Hasle · Lie · Quak *Editors*

Geometric Modelling, Numerical Simulation, and Optimization: Applied Mathematics at SINTEF

Geir Hasle · Knut-Andreas Lie · Ewald Quak
Editors

Geometric Modelling, Numerical Simulation, and Optimization:

Applied Mathematics at SINTEF

With 162 Figures, 59 in Color, and 34 Tables

Springer SINTEF

Editors

Geir Hasle
Knut-Andreas Lie
SINTEF ICT, Applied Mathematics
P. O. Box 124 Blindern
NO-0314 Oslo, Norway
Geir.Hasle@sintef.no
Knut-Andreas.Lie@sintef.no

Ewald Quak
SINTEF ICT, Applied Mathematics
P. O. Box 124 Blindern
NO-0314 Oslo, Norway
Ewald.Quak@sintef.no
and
Centre for Nonlinear Studies (CENS)
Institute of Cybernetics
Tallinn University of Technology
Akadeemia tee 21
EE-12618 Tallinn, Estonia
Ewald.Quak@cs.ioc.ee

Library of Congress Control Number: 2007927185

Mathematics Subject Classification (2000): 65-01; 65D17, 65D18, 65D05, 65D07, 65D10; 35F25, 35J05, 35Q55, 65M06, 65N06, 65N30, 65Y05, 65Y10, 68N19, 76B60, 76B70, 76D05, 76M10; 65K05, 90C11, 90C27, 90C59, 90C90

ISBN 978-3-540-68782-5 Springer Berlin Heidelberg New York

Springer is a part of Springer Science+Business Media

springer.com

Typesetting by the editors using a Springer TEX macro package
Production: LE-TEX Jelonek, Schmidt & Vöckler GbR, Leipzig
Cover design: WMXDesign GmbH, Heidelberg

Printed on acid-free paper 46/3180/YL - 5 4 3 2 1 0

Preface

This book presents current activities of the Department of Applied Mathematics at SINTEF, the largest independent research organisation in Scandinavia. The book contains fifteen papers contributed by employees and fellow partners from collaborating institutions.

The research and development work within the department is focused on three main subject areas, and the structure of the book reflects this clustering:

Part I Geometric Modelling
Part II Numerical Simulation
Part III Optimization

Addressing *Mathematics for Industry and Society*, each contribution describes a problem setting that is of practical relevance in one of the three areas and communicates the authors' own experiences in tackling these problems. The papers are written in a tutorial style, understandable for non-specialist researchers and students. The goal is to highlight the importance of mathematics for practical issues by providing material for educational purposes, such as a course or seminar on mathematical topics of industrial and societal relevance.

To ensure the scientific quality of the contributions to this volume, each paper was carefully reviewed by several international experts. Special thanks go to all authors and referees, without whom making this book would not have been possible. Last but not least, the friendly and effective collaboration with Springer Verlag through Martin Peters and Ute McCrory is kindly appreciated.

Oslo/Tallinn,
November 2006

Geir Hasle
Knut-Andreas Lie
Ewald Quak

Contents

List of Contributors

Jørg Espen Aarnes
SINTEF ICT, Applied Mathematics
P.O. Box 124 Blindern, NO–0314
Oslo, Norway
Jorg.Aarnes@sintef.no

Odd A. Andersen
SINTEF ICT, Applied Mathematics
Now: United Nations Statistics
Division, New York, USA
odd.andersen@gmail.com

Helge Avlesen
Parallab, UNIFOB, University of
Bergen
Thormøhlensgate 55, NO-5008
Bergen, Norway
avle@ii.uib.no

Jan Gunnar Bjaalie
Neural Systems and Graphics
Computing Laboratory, CMBN &
Institute of Basic Medical Sciences,
University of Oslo
P.O. Box 1105 Blindern, NO–0316
Oslo, Norway
j.g.bjaalie@medisin.uio.no

Marielle Christiansen
Department of Industrial Economics
and Technology Management
Norwegian University of Science and
Technology
Trondheim, Norway
Marielle.Christiansen@iot.ntnu.no

Geir Dahl
Center of Mathematics for Applications, University of Oslo
P.O. Box 1053 Blindern, NO–0316
Oslo, Norway
geird@ifi.uio.no

Tor Dokken
SINTEF ICT, Applied Mathematics
P.O. Box 124 Blindern, NO–0314
Oslo, Norway
Tor.Dokken@sintef.no

Truls Flatberg
SINTEF ICT, Applied Mathematics
P.O. Box 124 Blindern, NO–0314
Oslo, Norway
Truls.Flatberg@sintef.no

Marte Fodstad
SINTEF Technology and Society
Applied Economics and Operations
Research
S.P. Andersens vei 5, NO–7465
Trondheim, Norway
Marte.Fodstad@sintef.no

Simen Gaure
Scientific Computing Group
Center for Information Technology
Services, University of Oslo
P.O. Box 1059 Blindern, NO-0316
Oslo, Norway
Simen.Gaure@usit.uio.no

Tore Gimse
SINTEF ICT, Applied Mathematics
Now: Conax AS, Oslo, Norway
Tore.Gimse@conax.com

Roar Grønhaug
Department of Industrial Economics
and Technology Management
Norwegian University of Science and
Technology
Trondheim, Norway
Roar.Gronhaug@iot.ntnu.no

Trond Runar Hagen
SINTEF ICT, Applied Mathematics
P.O. Box 124 Blindern, NO–0314
Oslo, Norway
Trond.R.Hagen@sintef.no

Geir Hasle
SINTEF ICT, Applied Mathematics
P.O. Box 124 Blindern, NO–0314
Oslo, Norway
Geir.Hasle@sintef.no

Martin Ofstad Henriksen
SINTEF ICT, Applied Mathematics
Now: Scandpower Petroleum
Technology AS, Kjeller, Norway

Kjetil Midthun
Department of Industrial Economics
and Technology Management
Norwegian University of Science and
Technology
Trondheim, Norway
Kjetil.Midthun@iot.ntnu.no

Jon Mikkelsen Hjelmervik
SINTEF ICT, Applied Mathematics
P.O. Box 124 Blindern, NO–0314
Oslo, Norway
Jon.A.Mikkelsen@sintef.no

Vegard Kippe
SINTEF ICT, Applied Mathematics
Now: Norsk Hydro, Bergen, Norway
Vegard.Kippe@hydro.com

Oddvar Kloster
SINTEF ICT, Applied Mathematics
P.O. Box 124 Blindern, NO–0314
Oslo, Norway
Oddvar.Kloster@sintef.no

Knut–Andreas Lie
SINTEF ICT, Applied Mathematics
P.O. Box 124 Blindern, NO–0314
Oslo, Norway
Knut-Andreas.Lie@sintef.no

Jens Olav Nygaard
SINTEF ICT, Applied Mathematics
P.O. Box 124 Blindern, NO–0314
Oslo, Norway
Jens.O.Nygaard@sintef.no

Christian Pettersen
Neural Systems and Graphics
Computing Laboratory, CMBN &
Institute of Basic Medical Sciences,
University of Oslo
P.O. Box 1105 Blindern, NO–0316
Oslo, Norway
Christian.Pettersen@medisin.uio.no

Ewald Quak
SINTEF ICT, Applied Mathematics
Now: Institute of Cybernetics
Tallinn University of Technology
Akadeemia tee 21
EE-12618 Tallinn, Estonia
Ewald.Quak@cs.ioc.ee

Alf Birger Rustad
Statoil Research Centre, Rotvoll
Arkitekt Ebbellsvei 10
NO–7005 Trondheim, Norway
abir@statoil.com

Frode Rømo
SINTEF Technology and Society
Applied Economics and Operations Research
S.P. Andersens vei 5, NO–7465
Trondheim, Norway
Frode.Romo@sintef.no

Johan Simon Seland
Center of Mathematics for Applications, University of Oslo
P.O. Box 1053 Blindern, NO–0316
Oslo, Norway
johans@cma.uio.no

Vibeke Skytt
SINTEF ICT, Applied Mathematics
P.O. Box 124 Blindern, NO–0314
Oslo, Norway
Vibeke.Skytt@sintef.no

Martin Stølevik
SINTEF ICT, Applied Mathematics
P.O. Box 124 Blindern, NO–0314
Oslo, Norway
Martin.Stolevik@sintef.no

Asgeir Tomasgard
Department of Industrial Economics and Technology Management
Norwegian University of Science and Technology
Trondheim, Norway
Asgeir.Tomasgard@sintef.no

Karsten Trulsen
Mechanics Division, Department of Mathematics, University of Oslo
P.O. Box 1053 Blindern, NO-0316
Oslo, Norway
karstent@math.uio.no

Torbjørn Utnes
SINTEF ICT, Applied Mathematics
Sem Sælandsv 5,
NO–7465 Trondheim, Norway
Torbjorn.Utnes@sintef.no

George Vasilakis
ITI CERTH
Thermi–Thessaloniki, GR–57001
Greece
vasilak@iti.gr

Welcome to the World of SINTEF Applied Mathematics!

Tore Gimse

SINTEF is a contract research institute organized as a private non-profit foundation in Norway. For more than 50 years, SINTEF has performed basic and applied research in cooperation with Norwegian and international academic, industrial, and public partners. You are very welcome to learn more about the SINTEF group and our activities and aims at our home page `http://www.sintef.no`. SINTEF Applied Mathematics is a small, yet dynamic department within SINTEF. Currently we are about thirty full-time employees in addition to advisers, post docs, PhD students, and other affiliates.

The present book originated as an idea to present some of our activities and projects to our cooperating university groups and to potential master and PhD students in the field. Since its modest birth, it has developed and grown into the present publication. We certainly hope that the topics addressed and the level of detail in the presentations appeal to the international reader, who may be familiar and generally interested in applied mathematics, but not necessarily a specialist in each particular area. The book will also give insight in how mathematics and computational science enhance engineering, management, and industrial development. We trust that the book is suitable for teaching.

Applied mathematics has a long and fruitful history in Norway. It ranges from the fundamental, yet applicable works of Nils Henrik Abel and Sophus Lie, via mathematics applied to important physical processes by Vilhelm Bjerknes, as a pioneer of modern meteorology and weather forecasting, to Kristian Birkeland, who developed the first industrial process for nitrogen fixation. Birkeland was one of the founders of Norsk Hydro, today a global energy and aluminum supplier. The present achievements off-shore Norway in the North Sea, which have put our small country in a position as the world's third largest exporter of oil, are also due to elaborate applications of mathematics and computational methods.

The petroleum business has had a significant impact on the course of applied mathematical research and education during the past three decades in

Norway. Diverse fields of activity, ranging from construction analysis, marine modeling, geosciences, computational fluid dynamics, chemistry, materials, environmental issues, production design, communication, planning, scheduling, and control to financial modeling and impact on the society, all include mathematical challenges. Although the petroleum industry in Norway is mainly focused on production and upstream activities, an impressive and internationally leading expertise has been built, adding know-how and software tools as important Norwegian export articles.

SINTEF has been a part of this development since before the first exploitation of oil in the North Sea. The applied mathematics groups at SINTEF, and at the Central Institute for Industrial Research (SI) in Oslo before the merger of the two in 1993, were well positioned for the petroleum age when it started in the early 1970s. In addition to general and basic knowledge of computational mechanics and fluid dynamics, what later became the Department of Applied Mathematics was a key actor in three important fields. First, the group in question had intrinsic knowledge of the development of computers and computational resources. Second, there was a successful activity related to computer aided design (CAD) and modeling. The primary focus was ship hull design. Novel and innovative computer programs were developed, clearing the ground for numerous CAD-related projects and important industrial achievements. Third, an area related to wave analysis was launched in the late sixties. This was later to be concentrated on two activities: wave focusing and wave impacts on marine structures.

With these three focal areas as a starting point, new activities have been built up, in close cooperation with our clients and partners. The petroleum industry and petroleum-driven business are still the single most important sector for the department. However, we have successfully mastered other areas, either as a sub-contractor for other SINTEF entities, or by operating in the market by ourselves. As examples of the latter, we are especially proud of our activities within stratified air flows, and our work on optimization and operations research applied to logistics and planning. In both areas we are running projects of significant positive impact, both commercially and to the society. Over the years, SI and SINTEF have spawned several successful companies from results of research activities in applied mathematics.

Today, the Department of Applied Mathematics at SINTEF is focused on three main areas: geometrical modelling, numerical simulation, and optimization. The structure of the current book reflects this clustering, with contributions from employees and fellow partners from cooperating institutions. Moreover, the book shows the dualism with which we face the world. On one side we conduct short-time research and development to answer questions posed by our clients or to improve their software products. Secondly, we develop software solutions that we bring to the market ourselves, as reflected in the contributions on intersection algorithms and CAGD, and the chapter on industrial vehicle routing.

On the other hand, we also perform long-term, fundamental research to help develop Norwegian industry and society and in particular to put ourselves in a position where we can answer what we believe will be tomorrow's questions from our clients. Some of the contributions therefore reflect current, leading-edge activities, such as development of multiscale methods for reservoir simulation, and work related to the use of graphics processors as high-performance computational resources. The latter activity has not yet attracted many industrial end users, but research groups and innovative software developers are highly interested.

The strategic research activities also illustrate the duality of increase in computational problem-solving ability very well. Roughly speaking, today's level of performance in scientific computing is due to equally important progress in hardware and software. Improvements in numerical methods have pushed the limits of computing in a similar way as has the increase in brute processor force, and will continue to do so. This potential of mathematical and computational methods inspires us to proceed along the line of developing and improving the mathematical side of computational applications. Our book reveals added motivation and new reasons for enjoying and continuing this exciting endeavor.

Tore Gimse,
Research Director (1995–2006)

Part I

Geometric Modelling

Overview

Geometric modelling is the classical, but still rapidly evolving branch of mathematics that investigates the measurements, properties and relationships of curves, surfaces and volumes, and provides the mathematical tools for describing these entities in natural and human-made shapes, as well as for designing and using them in applications.

Computer-based geometric models can in fact describe not only physical objects, but even nonexistent virtual objects. Such virtual objects play a central role in a lot of very different application settings, for example in engineering through Computer Aided Design (CAD) and Computer Aided Manufacturing (CAM), but also in entertainment for the production of animation movies and computer games. The study of the mathematical foundations of this subject area is often referred to as Computer Aided Geometric Design (CAGD).

Obviously 3D computer models form an important means for understanding the structures and properties of both real and virtual objects. 3D models are of course ubiquitous in large-scale Virtual Reality (VR) applications in industry, but the increasing computational power and graphics capacity of standard computers are driving forces in the current rapid evolution towards the creation and delivery of high-fidelity and high-performance 3D shape content for everybody.

In general, the current SINTEF activities within geometric modelling combine theory and practical experience from spline technology, applied algebraic geometry, and numerical methods with requirements from industry and society, based on what is computationally feasible on commodity and emerging hardware technologies.

Applied research related to geometry has been central for R& D activities in Oslo for almost 50 years. After the first Norwegian computer "Nusse" was built in 1953, the researchers at what was the former Central Institute for Industrial Research (SI), which is now part of SINTEF, looked for industrial applications of computers. The first problems addressed were within ship building. The introduction of mini-computers and graphical displays (around 1975) allowed SI's researchers to address a wide range of challenges within producing industries. UNIX workstations (from 1990) and later PCs with high-end graphics cards (from 2000) opened up possibilities to address challenges within medicine, geology and geophysics, and entertainment.

Piecewise polynomial spline functions, as used to describe 3D objects parametrically, form a cornerstone within geometric modelling and CAGD. The de facto CAD industry standard of NURBS (non-uniform rational B-splines) is based on these functions. For more than 30 years, researchers from Oslo have contributed to spline research on a prominent international level; there is even a widely used Oslo algorithm for the insertion of new breakpoints (knots) into such functions. In all this time, there has been a very fruitful symbiosis between the geometric modelling group at SI and later SINTEF, and the team of spline researchers at the University of Oslo, combining practical considerations with theoretical insights. As a consequence, an individual researcher from one group may have a part-time position at the partner organization, or shift his affiliation from one to the other, either for a limited period or indefinitely. Joint training of students on the master and doctoral levels is also an important benefit. This joining of forces continues to this day, where the geometry group of the University of Oslo is now part of the Norwegian Centre of Excellence "Centre of Mathematics for Applications" (CMA, `www.cma.uio.no`), in which SINTEF is an official partner.

Geographically located at the outskirts of Europe, the importance of maintaining an international presence is obvious. While the University of Oslo organizes a well-established series of international conferences on mathematical methods for curves and surfaces, with its seventh instalment coming up in 2008, the geometric modelling group at SI and then SINTEF has been involved in international projects, especially in the EU's framework programmes, since the 1980s. More recently, it has also taken over leading management roles in such projects. Two of them, AIM@SHAPE and GAIA II, will be discussed a little bit more below. A third one was the EU FP5 Marie Curie Research Training Network MINGLE (Multiresolution in Geometric Modelling, `http://www.cs.technion.ac.il/~vitus/mingle/`), coordinated by SINTEF. The project provided visiting positions for young doctoral students and post docs at nine European organizations (including a small Norwegian company), organized an international summer school and a workshop, and published two books on the topic.

Since the Oslo-based research in spline functions, including more recently also spline-based wavelets, is well documented in the scientific literature, with a fairly recent basic tutorial on splines and spline wavelets available in the book *Tutorials on Multiresolution in Geometric Modelling*, published by the MINGLE project consortium, the topics of the five geometric modelling contributions in this volume were chosen somewhat differently. Two of them discuss very general issues of CAD systems, one describing the basic setup of such a system, and the other the fundamental problem of how to find intersections of CAD models. In two other papers, the geometries investigated are represented not as parametric CAD models, but implicitly, with their visualization being a main concern. In one of them, the context is another important application domain, namely medicine. In the other, the main idea is the use of dedicated graphics hardware for the visualization of implicit surfaces by raytracing. The use of such graphics processors (GPUs) has recently become a focus of attention within SINTEF Applied Mathematics, and involves both the geometric modelling and numerical simulation groups. For that reason, a general introduction to this subject was included, giving also an overview of several applications pursued by the SINTEF teams. This topic is then expanded not only in some of the geometric modelling papers, but also in Part II of this book on numerical simulation.

The group's geometric modelling research and development experience of decades of person-years is contained in a variety of software libraries, developed and maintained over a long period of time as a sort of institutional memory. The SINTEF Spline Library SISL (`www.sintef.no/sisl`) is the premier example, also widely used outside of SINTEF, both for commercial and academic purposes.

Several of these SINTEF geometry software tools, including a version of SISL for non-commercial use under a GPL licence, have been incorporated in the software tools repository of the EU FP6 IST Network of Excellence on Shape Modelling AIM@SHAPE (`www.aimatshape.net`), in which SINTEF participates and for which this editor is the Technical Manager. Since the project start in 2004, thirteen leading European institutions in the field are cooperating to establish an integrated European infrastructure for shape modelling research, linking it to latest developments in the area of knowledge management. As a major technological collaboration tool, a Digital Shape Workbench (DSW) is being developed to provide an eScience framework available to researchers also beyond the network consortium. The DSW incorporates a shape model repository, the aforementioned software tools repository, a digital library, and a search engine for 3D shapes. Annual summer schools are held to train researchers from inside and outside the consortium in the use of the DSW and in the research topics of the network, the most recent one organized by this editor in Tallinn, Estonia, where he is currently a visiting senior research fellow on an EU FP6 Marie Curie Transfer of Knowledge project, whose support is graciously acknowledged.

Some of the work in the AIM@SHAPE project to join shape modelling and knowledge technology concepts is described here by the paper of Andersen and Vasilakis, which provides both an introduction to CAD modelling in general as well as an introduction-by-example to the concept of an ontology. It also exemplifies the collaboration with other European partners, in this case through a co-author from the Greek project partner, the Informatics and Telematics Institute (ITI).

Mathematics for CAD/CAM-systems, with a special focus on surface intersection, has been central in our activity for more than two decades, and still plays an important role within the research program, especially in relation to CAD-type intersection algorithms. The intersection technology for surfaces of NURBS (non-uniform rational B-spline) type in the commercial systems of the company CoCreate (`www.cocreate.com`) has been developed by our team and is still being continually improved.

Based on this long-standing experience, the paper by Dokken and Skytt on intersection algorithms in Computer Aided Geometric Design (CAGD) spans from an introduction into the basic problems of CAD-surface intersection to latest developments for singular and near-singular intersection that were addressed in the SINTEF-coordinated EU FP5 Future and Emerging Technologies project GAIA II (`http://www.sintef.no/content/page1___5917.aspx`) in close cooperation with the company think3 (`www.think3.com`). A new software tool is currently being tested and expected to be made available soon under the GNU GPL-licence.

The SINTEF geometry group has also been actively engaged in various applications in the medical sector for quite some time now, contributing in the past also to the establishment of a Norwegian company in computer-aided surgical training, (`www.simsurgery.com`). The contribution to this volume about some recent work in this field is the paper by Nygaard et al. It describes the development of a toolbox for implicit representation of 3D geometry, addressing some of the needs of neuro-scientists concerning 3D geometry reconstruction, modelling, and visualization.

Another recent development that has, as already mentioned, created a new link between the geometry and simulation activities at SINTEF is the use of the computational power of programmable graphics cards (GPUs) for general-purpose computing. The ongoing strategic research project, "GPGPU: Graphics hardware as a high-end computational resource" (`www.sintef.no/gpgpu`), is funded by the Research Council of Norway and coordinated by SINTEF. Some of its results are presented in the papers by Dokken et al. and Seland and Dokken in Part I, as well as in the paper by Hagen et al. in Part II of the book. The paper by Dokken et al. gives an introduction to the programming of graphics cards and, more generally, to the paradigm of data-stream-based computing. The paper by Seland and Dokken discusses raytracing of algebraic surfaces, defined implicitly as level sets of polynomials, and how one can obtain very efficient rendering of such surfaces by utilizing the programmable parts of a modern graphics card. These latter contributions document also some of the current cooperation with the geometry group at University of Oslo as part of the already mentioned Norwegian Centre of Excellence, CMA.

Ewald Quak
Senior Scientist
Editor, Geometric Modelling

Building an Ontology of CAD Model Information

Odd A. Andersen and George Vasilakis

Summary. The purpose of this tutorial paper is twofold. Firstly it can serve as an introduction to CAD shape modeling in general. Secondly it can be read as an introduction-by-example to the concept of an ontology. In computer science, an ontology is a rigorous conceptual model of a specific domain. Such models are useful in various contexts such as advanced information retrieval, knowledge sharing, web agents, natural language processing and simulation and modeling. There is an ongoing effort in the EU-funded Network of Excellence AIM@SHAPE to develop ontologies formalizing various aspects of digital shapes, by creating several domain-specific ontologies and by integrating them into a general common ontology. As a part of this effort, an ontology for CAD model information based on the STEP 42 standard has been proposed. This article presents the establishment of this particular ontology as an introductory example.

Key words: computer aided design, geometry, topology, ontology, RDFS, STEP, OWL

1 Introduction

In today's networked world, the quantity of generated information is growing at an ever increasing pace. This phenomenon, often referred to as the *information deluge*, makes it increasingly important to facilitate ways of processing and exchanging information that can be automatically carried out by the computer. *Ontologies*, introduced in Artificial Intelligence in the field of Knowledge Management, provide the means to structure and manage all this information and make it easily accessible. In this context, an ontology is a rigorous conceptualization of some knowledge domain, where all concepts and relations are clearly and unambiguously defined. By structuring knowledge according to a defined ontology, computers can semantically interoperate when it comes to processing, exchanging or presenting information. Ontologies can be domain-specific, modeling the information used in a specific setting, or they can be general, in which case they might serve to integrate several domain-specific ontologies. Such ontologies are referred to as *upper-level ontologies*. One well-known example of an extensive general ontology is the CYC project [2]. (The expression "CYC" is derived from "encyclopaedia").

Nowadays there are several languages and tools for formally defining ontologies on arbitrary knowledge domains. However, the use of conceptual models in computer science is as old as the discipline itself. Such models have always been used in software development. We can recognize ontology-like concepts in the object-oriented software design, database structures, exchange formats, etc. Obviously, there is a lot of information out there that is already structured, and whose potential for wider use is great if we can integrate them with ontologies from other domains. As an example, the product exchange format STEP (ISO 10303), widely used in industry for more than a decade, uses its own modeling language EXPRESS to define its data models, which can be seen as ontologies for the representation of product model information.

The European Network-of-Excellence project AIM@SHAPE [1], which is part of the EU Sixth Framework Programme, is directed towards research in representing and processing knowledge related to shapes (where the word *shape* is used in a general sense, referring to any object that is characterized by its visual description). An important part of this work consists of associating semantics with shapes, which again has led to the development of ontologies for some shape-related domains, as well as an effort to integrate them into a larger whole. As a part of this effort, it was decided to write a tentative ontology for computer aided geometric design and to base the work on part 42 of the STEP standard, which deals with geometric and topological representations. This article intends to present some key aspects of this work, using it as an illustration of what a domain-specific ontology might look like. The article can also be read as an brief introduction to the usual way of modeling shapes in the CAD community.

In the first part, we present an introduction to ontologies and their use in computer science and knowledge management. In the second part, we look at the basic concepts in computer aided geometric design today. In the third part, we discuss how these concepts are represented in the STEP EXPRESS language. Then in a fourth part we present how this structure was partially recycled in our sample ontology. Finally, we give some concluding remarks regarding the direction this work is taking concerning AIM@SHAPE.

2 The Ontology Concept and Why It Is Useful

2.1 Introduction

Since its conception the World Wide Web has grown and expanded considerably. It also has undergone continuous development and transition. Web technologies have provided frameworks and solutions for using and exploiting the vast resources made available by the Web. The information available on the Web has also grown and reached the point where management and searching facilities have become more crucial than the information itself. After all, what good is it to have the information stored securely in a database server if it is so difficult to discover and access it?

Until recently the focus regarding Web content has been targeted on how to display information rather than the information itself. The hypertext markup language (HTML), on which the Web is still based, describes how information should be laid out on a Web page for humans to read. The problem of semantic search through the Web is perhaps the most critical limitation of the current state of the Web. In effect, there are no methods of gaining broader information about web documents.

A few years ago, the W3C [3] issued the Semantic Web initiative in order to address the current limitations of the Web. The objective of the Semantic Web is to provide an efficient way of representing content and information on the World Wide Web, so that it can be used not just for display purposes, but for automation, integration and reuse of data across various applications [7, 13]. For the Semantic Web to function, computers must have access to structured collections of information and sets of inference rules that they can then use to conduct automated reasoning [13].

Two important elements required for the Semantic Web can be identified in the above scenario: a means to represent information in a structured and formal manner, and tools to process this knowledge and derive resulting conclusions.

Knowledge management is a field in Artificial Intelligence which deals with the first element. Ontologies are a key enabling technology for the Semantic Web and are used to capture information structures available in the Web. An Inference Engine is a software tool which can process knowledge structures in order to deduce the required results.

The most basic ontology language for the Semantic Web can be identified with the Resource Description Framework (RDF) and the RDF Schema [8, 9, 15, 14]. RDF Schema provides some classification and basic rules but it is otherwise a very simple ontology language. The Web Ontology Working Group of W3C [11, 12] identified a number of characteristic use-cases for the Semantic Web which would require much more expressiveness than that offered by RDF and RDF Schema. The Web Ontology Language - OWL [5, 20], is perhaps the prevailing standard today in ontology design and provides a much richer vocabulary than that supported by RDF Schema. Furthermore, it provides support for reasoning and has a formal semantics.

2.2 What Is an Ontology?

A specification of a representational vocabulary for a shared domain of discourse – definitions of classes, relations, functions, and other objects – is called an ontology [19]. In general, ontologies describe formally a domain of discourse. An ontology embodies some sort of world view with respect to the given domain. The world view is often conceived as a set of terms (e.g. entities, attributes, and processes), their definitions and inter-relationships; terms denote important concepts (classes of objects) in the domain. This is referred to as a conceptualization. Recording such a conceptualization with an ontology is referred to as ontology development. The subject of Section 5 outlines an example of such a development.

In the context of the Web, ontologies provide a shared understanding of a domain. Such a shared understanding is necessary to overcome differences in terminology. Such differences can be overcome by mapping the particular terminology to a shared ontology, or by defining direct mappings between the ontologies. In either case, it is easy to see that ontologies support semantic interoperability. It is also evident that the ability to standardize the definition of real-world concepts becomes very powerful as we begin to investigate knowledge that spans across multiple domains.

In ontology development, classes are created to encapsulate the meaning of concepts in the target domain. A class describes all those features that define the concept in the knowledge domain. An individual, or an instance, of that class in the ontology represents an object whose type is characterized by the concept in the domain. For example the class Animation Sequence in an ontology describing virtual humans could describe the concept of an animation sequence that is used to show actions of virtual humans. An instance of the Animation Sequence class in this case would be a specific animation sequence, with all the necessary meta-data information (i.e. format, description, associated files, etc.), used for example to visualize a virtual human in the action of running.

Relations between classes are defined by properties and describe the kind of associations that are possible between classes. The relationships between concepts include typically hierarchies of classes, also called taxonomies.

Properties are also used to describe features of the class and represent attributes that have a specific type such as boolean, integer, string, etc. Examples of such properties are phone number, title, age, etc. Restrictions can be imposed both on classes and on properties in order to refine their semantics.

2.3 Uses and Benefits of Ontologies

The motivation behind the development of an ontology for a particular domain of interest falls into the following areas:

- sharing a common understanding of the information in a knowledge domain;
- improving interoperability among applications that use the domain knowledge;

- making domain assumptions explicit so that applying changes becomes easier as these assumptions evolve;
- enabling re-use of the domain knowledge: the ontology can be used by multiple applications.

Ontologies are useful for the organization and navigation of Web sites. Many Web sites today use a hierarchy of terms for navigation on the web site's contents.A The accuracy of Web searches can also be improved through the use of ontologies. Search engines can look for pages that refer to a precise concept in an ontology, instead of performing keyword-searching resulting in high recall but very low precision. Relations in the ontology can also be exploited to deduce functionality (i.e. relevant resources that have been visited, or bought, by the same customer).

Ontologies are extensible and scalable since they can grow independently of the applications that are using them. They also provide disambiguation of domain knowledge as the information captured by the ontology is well-defined and logically consistent, thus minimizing ambiguity.

Finally, knowledge captured by an ontology can be used by automated reasoners to deduce (infer) conclusions, thus making implicit knowledge explicit.

2.4 Ontology Development

Ontology development does not constitute a new approach to knowledge representation and management. Ontologies were developed in artificial intelligence to facilitate knowledge sharing and re-use. Nevertheless, their popularity has recently grown and more and more new technologies seem to revolve around them.

Ontology development, however, still remains a process that is not well understood. This is due to the sheer size of the knowledge we are usually faced with, for a specific scientific domain, and the complexity of the problem of storing and managing this knowledge in a flexible and efficient way. During the ontology creation process different perspectives of the domain must be combined and a consensus must be reached among a group of experts with different backgrounds. Also, because of the nature of the problem, ontology development is a rather subjective matter - there is not a single correct ontology for a specific domain. The resulting ontology structure depends on usage, scope, anticipated extensions, etc. If for example the ontology is foreseen to change frequently the emphasis should be put on following a basic and general approach so that changes can be easily applied.

Ontologies are not systems of logic but they may specify rules to support several logical inferences. For example, the equivalence of two concepts may be specified in an ontology but only an inference engine can deduce that instances belonging to these classes should support the same functionality. Creating an ontology that is well-structured, is free of contradictions, and accurately expresses the intent of its designers is not easy. It requires very good knowledge of the domain to be modeled, the ability to conceptualize the underlying ideas and a good knowledge of the syntax of the ontology language, so that the model is expressed correctly.

We can identify three basic elements in ontology engineering (and knowledge management in general):

- Knowledge acquisition
  This involves the automatic extraction of knowledge from unstructured or semi-structured sources of information and also the acquisition of expert knowledge from human domain experts through editing tools, like Protégé [10], which support the creation and population of ontologies
- Knowledge representation
  Once knowledge has been automatically extracted or acquired from human sources it is then recorded using an ontology language (including of course also a query language to provide access to the knowledge so stored).
- Knowledge use and reuse
  Perhaps the most important element to knowledge management is how to access the information that has been stored and to be able to reason on the acquired knowledge in order to draw some conclusions. Tools that enable this interaction with the acquired knowledge structures include functionality for discovering, browsing, visualizing, reasoning, maintaining and merging information captured by an ontology.

3 Basic Concepts in Computer Aided Geometric Design

In the following text, we aim to summarize the modeling concepts used in state- of-the-art CAD curve and surface modeling today. For good, state-of-the-art material, we can recommend [17] or [18]

3.1 Representations of CAD Models

A lot of the ground-breaking theoretical work in Computer Aided Geometric Design was carried out during the 1960s and 1970s. The first working demonstration of a CAD system was SKETCHPAD, presented in 1963, whereas the first exploitable systems arrived at the beginning of the following decade. The CAD world of the 1970s was characterized by systems that were curve-based and mainly 2D. This decade also saw the gradual appearance of 3D wire-frame based systems. At this time, discussions of volume modeling appeared at academic conferences. Volume modeling became state-of-the-art during the 1980s, the driving force being the automotive and aerospace industry. This also increased the importance of *parametric, polynomial surfaces*. The two main approaches within volume modeling were *constructive solid geometry* (CSG) and *boundary representations* (B-reps). The former specifies 3D shapes by

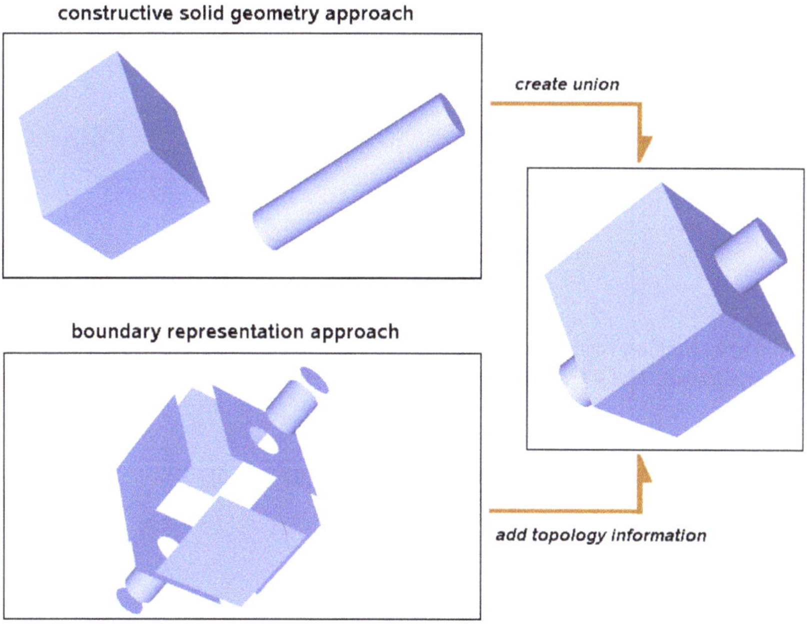

Fig. 1. Constructive solid geometry versus boundary representation approach.

using boolean operations on geometrical primitives (prisms, cylinders, etc.) whereas the latter stitches together individual curve segments and surface patches to represent the complete boundary of a 3D model[1].

Boundary representations constitute the basis of most of today's CAD systems. The number of representable boundary surfaces increased substantially when systems were extended to handle *rational polynomial surfaces* (NURBS) at the start of the 1980s. Work on the STEP standard was started later the same decade, and the current tools for geometry and topological representations were integrated into part 42 of this standard. STEP had its first official release in 1994. Since then, the standard tools for representing geometry in the context of CAD have changed very little. As for geometry, the building blocks are mainly 1st and 2nd order algebraic primitives (lines, planes, spheres, cones, conical sections, toruses, etc.) and parametric higher-order polynomial representations (Bézier and NURBS curves and surfaces).

In the following subsections we will have a closer look at curve and surface representations, as well as how they are combined to form a complete model by specifying topological information.

---

[1]It should be mentioned, however, that although the *definition* of shapes in solid modeling is based on boolean operations on primitives, their actual *representation* is typically based on B-reps-type data structures.

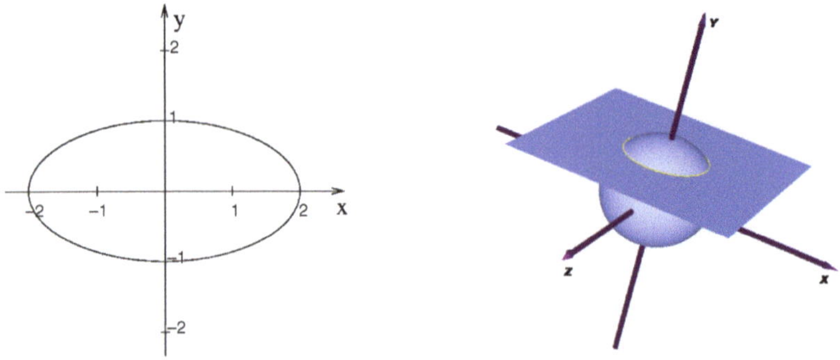

2D curve (ellipse) defined by the equation $\frac{1}{2}x^2 + y^2 - 1 = 0$

3D curve (circle) defined by the two equations:
$x^2 + y^2 + z^2 - 1.5 = 0$ (3D sphere)
$y - 1 = 0$ (3D plane)

Fig. 2. Algebraic curves in 2D and 3D.

3.2 Curve Representations

There are two common ways of expressing curves in Euclidean space: the *algebraic* representation and the *parametric* representation.

Fundamental Algebraic Curves

In the algebraic representation, a 2D curve is defined as all points (x, y) satisfying an algebraic equation $f(x, y) = 0$. A 3D curve is defined as all points (x, y, z) simultaneously satisfying two algebraic equations $f(x, y, z) = 0$ and $g(x, y, z) = 0$. (Note: such a curve could be interpreted as the *intersection* of the two algebraic surfaces defined by f and g).

If a 2D curve can be defined by a function that is a polynomial, the *degree* of the curve is equal to that of the polynomial. Likewise, a 3D curve expressible by the intersection of two polynomial functions has a degree that is the product of the degrees of the polynomials.

Algebraic curves of degree higher than two are not guaranteed to have a *rational parametric* representation. For this reason, they are rarely used in today's CAD systems, and are not supported by STEP.

Parametric Curves

Parametric curves are defined as being differentiable maps of (an interval of) the real line into Euclidean space. Parametric representations are popular in CAD for a number of reasons. They are easy to draw and manipulate, and the underlying theory is well known by engineers and mathematicians alike. In the field of differential geometry, curves are almost exclusively represented

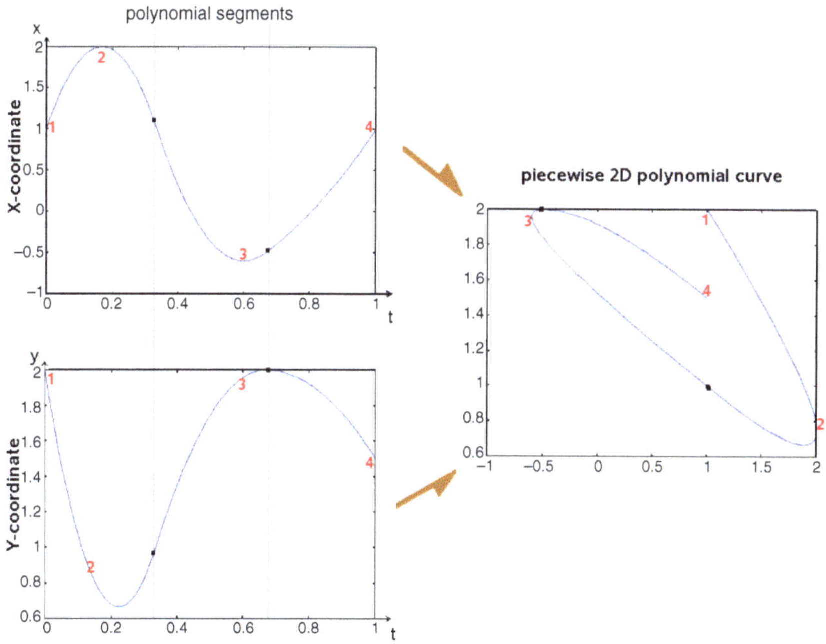

Fig. 3. A 2D curve constructed by piecewise polynomial functions for its x and y coordinate. Some corresponding points are numbered, and the transitions between polynomial segments are marked with dots.

in parametric form. Moreover, to have parameter values associated with each and every point on a curve or surface makes it easy to subdivide the object into smaller parts, as well as mapping textures.

In CAD, parametric curves are represented by *piecewise, rational polynomials*. Each of the spatial coordinates of the curve is thus evaluated as a rational polynomial of its parameter domain, with the restriction that the rational polynomials must share the same denominator. By *piecewise*, we mean that the parameter domain is divided into several subintervals, and that the defining polynomials change from one such piece to the next. However, the polynomials are constructed so that the curve stays continuous[2] even across subintervals.

[2] *Continuity* is a central issue in curve and surface design: If we say that a parametric curve is C^n continuous at a certain point, it means that its parametric description is n times differentiable at that point. Geometrically, a C^1 curve therefore has a uniquely defined tangent, and a C^2 curve has both a unique tangent and unique curvature. If a curve becomes n times differentiable at a given point when (re)parameterized by arc length, it is called G^n continuous at that point. This is called *geometric continuity*.

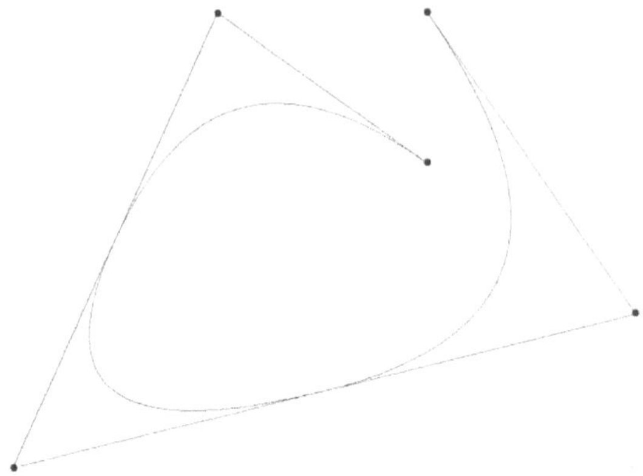

Fig. 4. A 2D NURBS curve, its control points (dots) and control polygon (line segments). By moving the control points, we can change the shape of the curve in an intuitive way.

Parametric, rational, piecewise polynomial curves of a given degree can be represented in several ways, but they are *always* convertible to a sequence of rational *Bézier curves* of the same degree. The Bézier representation was first studied by Paul de Casteljau in 1959, and independently by Pierre Bézier during the 1960s. The advantage of this representation is that the coefficients defining the curve have an intuitive geometric meaning. These coefficients can be seen as a sequence of points in 2D (or 3D) space, defining a piecewise linear curve that outlines the "general shape" of the polynomial curve. By moving these points-coefficients, the curve can be modified in a predictable way. Moreover, the computational algorithm[3] used to evaluate points and derivatives is based on linear interpolation, and inherently stable.

The problem with sequences of Bézier curves is that their control points cannot be freely manipulated without loss of continuity from one curve segment to another. In practice, a different representation is often preferred: the *non-uniform, rational B-spline* (NURBS). Like the Bézier curve, the NURBS-curve is also defined by a sequence of control points, but these can be freely manipulated without any risk of breaking continuity.

Rational polynomial curves can describe a wide range of curve shapes. The algebraic curves of degree 2 (conic sections) can all be expressed in parametric form as rational polynomial curves (in particular: they can be expressed as rational Bézier curves).

[3]known as the *de Casteljau algorithm*

Procedurally Defined Curves

Procedurally defined curves are curves that might not be directly representable in a simple algebraic or rational polynomial way, but whose points and derivatives can be evaluated by some user-defined procedure. In 2D, this is a good way of dealing with *offset curves*. An offset curve is specified by departing from a base curve and moving each of its points a specific distance along the direction perpendicular to the curve tangent at that point.

3.3 Surface Representations

As with curves, surfaces can be represented algebraically, parametrically, or defined procedurally. In practice, surfaces in CAD are usually designed by means of curves in some way; for instance by interpolating a grid of curves or by sweeping a curve along a path. In this section, we will have a brief look at some common surface representations.

Fundamental Algebraic Surfaces

An algebraic surface is (for our purposes) a 2-manifold embedded in 3D Euclidean space. The surface can be described as the set of points (x, y, z) that satisfy an equation $f(x, y, z) = 0$. If the point (x_0, y_0, z_0) satisfies this equation, we also know that the normal of the surface at this point is directed along $\nabla f(x_0, y_0, z_0)$. As for curves, the degree of the algebraic surface is equal to be the degree of the polynomial defining f. (Although f can be defined with any differentiable function, we usually stick to polynomials for CAD applications).

Example 1. Planes are defined by the linear equation:

$$f(x, y, z) = ax + by + cz + d = 0$$

Example 2. Quadrics are defined by the implicit equation:

$$q(x, y, z) = 0$$

where q is a quadratic polynomial in x, y, and z. This is the proper generalization of a conic section to the case of surfaces. Some important quadric surfaces are the *ellipsoids* (with the spheres as a special case), the *cones*, the *cylinders*, the *paraboloids* and the *hyperboloids*.

As in the curve case, algebraic surfaces of degree higher than two are rarely used in a CAD context. (An important exception is the *torus*, which is a *quartic* that can be handled fairly easily).

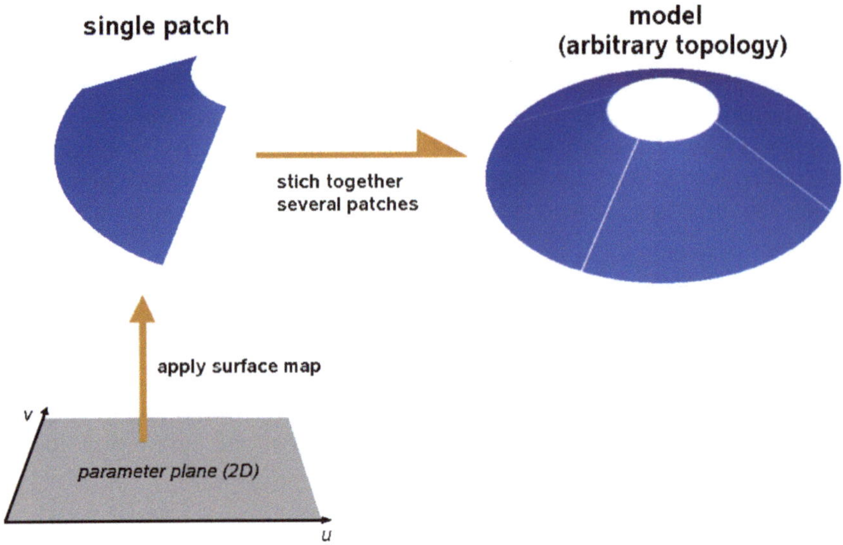

Fig. 5. Stitching together of several patches to generate a surface with arbitrary topology.

Parametric Surfaces

In differential geometry, a curve can always be defined as a differentiable map between the real line and Euclidean space. Analogous to this line of thought, we might imagine that a surface should be expressible as a differentiable map between $\mathbb{R}^2$ and Euclidean space. This is, however, not true in general. A given surface can have arbitrary topology, which makes the existence of differentiable maps from the real plane true only locally. In the case of boundary representation models in CAD, though, the topology issue is handled separately by piecing together smaller surface patches along boundaries into a structure with the correct topology. This will be discussed in more detail in Section 3.4. The patches are surfaces that *are* expressible as a map from a region of the real plane (usually a rectangle) into Euclidean space.

In Section 3.2 we saw how a parametric curve in CAD is defined by a mapping from its parameter domain into Euclidean space by means of *piecewise, rational polynomial functions*. In practice, the numerators and denominators of the rational functions are each specified as linear combinations of a limited number of *basis functions*, where each basis function is a specified univariate piecewise polynomial.

To extend this to the surface case, we need basis functions that can be used to describe a mapping from a rectangular domain in $\mathbb{R}^2$ into Euclidean space. This means that our basis functions must be piecewise polynomial *bivariate* functions. The most frequent way of defining the set of basis functions for a parametrical surface is by specifying two *univariate* sets of basis functions and taking their *tensor product.* In other words, if $\{\phi_i\}_{i=1..m}$ and $\{\theta_j\}_{j=1..n}$ are two sets of univariate basis functions, then we obtain a bivariate set of basis functions $\{\psi_{ij}\}_{i=1..m,j=1..n}$ by defining $\psi_{ij}(u,v) = \phi_i(u)\times\theta_j(v)$. A ratio of linear combinations of these basis functions will provide our mapping from the parameter plane into Euclidean space.

Remember that the univariate sets of basis functions $\{\phi_i\}_{i=1..m}$ and $\{\theta_j\}_{j=1..n}$ consist of functions that are piecewise polynomials. By taking their tensor products, we obtain functions that are also piecewise polynomials, and where the parameter domain of each such piece is a distinct rectangular part of the parameter domain of the complete surface. If the functions in $\{\phi_i\}_{i=1..m}$ and $\{\theta_j\}_{j=1..n}$ have maximal degrees of respectively m and n, we say that the basis functions $\{\psi_{ij}\}_{i=1..m,j=1..n}$ (and by consequence also the resulting surface) are of degree (m,n).

The most popular means of expressing piecewise, rational, polynomial surfaces are by using the *tensor-product NURBS representation.* In this form, the coefficients of the basis functions (called *control points*) have intuitive geometric meaning, and can be moved freely around to modify the geometry of the surface. The continuity of the surface will not suffer from such modifications. A NURBS surface can be split up into an array of rational *Bézier patches*, where each such patch represents a polynomial piece of the original surface. This representation is rarely preferred though, as it leads to a substantial increase in the total number of control points.

The parametric domain of a tensor product surface, being a rectangle in $\mathbb{R}^2$, can be interpreted as a part of 2D Euclidean space, and we can define 2D parametric curves from the real line *into* this domain. By subsequently applying the surface map we obtain a curve in 3D Euclidean space, located on the surface. This curve is called a *curve on surface* (CONS), and is an important concept in geometric modeling. A *closed* curve has coincident start and end points. A CONS that is also a closed curve can be used to *trim* a surface. The trimmed surface is defined to be the part of the surface that is mapped from the inside (or outside) of the closed curve in the parameter domain. This makes it possible to model surfaces with holes (illustrated in Figure 6). Another use of CONS is to approximate the intersection between a parametric surface and a second geometric object.

Note that if the CONS is specified in a tensor product surface's parametric domain by a Bézier curve of degree p, and if the surface itself is of degree (m,n), then the resulting CONS representation in 3D has a total degree of $p(m+n)$, which can rapidly become very high.

Rational polynomial tensor product surfaces can represent a wide range of surface types. A *bilinear surface*, which fits the simplest surface between four given points, is a tensor product surface of degree 1 in both parameters. The aforementioned *quadrics*, including spheres, are another important class of surfaces that can be represented exactly. *Ruled surfaces* (or "lofted" surfaces), which are produced by linear interpolation of two curves, can be seen as rational polynomial tensor product surfaces if the interpolated curves

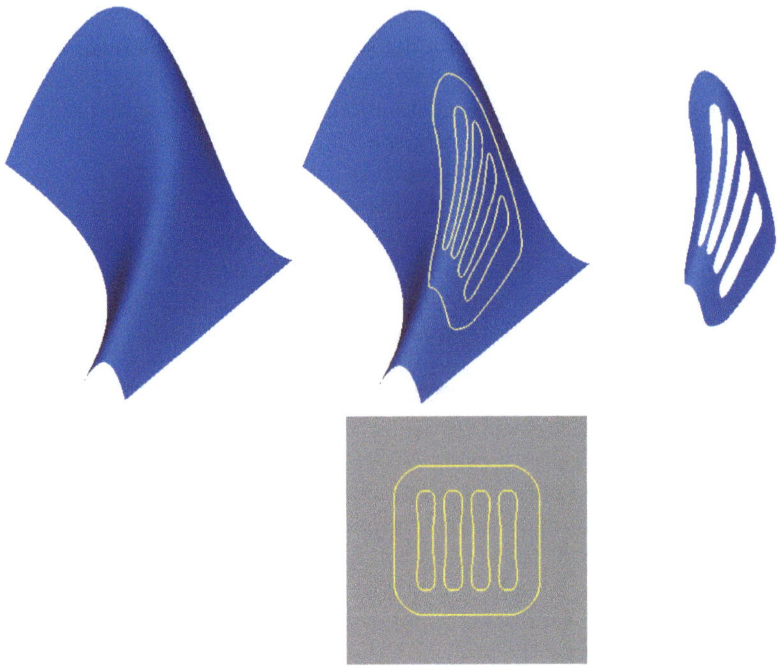

Fig. 6. The top row shows a spline surface (left), the same surface with curves on it (middle), the surface after trimming with the same curves (right). The bottom row shows a 2D representation of the curves in the surface's parameter plane.

are also rational polynomial. The same goes for *translational surfaces*, where the iso-parametric curves for each parameter are translated versions of each other, and the *Coons patches*, which interpolate a loop of four curve segments. Other surfaces that can be represented with rational polynomial tensor product surfaces are *surfaces of linear extrusion*, where the surface is obtained by sweeping a curve along a line, and *surface of revolution*, where the curve is swept along a circle. Note that most of the above mentioned surfaces are defined by a procedure applied on some given data (curves). Therefore, these surfaces are procedurally defined, even though their actual representation can be done with rational polynomial surfaces.

An example of a procedurally defined surface that cannot in general be expressed as a rational polynomial tensor product surface is the *offset surface*. Analogous to the offset curve, an offset surface is defined by moving every point of a base surface a specified distance along the surface normal at that point.

As a concluding remark it can be mentioned that there exist many other surface representations that, partly due to a lack of support in exchange formats like STEP, are not much used in the CAD community. *Subdivision surfaces* and *Bézier triangles* are two examples.

3.4 Binding Things Together: the Representation of Topology

As mentioned in Section 3.1, the boundary representation model (B-rep) is the representation of choice for most of today's CAD systems. A B-rep is obtained by sewing together individual curves and surface patches into a coherent surface representing a full 3D object with arbitrary topology (which means that it can have any number of handles and holes). Of course, the appearance of such an object could be obtained by merely placing surface patches side by side in the 3D scene. However, there would be no semantic information available that expressed the fact that these surfaces were parts of a whole, and that they were considered to be connected. The 'seam', the semantic thread binding together all involved parts of an object, is provided using topological primitives. These topological primitives have no geometric representation by themselves, but associate with each other and with geometric objects to create the underlying structure of a model.

The simplest topological entity is the *vertex*, which can associate with a point in Euclidean space, but carries no information of its own. Then we have the *edge*, which is defined by two vertices and expresses a connectivity relation between them. Geometrically, an edge may associate with a curve (which is required to be non-self-intersecting). If we define an ordering of an edge's two vertices, we have an *oriented edge.*

A chain of unique, oriented edges, connected head-to-tail, forms a *path.* If the first and last vertex of the path is the same, it is called a *loop.*

A *face* is a topological entity of dimensionality 2 which is defined by one or more non-intersecting loops, and which can associate geometrically with a surface. The loops represent the face's outer (and possibly inner) boundaries. Different faces might have boundary loops that share certain edges, which means that those faces are connected along those edges.

A set of faces connected along common edges in a way such that they form a 2-manifold is called a *shell.* A shell may be *open* or *closed.* An open shell has a outer boundary consisting of all those edges that are only participating in a single face, whereas all edges in a closed shell participate in exactly two faces and thus the closed shell has no boundary. If the faces of a closed shell are associated with surfaces in Euclidean space, the resulting, combined surface divides the space into two regions, one bounded and one unbounded. For this reason, closed shells are used to delimit a volume (a 3D model) in Euclidean space.

4 The STEP Standard and Part 42

STEP stands for "Standard for the Exchange of Product Model Data", and became an international standard (ISO 10303) in 1994. It provides system-

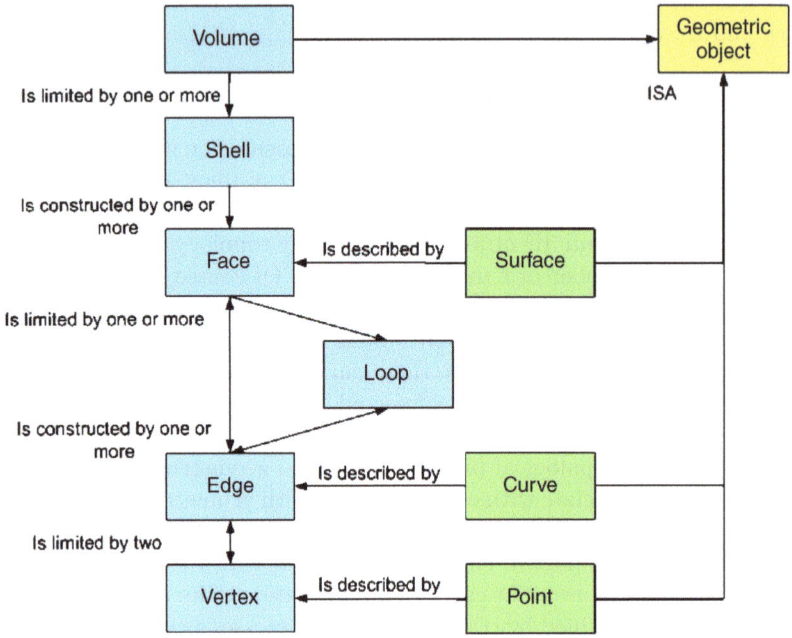

Fig. 7. A schematic view of the relationships between topological entities, geometric entities and model.

independent formats for describing product model data, and addresses the key engineering problems of portability, interoperability, longevity and extensibility. The term "product data" is used in a wide sense, and refers to all information generated for a product or a process during its life cycle, from design and manufacture to use, maintenance and disposal. The STEP standard is organized as a series of separately published parts, each of which falls into one of the categories:

- integrated resources
- application protocols
- abstract test suites
- implementation methods
- conformance testing

The STEP standard was written to obtain data exchangeability between systems, and has become extensively used by industry, particularly in the CAD/CAM world.

4.1 Modeling Language

The data modeling language used by STEP is called EXPRESS, and is part of the standard itself (ISO 10303-11). It is a formal, object-oriented data specification language, based on an entity relationship attribute model with generalization/specialization and specification of constraints. It can be presented in graphical or written form. The language is readable to humans and fully computer interpretable.

The fundamentals for building a data model in EXPRESS are *schemas*, *entities*, *attributes*, *types* and *rules*. A *schema* defines an area of interest and contains all relevant specifications for that particular area. It can be regarded as a domain-specific ontology, whose definitions and concepts can be re-used elsewhere by means of an interface between *schemas*. For readers familiar with object-oriented design, the concept of *entities* should be quite straightforward, as an *entity* specifies a class of objects with common properties. It defines a real-world concept, from which specific *instances* can be generated. A supertype/subtype mechanism makes it possible to let entities be generalizations/specializations of other entities. *Attributes* are properties that assign semantics-laden values from some value domain to entities, or representing relations between entities or instances of entities. The value domain that an attribute can take is defined by its *type*. Finally, *rules* are provided to express constraints that must be satisfied by the data model, either on the global level or for a specific entity.

EXPRESS provides many of the common ideas used in ontology design: *classes* and *instances of classes* (by using entities), *multiple inheritance of classes* (by the supertype/subtype mechanism of entities), *properties* and *class relations* (by using attributes), *value domains* (of attributes), *cardinality* (number of values that an entity may possess for a given attribute), *part-of relations* (again using attributes) and *asserted conditions* (by using rules).

There is no mechanism in EXPRESS for modeling inheritance relationships between attributes, only between entities. Neither does EXPRESS support polymorphism or encapsulation.

More details on the EXPRESS language will be presented in the comparison with RDFS/Protégé in Section 5.

4.2 Part 42: Geometric and Topological Representation

This part of the STEP standard belongs to the "integrated resources" category, which signifies that its data constructs are independent of any specific application context and that its schemas are used for all application protocols in the standard. The STEP Part 42 contains schemas for representing the shape of product models, and is the part of interest to us for specifying our ontology of CAD model information. In addition to a list of word and symbol definitions, normative references and abbreviations, Part 42 contains

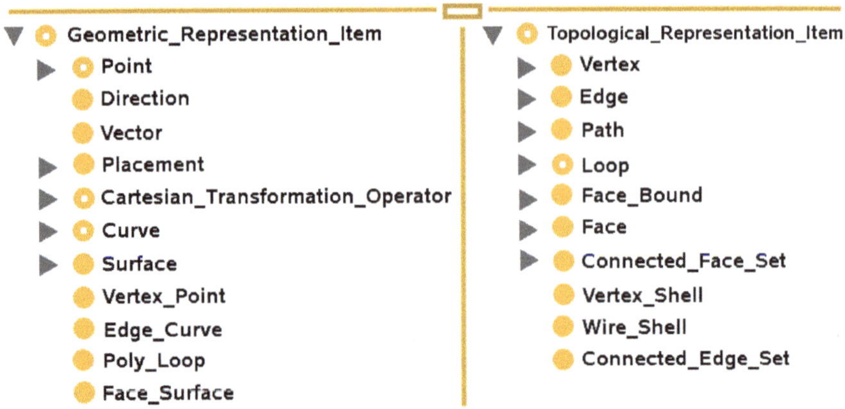

Fig. 8. Names of main base classes in the EXPRESS Geometry and Topology Schema, visualized in Protégé. Classes with a grey triangle on their left have additional subtypes.

three schemas. The two first schemas specify the data models used for representing *geometry* (in Euclidean space) and *topology*, respectively. Together, these schemas provide the basic structure necessary for specifying boundary representations, as mentioned in Section 3.1. The third schema provided by Part 42 is called "Geometric Shape Models", and provides resources for communicating the size, position and shape of objects. A large part of this schema is dedicated to the specification of constructive solid geometry objects (as opposed to boundary representations). Although it also contains more specific constraints and entities related to boundary representations, we consider this schema as out of scope for the simpler ontology that we aim to specify in Section 4.3.

Neither of the schemas live in a vacuum. They refer to each other, as well as to types found in schemas from other STEP parts. Concepts related to various measurements are imported from the *Measure* schema (STEP part 41), where basic, self-explanatory types such as 'length_measure', 'plane_angle_measure' and 'parameter_value' are found. Concepts related to a particular representation are imported from the *Representation* schema, which provides entities such as 'representation_context' and 'functionally_defined_transformation'.

An exhaustive description of the schemas requires a manual of its own[6]. In the following paragraphs we will give the reader an idea about their structure by using some relevant examples.

4.3 Outline of the STEP 42 Geometry Schema

The Geometry schema contains the data models for concepts that directly relate to 2D or 3D Euclidean space. Most fundamentally, it specifies an entity called 'geometric_representation_context', which derives from the imported 'representation_context' mentioned above. This entity is important, as it designates the 1D, 2D or 3D world in which a particular geometric model lives. Different geometrical items cannot relate to each other (e.g. distances become meaningless) unless they are part of the same context.

In addition to the 'geometric_representation_context', the Geometry schema specifies another fundamental entity called the 'geometric_representation_item'. This is the superclass for all entities that have geometric (as opposed to topological) meaning. It has the property that it is founded in one or more 'geometric_representation_context' of the same dimension.

All the rest of the entities specified in the Geometry schema are specializations (subtypes) of the 'geometric_representation_item'. This includes, for example, 'point', 'direction', 'vector', 'placement', 'curve', 'surface' and 'cartesian_transformation_operator'.

The 'point' entity represents a specific position in space. Its subtypes represent different ways of defining this position. For instance, the 'cartesian_point' uses a list of coordinates, whereas the 'point_on_curve' specifies its location by referring to a specific curve (instance of the 'curve' entity) as well as the parameter value of the point on the curve.

The difference between the 'direction' and the 'vector' entities in STEP, is that the former solely models a direction in space, whereas the latter also has a length. Thus, the vector has two attributes, its *orientation* (taking a 'direction' as value) and its *magnitude* (taking a 'length_measure' as value).

A 'placement' is similar to a point in that it refers to a position in space, but differs in that it is not used to represent that point per se, but rather to specify the reference location of another geometric item. Subtypes of 'placement' can in addition specify a local coordinate system at this point.

The Geometry schema contains many entities describing curves or surfaces. All are specializations of respectively the 'curve' or the 'surface' supertypes. Algebraic representations of curves and surfaces are limited to elementary kinds: 'line' and 'conic' in the curve case, 'plane', 'cylindrical_surface', 'conical_surface', 'spherical_surface' and 'toroidal_surface' in the surface case.

All curves and surfaces defined in the Geometry schema have a defined parameterization, even those who are algebraically or procedurally defined. This ensures that it is always possible to refer directly to a point on a given curve or surface using parameter values. For instance, the 'circle' entity represents a circle defined by the attributes *radius* (taking a 'length_measure' as value) and *position* (taking for value a subclass of 'placement', defining the center position and a local coordinate system). If we denote the radius R, the center C and the local coordinate system $(\mathbf{x}, \mathbf{y}, \mathbf{z})$, the circle's parameterization is defined by:

$$\lambda(u) = C + R((\cos u)\mathbf{x} + (\sin u)\mathbf{y})$$

where $0^o \leq u \leq 360^o$, and $\lambda(u)$ is the corresponding point on the circle.

Curve types with identifiable end points are represented by the type 'bounded_curve', which contains subtypes such as 'b_spline_curve', 'trimmed_curve' and 'composite_curve'. The 'b_spline_curve' entity models the general representation of a piecewise (possibly rational) polynomial curve, as mentioned in Section 3.2. Polynomial curves and surfaces are in general characterized by their *degree*, their *control points* and their *parameterization*, as well as whether they are *rational* or not. The first two of these properties are modeled using attributes, the last two by using subtypes. Likewise, surfaces with finite area and identifiable boundaries are all specializations of the 'bounded_surface' entity, where we find subtypes such as 'b_spline_surface' (tensor product surfaces, modeled much the same way as 'b_spline_curve', 'rectangular_composite_surface' (assembled from an array of rectangular surfaces with transition continuity information), and 'curve_bounded_surface', which defines a *trimmed* surface whose boundaries are defined by one or more 'boundary_curves'. These again are composed of segments that are *curves on surfaces* (CONS - as mentioned in 3.3).

STEP provides two ways of representing CONS, either by using the 'pcurve' entity or by using the 'surface_curve' entity. The former has the simplest representation (in terms of attributes and constraints). It represents a curve on a surface, defined by an attribute referring to the actual surface, as well as an attribute referring to a 2D curve in the parametric domain of the surface. The 'surface_curve' contains more information, which must be properly constrained using *rules*, and is well suited for representing curves lying on more than one surface (*intersection curves*) or which have two distinct representations in the parametric plane of a single surface(*seam curves*). For a detailed treatment of the intersection problem in CAD, see the next chapter [16] in this book.

Procedurally defined curves in STEP are 'offset_curve_2D', 'offset_curve_3D' and 'curve_replica'. Procedurally defined surfaces are 'swept_surface' (with the subtypes 'surface_of_linear_extrusion' and 'surface_of_revolution'), 'offset_surface' and 'surface_replica'.

Other forms of procedurally defined surfaces, or curves and surfaces without any parametric form of representation, are not covered. Neither are any other algebraically defined curves or surfaces nor Bézier triangles.

4.4 Outline of the STEP 42 Topology Schema

This schema defines fundamental topological entities, as described in Section 3.4. The fundamental concept here is the entity called 'topological_representation_item'. This is the supertype for all other entities in the topology schema. The fundamental entities, in order of increasing complexity, are 'vertex', 'edge', 'path', 'loop', 'face' and various kinds of *shells*. In addition, entities are also defined for more general high-level structures (not bound by the constraints of a *shell*) such as 'connected_face_set' and 'connected_edge_set'.

We already mentioned that topological vertices may associate with geometric points, edges with curves and faces with surfaces. This is modeled in STEP by using the entities 'vertex_point', 'edge_curve' and 'face_surface' (which are subtypes of 'vertex', 'curve' and 'face' respectively). These add the semantics to their supertypes of being associated with geometric entities. Since they refer to geometry, they are at the same time subtypes of 'geometric_representation_item', and are an example of multiple inheritance in EXPRESS. There is a *rule* stating that any part of a curve or surface that corresponds to the domain of a topological entity (such as 'edge' or 'face') must be a *manifold*, which entails that they are not self-intersecting.

Edges are defined by referencing two vertices. An edge may in addition specify an *explicit ordering* of its vertices, and thus be of the subtype 'oriented_edge'. An edge needs to be oriented in order to participate in the definition of a 'path'. A special subtype of 'path' is 'open_path', which has the additional constraint that each 'vertex' participating in the definition of the part is referred to only once. (In particular, this means that the start and end vertices are different).

On the other side, a 'loop' is constrained to have coincident start and end vertices. The loop may be defined by a sequence of 'oriented_edge'. In this case, it is of the subclass 'edge_loop', which is coincidentally also a subtype of 'path'. It may also be defined by a single 'vertex', in which case its subtype is 'vertex_loop'. It represents a loop that is degenerated into a single point. Finally the loop may be specified by a set of (geometric) 'cartesian_point', which refer to the loop around a planar polygon in 2D or 3D space. This representation has the subtype 'poly_loop' and is simultaneously a subtype of 'geometric_representation_item'.

A 'face' is defined by its one or more 'face_bound' which again are defined by a set of 'loop'. A 'face' may have an *orientation* which is given by the orientation of its defining 'loop's and possibly a flag to invert it.

Shells are in STEP defined to be structures that are connected objects of fixed dimensionality 0, 1, or 2, typically used to bound a region. The 0D entity is called 'vertex_shell', and is defined by a single 'vertex'. The 1D entity is called 'wire_shell' and is defined by a collection of 'loop'. In the 2D case, a shell can be represented as an 'open_shell' or a 'closed_shell', which are defined by collections of 'face'. If the faces of an 'open_shell' or a 'closed_shell' refer to geometric surfaces, the combined surface should be a *connected, oriented 2-manifold.* This requires the participating faces to be consistently oriented (the orientation should not flip between two neighboring faces). The manifold should be *open* for an 'open_shell', meaning that it has a nonempty boundary and does not divide the space into two distinct regions. On the other hand, it should be 'closed' for a 'closed_shell', meaning that it has an empty boundary and that it divides the space into one bounded and one unbounded region. In this context, the boundary is defined by all participating edges that are not shared by two faces, as well as the vertices they are connecting. Moreover, instances of 'open_shell' or 'closed_shell' must satisfy certain topological conditions formalized by the *Euler equations.*

More general sets of faces, not necessarily satisfying the conditions required for 'open_shell' or 'closed_shell', can be represented by the entity 'connected_face_set'.

5 Building a STEP-Based Ontology in Protégé

The original goal of the AIM@SHAPE ontology initiative was to specify four domain-specific ontologies and to place them in a common framework by integrating them into a common upper-level ontology. The four domains originally chosen were "*Virtual Humans*", "*Geometry/Topology*", "*Design*" and "*Acquisition and Reconstruction Pipeline*". Obviously, in order for this to work, the ontologies in question would have to be expressed in a common language, which was originally chosen to be RDFS, using the Protégé ontology editor [10] provided by Stanford University. At a later stage, the common format was changed to OWL [5], which represents the latest recommendation in the Semantic Web initiative, and poised to become a general web standard.

Our task was to make a draft proposal for the second of the above mentioned ontologies. As there already existed a heavily used industrial format for product model information, that formulated a similar ontology, it was decided to base our work as much as possible on the existing standard.

The basis for our work was the original 1994 edition of STEP 42.

5.1 The Scope of Our Ontology

We wanted to write an ontology of CAD model information that should be

- expressible using RDFS for the ontology schema (and RDF for instances);
- simple (since this would be a first attempt later to be integrated in a common framework);
- as close to the industry standard as possible.

Even though the STEP standard was developed to obtain data exchangeability between systems in the product development world, the ultimate goal of the AIM@SHAPE initiative would be to allow information to be exchanged between systems of very different natures. This is however an ongoing work and beyond the scope of this article.

5.2 Comparing EXPRESS to Modeling in RDFS Using Protégé

Like EXPRESS, the RDFS modeling language, built on XML, adheres to an object-oriented philosophy, and there are relatively close correspondences to the fundamentals in EXPRESS, as can be seen in Table 1.

The RDFS *schema* is similar to its EXPRESS counterpart in that it models a certain knowledge domain whose concepts also can be re-used by other

Table 1. Comparison of concepts between EXPRESS and RDFS

EXPRESS concept	**similar RDFS concept**
schema	*schema*
entity	*class*
attribute	*slot*
type	*XML type definitions, for defining basic/primitive types*
rule	*none (the Protégé environment does, however, provide support for its own, non-standard, PAL-constraints)*
instance	*instance*

domains. Unlike the EXPRESS *schema*, however, it cannot be included partially.

The concept of *class* in RDFS is more or less equivalent to the EXPRESS *entity*; it specifies a category of objects with *common properties* that can also be *instantiated.* (Instances are similarly used in the two modeling languages). Like EXPRESS subtypes/supertypes, classes can *inherit* from each other, and we also have the possibility of *multiple inheritance.* Both EXPRESS and RDFS differentiate between *abstract* and *concrete* classes/entities. An *abstract* class cannot have instances, it can only be used as a parent class to other classes. Protégé refers to this as the *role* of the class.

The EXPRESS *attribute*, which is used to model the properties of entities, has its counterpart in the RDFS *slot.* EXPRESS *attributes* have value domains specified by their *type.* The concept of *type* in EXPRESS is more general than *entity.* Entities *are* types, but EXPRESS also specifies other, built-in types, such as *NUMBER* and *STRING*, as well as types that can be constructed by the user (*ENUMERATION*, *SELECT*), collections of types (*ARRAY*, *SET*...), etc.

Like EXPRESS *attributes*, RDFS *slots* also have value domains, but they are specified somewhat differently. Like in EXPRESS, the value domain can be set to instances of a specific class, or to a built-in primitive type (*Boolean*, *Float*, *Integer*, *String*) or user-specified enumeration (*Symbol*). Unlike EXPRESS, *slots* can also have a value domain that is the set of *names of classes.* Rather than indicating a specific *instance* of a given class, such a slot indicates a *class* in itself.

As just stated, EXPRESS must use a *collection type* (ARRAY, SET, etc.) in order to specify an *attribute* taking multiple *values.* This is modeled in Protégé using *cardinality*, which is a property assigned to each specified *slot* in addition to its value domain. A cardinality specifies a restriction on the number of times that the specific property can occur for a specific instance of a class. (Note: this is a feature provided by Protégé but which is not formally a part of RDFS).

There are a few other differences to note between *slots* and *attributes*. Firstly, EXPRESS supports *derived attributes*, which is a feature of which RDFS has no direct equivalent. A *derived attribute* is an attribute whose value for a particular instance is not explicitly given, but calculated by some formula, based on other information about the instance. Secondly, in RDFS, *slots* can also inherit from each other, in much the same way classes do.

Both RDFS and EXPRESS support domain restrictions for slots/attributes for a specific class/entity. This means for instance that if the general value domain for the slot S is V, and $Vsub$ is a class inheriting from V, then the value domain for S when used in a particular class can be *restricted* to only refer to instances of $Vsub$, not V in general. Another feature provided by EXPRESS is the possibility of two attributes to be *inverses* of each other. If $S1$ and $S2$ are inverse slots and class instance A has the slot $S1$ taking class instance B as its value, then B must have the slot $S2$ taking A as its value. Although this functionality is also available in the Protégé environment, it is not part of RDFS as such.

Constraints specified in EXPRESS are called *rules*. There are various kinds of rules in EXPRESS, which might be global, entity-specific or related to the inheritance hierarchy itself. RDFS by itself offers no constraint specification mechanism, but Protégé provides it with a plug-in called PAL ("Protégé Axiom Language"). The purpose of PAL is to support arbitrary logical constraints, and it offers functionality comparable to the global and entity-specific rules of EXPRESS. As for the EXPRESS rules pertaining to the inheritance hierarchy itself, there is no equivalent either in RDFS or Protégé PAL, simply because the inheritance structure is not complex enough to warrant one. That is, in EXPRESS it is possible to *combine* subtypes of an entity into additional, not explicitly defined, subtypes that can be instantiated just like any other subtype. For example, in STEP42, the entity 'b_spline_curve' has the subtypes 'bezier_curve' and 'rational_b_spline_curve'. There is nothing to prevent a user for instantiating a 'b_spline_curve' that is *both* a 'bezier_curve' and 'rational_b_spline_curve'. In this case, such a behavior is desirable, but in other cases, where subtypes should be mutually exclusive, constraints on the hierarchy must be defined. As RDFS subclasses cannot be combined in this fashion anyway, no such constraints are necessary.

Both EXPRESS and RDFS/Protégé provide additional features as well, but not being particularly relevant to our ontology example, we will not go into details.

5.3 A Short Comment on OWL

Although OWL was not used in the work described in this paper, it was later chosen by the AIM@SHAPE members to be the language of choice for later work. Moreover, it is a more expressive language than RDFS and is the latest recommendation in the Semantic Web initiative. For this reason, we find it natural to give in this paper a short introduction of how OWL compares to and improves on RDFS.

OWL builds on top of RDF Schema. The purpose of OWL is identical to that of RDF Schema: to provide an XML vocabulary for defining classes, properties and their relationships. RDF Schema enables you to express very rudimentary relationships and has limited inferencing capability.

OWL provides a syntax for expressing statements about properties and classes, above and beyond what can be expressed with RDF Schema. OWL enables you to express much richer relationships, thus yielding a much enhanced inferencing capability.

In OWL, restrictions can be imposed both on classes and on properties in order to refine their semantics. Suppose for example that we wish to declare that a class C satisfies certain conditions, that is, all instances of C satisfy the necessary conditions. Using an OWL *Restriction* in an ontology describing a University, for example, we can easily describe the axiom that says that *all academic staff members must teach one graduate course.* This imposes a restriction on the class *Academic Staff*, thus making it more specific. In this example we state that every professor must teach a graduate course. In terms of logic we have a *universal quantification.*

Properties can also be defined to have special characteristics. For example, sometimes it is useful to say that a property is *transitive* (like "greater than"), *unique* (like "is mother of"), *symmetric* (like "is sibling of"), or the *inverse of* another property (like "eats" and "is eaten by").

In OWL it is possible to talk about boolean combinations (*union, intersection, complement*) of classes. In the University example, we can say that courses and staff members are *disjoint* and we can find all the people at the University by finding the union of the *Staff Members* and the *Student* classes.

In RDF Schema we cannot declare range restrictions that apply to some classes only. For example, we cannot say that cows eat only plants, while other animals may eat meat too.

Sometimes we wish to place restrictions on how many distinct values a property may or must take. For example, we would like to say that a person has exactly two parents, and that a course is taught by at least one lecturer. Such *cardinality* restrictions are impossible to express in RDF Schema.

5.4 Defining the 'Beyond Step' Ontology

Due to its close resemblance with the STEP 42 Geometry and Topology schemas, which have already been outlined in Sections 4.2, 4.3 and 4.4, we do not delve into details in the structure of our ontology, informally named "Beyond STEP". Instead, we will just point out some of the design decisions we had to make in order to overcome arising difficulties, as well as explaining our decision to extend the ontology with general algebraic curves and surfaces. The actual ontology can be viewed on and also downloaded from the AIM@SHAPE project website [1].

"Beyond STEP" was implemented as a single Protégé project, covering both of the concerned STEP 42 schemas. The names of classes and slots were always chosen to match their STEP entity/attribute equivalents as closely as possible, except for using capital initial letters for classes.

When implementing the ontology, the first question to address was how to deal with types and entities imported from *other* schemas. For each imported item, there were basically two options: insert its definition into the new ontology, or avoid using it. The first option was taken for the basic elements 'length_measure', 'parameter_value' and 'plane_angle_measure'. These all contain a single numeric value, but have semantically different meanings. Other items mentioned from the 'measure schema' did not play any active part in the geometry or topology schemas, and were ignored. Among the supertypes from the 'representation schema', only one was inserted in our new ontology. This was the 'representation_item' entity, which in our ontology became the superclass of the major classes 'geometric_representation_item' and 'topological_representation_item'. The other supertypes imported from 'representation schema' only had a single relevant specialization, and were considered unnecessary. The rest of the 'representation schema' types that were imported into the original schemas were either only relevant in the specification of functions and rules, or their roles could be substituted by adding information in other places and were thus left out.

Another matter that had to be settled from the beginning was how to deal with the *basic type definitions* in the STEP 42 schemas. These are the types that are not *entities* and cannot be instantiated, but which may be used in *rules* or to specify an *attribute*'s value domain. Some of the basic type definitions were not used in the Geometry or Topology schema, or were used only for rules and functions. These were ignored altogether. The remaining type definitions could be separated into EXPRESS INTEGERs, ENUMERATIONs and SELECTs. The only imported INTEGER type was 'dimension_count'. Used as domain value type for attributes/slots, we merely substituted its use with Protégé Integers (which are RDFS Literal values). Likewise, all imported ENUMERATION types were replaced with the use of the Protégé Symbol value type (again a RDFS Literal), which defines a list of allowed values. In the EXPRESS schemas, the SELECT types were used to define value domains that extend to more than one class. In RDFS, where there is no limit to the number of classes that constitute a slot's value domain, we could dispose of the intermediary SELECT type.

The implementation of EXPRESS rules and constraints in terms of Protégé PAL-constraints was considered out-of-scope for the development of this tentative ontology. They could however be included at a later iteration of the development.

One type of EXPRESS constraints could not be ignored, namely that related to the inheritance hierarchy itself. As explained earlier, the inheritance mechanism in Protégé/RDFS lacks the degree of freedom of the EXPRESS framework, where sub-entities are not mutually exclusive. Much of the time

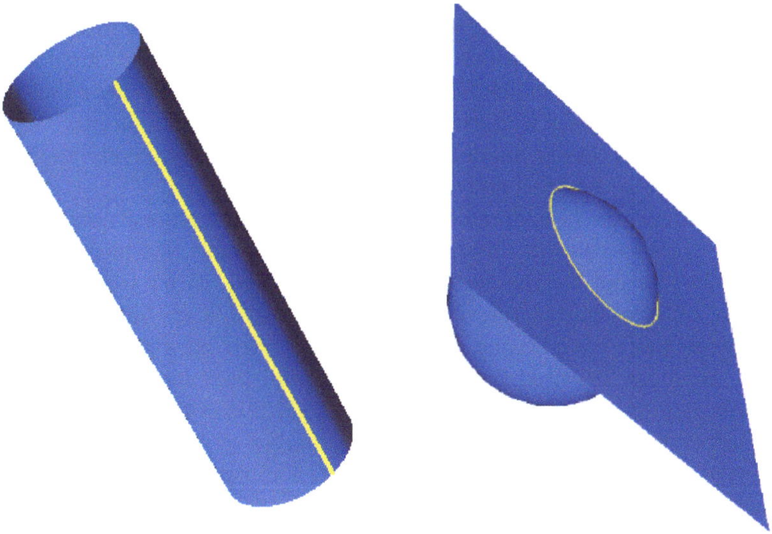

Fig. 9. Surface curves in the roles of a seam curve (left) and an intersection curve (right).

this did not really pose a problem, because this exclusivity constraint was imposed anyway for most concerned inheritance relations. However, in some cases it had to be dealt with. In cases where the Geometry or Topology schema allowed two or more sub-entities to combine when instantiating an object, the classes corresponding to each such combination had to be explicitly introduced to our Protégé ontology.

One such case was for the EXPRESS 'surface_curve', which we have already briefly discussed earlier. This entity represents a curve on a surface, and can more specifically represent the curve of intersection between two surfaces. On the other hand, it can also represent a "seam curve", where a surface meets itself along two of its boundaries (like a cylindrical surface). However, it cannot represent both at the same time.

For that reason, 'surface_curve' has two subtypes, 'intersection_curve' and 'seam_curve', which are made mutually exclusive. In addition, a surface curve can be *bounded*, giving it the property of having finite arc length and identifiable end points. Therefore 'surface_curve' also has a subtype called 'bounded_surface_curve'. This type is *not* mutually exclusive with the other subtypes, so it is possible to create a curve instance that is *both* a 'intersection_curve' *and* a 'bounded_surface_curve', and the same goes for 'seam_curve'. In order to allow this in Protégé, we had to create the specific classes 'bounded_intersection_curve', 'bounded_seam_curve', which simultaneously inherits from 'bounded_surface_curve' and respectively 'intersection_curve' and 'seam_curve'.

The problem was resolved in a similar way for the few other cases. For the entity 'b_spline_curve', its various specializations all had the additional ability of being *rational* or not, which in EXPRESS was modeled by their ability to create common instances with the 'rational_b_spline_curve' subtype. We had to add the possible classes manually to our ontology. A completely analogous situation also arose for the entity 'b_spline_surface'.

The last such case to be handled was for 'topological_representation_item'. It is the direct supertype of all topological concepts, most of which are mutually exclusive, except for the two subtypes 'loop' and 'path'. It is allowed to instantiate objects that represent at the same time a 'loop' and a 'path'. To make matters more complicated, both 'loop' and 'path' have several subtypes: for 'path' these are 'oriented_path', 'open_path' and 'edge_loop', for 'loop' they are 'vertex_loop', 'poly_loop' and 'edge_loop' (notice that the 'edge_loop' is an explicit specialization of both entities). The number of possible combinations of subtypes might seem formidable. However, a closer look reveals that the entity-specific constraints eliminate the instantiation of most such combinations, and as the combination of a basic 'path' and a 'loop' is already explicitly present as 'edge_loop', the only new class we chose to add to our class hierarchy was the 'oriented_edge_loop', which inherits from 'oriented_path' and 'loop'.

As an extension, we added two classes without any equivalent in STEP, namely the 'general_algebraic_curve' (as a direct subtype of 'curve', and the 'general_algebraic_surface' (as a direct subtype of 'surface'). Today, the support of algebraic curves and surfaces in STEP 42 is limited to some basic types. However, due to current academic developments, as well as the ever-increasing power of computer graphics cards, it can be argued that future CAD systems might increasingly orient themselves towards expressing curves and surfaces by means of complex algebraic representations. This was our motivation for introducing these classes into the ontology.

6 Conclusion; Where Do We Go from Here?

Later developments in the AIM@SHAPE work on ontologies led to the decision that further development of the "Beyond STEP" ontology would have to be abandoned for the moment, while the other three domain-specific ontologies, which focus on the higher-level fields of shape acquisition, the design process and virtual human modeling, were to be further developed, and are expected to be integrated into a common upper-level ontology as work proceeds.

In the spirit of the Semantic Web vision, a search engine for digital shapes is also being developed. This search engine will be able to search the AIM@SHAPE repositories for specific shapes, shape processing tools or relevant publications. Both the search engine and the repositories will work with information within the framework of the specified ontologies. The search engine will also provide inference capabilities, by utilizing a Description Logic(s) reasoner such as RacerPro [4] for example, in order to not only expose searching functionality but to also be able to implicitly deduce knowledge from explicit information residing in meta-data repositories.

The AIM@SHAPE resource repositories are expected to grow to a considerable size over the next years. Although it is very hard to give estimates, it is expected that the number of available shapes will reach one thousand, the number of available software tools to be around two hundred, and the number of referenced publications to be in the thousands. In addition to providing the academic (and industrial) community with valuable and easily-accessible resources, the AIM@SHAPE ontology framework will also serve as a good example of the potential that the Semantic Web holds for these communities.

Although the "Beyond Step" ontology does not play an active part in this current development, it is clear that the widespread use of the STEP standard in the industrial community warrants closer examination on how this information can be integrated on a higher level, thus becoming available for use in other settings. Whether integration in some upper ontology can be done directly, or whether a reformulation (like our informal example) will prove to be necessary is a question that will have to be addressed more rigorously at some point in the future.

Acknowledgement. This research was financially supported by the EU through the FP 6 IST Network of Excellence AIM@SHAPE - Contract No. 506766.

References

1. `http://www.aimatshape.net`
2. `http://www.cyc.com`
3. `http://www.w3.org`
4. `http://www.racer-systems.com`
5. `http://www.w3.org/tr/owl-features/`
6. Technical Committee ISO/TC 184. *Industrial automation systems and integration - Product data representation and exchange - Part 42: Integrated generic resources: Geometric and topological representation*, 1994.
7. `http://www.w3.org/2001/sw/`
8. `http://www.w3.org/rdf/`
9. `http://www.w3.org/tr/rdf-schema/`
10. `http://protege.stanford.edu`
11. `http://www.w3.org/sw/webont/`
12. S. Bechhofer, M. Dean, F. van Harmelen, J. Hendler, I. Horrocks, D. McGuinness, P. Patel-Schneider, and L. Stein. Web ontology language (OWL) reference version 1.0. W3c recommendation, World Wide Web Consortium, 2004.
13. T. Berners-Lee, J. Hendler, and O. Lassila. The semantic web. *Scientific American*, May 2001.
14. D. Brickley and R. V. Guha. RDF vocabulary description language 1.0: RDF schema. W3c recommendation, World Wide Web Consortium, February 2004.

15. J.J. Carroll and G. Klyne. RDF concepts and abstract syntax. `http://www.w3.org/TR/rdf-concepts/`, February 2004.
16. T. Dokken. Intersection algorithms and CAGD. *This book...*, 1(0–0):123–456, 2006.
17. G. Farin. *Curves and Surfaces for CAGD.* Morgan Kaufmann Publishers, fifth edition, 2002.
18. G. Farin, J. Hoschek, and M.-S. Kim. *Handbook of Computer Aided Design.* Elsevier Science B.V., first edition, 2002.
19. T.R. Gruber. A translation approach to portable ontology specifications. *Knowl. Acquis.*, 5(2):199–220, 1993.
20. D.L. McGuinness and F.van Harmelen. Owl web ontology language overview. Technical report, World Wide Web Consortium, February 2003.

Intersection Algorithms and CAGD

Tor Dokken and Vibeke Skytt

Summary. An approach for calculating intersections between parametric surfaces based on long experience in developing intersection algorithms for industrial use, is presented. In addition two novel methods that help improve quality and performance of intersection algorithms are described:

- **An initial assessment of the intersection complexity** to identify most transversal intersections, and to identify surface regions with possible complex intersection topology. To find regions where the surfaces possibly intersect, and regions where surface normals possibly are parallel, the computational power of multi-core CPUs and programmable graphics processors (GPUs) is used for subdivision of the surfaces and their normal fields.
- **Approximate implicitization of surface regions** to help analyse singular and near singular intersections.

Key words: Intersection algorithm, recursive subdivision, approximate implicitization, programmable graphics cards.

1 Introduction

The work within SINTEF on intersection algorithms started in 1979 [7, 23] with recursive intersection algorithms for curves, moving on to recursive surface-surface intersection algorithms for transversal intersections in the start of the 1980s [8]. A complete solution for transversal intersections was ready around 1989 in the SINTEF Spline Library [10]. In 2005 a new generation of algorithms was prototyped addressing singular and near-singular intersections; intersections where the surfaces are parallel or near parallel on or close to the intersection, and in 2006 the first attempts of GPU acceleration of NURBS surface intersections and self-intersections have been made.

This is not a survey of intersection algorithms, as the published approaches to intersections algorithms are numerous, and a survey would require a much longer paper. However, we have tried to provide references to published papers to give the reader access to more information. Performance and quality of intersection algorithms have been and are still important competitive factors within CAD/CAM. Thus, a lot of work on intersection algorithms performed in industry, R&D-institutes and universities has never been published. Some of these results were lost when CAD-systems were phased out, other results are well hidden within the current commercial CAD-systems.

This contribution is organised as follows:

- In Section 2 the difference between CAD-intersections and set intersections is addressed along with intersection complexity, the relation of CAD-intersections to the STEP-standard and other CAD-intersection challenges.
- An overview of different classes of surface intersections is given in Section 3.
- To improve quality and performance of intersection algorithms we discuss in Section 4 how multi-core CPUs and programmable graphics cards can be used early in the intersection process to detect intersection configurations that only result in transversal intersection curves. Employing this type of approach it is possible to avoid the use of more complex intersection algorithms on what can be described as simple intersection situations.
- Many different algorithms or tools can be used in intersection calculations. In Section 5 we address those considered to be most important including the novel approach of approximate implicitization, and the potential of combining algebraic and parametric descriptions.
- In the Appendix we discuss different approaches for the representation of piecewise polynomials and their influence on the quality of the intersection calculation. The Appendix also addresses the importance of consistence between different algorithms.

CAD-systems play a central role in most producing industries. The investment in CAD-model representation of current industrial products is enormous. CAD-models are important in all stages of the product life-cycle, some products have a short life-time; while other products are expected to last at least for one decade. Consequently backward compatibility of CAD-systems with respect to functionality and the ability to handle "old" CAD-models is extremely important to the industry. Transfer of CAD-models between systems from different CAD-system vendors is essential to support a flexible product creation value chain. In the late 1980s the development of the **STEP standard** (ISO 10303) *Product Data Representation and Exchange* [29] started with the aim to support backward compatibility of CAD-models and CAD-model exchange. STEP is now an important component in all CAD-systems and has been an important component in the globalisation of design and production. The drawback of STEP is that it has standardised the geometry processing technology of the 1980s, and the problems associated with that generation of technology. Due to the CAD-model legacy (the huge bulk of existing CAD-models) upgraded CAD-technology has to handle existing models to protect the resources already invested in CAD-models. Consequently the CAD-customers and CAD-vendors are conservative, and new technology has to be backward compliant. Improved intersection algorithms have thus to be compliant with STEP representation of geometry and the traditional approach to CAD coming from the late 1980s. For research within CAD-type intersection algorithms to be of interest to producing industries and CAD-vendors backward compatibility and the legacy of existing CAD-models have not to be forgotten.

2 CAD Intersection $\neq$ Set Intersection

2.1 STEP Constrains CAD Intersection Representation and Functionality

Finding the intersection of two or more objects can at a first glance seem to be a simple problem. Given two discrete sets A and B, then the intersection $A \cap B = \{x \mid x \in A \wedge x \in B\}$ is theoretically easy to determine. However, when addressing continuous sets the situations is more complex. Given two bounded surfaces in 3D then the correct intersection can be

- Empty.
- Points.
- Curves.
- Surface regions.
- Combinations of these.

However, when intersecting two surfaces in CAD-systems the intersection result has to be conform to what can be represented in the data structures of CAD-systems. CAD data structures are constrained by the STEP-standard, and are based on boundary representation of geometry, see Figure 1. The connectivity of the CAD-model is stored in a "topology" structure, and the topological elements are assigned a geometric description:

- A volume is limited by an outer shell and possibly a number of inner shells.
- A shell is described by a structure of faces.
- A face is described by a surface and limited by one or more loops. (One outer loop, and possibly numerous inner loops).
- A loop is an ordered collection of edges.
- An edge is described by a curve and limited by two vertices.
- A vertex is described by a point.

Figure 1 shows that there is not any direct link between a point and a surface. Thus the storage of surface intersections resulting in distinct intersection points is not straight forward. One of the reasons for this is that when STEP was agreed upon most surface intersection methods could only be relied

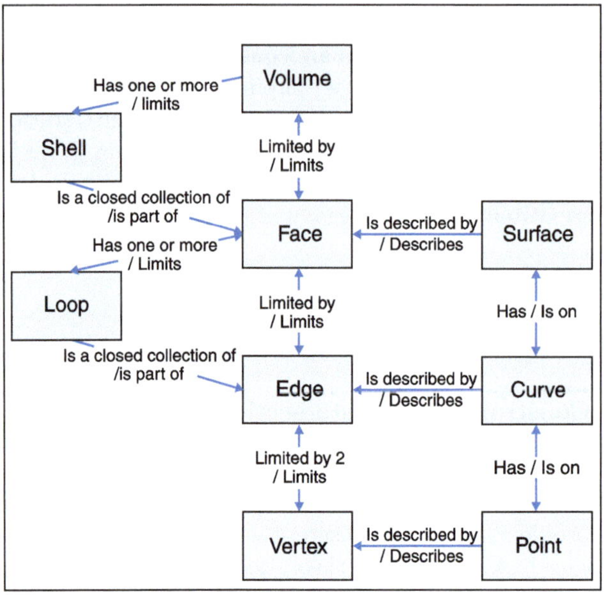

Fig. 1. Example of entities and relationships in a CAD-type data structure.

upon to produce high quality transversal intersection curves, distinct intersection points would not be detected. Also single intersection points in most cases will be removed as they somehow reflect a non-manifold structure of the geometric object. STEP assumes that the volume represented is manifold. Consequently current CAD-technology assumes that the intersection of two surfaces is either:

- Empty.
- Curves.

A first choice of an approach for restructuring intersection results could be to discard discrete intersection points and represent intersection regions with a curve describing the boundary of the intersection region. In some cases this would be the right approach. However, frequently when two surfaces are closer than the specified *point equality tolerance* without a real intersection occurring the surfaces are expected to be identified as intersecting. Consequently there are no exact rules for how to classify the intersection using empty sets and CAD-curves.

Another constraint from STEP on intersection algorithms is the representation offered for curves. Curves are in STEP represented as:

- First and second order algebraic curves (line, circle, ellipse,....).
- NURBS-curves.

Only special cases of intersections can be represented exactly by these representation formats, e.g., the intersection of two planes, the intersection of a plane and a quadric surface (sphere, cylinder,...). The general intersection of two quadric surfaces cannot be represented by the STEP type curves. Consequently most intersection curves have to be approximated.

Thus when representing the intersection of two surfaces we have in general to accept approximation tolerances:

- The approximated intersection curve should be as "simple" as possible.
- It should be within specified tolerances of the surfaces intersected (distance from surfaces).
- It should be within tolerances of the faces that it bounds, (distance from the faces).

2.2 Challenges of CAD-type Intersections

If the faces of the volume are all planar, then it is fairly straightforward to represent the curves describing the edges with minimal rounding error. However, if the faces are sculptured surfaces, e.g., bicubic NURBS - NonUniform Rational B-splines, the edges will in general be free form space curves with no simple closed mathematical description. As the tradition (and standard) within CAD is to represent such curves as NURBS curves, approximation of edge geometry with NURBS curves is necessary. For more information on the challenges of CAD-type intersections consult [48].

When designing within a CAD-system, *point equality tolerances* are defined that determine when two points should be regarded as the same. A typical value for such tolerances is 10^{-3}mm, however, some systems use tolerances as small as 10^{-6}mm. The smaller this tolerance is, the higher the quality of the CAD-model will be. Approximating the edge geometry with e.g., cubic spline interpolation that has fourth order convergence using a tolerance of 10^{-6} instead 10^{-3} will typically increase the amount of data necessary for representing the edge approximation by a factor between 5 and 6. Often the spatial extent of the CAD-models is around 1 meter. Using an approximation tolerance of 10^{-3}mm is thus an error of 10^{-6} relative to the spatial extent of the model.

The intersection functionality of a CAD-system must be able to recognise the topology of a model in the system. This implies that intersections between two faces that are limited by the same edge must be found. The complexity of finding an intersection depends on relative behaviour of the surfaces intersected along the intersection curve:

- **Transversal intersections** are intersection curves where the normals of the two surfaces intersected are well separated along the intersection curve. It is fairly simple to identify and localise the branches of the intersection when we only have transversal intersection.

Fig. 2. Transversal intersection between two sculptured surfaces

Fig. 3. Tangential intersection between two surfaces

- **Singular and near singular intersections** take place when the normals of the two surfaces intersected are parallel or near parallel in single points or along intervals of an intersection curve. In these cases the identification of the intersection branches is a major challenge.

Figures 2 and 3 respectively show transversal and near-singular intersection situations. In Figure 2 there is one unique intersection curve. The two surfaces in Figure 3 do not really intersect, there is a distance of 10^{-7} between the surfaces, but they are expected to be regarded as intersecting. To be able to find this curve, the point equality tolerance of the CAD-system must be considered. The intersection problem then becomes: Given two sculptured surface $f(u, v)$ and $g(s, t)$, find all points where $|f(u, v) - g(s, t)| < \varepsilon$ where ε is the point equality tolerance.

2.3 The Algebraic Complexity of Intersections

The simplest example of an intersection of two curves in $\mathbb{R}^2$ is the intersection of two straight lines. Let two straight lines be given:

- A straight line represented as a parametric curve

$$\mathbf{p}(t) = \mathbf{P}_0 + t\mathbf{T}_0, t \in \mathbb{R},$$

  with $\mathbf{P}_0$ a point on the line and $\mathbf{T}_0$ the tangent direction of the line.
- A straight line represented as an implicit equation

$$q(x, y) = ((x, y) - \mathbf{P}_1) \cdot \mathbf{N}_1 = 0, (x, y) \in \mathbb{R}^2,$$

  with $\mathbf{P}_1$ a point on the line, and $\mathbf{N}_1$ the normal of the line.

Combining the parametric and implicit representation the intersection is described by $q(\mathbf{p}(t)) = 0$, a linear equation in the variable t. Using exact arithmetic it is easy to classify the solution as:

- An empty set, if the lines are parallel.
- The whole line, if the lines coincide.
- One point, if lines are non-parallel.

Next we look at the intersection of two rational parametric curves of degree n and d, respectively. From algebraic geometry [41] it is known that a rational parametric curve of degree d is contained in an implicit parametric curve of total degree d.

- The first curve is described as a rational parametric curve

$$\mathbf{p}(t) = \frac{\mathbf{p}_n t^n + \mathbf{p}_{d-1} t^{n-1} + \ldots + \mathbf{p}_0}{h_n t^n + h_{n-1} t^{n-1} + \ldots + h_0}.$$

- The second curve is described as an implicit curve of total degree d

$$q(x, y) = \sum_{i=0}^{d} \sum_{j=0}^{d-i} c_{i,j} x^i y^j = 0.$$

By combining the parametric and implicit representations, the intersection is described by $q(\mathbf{p}(t)) = 0$. This is a degree $n \times d$ equation in the variable t. As even the general quintic equation cannot be solved algebraically, a closed expression for the zeros of $q(\mathbf{p}(t))$ can in general only be given for $n \times d \leq 4$. Thus, in general, the intersection of two rational cubic curves cannot be found as a closed expression. In CAD-systems we are not interested in the whole infinite curve, but only a bounded portion of the curve. So approaches and representations that can help us limit the extent of the curves and the number of possible intersections will be advantageous.

We now turn to intersections of two surfaces. Let $\mathbf{p}(s,t)$ be a rational tensor product surface of bi-degree (n_1, n_2),

$$\mathbf{p}(s,t) = \frac{\sum_{i=0}^{n_1}\sum_{j=0}^{n_2} \mathbf{p}_{i,j} s^i t^j}{\sum_{i=0}^{n_1}\sum_{j=0}^{n_2} h_{i,j} s^i t^j}.$$

From algebraic geometry [41] it is known that the implicit representation of $\mathbf{p}(s,t)$ has total algebraic degree $d = 2n_1n_2$. The number of monomials in a polynomial of total degree d in 3 variables is $\binom{d+3}{3} = \frac{(d+1)(d+2)(d+3)}{6}$. So a bicubic rational surface has an implicit equation of total degree 18. This has 1330 monomials with corresponding coefficients.

Using this fact we can look at the complexity of the intersection of two rational bicubic surfaces $\mathbf{p}_1(u,v)$ and $\mathbf{p}_2(s,t)$. Assume that we know the implicit equation $q_2(x,y,z) = 0$ of $\mathbf{p}_2(s,t)$. Combining the parametric description of $\mathbf{p}_1(u,v)$ and the implicit representation $q_2(x,y,z) = 0$ of $\mathbf{p}_2(s,t)$, we get $q_2(\mathbf{p}_1(u,v)) = 0$. This is a tensor product polynomial of bi-degree $(54,54)$. The intersection of two bicubic patches is converted to finding the zero of

$$q_2(\mathbf{p}_1(u,v)) = \sum_{i=0}^{54}\sum_{j=0}^{54} c_{i,j} u^i v^j = 0.$$

This polynomial has $55 \times 55 = 3025$ monomials with corresponding coefficients, describing an algebraic curve of total degree 108. As in the case of curves, the surfaces we consider in CAGD are bounded, and we are interested in the solution only in a limited interval $(u,v) \in [a,b] \times [c,d]$. This illustrates that the intersection of seemingly simple surfaces can result in a very complex intersection topology.

2.4 Intersection Stability and Floating Point Arithmetic

For the intersection of sculptured surfaces one source of accuracy problem is geometries constructed using approximation methods. However, there are also other sources. For industrial applications most often a floating point representation for numbers is chosen. In most such systems exact arithmetic is regarded as too slow to be of interest. Further the complexity of the problems to be solved is outside the domain of problems current computer algebra systems can handle.

For double precision IEEE arithmetic [28], *machine precision* is $\epsilon = 2^{-53} \approx 1.11 \times 10^{-16}$. The floating point approximation to a, denoted float (a), satisfies float $(a) \in [a(1-\epsilon), a(1+\epsilon)]$. The values of the closest neighbours to a floating point number a' are approximately $(1-2\epsilon)\,a'$ and $(1+2\epsilon)\,a'$.

The theory and algorithms being the basis of B-splines and NURBS assume that the actual numbers used are real numbers. However, when implementing the algorithms on computers using floating point arithmetic, the coefficients come from the discrete set of floating point numbers. When calculating intersections this does not create severe problems as long as the curves and surfaces are transversal (not near parallel) close to the intersection. However, an unstable situation arises when e.g., two curves intersect and are near parallel at the intersection point. Such intersections we call *near-singular*. Very often curve or surface intersections are produced by a lengthy computation using floating point arithmetic. During these processes rounding errors can accumulate extensively if care is not taken. Dependent on the actual algorithmic implementations the rounding error accumulation can evolve in different ways, and the same input given to supposedly equivalent algorithms can produce curves and surfaces that differ significantly. When such curves and surfaces are intersected later in the process chain, the intersection results produced will be inflicted with uncertainty. This is illustrated by the examples below.

Example 1 (Near singular intersection of straight lines). In spite of the simplicity of this intersection, the accuracy problem is highly relevant. We consider the intersection of two near parallel lines. The first represented as a parametric straight line $\mathbf{l}_1^j(s)$, $s \in [0,1]$, $j \in Z$,

$$\mathbf{l}_1^j(s) = (-1, 1 + j\epsilon)(1-s) + (1, 1-\epsilon)s.$$

The second represented with an implicit equation $l_2(x,y) = 0$,

$$l_2(x,y) = y - 1.$$

We want to see what happens when the left end point of the first line is changed in steps proportional to the smallest possible relative distance between adjacent floating point numbers, $\epsilon = 2.2 \times 10^{-16}$. We combine parametric and implicit representations to reduce the problem to one linear equation in one variable. If both lines had been represented by implicit equations, or both lines were represented as parametric lines, the intersection problem would be represented as two linear equations in two variables. Combining parametric and implicit representations is a very useful tool in intersection algorithms, and will be discussed in more detail in Section 5.3. The intersection of $\mathbf{l}_1^j(s)$ and $l_2(x,y)$ is

- $(x,y) = \left(\frac{j-1}{j+1}, 1\right)$ for $j \neq -1$.
- $y = 1$, for $j = -1$.

For increasing values of j, starting from $j = 0$ the intersection points are

$$\left\{(-1,1), (0,1), (\frac{1}{3},1), (\frac{1}{2},1), (\frac{3}{5},1), (\frac{2}{3},1), (\frac{5}{7},1), \ldots\right\}.$$

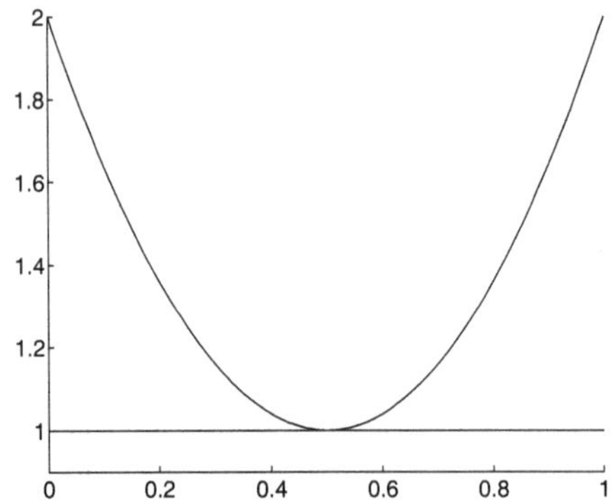

Fig. 4. A near singular intersection between a straight line and a parabola.

A change of magnitude $\epsilon = 2.2 \times 10^{-16}$ in the y-coordinate of the left end point of the straight line gives a large jump (16 orders of magnitude higher) in the x-coordinate of the intersection point.

Example 2 (Near singular intersection of straight line and parabola). Given a straight lines $l_1^j(x,y) = 0$, $j \in Z$,

$$l_1^j(x,y) = y - (1 + j\epsilon),$$

and a parabola $\mathbf{p}(s)$,

$$\mathbf{p}(s) = (s, 4s^2 - 4s + 2).$$

The intersection of the line and the parabola, see Figure 4, is:

- Empty for $j < 0$.
- One point $(x,y) = \left(\frac{1}{2}, 1\right)$ for $j = 0$.
- Two points $\left(\frac{1}{2} - \frac{1}{2}\sqrt{j\epsilon}, 1 + j\epsilon\right)$ and $\left(\frac{1}{2} + \frac{1}{2}\sqrt{j\epsilon}, 1 + j\epsilon\right)$ for $j \geq 1$.

When we move the straight line from parameter value j to parameter value $j+1$, the first intersection point moves from $\left(\frac{1}{2} - \frac{1}{2}\sqrt{j\epsilon}, 1 + j\epsilon\right)$ to $\left(\frac{1}{2} - \frac{1}{2}\sqrt{(j+1)\,\epsilon}, 1 + (j+1)\,\epsilon\right)$. This is a translation described by the vector $\left(-\frac{\sqrt{\epsilon}}{2}\left(\sqrt{(j+1)} - \sqrt{j}\right), \epsilon\right)$. At first glance this might not seem such a large number. However, $\epsilon = 2.2 \times 10^{-16}$, results in $\sqrt{\epsilon} = \sqrt{2.2 \times 10^{-16}} \approx 1.483\,2 \times 10^{-8}$. The movement of the straight line from $j = 0$ to $j = 1$, from $j = 1$ to $j = 2$ and from $j = 2$ to $j = 3$, gives the following changes in the location of the x-coordinate of the intersection point: -7.416×10^{-9}, $-3.071\,8 \times 10^{-9}$, and $-2.357\,1 \times 10^{-9}$, while the y-coordinate only changes 2.2×10^{-16}. The change in location of the x-coordinate of the intersection point is 7 orders of magnitude larger than the movement in the y-direction. This illustrates how unstable a near-singular intersection can be.

From the above examples we can conclude that rounding errors must be kept to a minimum in intersection algorithms to ensure the highest possible quality of the intersection result.

2.5 Floating Point Arithmetic and Choice of Polynomial Bases

In the STEP standard and in CAD-systems the NonUniform Rational B-splines are chosen as the representation format for rational piecewise rational curves. This is a very good choice as NURBS and a proper choice of algorithms reduce the size of rounding errors compared to alternative bases, e.g., the power bases. However, if the B-splines are converted, e.g., to a power basis in an algorithm, there is a great risk of introducing unnecessary rounding errors. It should also be noted that the algorithms chosen should be consistent. E.g., when a NURBS-curve is cut in two pieces by using a subdivision algorithm the B-spline vertex representing the subdivision point should be identical to the point resulting when evaluating the original curve at the subdivision point. If inconsistent algorithms are used, unnecessary noise is introduced into the intersection algorithm. More details on different polynomial bases can be found in the Appendix.

2.6 Representing the Intersection

The exact intersection of two surfaces will be a combination of points, curves and surface regions. However, when calculating intersections using tolerances, a neighbourhood of the exact intersection will be within the intersection tolerance as well. In addition to these neighbourhoods there may be points, curves and regions that are considered intersections within the given tolerance, but which have no corresponding exact intersection. A good description of the intersection should combine the description of regions resulting from using intersection tolerances with a description of the exact intersections if they exist.

For the intersection of two 2D curves it is fairly straightforward to describe tolerance dependent intervals in addition to the description of the exact intersection points and intersection intervals. The description of the tolerance dependent regions is more challenging when intersecting surfaces. The boundaries of intersection regions around transversal intersection curves will be closely related to the geometry of the intersection curve. However, the boundaries of intersection regions around intersection curves where the surfaces are near parallel or parallel, will result in intersection regions with a shape deviating significantly from the shape of the exact intersection curve. The shape of the intersection regions in singular or near-singular situations will be unstable, as minor changes to the surfaces will result in significant changes in the shape of the intersection regions. When using floating point arithmetic the choice of polynomial bases and algorithms will to a great extent influence the size and shape of the regions produced, see the Appendix.

In CAD-systems the representation of surfaces implicitly supposes that there is an edge between adjacent faces. Consequently information that two surfaces are near parallel and intersect within a tolerance in a large region, has to be translated to well defined edges. These can then be used to trim the surfaces to establish well defined adjacent surfaces. It would be preferable to be able to store and process the intersection regions in a proper way. If a CAD system is only based on boundary structures as is described in STEP (ISO 10303) *"Product Data Representation and Exchange standard"* [29], storage of such regions of intersection is normally not supported. In future intersection algorithms proper handling of intersection regions resulting from intersection tolerances, and the use of floating point arithmetic should be introduced. In such algorithms Interval Arithmetic is one of the approaches that should be considered [1, 19, 27, 35].

Instead of a detailed description of the intersection regions, an alternative is to assign points and curves the role of placeholders for the intersection regions. Additional information can be attached to the points and curves to describe the type of intersection region they represent. How to represent intersection curves geometrically will be addressed in Section 5.6.

3 Traditional Classes of Surface Intersection Methods

Over the years many different strategies for the intersection of surfaces have been developed in CAD, the most used methods are shown in Table 1. Simple and fast methods with limited quality guarantees are often used although more advanced methods with quality guarantees, e.g., recursive subdivision, are available. The main reason for this is that the dominant part of intersections are transversal, and that for transversal intersections the simpler methods most often produce a satisfactory result.

Figures 5 and 6 illustrate the difference for near-singular situation between a recursive subdivision intersection algorithm working on the NURBS descriptions of surfaces, and an intersection algorithm that first makes triangulations of the surfaces and then intersect the triangulations. In Figure 5 the triangulations are rough and the intersection of the triangulations have the wrong topology. The intersection curves are more oscillating than the correct intersection. In Figure 6 the intersection is closer to the correct intersection, however, the topology of the intersection is still not correct.

Intersection algorithms describe in table 1 can be divided into two types:

- Identification of intersection branches.
- Tracing of intersection curves to satisfy some tolerance requirements when one or more points on the curve are identified.

Methods from both categories can be combined to build a fast and reliable intersection procedure, and it is natural to split surface intersection algorithms into two phases:

Table 1. Different CAD-intersection methods and their properties. Note that the only method with a guarantee on the intersection topology is recursive subdivision. All methods have problems related to singular and near singular intersection situations.

	Guarantee	**Quality**	**Strategy**	**Problems**
Triangulation	No	Proposes intersection topology with possible missing and false branches	Intersects triangulated surfaces	Near singular intersections
Lattice evaluation	No	Proposes intersection topology, can miss branches	Intersects curve network in each surface with other surface	Near singular intersections
Marching	No	Works on one intersection branch	Traces intersection curve from a point supplied to the algorithm	Singular points
Refinement	No	Works on one intersection branch	Inserts points in interior of segment defined by two points supplied to the algorithm	Singular points
Recursive subdivision	Yes	Tolerance dependent complete topology	Chops surfaces into sub-piece until reliable intersection	Unstable in near singular situations

- Phase 1: Identification of the intersection topology, as intersection regions, intersection curves and intersection points.
- Phase 2: Refinement of intersection curves to satisfy tolerance requirements, and describing the boundary of intersection regions.

We will in the subsequent Sections address these phases. First we will look at methods for computing the topology of an intersection problem.

As described in Section 2.3, to find the mathematically correct solution of the intersection of two rational bicubic parametric surfaces is an extremely challenging problem. Two basic approaches for "solving" the problem are:

- Develop an algorithm that works in most cases, but with no guarantee of a complete solution. Two such methods are addressed, namely surface intersection by triangulation in Section 3.1, and lattice evaluation in Section 3.2.

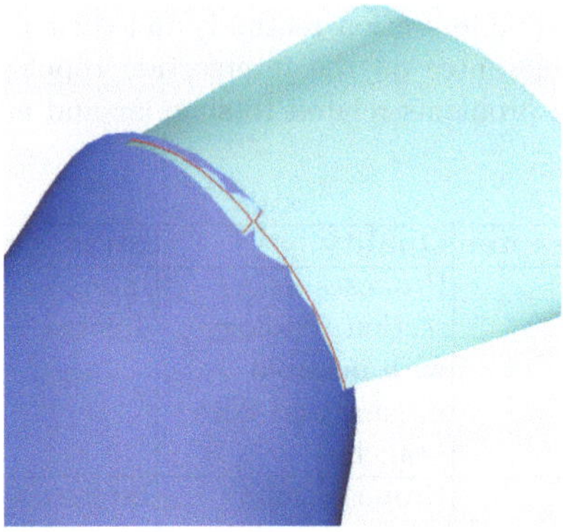

Fig. 5. The exact intersection curve between two surfaces are displayed, along with rough triangulation of the surfaces. The intersection between the two triangulations is visually produced by the graphics package. We see that the exact intersection curves and the intersection between the triangulations have different branch structure.

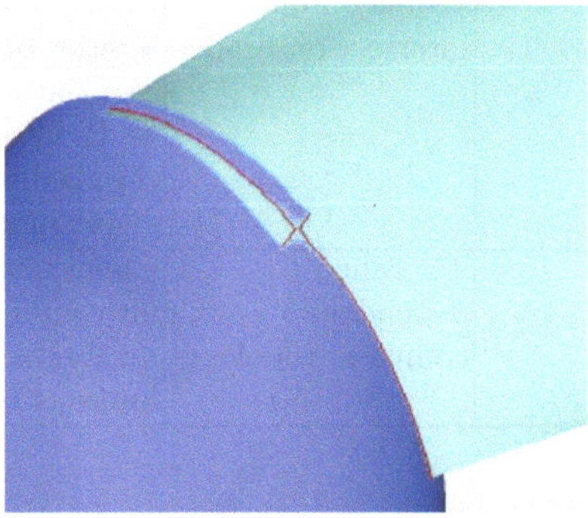

Fig. 6. The surfaces are sampled approximately $8\times$ denser in both parameter directions thus producing approximately $64\times$ more triangles than in figure 5, however, still the intersection topology produced by intersecting the triangulations is not correct.

- Develop an algorithm that provides a "correct" solution within specified quality tolerances. The recursive subdivision approach is discussed in Section 3.5.

Experience shows that algorithms derived by the first approach are faster than those using the second approach. Algorithms that provide a solution within specified quality tolerances have to use computational resources in the verification process, and will thus be slower than the simplistic approaches. In some cases the verification process of seemingly simple intersection problems takes more resources than seems necessary.

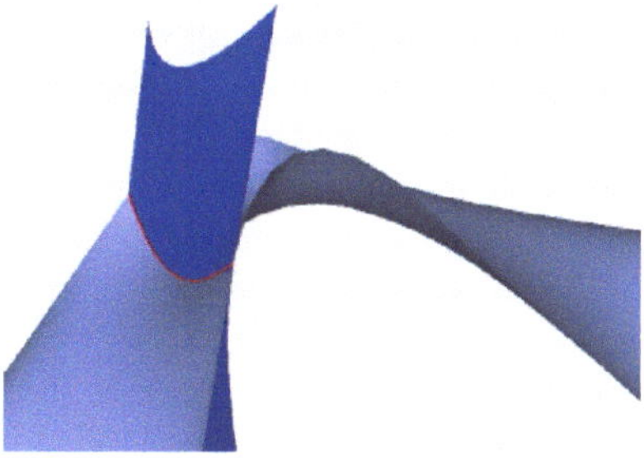

Fig. 7. A transversal intersection between two surfaces. At one end of the intersection curve the surfaces are near parallel.

Figure 7 shows two surfaces that intersect along constant parameter curves in both surfaces. By a constant parameter curve in a parametric surface we mean a curve in the parametric surface generated when assigning one of the parameter values a constant. The intersection is transversal, but the angle between the surface normals along the intersection curve is small. Thus the intersection region corresponding to the curve has a significant size, and to guarantee that no additional small intersection loops exist in the vicinity of the curve is a non-trivial task.

In the subsections following we will shortly describe the methods listed in 1. Marching and refinement differ from the other methods by the need of information on the location of points on the intersection tracks. The other methods do not need such input.

A number of papers are referenced in the following subsection, other papers of interest are [32, 33, 34, 36, 39, 40, 45].

3.1 Surface Intersection by Triangulation

To avoid the algebraic complexity of computing the true intersection of two sculptured surfaces, the surfaces can be replaced by piecewise linear approximations. These are then used as in the intersection calculation. At first glance this approach appears straightforward, however, among the problems of this approach are:

- The triangulations can introduce additional intersections. In Figure 8 we illustrate this by 2D circle intersections.
- The triangulations can lose intersections. In Figure 9 we illustrate this by 2D circle intersections.

It can be argued that finer approximations with more linear segments would solve the problems above. However, even with denser approximations we can always make examples similar to Figures 8 and 9.

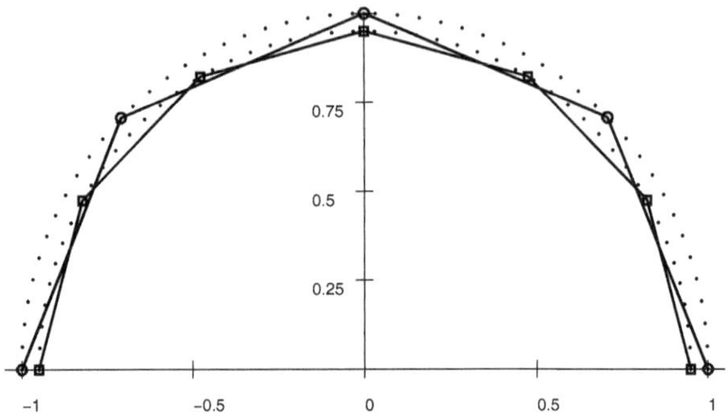

Fig. 8. The original geometries are two half circles around the origin, of radius 1 and 0.95 respectively. These do not intersect. The first half circle is approximated with 4 straight line segments, the second circle with 6 straight line segments. The resulting linear approximations intersect in 8 points.

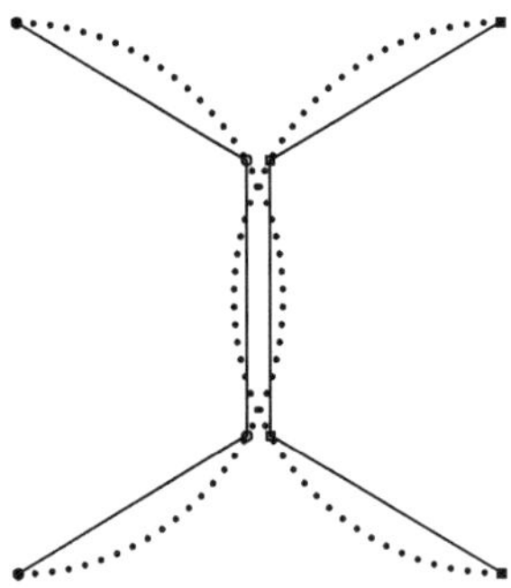

Fig. 9. The original geometries are two half circles of radius 1.1 with centres at (-1,0) and (1,1). The half circles intersect in two points. However, the linear approximations, each consisting of 3 straight lines, do not intersect.

Table 2. Properties of intersection methods based on triangulation.

	Intersection by triangulation
Distinct intersection points	Method assumes all intersections are curves
Transversal intersection curves	Will perform well, but risk of incorrect topology
Near singular intersections	Random behaviour
Singular intersections	Random behaviour
Intersection topology quality	Risk to both miss and get false branches
Computational performance	Good

- Intersection of triangulations is in itself a challenging task. When generating dense triangulations of surfaces that have near-singular intersections, then situations like those described in Section 2.4, and illustrated in Figures 5 and 6 arise, and the instabilities associated with them have to be handled.

If the results of the intersection of two triangulations were always sequences of straight line segments, with a gradual variation of the length of the line segments, then using these results would be fairly straightforward. However, the result of the intersection of two triangulations is often not so well behaved:

- Some of the straight line segments can be extremely short. This is the result if a triangle intersects another triangle close to a vertex.
- Triangles might partly coincide.
- The triangulations of surfaces that have only transversal intersections, might have coincident or near coincident triangles. For example a small translation of the circles in Figure 9 will make the middle line segments coincide.

Consequently, intersection of the triangulations of surfaces has its role when generating a conjecture of the intersection situation. However, the quality of the conjecture needs to be assessed before any further use of the result, e.g., by using Sinha's theorem [47] described in Section 5.2.

3.2 Surface Intersection by Lattice Evaluation

In this approach, one of the surfaces is replaced by a grid of curves lying in the surface. Thus the computationally demanding intersection of two surfaces is replaced by a series of curve-surface intersections. In Section 2.3 we showed that the exact solution of the intersection of two bicubic patches can be converted to finding the zero set of a polynomial equation in two variables of bi-degree (54,54). Lattice evaluation removes one variable, and consequently we are left with a problem that can be converted to finding the zeroes of one univariate polynomial of degree 54.

To ensure symmetry in the algorithm both surfaces should be approximated by a grid of curves lying in the surface and then intersected with the other surface. The result of lattice evaluation will be a set of points lying in the surface. Using the structure of the grid can help sort out how to connect the points. However, there will be situations where marching or bisection will be necessary to build an intersection topology.

It is obvious that small intersection loops can be missed out by lattice evaluation. Thus, also the result of lattice evaluation can only be regarded as a conjecture of the correct intersection topology. In many situations the recursive approaches in the following sections can be used to verify if the intersection conjecture is complete.

Table 3. Properties of intersection by lattice evaluation.

	Lattice based intersection
Distinct intersection points	Method assumes all intersections are curves
Transversal intersection curves	Will find transversal intersections
Near singular intersections	Random behaviour
Singular intersections	Random behaviour
Intersection topology quality	Risk to miss intersection branches
Computational performance	Good

If the curve-surface intersection algorithms used in lattice evaluation are properly implemented, lattice evaluation will not produce false intersection points as methods for surface intersection based on triangulations will do.

3.3 Intersection by Refinement

Intersection by refinement, also denoted by *bisection*, is very useful when the end points of an intersection branch are known, and additional points on the intersection track are required. When calculating new points on an intersection track, it is important to ensure that a new point really is on the intersection curve being refined. If care is not taken, a new point can belong to an adjacent intersection curve. This can easily happen in near singular situations.

Some algorithmic components in intersection by refinement are:

- Predicting an intersection point between the already existing points.
- Correcting the predicted point to be on the intersection.
- Validating a new point, e.g., by checking that the two adjacent points on the intersection curve have tangents pointing in the same direction as the tangent of the new point by ensuring that the tangent scalar products are positive. If the tangents point in different directions, the point might belong to a different intersection branch. The check should be performed both in 3D and in the parameter domains of the surfaces.
- Checking the quality of the intersection curve approximation, e.g., by calculating intermediate points on the intersection tracks, and by checking that the tangent direction of the new curve segment is not varying too much.

3.4 Intersection by Marching

Intersection by marching only requires that one point on the intersection branch is known. If properly implemented it will be sufficient to just know a point sufficiently close to the intersection branch to be able to iterate onto the intersection branch.

Some algorithmic components in intersection by marching are:

Table 4. Properties of intersection by refinement.

	Intersection by refinement
Distinct intersection points	Method assumes all intersections are curves
Transversal intersection curves	Will trace out the curve on which it is applied
Near singular intersections	Unknown
Singular intersections	Implementation dependent
Intersection topology quality	Works only on one branch
Computational performance	Good
Comment	Two points representing the ends of the segment to be refined to be supplied to the algorithm

Table 5. Properties of intersection by marching

	Intersection by marching
Distinct intersection points	Method assumes all intersections are curves
Transversal intersection curves	Will trace out the curve on which it is applied
Near singular intersections	Unstable in most cases
Singular intersections	Implementation dependent
Intersection topology quality	Works only on one branch
Computational performance	Good
Comment	Start point to be supplied to the algorithm

- Predicting a next point on the intersection branch.
- Correcting the predicted point to be on the intersection.
- Validating a new point, e.g., by checking that the new and existing points on the intersection branch have tangents pointing in the same direction as the tangent of the new point by ensuring that the tangent scalar products are positive. If the tangents point in different directions, the point might belong to a different intersection branch. The check should be performed both in 3D and in the parameter domains of the surfaces.
- Checking the quality of the intersection curve approximation, e.g., by calculating intermediate points on the intersection tracks, and by checking that the tangent direction of the new curve segment is not varying too much.

3.5 Recursive Intersection Algorithms

Around 1980 algorithms for subdividing polynomial and piecewise polynomial curves and surfaces were introduced.

- The Lane–Riesenfeld Algorithm for subdivision of Bezier curves [31].
- de Casteljau's algorithm [17].
- The Oslo Algorithm [4] for subdivision of B-spline represented curves and surfaces.

Table 6. Properties of intersection by subdivision.

	Subdivision based intersection
Distinct intersection points	Are detected
Transversal intersection curves	Are detected
Near singular intersections	Will often work
Singular intersections	Will often work
Intersection topology quality	In general good
Computational performance	Good in transversal cases, often slow in singular cases

Triggered by these algorithms work on recursive subdivision based curve and surface intersection algorithms started [8, 9, 11, 12, 15, 30, 47].

In [5] (from 1973) we find the following description of recursive programming: *"To make a recursive procedure is a process that is very similar to a mathematical proof by induction. First the outward effect of the procedure is established. Then the content of the procedure is described as a program with two branches: one branch with no recursive call, the induction basis; the other branch with a recursive call, the induction step. The second branch cannot be constructed without precisely knowing beforehand what the consequences of the recursive call will be."*

The consequences of the recursive step in intersection algorithms are, as in all recursive programming, dependent on the cut-off strategies employed, strategies to reduce the total number of recursive calls when running the algorithm. In an intersection algorithm we have no control over the surfaces to be intersected. Thus an algorithm designed for the identification of intersection tracks can be given the task of intersecting two coincident or partially coincident surfaces. For most recursive intersection algorithms the intersection of coincident surfaces represent an extreme challenge.

Example 3. Assume that we have a recursive surface-surface intersection algorithm for the intersection of two Bezier surfaces that uses the convex hulls of the subpatches to be intersected as the only tool for checking the possibility of intersections. No other strategies for terminating the recursion are used. Let the surfaces to be intersected have parameter domains $[0,1] \times [0,1]$. Instead of intersecting two different surfaces, we intersect the surface with itself. We want to stop the subdivision process when the area of the sub-patch domains are less or equal to $10^{-7} \times 10^{-7}$(floating point single precision), thus giving 7 digits of accuracy for the parameter values generated for the intersection track. Since we intersect a surface with itself, we will subdivide the surface into approximately $10^7 \times 10^7 = 10^{14}$ subpatches. Assuming that we can produce 10^9 subpatches a second, the subpatch generation will take 10^5 seconds $= 27.7$ hours. If we want double precision accuracy the subpatches will be of size $10^{-16} \times 10^{-16}$, and we get 10^{32}subpatches. Once more assuming we can produce 10^9 subpatches a second, the subpatch generation will take 10^{23} seconds $= 3.17\ 10^{15}$ years.

The example above describes a recursive intersection algorithm that is inadequate for the problem. It could be argued that it is stupid to generate surfaces that coincide or oscillate through each other, and that intersection algorithms should not be forced to handle such situations. However, a number of seemingly simple operations in CAD-systems produce such surfaces, e.g., a rolling ball blend between two surfaces. Consequently the surface intersection algorithms must handle for instance intersections between a blend surface and one of its generating surfaces. Within a given CAD-system, the history of surface creation is often stored, and such intersections can be avoided. However, when exporting the geometry in a neutral format, such history information is in general lost, and the receiving system has no reasons to avoid intersections between surfaces oscillating through each other.

For a subdivision based surface intersection algorithm to perform well, different strategies have to be employed to avoid too deep levels of subdivision. For transversal intersections these work well, and ensure a good computational performance. However, for near singular intersections the recursion will often be deep and the performance not as industrial systems require. In the subsequent sections a number of such strategies will be discussed. It should be noted that it is important always to subdivide from the original surface. In different parts of the surface the necessary level of subdivision will often vary. Adjacent subpatches at the lowest necessary level will thus often be at different subdivision granularity. By subdividing the subdivided subpatches the history of rounding errors will not be the same, and small discrepancies in coefficients must be expected. By always using the original surface when subdividing and evaluating, such discrepancies are avoided.

4 Early Detection of Transversal Intersections Followed by Detailed Investigation of Problematic Regions

The overview in the previous section indicates that there is a conflict between performance and guarantee of quality in CAD surface intersection algorithms. However, this is based on the assumption that it is computationally too expensive to establish the quality guarantees. The computational performance of CPUs (2006) has grown by 3 orders of magnitude since 1990 (Moore's law). Until recently the increased computational performance has resulted in a higher clock speed on the CPU and direct boost of existing algorithms. However, in 2006 multi-cores are introduced in the CPU. As most existing intersection algorithms have been implemented based on a sequential paradigm they will not benefit from the multi-core CPUs without being rewritten. Current (2006) programmable graphics cards (GPUs) offer a computational performance that is an additional 1 to 2 orders higher than the multi-core CPUs. Consequently approaches that can use the parallelism of multi-core CPUs and programmable GPUs have the potential of improving both quality and performance of intersection algorithms.

With the above evolution of computational performance in mind the following structuring of intersection algorithms is possible:

1. **Establish if only transversal intersections are possible.** To do this we will naively subdivide both surfaces and their corresponding normal fields to a predefined number of sub-regions. Then the sub-regions are checked both for spatial overlap, and overlap of normal fields, using the approaches of Sections 5.1 and 5.2. How these approaches can be improved by using the extreme computational performance of programmable graphics cards is discussed in Section 4.1.
2. **Detailed investigation of possible singular or near singular intersections.** Stage 1 will provide information on which regions of the surface can intersect, and regions where the normal fields can overlap. In singular or near singular surface intersection situations stage 1 above will identify the location of the possible singularities or near singularities in small sub-regions of the surfaces. Consequently a recursive subdivision algorithm can focus on these areas rather than on the total surfaces. When possible singular or near singular intersection regions are identified, the use of approximate implicitization can be advantageous when sorting out which cases are clearly not singular. Approximate implicitization is also a good tool to use when trying to find the intersection topology in the near singular or singular situations. More details are given in Section 4.2.

When the two stage topology search have been performed the surface intersection has been classified as either:

- **Pure transversal intersections with all branches touching the boundary.** This situation arises when the approximated normal fields of the surfaces do not overlap. The naive subdivision will reduce the extent of the approximated normal fields and consequently classify more intersections to be without normal field overlap. The intersection topology can be built from points found by intersecting the boundary of both surfaces with the other surface.
- **Pure transversal intersections with identified internal loops.** This situation arises when the approximated normal fields of the surfaces overlap, but no selection of possibly intersecting subsurfaces of the two surfaces have approximated normal fields that overlap. Consequently there will be no internal loops in the subsurface intersections, and points on all intersection branches, also the loops, can be found by the approach above, now employed at subsurface level.
- **Singular or near singular intersections:** Information found will be intersection points, start end points of intersection branches, and intersection regions where the intersection topology is too unclear to provide a clear point or branch structure.

A geometric approximation of intersection curves can be by marching or other refinement strategies such as bisection as described in Section 5.5.

4.1 Brute Force Subdivision to Find Transversal Intersections and Self-Intersection Free Surfaces

We propose to use the parallel processing power of multi-core CPUs and GPUs for brute force subdivision of the surfaces and their normal fields into a restricted number of levels. More details on this approach can be found in [3]. By brute force subdivision of the surface we mean a process that splits the NURBS surfaces into respectively $l_1 \times l_2$ and $\hat{l}_1 \times \hat{l}_2$ Rational Bezier surfaces. The actual implementation will depend on the number of rational Bezier patches in each surface and if the brute force subdivision is to be performed on the CPU or the GPU. One approach is to first increase the knot multiplicity of the NURBS surfaces to describe each polynomial segment as a rational Bezier patch, and then subdivide each rational Bezier patch into $2^n \times 2^n$ pieces. Another approach is to go directly to the subdivided Bezier patches by using the Oslo algorithm.

The brute force subdivision will identify most loop-free surface intersections and many transversal surfaces intersections with loops. To build the intersection topology of such transversal intersections can be done from identified intersection points on the surface boundaries and identified intersection points on the intersection loops. For possible singular intersections the information found related to spatial overlap of subparts, or the overlap of subpart normal fields can be used to guide a subsequent recursive algorithm.

The convergence of the control polygon of a NURBS surface to the surface is $O(h^2)$, with h the width of the largest B-spline knot interval. Consequently to base the comparison on the extent of the control polygons of the subdivided surfaces will yield more accurate results than comparing the extent of the control polygons of the original surfaces and/or their normal fields. This approach makes sense as most surface intersections in CAD do not contain loops. In a similar way most surfaces do not have self-intersections, so a brute force approach that can establish that no self-intersection takes place will also be extremely beneficial in algorithms for surface self-intersection.

To illustrate this approach we compare in Table 7 the execution of the following self-intersection calculation on a CPU and a GPU:

- Subdivision of a bicubic Bezier patch and the (quintic) patch representing surface normals into $2^n \times 2^n$ subpatches.
- Test for degenerate normals for subpatches.
- Computation of the approximate normal cones for subpatches.
- Computation of the approximate bounding boxes for subpatches.
- Computation of bounding box pair intersections for subpatches.

For details on the test see Table 7. The expected growth in computational performance of GPUs is expected to be significantly higher than the growth of computational performance of multi-core CPUs in the next years (form 2006). So a split of the brute force subdivision between CPUs and GPUs will be feasible for many years. However, GPUs are only a specialisation of

Table 7. Comparison of the execution of a brute force subdivision algorithm for surface self-intersection detection. The GPU used is a NVIDIA GeForce 7800GT, the CPU an AMD X2 4400+ without threading. Note that for low levels of subdivision the CPU is fastest, while for $n > 4$ the GPU is performing best. As the expected growth in performance of GPUs is higher that for CPUs, even with multi-cores, it makes sense to use GPUs for extensive subdivision. However, an initial subdivision should run on the CPU.

n	#subpatches	GPU	CPU	Speedup
4	16×16	7.456×10^{-3}	6.831×10^{-3}	0.9
5	32×32	1.138×10^{-2}	7.330×10^{-2}	6.4
6	64×64	7.271×10^{-2}	1.043×10^{0}	14.3
7	128×128	9.573×10^{-1}	1.607×10^{1}	16.8
8	256×256	1.515×10^{1}	2.555×10^{2}	16.9

general stream processors, so in general the approach is to perform the first levels of subdivision on the multi-core CPUs, and then perform the finer level subdivision on a stream processor such as a GPU or a Cell processor.

4.2 Guided Search for Singular and Near-Singular Intersections

The traditional approach to recursive intersection algorithms has been to regard all intersections as challenging and use a very complex intersection approach for all intersections to be calculated. As most intersections are simple this significantly reduces the algorithmic performance. The brute force subdivision, proposed in Section 4.1 will identify most transversal intersections (both loop free and those with loops), and can also provide information on which regions of the surfaces do not satisfy the criteria for transversal intersection, and which regions of the two surface cannot intersect. Consequently a recursive intersection algorithm can use this information to guide the advanced strategies to work only on the regions of the surface intersection that cannot be identified as clearly transversal. This means that more demanding methods, e.g., approximate implicitization will only be used in the near singular or singular situations.

In the case of a self-intersection the brute force subdivision will identify which regions of the surface have vanishing gradients, and which regions of the surface intersect another region. This information can be used to guide the recursive self-intersection algorithm to use advanced strategies only when a potential non-trivial self-intersection is identified.

5 Useful Tools for Surface Intersection Algorithms

In industrial type intersection algorithms a number of different strategies are employed to decide that further subdivision is not necessary, and thus give acceptable computation times. Some approaches for reducing the number of subdivision in recursive intersection algorithms are and will be explained in more detail later.

- **Determining if spatial overlap is possible.** To make and compare pre-evaluated axis-oriented boxes around the surfaces is much faster than to compute and compare control polygon based convex hulls around the surfaces. To be less dependent on the orientation of the coordinate system rotated boxes can be precomputed. For surfaces or subsurfaces that are not close, the box tests are very efficient for deciding that no intersection is possible. We will look in more detail at how to detect or rule out spatial overlap in Section 5.1.
- **Comparing the behaviour of surface normals.** Criteria ruling out the possibility of internal closed intersection loops are addressed in Section 5.2. If we can ensure that no internal intersection loop exists, then all intersection curves have to touch the boundary of one of the surfaces, and no further subdivision is needed. Thus we can find points on all intersection curves by just intersecting each surface with the boundary curve of the other surface. A controlled approximation of an intersection curve can be found using the methods of Section 5.5.
- **Subdividing surfaces in an "optimal" way.** In situations where we have to subdivide the surfaces to simplify the problem, the locations where we subdivide have a great influence on later need for subdivision. If the intersection has a closed internal loop, a subdivision that splits the intersection into two or more parts, simplifies the problem. On the other hand in a situation where the intersection oscillates around the subdivision line, a problem might become even more complex after subdivision, as one intersection curve is broken into subcurves, that later have to be aggregated to one curve. The subdivision strategy must be optimised towards producing situations where a recursion branch may be terminated [49].
- **Recognise total coincidence**. The previously mentioned strategies will not lead to a conclusion in coincident situations. Such situations have to be identified separately.
- **Combine the use of parametric and algebraic representations** can be used for reducing the number of parameter directions and thereby simplify the intersection problem. The exact algebraic description or an algebraic approximation of one of the surfaces can be applied for separating surfaces or to establish that the surfaces intersect within predefined tolerances. For NURBS surfaces the computational cost of making exact algebraic descriptions is high due to the high polynomial degrees of the rational parametric tensor product surface. However, the use of approximate implicitization, addressed in Section 5.4, with total degree as low as 4, 5 or 6 will often be efficient.

- **Stop recursion if going too deep.** Detect if the recursion is too extensive and provide information on the problematic situation that has arisen. This is a last opportunity option and must be used with extreme care, otherwise the quality guarantees of the method will be broken.

5.1 Determining if Spatial Overlap is Possible

Before the intersection algorithm performs a detailed analysis of the intersection of two NURBS surfaces it is important to establish if such an intersection can exist. In the following we assume that the weights of the NURBS surfaces are all positive to rule out asymptotic behaviour inside the surfaces intersected. Assume that $\mathbf{p}(s,t)$ is a NURBS surface with positive weights $w_i > 0$, $i = 1, \ldots, N$.

$$\mathbf{p}(s,t) = \frac{\sum_{i_1=1}^{N_1}\sum_{i_2=1}^{N_2} \mathbf{p}_{i_1,i_2} B_{i_1,i_2}(s,t)}{\sum_{i_1=1}^{N_1}\sum_{i_2=1}^{N_2} w_{i_1,i_2} B_{i_1,i_2}(s,t)} = \sum_{i_1=1}^{N_1}\sum_{i_2=1}^{N_2} \frac{\mathbf{p}_{i_1,i_2}}{w_{i_1,i_2}} \alpha_{i_1,i_2}(s,t)$$

$$= \sum_{i_1=1}^{N_1}\sum_{i_2=1}^{N_2} \mathbf{c}_{i_1,i_2} \alpha_{i_1,i_2}(s,t), \text{ where} \tag{1}$$

$$\alpha_{i_1,i_2}(s,t) = \frac{w_{i_1,i_2} B_{i_1,i_2}(s,t)}{\sum_{j_1=1}^{N_1}\sum_{j_2=1}^{N_2} w_{j_1,j_2} B_{j_1,j_2}(s,t)}, \tag{2}$$

$$\text{and } \mathbf{c}_{i_1,i_2} = \frac{\mathbf{p}_{i_1,i_2}}{w_{i_1,i_2}}, i_1 = 1, \ldots, N_1, i_2 = 1, \ldots, N_2. \tag{3}$$

The basis $\{\alpha_{i_1,i_2}(s,t)\}_{i_1,i_2=1,1}^{N_1,N_2}$ is a partition of unity since

$$\alpha_{i_1,i_2}(s,t) = \frac{w_{i_1,i_2} B_{i_1,i_2}(s,t)}{\sum_{j_1=1}^{N_1}\sum_{j_2=1}^{N_2} w_{j_1,j_2} B_{j_1,j_2}(s,t)} \geq 0, \text{ and}$$

$$\sum_{i_1=1}^{N_1}\sum_{i_2=1}^{N_2} \alpha_{i_1,i_2}(s,t) = \sum_{i_1=1}^{N_1}\sum_{i_2=1}^{N_2} \frac{w_{i_1,i_2} B_{i_1,i_2}(s,t)}{\sum_{j_1=1}^{N_1}\sum_{j_2=1}^{N_2} w_{j_1,j_2} B_{j_1,j_2}(s,t)} = 1.$$

Thus these surfaces can be viewed as convex combinations of the projection of the vertices from $\mathbb{RP}^3$ (the real projective space) to $\mathbb{R}^3$, performed by dividing the 3 first coordinates of the coefficient in $\mathbb{RP}^3$ by the fourth coordinate.

An intersection algorithm can use combinations of different approaches for testing if two surfaces overlap:

Fig. 10. Two sculptured surfaces lie very close , but do not intersect.

- **Axes-parallel boxes.** Trying to decide if two surfaces overlap by comparing if axes-parallel boxes each surrounding one surface intersect, is computationally inexpensive. For NURBS-surfaces it is computationally inexpensive to make such an axes-parallel box containing the surface, for other types of surfaces, e.g., procedural surfaces, making such boxes will be more expensive.
- **Improved fixed orientation boxes.** By creating boxes with all possible 45 degree rotations around the coordinate axes, we produce a set of boxes that fit fairly well to most surfaces with a distinct orientation. All these rotations involves coordinates multiplied by $\pm\cos(\frac{\pi}{4})$ or $\pm\sin(\frac{\pi}{4})$. As $\left|\cos(\frac{\pi}{4})\right| = \left|\sin(\frac{\pi}{4})\right| = \frac{\sqrt{2}}{2}$, we do not need to scale by $\frac{\sqrt{2}}{2}$ during the rotation, the scaling can be done after the rotation, thus improving computational performance.
- **Rotated boxes.** Both axes-parallel boxes and other improved boxes can be made tighter by rotating the coordinate system according to the particular situation.
- **Convex hulls.** In situations where the surfaces are curved and near parallel, high levels of subdivision are necessary to establish separation. However, establishing convex hulls is computationally more expensive than making and comparing boxes.
- **SLEVEs** represent a tighter hull to a surface than the convex hull [37]. As for convex hulls, the computational cost of establishing the SLEVEs, and comparing these, are significantly higher than for axes-parallel and rotated boxes.
- **Separation by algebraic surface.** Let us assume that we have two parametric surfaces $\mathbf{p}(s,t), (s,t) \in \Omega_1$, and $\mathbf{q}(u,v)$, $(u,v) \in \Omega_2$, that we want to separate. If we can find an algebraic surface $q(x,y,z) = 0$ that satisfies $q(\mathbf{p}(s,t)) > c$, $(s,t) \in \Omega_1$, and $q(\mathbf{q}(u,v)) < c$, $(u,v) \in \Omega_2$, then the two surfaces do not intersect.

The main challenge of using the last technique is to find the proper algebraic surfaces. In the items above, where we use axes-parallel or rotated boxes, we have predetermined the orientation of the planes we want to employ for testing. By allowing the orientation of the planes to be influenced by the actual geometries we can get better results than just using axes-parallel or rotated boxes. However, the penalty is increased computation time. By allowing the separating algebraic surface to have higher degree, it can follow the shape of the surfaces better and thus be more efficient as a tool. One way of creating such a surface is to use approximate implicitization of the surfaces to produce the separating algebraic surface, see Section 5.4. Dependent on the relative spatial relation between the two surfaces we want to check, requirements will vary as to what is a "good" algebraic surface.

- **Surfaces that are close and near parallel in a spatial region.** By approximating the whole of one of the surfaces or a subregion by an algebraic surface, a candidate for establishing separation can be found. Figure 10 shows a situation where the surfaces obviously do not intersect to the human eye, but where the recursive subdivision based intersection tools need a long time to come to the same conclusion. This case is easily handled by approximating one surface by an algebraic surface and use this algebraic surface to separate the initial surfaces.
- **Surfaces that are near touching along boundaries with a complex shape.** In this case there is possibly a near smooth transition between the surfaces. Thus an algebraic surface that approximates one of the surfaces would possibly have multiple intersections with the other surface. To achieve separation it is be better to approximate the boundary of one of the surfaces and simultaneously enforce the algebraic surface to be near orthogonal to the surfaces to be tested for separation.

5.2 Comparing the Behaviour of Surface Normals

After establishing the fact that two surface may intersect, our aim is to be able to make the conclusion that the topology of the current intersection problem is defined uniquely from intersection points found at the surface boundaries. This is equivalent to concluding that no inner intersection loop exist for the current surfaces. Then we can terminate the recursion.

The paper of Sinha [47] from 1985 provides a theoretical foundation for deciding that there is no possibility for internal intersection loops in the intersection between two smooth surfaces. We use the formulation of the theorem from [24]. The theorem considers the intersection of two surfaces S_1 and S_2 that are smoothly embedded in $\mathbb{R}^3$. Since it is the behaviour inside of the intersection curve that is of interest, the surfaces are restricted such that they intersect along their common boundary. Thus we require that $S_1 \cap S_2 = \partial S_1 = \partial S_2$, where ∂S_1 is the boundary of S_1 and ∂S_2 is the boundary of S_2. We also assume that the surfaces are transversal to each other along their common boundary $\partial S_1 = \partial S_2$, meaning that the normals are not parallel along the common boundary. When these conditions are satisfied, there are points $\mathbf{x}_1 \in S_1$ and $\mathbf{x}_2 \in S_2$,where the normals $\mathbf{n}_1(\mathbf{x}_1)$ and $\mathbf{n}_2(\mathbf{x}_2)$ are parallel.

This theorem is well suited for the most common situation, where the surfaces have no vanishing first partial derivatives, and the surfaces intersect transversely. In most intersection cases the intersection will be transversal along the intersection curve, and sufficient levels of recursive subdivision will produce subsurfaces where the normal fields of the two subsurfaces do not overlap. Thus the intersection topology can be resolved [42, 46, 50, 51]. As CAD-systems have limited tests for surface degeneracy and surface self-intersections, vanishing first partial derivatives will be encountered sometimes. Intersections that are singular or near-singular are even more frequent, e.g., the rolling ball blend surface between two smooth surfaces will touch the two parent surfaces tangentially and thus the intersection curves will be singular or near-singular.

How can Sinha's theorem be applied to singular intersection points/curves and points/curves with vanishing partial derivatives?

- **Vanishing partial derivative in distinct points.** If the vanishing partial derivative is located in separate points, then recursive subdivision will place the points in questions in small subsurfaces, and the theorem can be applied for the subsurfaces where the normals are transversal. If the points in question are located sufficiently far from the actual intersection curve, recursive subdivision will after sufficient levels of recursion resolve the situation. If the point with vanishing first partial derivatives is located on the intersection curve, recursive subdivision will produce a small subsurface where the intersection topology cannot be resolved only based on the behaviour of normal vectors using Sinha's theorem. Subdividing in a distinct point with vanishing first partial derivatives will bring the point with vanishing first partial derivatives to the boundary of a subsurface and simplify the analysis of the intersection topology.
- **Vanishing partial derivatives on curve segment.** If the curve with a vanishing first partial derivative is well separated from the intersection curve, sufficient subdivision will resolve the situation, as explained above. However, if the curve with a vanishing first partial derivative is close to the intersection curve, oscillating around the intersection curve, or coincides with the intersection curve, recursive subdivision combined with an analysis of surface normal behaviour cannot resolve the situation.
- **Intersection curves with parallel normals in distinct points.** Recursive subdivision will in principle locate the singular intersection points in small subsurfaces. However, close to the singular points on the intersection curve the subsurfaces will be near parallel and very close. As this type of algorithm is running in floating point, the consequence will be "noisy" subsurfaces, and it is thus often challenging to sort out the intersection

configuration. With intersection tolerances playing a role as well, there is a risk that false intersections can be identified in a region around the distinct singular intersection points if care is not taken.

- **Intersection curve segments with parallel normals.** In most industrial examples of such situations, the singular or near-singular intersection curve is parallel to a constant parameter line in one of the surfaces due to the design technique used, e.g., a blend surface (a surface making a smooth transition between two other surfaces) will touch the surfaces being blended along two opposite edges. Thus identifying this situation will prevent high levels of recursion. Relying only on recursive subdivision will in this situation identify a region around the actual intersection curve where the intersection topology is undetermined. If high levels of recursion are allowed, then the algorithm will be computationally expensive. Moreover, the exact intersection result in this region will be complex, and the application using the intersection algorithm is not necessarily able to handle the exact intersection.

To use Sinha's theorem, approximations to the normal sets of the surfaces have to be calculated. For a smooth NURBS surface $\mathbf{p}(s,t)$ we can find a closed expression for the cross product of the partial derivatives $\mathbf{n_p}(s,t) = \mathbf{p}_s(s,t) \times \mathbf{p}_t(s,t)$ represented as another NURBS-surface. The vertices of this NURBS normal surface can be used for estimating sets guaranteed to contain the normal set. As $\mathbf{n_p}(s,t)$ is oriented and does not describe unit normal vectors, care must be taken when comparing normal sets based on $\mathbf{n_p}(s,t)$. Both $\mathbf{n_p}(s,t)$ and the antipodal $-\mathbf{n_p}(s,t)$ have to be used, and the sectors of the unit sphere spanned by the vertices of $\mathbf{n_p}(s,t)$ and $-\mathbf{n_p}(s,t)$ must be compared rather than their geometric extents.

Ruling out the possibility of intersection loops in a recursive intersection algorithm reduces the need for recursive subdivision. However, to compute the normal surface corresponding to a NURBS surface and compare these surfaces is computationally expensive. Thus approximations to a normal set, which is guaranteed to contain the normal instead of the exact normal set, will be beneficial in many instances of surface-surface intersection, e.g., when the normal sets are well separated. When making these approximations to the normal sets there is the trade-off between tightness of the approximation to the actual normal set, and computational resources needed when generating the normals sets and subsequently comparing the normal sets. Some of the approaches proposed are:

- The actual sets of normals of two surfaces $\mathbf{p}$ and $\mathbf{q}$ can be compared by intersecting $\frac{\mathbf{n_p}}{\|\mathbf{n_p}\|_2}$ and its antipodal with $-\frac{\mathbf{n_q}}{\|\mathbf{n_q}\|_2}$ and its antipodal. However, this is in most cases not necessary and approximations containing the set of normals can be used.
- To produce cones that contain the normal set of a surface is not computationally expensive. Comparing cones is not computationally demanding either, thus tests based on "normal" cones have a natural role in recur-

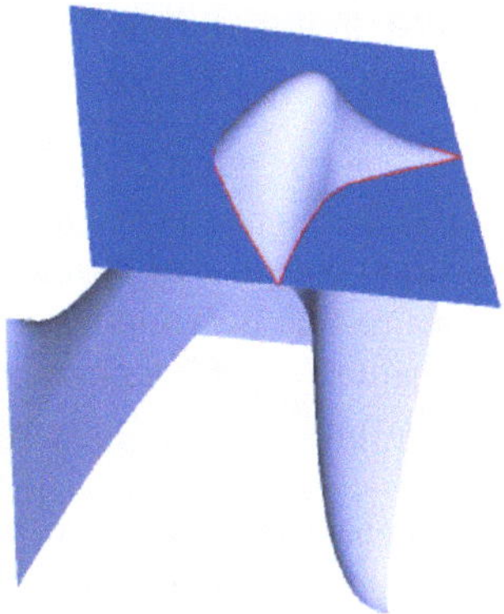

Fig. 11. Two surfaces intersect in two intersection curves which meet in singular branch points in both endpoints.

sive surface-surface algorithms. However, when the intersection curve has points with parallel normals, this is not sufficient. For NURBS surfaces such normal cones can be based on the vertices of the closed expression $\mathbf{n_p}(s,t)$.

- A more accurate approach is to compare the convex hulls of the sets of normals, however, still singular intersection points cannot always be handled properly. Ensuring that the singular intersection points lie in surface corners can lead to a solvable problem. In Figure 11 an intersection between two surfaces is shown. We subdivide the two surfaces in such a way that the branch point between the intersection curves lies at the corner of both surfaces.
- In [24] a fairly simple criterion is introduced. If the normal sets can be separated by a plane, then there is no possibility for a closed intersection loop. This observation based on Sinha's theorem can be generalised to the separation of normal sets by an algebraic surface. However, the calculation of the plane and especially of the algebraic surface can be computationally expensive.

5.3 Combining Algebraic and Parametric Expressions

All rational polynomial parametric curves and surfaces have an algebraic representation. The opposite is not true, many algebraic curves and surfaces of total degree higher than two do not have a rational parametrisation [2, 40, 41, 43]. However, all algebraic curves and surfaces of degree one and two have rational parametric representations.

Assuming that the curves and surfaces to be intersected have both an algebraic and rational parametric description, we have the following simplifications:

- Intersection of two 2D parametric curves can be reduced from two polynomial equations in two variables to one polynomial equation in one variable.
- Intersection of two algebraic curves can be reduced from two polynomial equations in two variables to one polynomial equation in one variable.
- Intersection of two 3D parametric surfaces can be reduced from three polynomial equations in four variables to one polynomial equation in two variables.
- Intersection of two 3D algebraic surfaces can be reduced from two polynomial equations in three variables to one polynomial equation in two variables.

The simplification, by reduction of the number of variables, has some drawbacks:

- The polynomial degree of the reduced system of equations will be the product of the parametric and algebraic degrees.
- When intersecting two parametric curves or surfaces, the solution of the reduced problem is only described in the parametrisation of one of the curves or surfaces. We can find the solution in the parametrisation of the other curve or surface by switching which curve or surface should be described as an algebraic curve or surface. However, to combine the solutions, the solutions of both alternatives have to be matched.
- The 3D tolerances are distorted when finding intersections by combining parametric and algebraic expressions.

Using existing methods it is feasible to find the exact algebraic representation of rational parametric 2D curves. No such exact methods are feasible for rational parametric surfaces. However, in many cases a good approximate algebraic surface of degree as low as 4 can replace the exact algebraic surface. An approach for finding such approximate algebraic curves and surfaces is addressed in Section 5.4. This approach is also well suited for piecewise curves and surfaces.

Distortion of 3D Tolerances for Implicit Descriptions

Let $q(x,y,z) = 0$ be an algebraic surface. We want to took at how the behaviour of $q(x,y,z)$ is related to 3D distances. Let $\mathbf{p}_0$ be a nonsingular point on the algebraic surface, and let $\mathbf{n}(\mathbf{p}_0) = \frac{\nabla q(\mathbf{p}_0)}{\|\nabla q(\mathbf{p}_0)\|_2}$ be the normalised gradient of q at $\mathbf{p}_0$. We want to look at points offset from q a distance ε, i.e., points $\mathbf{p}_0 + \varepsilon\mathbf{n}(\mathbf{p}_0)$. Taylor expanding $q(\mathbf{p}_0 + \varepsilon\mathbf{n}(\mathbf{p}_0))$ with respect to ε gives

$$q(\mathbf{p}_0 + \varepsilon\mathbf{n}(\mathbf{p}_0)) = q(\mathbf{p}_0) + \nabla q(\mathbf{p}_0)\cdot\mathbf{n}(\mathbf{p}_0)\varepsilon + \frac{1}{2}\left(\mathbf{n}(\mathbf{r})^T\left(\nabla^2 q(\mathbf{r})\right)\mathbf{n}(\mathbf{r})\right)\varepsilon^2, \quad (4)$$

with $\mathbf{r} = \mathbf{p}_0 + \theta\varepsilon\mathbf{n}(\mathbf{p}_0), 0 \leq \theta \leq 1$. Provided ε is small and $|\varepsilon| << \|\nabla q(\mathbf{p}_0)\|$, remember that $q(\mathbf{p}_0) = 0$ and use the definition of $\mathbf{n}(\mathbf{p}_0)$, then we have

$$q(\mathbf{p}_0 + \varepsilon\mathbf{n}(\mathbf{p}_0)) \approx \nabla q(\mathbf{p}_0) \cdot \mathbf{n}(\mathbf{p}_0)\varepsilon = \|\nabla q(\mathbf{p}_0)\|_2 \varepsilon.$$

Given a point $\mathbf{p}$ close to the algebraic surface, then the distance from $\mathbf{p}$ to the algebraic surface can be estimated by

$$\frac{|q(\mathbf{p})|}{\|\nabla q(\mathbf{p})\|_2}.$$

As a consequence the spatial tolerances have to be adjusted according to the spatial location of surfaces, when we use the algebraic parametric combination in intersection algorithms. If $\|\nabla q(\mathbf{p}_0)\| << |\varepsilon|$, and ε is small, then we have a near-singular point and the distance approximation has to be based on the second order term in (4) resulting in

$$|\varepsilon| \approx \sqrt{\left|\frac{2q(\mathbf{p})}{\mathbf{n}(\mathbf{p})^T\left(\nabla^2 q(\mathbf{p})\right)\mathbf{n}(\mathbf{p})}\right|}.$$

Although it is advantageous to simplify an intersection problem by a combination of algebraic and parametric representations, this combination does not keep sufficient information to calculate accurate distances. Thus it is also important to keep the original representation of the objects to be able to handle intersection tolerances properly.

5.4 Approximate Implicitization

In [11, 13, 14, 16, 44] the theory behind approximate implicitization is described. The idea is to approximate a rational parametric surface $\mathbf{p}(s,t)$, $(s,t) \in \Omega \subset \mathbb{R}^2$, Ω being a closed and bounded set, by an algebraic surface $q(x,y,z) = 0$ of degree $m > 0$. This is different from exact implicitization, where the aim is to find the exact algebraic representation of the parametric surface. The combination of the algebraic surface $q(x,y,z) = 0$ and the rational parametric surface $\mathbf{p}(s,t)$ can be expressed as

$$q(\mathbf{p}(s,t)) = (\mathbf{Db})^T \alpha(s,t).$$

Here $\mathbf{b}$ is a vector containing the unknown coefficients of the algebraic surface to be found, $\alpha(s,t)$ is a vector containing rational basis functions, and $\mathbf{D}$ is a matrix with constants that ensures that factorisation is correct. As $q(x,y,z)$ is a polynomial of total degree m we have that $q(\mathbf{p}(s,t))$ is a sum of products of a maximum of m coordinate functions of $\mathbf{p}(s,t)$. Consequently each entry of $\mathbf{D}$ will containing products of m coefficients of the coordinate functions of $\mathbf{p}(s,t)$. Provided that the basis $\alpha(s,t)$ is a partition of unity over Ω we have that

$$\max_{(s,t)\in\Omega} \|q(\mathbf{p}(s,t))\|_2 = \max_{(s,t)\in\Omega} \left\|(\mathbf{Db})^T \alpha(s,t)\right\|_2$$
$$\leq \|\mathbf{Db}\|_2 \max_{(s,t)\in\Omega} \|\alpha(s,t)\|_2 \leq \|\mathbf{Db}\|_2 .$$

If we can find $\mathbf{b} \neq \mathbf{0}$ such that $\mathbf{Db} = 0$, then we have found an exact implicitization of $\mathbf{p}(s,t)$. If we can find $\|\mathbf{b}\|_2 = 1$, such that $\|\mathbf{Db}\|_2 << 1$, then we have a possible approximation to $\mathbf{p}(s,t)$. We say possible approximation as a small value of $\|\mathbf{Db}\|_2$ is related to approximation in projective space, not affine space. In affine space the approximation might have unwanted singularities, near-singularities or multiple branches in the region of interest.

By using singular value decomposition on $\mathbf{D}$ and selecting the vectors corresponding to the smallest singular values of $\mathbf{D}$, we can calculate candidate algebraic approximations to $\mathbf{p}(s,t)$, and then check these for singularities or near-singularities. If the chosen algebraic degree is too small, all candidate approximations can be singular or near-singular in the region of interest. Experience shows that an algebraic degree as low as 4 often gives algebraic approximations well suited for separation purposes. If the implicit surface $q(x,y,z) = 0$ is described in a Bernstein basis defined over a tetrahedron containing the parametric surface being approximated, and the parametric surface is described in a basis that is a partition of unity, then all elements of the matrix $\mathbf{D}$ are non-negative and the rows sum to 1. In this case very stable methods exist for building the matrix $\mathbf{D}$ [13]. Constraints enforcing a specific behaviour of position, tangents and normal in points or along curves can be added, thus tailoring the approximation to a specific purpose.

Approximate implicitization is well suited for implementation in floating point arithmetic. Minimal propagation of rounding errors is guaranteed if bases that are a partition of unity (e.g., the Bernstein basis or B-spline basis) are used both for the representation of the implicit surface and the parametric surface. The convergence of approximate implicitization of surfaces is

$$O(h^{\left\lfloor \frac{1}{6}\sqrt{(9+12m^3+72m^2+132m)}-\frac{1}{2}\right\rfloor}),$$

with m the degree of the algebraic surface used in the approximation and h the radius of the smallest possible circle within the parameter domain Ω containing the part of the surface being approximated. This gives convergence rates as follows: Total degree 3: $O(h^{\lfloor 5.6847 \rfloor}) = O(h^5)$; Total degree 4: $O(h^{\lfloor 7.7614 \rfloor}) = O(h^7)$; Total degree 5 is exactly: $O(h^{10})$; Total degree 6: $O(h^{\lfloor 12.394 \rfloor}) = O(h^{12})$. Thus halving the radius h for total degree 5 reduces the error in projective space by a factor of at least $(\frac{1}{2})^{10}$ corresponding to 3 orders of magnitude, for total degree 3 halving of the radius h reduces the error by $(\frac{1}{2})^5$ corresponding to approximately 1.5 orders of magnitude. The good convergence rate implies that one extra recursion step can improve the algebraic approximation of a sub-surface in a recursive algorithm for computing intersections drastically. A combination of recursion and approximate implicitization can be beneficial.

5.5 Refinement of Intersection Curves

When the topology of the intersection has been computed, the actual tracking of the intersection curves found can be done with methods tailored for this task. To find the exact description of an intersection curve as a closed expression is in most cases not possible. As explained in Section 2.3 the intersection of two bicubic parametric surfaces is a degree (54,54) algebraic curve in the parameter domain of each of the surfaces. Consequently the intersection curve has to be approximated. However, there are three descriptions of the intersection curve: A 3D curve and a curve in the parameter domain of each surface. Thus there will be three approximated versions of the intersection curve, these will most often not coincide.

The classification of a good approximation of an intersection curve is to a great extent dependent on the planned use of the intersection curve as the following list illustrates:

- **Intersection curves used for visual inspection** can be adapted to the actual resolution of the display used. As a typical computer screen has a resolution of between 1000x1000 pixels and 2000x2000 pixels, an error of 0.001 of the screen width will hardly be visible. However, as visualisation of sculptured surfaces in many cases is based on triangulations of the surfaces to display, an intersection curve together with the tessellated surfaces can give unwanted visual effects. The 3D intersection curve itself will often be converted to a polyline before visualisation. Thus what is actually visualised are triangulations and polylines, in some locations the polyline will be on one side of a surface, in other locations on the other side. Thus care must be taken to produce the wanted visual effect. To ensure that the intersection curve is visible when the surfaces are displayed separately, the intersection curve can be displayed using texture mapping. A 2D texture in the parameter domain of the surface is glued on to the surface. However, displaying two triangulated surfaces sharing an intersection curve by this approach can result in the intersection curve in one surface being hidden by the other surface, or we can experience that the intersection is displayed as two visual curves that do not coincide.
- **Intersection curves in a boundary structure volume model** most often describe the geometry of the edges between surfaces. Although such intersection curves should have a very accurate description, the data volume of such accurate approximations will often be larger than desired, thus the approximations used are fairly crude, but accurate enough to allow for later calculations of more accurate points and representations. Most often it is sufficient to deal with C^1-continuous curves and the parameterisations of the intersection curves do not have to be of high quality.
- **Intersection curves intended for designing new surfaces are** a great challenge. As varying speeds of the parameterisations and small oscillations in the curves often result in artifacts in the resulting surface, such curves should be at least C^2-continuous and have a parametrisation that is not

too varying. In many cases an approximate curve length parametrisation is desirable. In other cases one wants to be able to model the parametrisation of the intersection curves according to the requirements of the surfaces to be designed. Often the intersection curves are the geometric representation of edges in the volume model. In these cases a close geometric fit to the exact intersection curve is important.

- **Intersection curves used as input to production processes** have to be tailored to the actual needs of the production process. The parametrisation of the intersection curve will often correspond to a movement of the production tools. If the intersection curve follows a nearly curve-length parametrisation, movements will be smooth, if the tangent length of the intersection curve varies, the result can be oscillatory behaviour of the production machinery, and thus bad or slow production processes.

In many cases the intersection curves to be refined are defined by surfaces intersecting transversely and thus simple strategies will suffice. However, in other cases the surfaces intersected will be near parallel or parallel on or close to the intersection curve, resulting in many possibly closed intersection curves that are not well separated spatially. To handle these cases, fairly advanced strategies have to be employed to ensure that there is no jumping between different intersection curves, and that the refinement of a closed intersection curve does not result in an infinite loop. Other challenges are posed by singular intersection points on the intersection curves and situations where the normals of the surfaces coincide. Further challenges result from intersecting adjacent surfaces that meet in a singular or near-singular way, e.g., once a blending surface is intersected with one of its parent surfaces. Another challenge arises if the parametrisation of the surfaces is nonregular. For a surface parametrisation to be regular the first partial derivatives have to be linearly independent for the domain of interest. In most methods for refining intersection curves, the first partial derivatives of the surface are used, thus in points where the surface parametrisation is nonregular, simple strategies will fail.

All strategies related for the handling of singular points, avoiding of jumping between intersection tracks, quality of the approximated intersection curve, etc., essentially leave us with one task: Given a point on the intersection curve, how can we find an additional point on the intersection curve?

Tracing the Intersection Curve by a System of ODEs

Assume, as in [38] and [22] that the two parametric surfaces $\mathbf{p}(s,t)$ and $\mathbf{q}(u,v)$ both have a regular parametrisation. Assume that the intersection curve has a 3D parametric description $\mathbf{c}(w)$, and let $\mathbf{p} = \mathbf{q} = \mathbf{c}$ be a point on the intersection curve between the two surfaces. To find additional points on the trace of the intersection curve in the parameter domains of both surfaces, we want to express the intersection curve as a set of ordinary differential equations in the parameter domains, and then apply methods for finding numerical solutions of differential equations to trace the curve.

Let $\mathbf{p}_s$, $\mathbf{p}_t$ and $\mathbf{n}_p$ respectively be the first order partial derivatives and normal for $\mathbf{p}$, and let $\mathbf{q}_u$, $\mathbf{q}_v$, and $\mathbf{n}_q$ respectively be the first partial derivatives and normal for $\mathbf{q}$. We now focus on the surface $\mathbf{p}(s,t)$. As $\mathbf{c}$ is a point on the intersection curve, the tangent at $\mathbf{c}$, denoted $\frac{d\mathbf{c}}{dw}$ will lie in the tangent plane of $\mathbf{p}(\mathbf{u},\mathbf{v})$ at $\mathbf{p}$, thus

$$\frac{d\mathbf{c}}{dw} = \mathbf{p}_s \frac{ds}{dw} + \mathbf{p}_t \frac{dt}{dw}.$$

As the normals of both surfaces are also normal to $\frac{d\mathbf{c}}{dw}$ we have that

$$\frac{d\mathbf{c}}{dw} \cdot \mathbf{n}_q = 0 \implies (\mathbf{p}_s \cdot \mathbf{n}_q)\frac{ds}{dw} + (\mathbf{p}_t \cdot \mathbf{n}_q)\frac{dt}{dw} = 0.$$

Thus $\frac{ds}{dw}$ and $\frac{dt}{dw}$ have to satisfy

$$\frac{ds}{dw} = \pm(\mathbf{p}_t \cdot \mathbf{n}_q) f(w) \tag{5}$$

$$\frac{dt}{dw} = \mp(\mathbf{p}_s \cdot \mathbf{n}_q) f(w). \tag{6}$$

The function $f(w)$ controls the behaviour of the parametrisation of the intersection curve. A natural choice for the parametrisation is curve-length parametrisation.

If we have $q(x,y,z) = 0$, as an exact implicit representation for $\mathbf{q}(u,v)$, the normal is given by $\mathbf{n}_q = \frac{\nabla q(\mathbf{p})}{\|\nabla q(\mathbf{p})\|_2}$, and we have a system of two ordinary differential equations for $\frac{ds}{dw}$ and $\frac{dt}{dw}$. If no exact implicit equation for $\mathbf{q}(u,v)$ exists, then $\mathbf{n}_q = \frac{\mathbf{q}_u \times \mathbf{q}_v}{\|\mathbf{q}_u \times \mathbf{q}_v\|_2}$, and we have a system of 4 ordinary differential equations as the equations $\frac{du}{dw}$ and $\frac{dv}{dw}$ have to satisfy

$$\frac{du}{dw} = \mp(\mathbf{q}_v \cdot \mathbf{n}_p) g(w) \tag{7}$$

$$\frac{dv}{dw} = \pm(\mathbf{q}_u \cdot \mathbf{n}_p) g(w), \tag{8}$$

with $g(w)$ controlling the parametrisation.

Predicting a Next Intersection Point

The simplest approach for solving the system of ordinary differential equations would be to approximate $\frac{du}{dw}$ and $\frac{dv}{dw}$ by a first order forward difference yielding in the parameter domain of $\mathbf{p}$

$$\frac{ds}{dw} = \frac{s_{i+1} - s_i}{\triangle w}$$

$$\frac{dt}{dw} = \frac{t_{i+1} - t_i}{\triangle w}$$

and

$$
\begin{aligned}
s_{i+1} &= s_i \pm \triangle w\,(\mathbf{p}_t \cdot \mathbf{n}_q)\, f(w) \\
t_{i+1} &= t_i \mp \triangle w\,(\mathbf{p}_s \cdot \mathbf{n}_q)\, f(w).
\end{aligned}
$$

Similarly in the parameter domain of the second surface we obtain

$$
\begin{aligned}
\frac{du}{dw} &= \frac{u_{i+1} - u_i}{\triangle w} \\
\frac{dv}{dw} &= \frac{v_{i+1} - v_i}{\triangle w}
\end{aligned}
$$

and

$$
\begin{aligned}
u_{i+1} &= u_i \mp \triangle w\,(\mathbf{q}_v \cdot \mathbf{n}_p)\, g(w) \\
v_{i+1} &= v_i \pm \triangle w\,(\mathbf{q}_u \cdot \mathbf{n}_p)\, g(w).
\end{aligned}
$$

Since we have formulated the prediction step as a set of ordinary differential equations, better prediction can be achieved by using higher order methods for ordinary differential equations, e.g., an adaptive 4-5 order Runge-Kutta method. Note that the high convergence rate of such methods can be disturbed when crossing knot lines in NURBS-surfaces.

Correcting Predicted Intersection Points

If both surfaces are planar, then the new point is $\mathbf{c}_{i+1} \equiv \mathbf{p}(s_{i+1}, t_{i+1}) \equiv \mathbf{q}(u_{i+1}, v_{i+1})$. In most other cases (except when the intersection curve is a straight line, or by sheer coincidence) $\mathbf{p}(s_{i+1}, t_{i+1}) \neq \mathbf{q}(u_{i+1}, v_{i+1})$. Let $\mathbf{c}_i \equiv \mathbf{p}(s_i, t_i) \equiv \mathbf{q}(u_i, v_i)$ be the current point on the intersection curve and $\mathbf{c}_i^{'}$ the tangent of the intersection curve at $\mathbf{c}_i$. Provided that $\triangle w$ is not too large, then an exact point on the intersection curve is defined by the intersection of $\mathbf{p}$, $\mathbf{q}$ and the plane A through the point $\mathbf{c}_i + \triangle w \mathbf{c}_i^{'}$ with $\mathbf{c}_i^{'}$ as normal

$$
A : \Big((x, y, z) - (\mathbf{c}_i + \triangle w \mathbf{c}_i^{'}) \Big) \cdot \mathbf{c}_i^{'} = 0.
$$

If $\triangle w$ is not too large, then the predicted points $\mathbf{p}(s_{i+1}, t_{i+1})$ and $\mathbf{q}(u_{i+1}, v_{i+1})$ will both be approximations to the intersection of of $\mathbf{p}$, $\mathbf{q}$ and A. Thus we can start to iterate from the parameter pairs (s_{i+1}, t_{i+1}) and (u_{i+1}, v_{i+1}) to find a more accurate approximation to the intersection of $\mathbf{p}$, $\mathbf{q}$ and A. We iterate until we find parameter pairs (σ_n, τ_n) and (υ_n, ω_n) such that either $\|\mathbf{p}(\sigma_n, \tau_n) - \mathbf{q}(\upsilon_n, \omega_n)\|_2 \leq \varepsilon$ or until we detect divergence. At each iteration step we approximate $\mathbf{p}$ by the tangent plane at (σ_n, τ_n) and $\mathbf{q}$ with the tangent plane at (υ_n, ω_n).

5.6 Representing the Intersection Curve

Representation of intersection curves by parametric cubic B-spline curves is a natural choice both in 3D and in the parameter domains of the intersected surfaces. The curve is built segment-by-segment by Hermite interpolation (interpolation of position and tangent at the start and end of each curve segment), and the continuity between adjacent segments is chosen to fit best with the expected use of the approximated intersection curves. In cubic Hermite interpolation of a function, a sequence of points with associated tangents are interpolated, and the resulting error term is $O(h^4)$, with h the smallest interval between points in the curve approximation. When using this approach for parametric curves, interpolating each coordinate independently, the resulting error term remains $O(h^4)$ as for Hermite interpolation of univariate functions.

For parametric curves, a proper choice of the tangent lengths can both improve quality and convergence rates of parametric cubic Hermite interpolation. This approach is often referred to as geometric Hermite interpolation. For interpolation of 2D curves using geometric cubic Hermite interpolation, it is shown in [6] that the error term can be improved to $O(h^6)$ provided the segment approximated is convex. In [25, 26] geometric cubic Hermite interpolation of 3D curves is shown to be $O(h^5)$ if curvature and torsion do not vanish. The tangent lengths can be used to improve the approximation order or alternatively to model other properties of the curve, and the correct choice of tangent lengths will depend on the purpose of the curve:

- **Intersection curves used for visual inspection** can be G^1 continuous without creating any unwanted visual effect, and the tangent lengths can be used to produce the best possible fit to the intersection curve. By G^1 continuity we mean that the position and tangent directions at the common point of adjacent curve segments are the same, this is different from C^1 where also tangent lengths have to be the same.
- **Intersection curves in a boundary structure based volume model** should be at least C^1, to have consistency of the evolution of tangent lengths as we move along the intersection curve. The intersection curve should be sufficiently accurate to be a starting point for later refinement if necessary.
- **Intersection curves intended for designing new surfaces** should have a well-behaved parametrisation. When a number of different curves are used as a basis for defining surfaces, the intuitive correspondence between points on the curves is related to the curve lengths of the different curves. Thus an approximately curve-length parametrisation is a good choice. Cubic Hermite interpolation will at most produce C^1 continuous curves so in order to have higher order continuity, the degree of the approximation must be increased.
- **Intersection curves used as input to production processes** have to be adapted to the formats used by the actual production processes. Consequently it is difficult to give detailed guidelines. Geometric cubic Hermite

interpolation will in many cases not have sufficient degrees of freedom to take care of the quality requirements of production processes. One alternative representation can be Pythagorean-hodograph polynomials [18].

6 Conclusion

We hope that this paper has managed to show that intersection algorithms for parametric surfaces is a challenging task with a considerable potential for further improvements. With the wide use of CAD in industry the economical gain from improved algorithms is large. However there are a lot of factors that slow down the introduction of new approaches to intersection algorithms:

- New intersection technology has to be compatible with existing CAD-models, as industry has invested heavily in the use of current generation CAD-systems.
- There has been continuous flow of buy ups and mergers in the CAD-industry over the last decade. Most CAD-vendors will not invest into improved technology if they feel confident in their market position.
- For industries with not-too-complex geometries, current technology seems to work, it is the high technology industries such as aerospace and automotive industry that are most severely affected by the shortcomings of current intersection algorithms.
- The STEP standard has standardised the ideas of the late 1980s' for the transfer of CAD-models. With current CAD focus on STEP it is difficult to achieve a renewal of CAD-model representation and associated intersection algorithms.

Based on the above observation the research on intersection algorithms addressing industrial needs will to a great extent be focused on incremental improvements. However, with the wide range of possibilities of improvements in intersection algorithms there is still a very large area of research open. Our hope is, however, that CAD-model representation is renewed and the full potential of interval arithmetic and combination of algebraic and parametric representations can be investigated.

Acknowledgement. This work was supported by:

- The European Commission through the RTD-project: GAIA II – *Intersection algorithms for geometry based IT-applications using approximate algebraic methods*, contract number FP5 IST-2001-35512.
- The European Commission through the NoE (Network of Excellence): AIM@SHAPE – *Advanced and Innovative Models And Tools for the development of Semantic-based systems for Handling, Acquiring, and Processing knowledge Embedded in multidimensional digital objects*, contract number FP6 IST NoE 506766.
- The Research Council of Norway through the project: GPGPU – *Graphics hardware as a high-end computational resource*, contract number 158911/I30.

References

1. S. L. Abrams, W. Cho, C. Y. Hu, T. Maekawa, N. M. Patrikalakis, E. C. Sherbrooke, and Ye X. Methods for rounded interval arithmetic. *Computer Aided Design*, 30(8):657–666, 1998.
2. C . Bajaj. The emergence of algebraic curves and surfaces in geometric design. In *Directions in Geometric Computing*, pages 1–28. Information Geometers, 1993.
3. S. Briseid, T. Dokken, T. R. Hagen, and J. O. Nygaard. Spline surface intersections optimized for GPUs. In V. N. Alexandrov, G. D. van Albada, P. M. A. Sloot, and J. Dongarra, editors, *Computational Science – ICCS 2006: 6th International Conference, Reading, UK, May 28–31, 2006, Proceedings, Part IV*, volume 3994 of *Lecture Notes in Computer Science*, pages 204–211, Berlin/Heidelberg, 2006. Springer.
4. E. Cohen, R. F. Riesenfeld, and G. Elber. *Geometric modeling with splines.* A K Peters Ltd., Natick, MA, 2001. An introduction, With a foreword by Tom Lyche.
5. O.J Dahl and D. Belsnes. *Algorithms and Data Structures.* Studentlitteratur, Lund, Sweden, 1973.
6. C. de Boor, K. Höllig, and M. Sabin. High accuracy geometric Hermite interpolation. *Comput. Aided Geom. Design*, 4(4):269–278, 1987.
7. T. Dokken. *Reference Manual, B-spline Library.* SINTEF (Senter for Industriforskning), Oslo, Norway, 1983.
8. T. Dokken. Finding intersections of b-spline represented geometries using recursive subdivision techniques. *Comput. Aided Geom. Design*, 2:189–195, 1985.
9. T. Dokken. *APS-SS: A subrouting package for modelling of sculptured surfaces.* SINTEF (Senter for Industriforskning), Oslo, Norway, 1987.
10. T. Dokken. *The SINTEF Spline Library, version 4.0.* SINTEF, Oslo, Norway, 1994.
11. T. Dokken. *Aspect of Intersection algorithms and Approximation, Thesis for the doctor philosophias degree.* PhD thesis, University of Oslo, 1997.
12. T. Dokken. Aspects of algorithms for manifold intersection. In *Numerical methods and software tools in industrial mathematics*, pages 365–380. Birkhauser, Boston, MA, 1997.
13. T. Dokken. Approximate implicitization. In *Mathematical methods for curves and surfaces (Oslo, 2000)*, Innov. Appl. Math., pages 81–102. Vanderbilt Univ. Press, Nashville, TN, 2001.
14. T. Dokken, H. K. Kellermann, and C. Tegnander. An approach to weak approximate implicitization. In *Mathematical methods for curves and surfaces (Oslo, 2000)*, Innov. Appl. Math., pages 103–112. Vanderbilt Univ. Press, Nashville, TN, 2001.
15. T. Dokken, V. Skytt, and A.-M. Ytrehus. Recursive subdivision and iteration in intersections and related problems. In *Mathematical methods in computer aided geometric design (Oslo, 1988)*, pages 207–214. Academic Press, Boston, MA, 1989.
16. T. Dokken and J. B. Thomassen. Overview of approximate implicitization. In *Topics in algebraic geometry and geometric modeling*, volume 334 of *Contemp. Math.*, pages 169–184. Amer. Math. Soc., Providence, RI, 2003.
17. G. Farin, J. Hoschek, and M.-S. Kim, editors. *Handbook of computer aided geometric design.* North-Holland, Amsterdam, 2002.

18. R. T. Farouki. Pythagorean-hodograph curves. In *Handbook of computer aided geometric design*, pages 405–427. North-Holland, Amsterdam, 2002.
19. R. T. Farouki, C. Y. Han, J. Hass, and T. W. Sederberg. Topologically consistent trimmed surface approximations based on triangular patches. *Comput. Aided Geom. Design*, 21(5):459–478, 2004.
20. R. T. Farouki and V. T. Rajan. On the numerical condition of polynomials in Bernstein form. *Comput. Aided Geom. Design*, 4(3):191–216, 1987.
21. R. T. Farouki and V. T. Rajan. Algorithms for polynomials in Bernstein form. *Comput. Aided Geom. Design*, 5(1):1–26, 1988.
22. A. Gálvez, J. Puig-Pey, and A. Iglesias. Differential method for parametric surface intersection. In *Computational Science and Its Applications – ICCSA 2004: International Conference, Assisi, Italy, May 14-17, 2004*, volume 3044 of *Lecture Notes in Comput. Sci.*, pages 651–660. Springer, Berlin, 2004.
23. K. Hasund, T. Dokken, and S. Ulfsby. System specifications, sculptured surfaces. Technical Report 16, SINTEF (Sentralinstitutt for industriell forskning), Oslo, Norway, 1980.
24. M. E. Hohmeyer. *Robust and Efficient Surface Intersection for Solid Modelling.* PhD thesis, Computer Science Division, University of California, 1992.
25. K. Höllig and J. Koch. Geometric Hermite interpolation. *Comput. Aided Geom. Design*, 12(6):567–580, 1995.
26. K. Höllig and J. Koch. Geometric Hermite interpolation with maximal order and smoothness. *Comput. Aided Geom. Design*, 13(8):681–695, 1996.
27. C.-Y. Hu, T. Maekawa, N. M. Patrikalakis, and X. Ye. Robust interval algorithm for surface intersections. *Computer-Aided Design*, 29(9):617–627, 1997.
28. IEEE. Specification for binary floating point arithmetic for microprocessor systems, international standard, 1989.
29. ISO. Industrial automation systems and integration - product data representation and exchange – part 42: Integrated generic resources: Geometric and topological representation, 1994.
30. S. Krishnan and D. Manocha. An efficient surface intersection algorithm based on lower-dimensional formulation. *ACM Trans. Graph.*, 16(1):74–106, 1997.
31. J. M. Lane and R. F. Riesenfeld. Bounds on a polynomial. *BIT*, 21(1):112–117, 1981.
32. T. Maekawa, W. Cho, and N. M. Patrikalakis. Computation of self-intersections of offsets of bézier surface patches. *Journal of Mechanical Design, ASME Transactions*, 119:275–283, 1997.
33. H. Mukundan, K. H. Ko, T. Maekawa, T. Sakkalis, and N. M. Patrikalakis. Tracing surface intersections with a validated ode system solver. In G. Elber, N. Patrikalakis, and P. Brunet, editors, *Proceedings of the Ninth EG/ACM Symposium on Solid Modeling and Applications, Genoa, Italy, June 2004*, pages 249–254. Eurographics Press, 2004.
34. N. M. Patrikalakis and T. Maekawa. Intersection problems. In *Handbook of computer aided geometric design*, pages 623–649. North-Holland, Amsterdam, 2002.
35. N. M. Patrikalakis and T. Maekawa. *Shape interrogation for computer aided design and manufacturing.* Springer-Verlag, Berlin, 2002.
36. N. M. Patrikalakis, T. Maekawa, K. H. Ko, and H. Mukundan. Surface to surface intersections. *Computer-Aided Design & Applications*, 1:449–458, 2004.
37. J. Peters and X. Wu. Sleves for planar spline curves. In *Computer Aided Geometric Design*, pages 615–635, 2004.

38. J. Puig-Pey, A. Gálvez, and A. Iglesias. A new differential approach for parametric-implicit surface intersection. In *Computational science–ICCS 2003. Part I*, volume 2657 of *Lecture Notes in Comput. Sci.*, pages 897–906. Springer, Berlin, 2003.
39. T. Sederberg, R. Goldman, and H. Du. Implicitizing rational curves by the method of moving algebraic curves. *J. Symbolic Comput.*, 23(2-3):153–175, 1997. Parametric algebraic curves and applications (Albuquerque, NM, 1995).
40. T. W. Sederberg. *Implicit and parametric curves and surfaces for computer aided geometric design.* PhD thesis, Purdue University, 1983.
41. T. W. Sederberg. Planar piecewise algebraic curves. *Computer Aided Geometric Design*, 1(3):241–255, 1984.
42. T. W. Sederberg and R. J. Meyers. Loop detection in surface patch intersections. *Comput. Aided Geom. Design*, 5:161–171, 1988.
43. T. W. Sederberg and Anderson D. C.and Goldman R. N. Implicit representation of parametric curves and surfaces. *Computer Vision, Graphics, and Image Processing*, 28:72–84, 1984.
44. T. W. Sederberg, J. Zheng, K. Klimaszewski, and T. Dokken. Approximate implicitization using monoid curves and surfaces. *Graphical Models and Image Processing*, 61(4):177–198, 1999.
45. T. W. Sederberg, J. Zheng, and Song X. A conjecture on tangent intersections of surface patches. *Comput. Aided Geom. Design*, 21(1):1–2, 2004.
46. T. W. Sederberg and A. K. Zundel. Pyramids that bound surface patches. *CVGIP: Graphical Model and Image Processing*, 58(1):75–81, 1996.
47. P. Sinha, E. Klassen, and K.K. Wang. Exploiting topological and geometric properties for selective subdivision. In *Symposium on Computational Geometry*, pages 39–45. ACM Press, 1985.
48. V. Skytt. Challenges in surface-surface intersections. In *Computational methods for algebraic spline surfaces*, pages 11–26. Springer, Berlin, 2005.
49. V. Skytt. A recursive approach to surface-surface intersections. In *Mathematical methods for curves and surfaces: Tromsø2004*, Mod. Methods Math., pages 327–338. Nashboro Press, Brentwood, TN, 2005.
50. A. K. Zundel and T. W. Sederberg. Surface intersection loop destruction. In *The mathematics of surfaces, VII (Dundee, 1996)*, pages 463–478. Info. Geom., Winchester, 1997.
51. A.K. Zundel. *Surface-Surface Intersection: Loop Destruction Using Bézier Clipping and Pyramidal Bounds. Thesis for Doctor of Philosophy.* PhD thesis, Brigham Young University, 1994.

A Appendix: Choice of Polynomial Basis

In CAD-systems curves and surfaces are represented by low degree algebraic representations (line, circle, ellipse, sphere, torus,...), or as piecewise rational parametric curves and surfaces represented by NonUniform Rational B-splines. The results of intersection algorithms are very sensitive to the accuracy of the geometries intersected. We know that different polynomial bases have different properties. When implementing algorithms in floating point arithmetic the behaviour of the different polynomial basis with respect to rounding errors is essential for the quality of the results of the algorithms, e.g., for intersection. Within CAD we address geometric objects that are finite in size. Thus polynomial bases with attractive properties on a limited interval can be the proper choice.

If the result of cubic spline interpolation is converted to a piecewise representation, using a power basis for each segment, the resulting composite curve will have small gaps between adjacent segments, while the use of the basis for Hermite interpolation or the Bernstein basis will not introduce such holes. In the following subsections we will look at the behaviour of rounding errors for three choices of polynomial bases:

- The power basis.
- The basis used in cubic Hermite interpolation.
- The Bernstein basis.

The discussion shows that among these bases, the best-behaved basis with respect to rounding error, is the Bernstein basis. The Bernstein basis is well known within CAGD - Computer Aided Geometric Design, but has drawn little attention outside of CAGD. The crucial property of the Bernstein basis in this respect is that the basis provides a partition of unity. By partition of unity we mean that the basis functions are all non-negative and sum to 1 on the interval of interest, say $[0, 1]$.

In CAD-systems the B-spline basis plays a central role when representing free form curves and surfaces within CAD-systems. This basis is used to create the rational B-spline basis (NURBS - NonUniform Rational B-splines) used when representing parametric curves and surfaces. The B-spline basis is a piecewise polynomial basis with explicit control of the continuity between polynomial segments. The B-spline basis and the NURBS basis with positive weights, are bases that are partitions of unity. These bases consequently have similar attractive properties with respect to rounding errors as the Bernstein basis. A B-spline basis is described by

- A polynomial degree.
- A number of basis functions.
- A knot vector, a non-decreasing sequence of real numbers that both describe the location of the joints and continuity between adjacent polynomial segments.

When working with B-splines and NURBS it is important to use evaluation and subdivision algorithms that are harmonised. Subdividing at a given point should give exactly the same point value as an evaluation. By harmonising evaluation and subdivision algorithms we make the behaviour of the algorithms more predictable. The methods that we have chosen for evaluation and subdivision of B-splines are respectively:

- de Boor Cox algorithm [17] for evaluation of the B-spline basis functions,
- the Oslo Algorithm [4] for evaluating the discrete B-splines used when subdividing a NURBS curve or surface.

A.1 The Power Basis

We are all familiar with the power basis representation of polynomials.

$$p(t) = a_n t^n + a_{n-1} t^{n-1} + \ldots + a_1 t + a_0.$$

In the power basis the coefficients of the polynomial are related to the derivatives of the polynomial for $t = 0$.

$$a_i = \frac{p^{(i)}(0)}{i!}.$$

Assume that we represent the polynomial in floating point arithmetic and thus that all coefficients of the polynomial have a relative rounding error $\varepsilon_i, i = 0, \ldots n$. Instead of the original polynomial $p(t)$ we consider the polynomial $\tilde{p}(t)$ with rounding errors,

$$\begin{aligned}\tilde{p}(t) &= a_n(1+\varepsilon_n)t^n + a_{n-1}(1+\varepsilon_{n-1})t^{n-1} + \ldots + a_1(1+\varepsilon_1)t + a_0(1+\varepsilon_0).\\ &= p(t) + a_n\varepsilon_n t^n + a_{n-1}\varepsilon_{n-1}t^{n-1} + \ldots + a_1\varepsilon_1 t + a_0\varepsilon_0.\end{aligned}$$

It is obvious that as we move away from $t = 0$, we must expect the rounding error to grow. To simplify the later discussion and be compatible with the Bernstein basis, we restrict the interval of interest to $t \in [0,1]$. Let us assume that all the relative rounding errors satisfy $|\varepsilon_i| = \varepsilon, i = 0, \ldots, n$, and the sign of each ε_i is chosen such that either $a_i\varepsilon_i \geq 0, i = 0, \ldots n$, or $a_i\varepsilon_i \leq 0, i = 0, \ldots n$. Now we define the polynomials

$$\begin{aligned}\tilde{p}_+(t) &= p(t) + \left(|a_n|\, t^n + |a_{n-1}|\, t^{n-1} + \ldots + |a_1|\, t + |a_0|\right)\varepsilon,\\ \tilde{p}_-(t) &= p(t) - \left(|a_n|\, t^n + |a_{n-1}|\, t^{n-1} + \ldots + |a_1|\, t + |a_0|\right)\varepsilon.\end{aligned}$$

These satisfy $\tilde{p}_-(t) \leq \tilde{p}(t) \leq \tilde{p}_+(t)$, and limit $p(t)$ when influenced by rounding errors. In double precision floating point ε is typically 1.1×10^{-16}, and in single precision 3.2×10^{-8}.

In the example below we use these limiting polynomials to show the possible location of a parametric curve with rounding error. To emphasis the effect of the rounding error, we have chosen $\varepsilon = 0.01$ to visualise this effect.

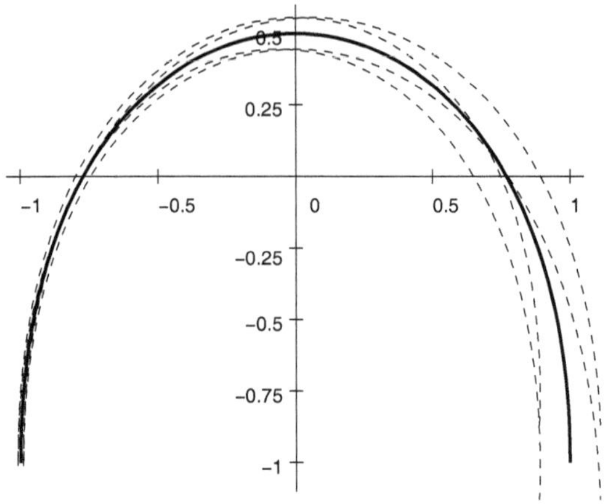

Fig. 12. The curve in Example 4 and the maximal rounding errors for the relative rounding error of 0.01 when the power basis is used. Note that the rounding error grows from the start to the end of the segment, and is a factor 13 larger at the end of the segment compared to the start of the segment.

Example 4 (Effect of rounding error for the power basis). Given a parametric curve $\mathbf{p}(t) = (x(t), y(t)) = \left(-4t^3 + 6t^2 - 1, -6t^2 + 6t - 1\right), t \in [0, 1]$, and assume that all coefficients of $\mathbf{p}(t)$ are assigned a relative error of 0.01. Then the extreme rounding error cases for $x(t)$ and $y(t)$ are

$$\begin{aligned}
x_+(t) &= x(t) + 0.01 \times (4t^3 + 6t^2 + 1) \\
x_-(t) &= x(t) - 0.01 \times (4t^3 + 6t^2 + 1) \\
y_+(t) &= y(t) + 0.01 \times (6t^2 + 6t + 1) \\
y_-(t) &= y(t) - 0.01 \times (6t^2 + 6t + 1).
\end{aligned}$$

In Figure 12 the curve is plotted along with the curves $(x_-(t), y_-(t))$, $(x_-(t), y_+(t))$, $(x_+(t), y_-(t))$ and $(x_+(t), y_+(t))$. Note that the rounding error grows with increasing parameter value. If the curve had been parametrised from right to left instead of from left to right the behaviour of the rounding error would be mirrored.

A.2 The Hermite Interpolation Polynomials

Cubic spline interpolation is traditionally described in a piecewise cubic Hermite basis to ensure C^1–continuity between adjacent segments in a simple way. The coefficients of the cubic Hermite bases are position and tangent at both segment ends. By using the position and tangent at the end of one segment as the position and tangent of the start of the next segment, C^1-continuity is not destroyed by rounding errors as would be the case when using a piecewise power basis.

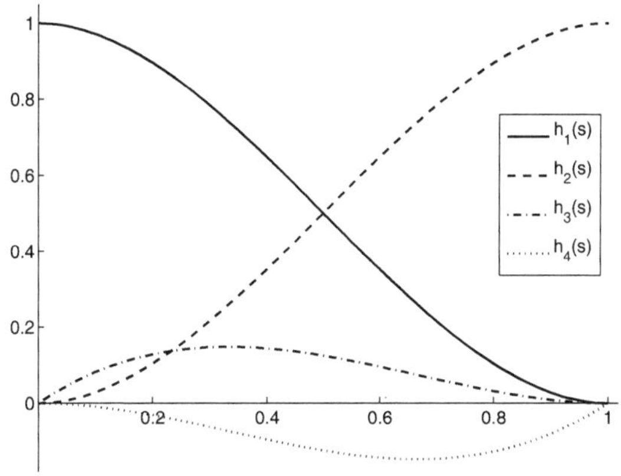

Fig. 13. The four basis functions used in cubic Hermite interpolation.

We will look at the same example as in Section A.1, but now represent the polynomial in a cubic Hermite basis. The basis functions are

$$\begin{aligned}
h_1(s) &= 2s^3 - 3s^2 + 1 \\
h_2(s) &= -2s^3 + 3s^2 \\
h_3(s) &= s^3 - 2s^2 + s \\
h_4(s) &= s^3 - s^2.
\end{aligned}$$

They satisfy:

	$h_1(t)$	$h_1'(t)$	$h_2(t)$	$h_2'(t)$	$h_3(t)$	$h_3'(t)$	$h_4(t)$	$h_4'(t)$
$t=0$	1	0	0	0	0	1	0	0
$t=1$	0	0	1	0	0	0	0	1

Example 5 (Effect of rounding error for the cubic Hermite basis). The description of the parametric polynomial in Figure 12 using the basis functions of Hermite interpolation, shown in Figure 13 is

$$\begin{aligned}
p(t) &= (x(t), y(t)) \\
&= (-1,-1)h_1(s) + (1,-1)h_2(s) + (0,6)h_3(s) + (0,-6)h_4(s) \\
&= (-h_1(s) + h_2(s), -h_1(s) - h_2(s) + 6h_3(s) - 6h_4(s)).
\end{aligned}$$

Now adding relative rounding errors with absolute value ε to all coefficients, and adjusting the sign of the rounding errors to produce extremal cases for the rounding errors we get

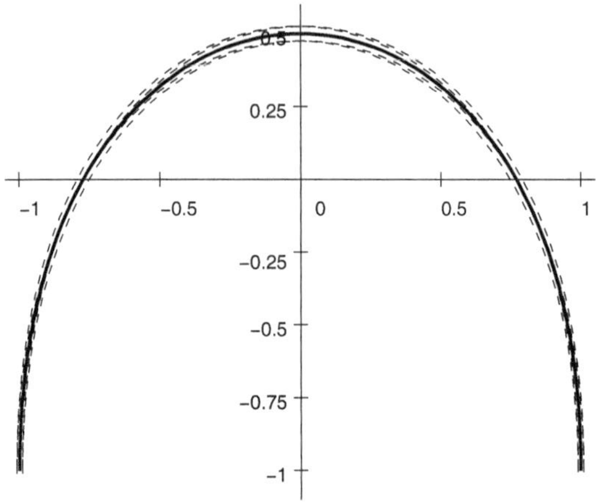

Fig. 14. The curve in Example 5 and the maximal rounding errors for the relative rounding error of 0.01 when cubic Hermite interpolation is used. Note that the rounding error is small at the segment ends and 2.5×larger in the middle.

$$
\begin{aligned}
x_-(s) &= x(s) - \varepsilon h_1(s) - \varepsilon h_2(s) \\
x_+(s) &= x(s) + \varepsilon h_1(s) + \varepsilon h_2(s) \\
y_-(s) &= y(s) - \varepsilon h_1(s) - \varepsilon h_2(s) - 6\varepsilon h_3(s) + 6\varepsilon h_4(s) \\
y_+(s) &= y(s) + \varepsilon h_1(s) + \varepsilon h_2(s) + 6\varepsilon h_3(s) - 6\varepsilon h_4(s).
\end{aligned}
$$

In Figure 14 the curve is plotted along with the curves $(x_-(t), y_-(t))$, $(x_-(t), y_+(t))$, $(x_+(t), y_-(t))$ and $(x_+(t), y_+(t))$. Note that the rounding error is largest in the middle of the curve and small at both end. The rounding error is independent of the direction in which the curve is parameterised.

A.3 The Bernstein Basis

The control of the relative rounding error when using cubic Hermite interpolation is much better compared to using the power basis [20, 21]. One way of explaining this is that the position of the curve is much more closely related to the location of the coefficients in the cubic Hermite basis than for the cubic power basis. A basis with an even closer relationship between values of coefficients, and the location of the curve, is the Bernstein basis. The Bernstein basis of degree n on the interval $[0, 1]$ is defined as follows

$$B_i^n(s) = \binom{n}{i}(1-s)^{n-i}s^i, \; i = 0, \ldots, n.$$

A polynomial described in the Bernstein basis has coefficients that we call control points c_i, $i = 0, \ldots, n$. When plotting a function described in the degree n Bernstein basis, we can assign a location to the control points by assigning

$$(\frac{i}{n}, c_i).$$

The Bernstein basis on $[0, 1]$ is a partition of unit on this interval, i.e.,

$$0 \le B_i^n(s) \le 1,\ s \in [0,1],\ i = 0, \dots, n$$
$$\sum_{i=0}^{n} B_i^n(s) = 1.$$

Now let $p(s) = \sum_{i=0}^{n} c_i B_i^n(s)$, be a polynomial represented in a degree n Bernstein basis. By assigning relative rounding errors to the coefficients we define the polynomial

$$\tilde{p}(s) = \sum_{i=0}^{n} c_i(1+\varepsilon_i)B_i^n(s)$$
$$= p(s) + \sum_{i=0}^{n} c_i \varepsilon_i B_i^n(s).$$

For $s \in [0, 1]$,we have

$$|\tilde{p}(s) - p(s)| \le \left| \sum_{i=0}^{n} c_i \varepsilon_i B_i^n(s) \right| \le \max_{0 \le i \le n} |c_i \varepsilon_i| .$$

Thus the rounding error for polynomials represented in the Bernstein basis is less than the largest coefficient rounding error. In Figure 15 the cubic Bernstein basis is displayed.

Example 6 (Effect of rounding error for the Bernstein basis). The description of the parametric polynomial in Figure 12 using the Bernstein basis is

$$p(s) = (x(s), y(s))$$
$$= (-1,-1)(1-s)^3 + (-1,1)3(1-s)^2 s + (1,1)3(1-s)s^2 + (1,-1)s^3.$$

Now adding relative rounding errors with absolute value ε to all coefficients and adjusting the sign of the rounding errors to produce extremal cases for the rounding errors we get

$$x_-(s) = x(s) - \varepsilon$$
$$x_+(s) = x(s) + \varepsilon$$
$$y_-(s) = y(s) - \varepsilon$$
$$y_+(s) = y(s) + \varepsilon.$$

In Figure 16 the curve is plotted along with the curves $(x_-(t), y_-(t))$, $(x_-(t), y_+(t))$, $(x_+(t), y_-(t))$ and $(x_+(t), y_+(t))$. Note that the rounding error is small all along the curve and smaller than the rounding error of either power basis or the cubic Hermite interpolation basis. The rounding error is independent of the direction in which the curve is parameterized.

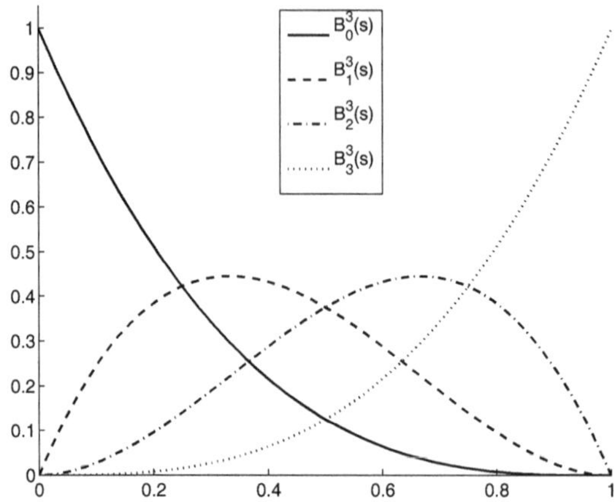

Fig. 15. The four basis functions used in the cubic Bernstein basis.

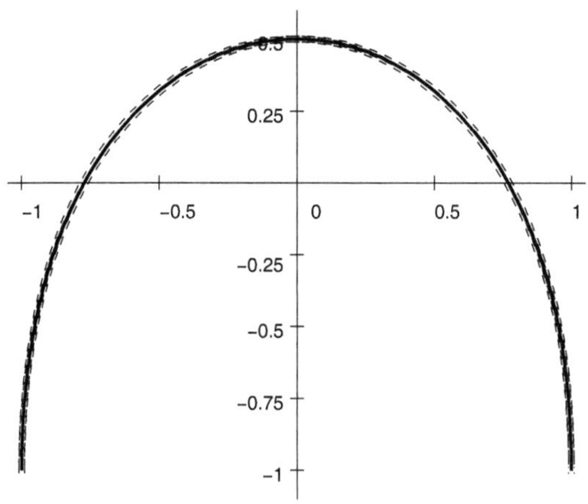

Fig. 16. The relative rounding error of 0.01 when the cubic Bernstein basis is used. The relative rounding error is the same all along curve.

Surface Modelling of Neuronal Populations and Brain Structures: Use of Implicitly Represented Geometries

Jens O. Nygaard, Jan G. Bjaalie, Simen Gaure, Christian Pettersen, and Helge Avlesen

Summary. We discuss some challenges faced by neuroscientists with respect to 3D geometry reconstruction, modelling and visualization. We have developed a toolbox based on implicit representation of geometry, distributed computing and an easy to deploy and maintain graphical Java3D based interface. We describe the principles underlying this toolbox and provide an outline of the problems and suggested solutions related to a specific project, `Neuroinf` [18], which is a collaboration between research groups in biomedical science, informatics, and mathematics at the participating institutions. Public access to these tools will be announced on the web page.

1 Introduction

Novel approaches in computational neuroanatomy include the simultaneous visualization and analysis of a large number of potentially huge data sets describing the distribution of nerve cells and other identified elements in brain tissue, collected with computerized microscope systems. Much of the analysis requires real-time comparison of the spatial patterns formed by each of the data sets.

In this chapter we discuss the development of tools for modelling, visualization and analysis of such neuroscience data. In particular, we describe an implemented web client based on Java3D, with full 3D viewing capability and interactive geometry editing. This is accompanied by a server dedicated to perform compute-intensive tasks, and to communicate with a database for easy data sharing.

Advantages of this client/server model include both a reduction in complexity in the client, making a Java implementation feasible, as well as giving the possibility of tailoring the compute-intensive server to parallel hardware and larger databases for number crunching and easy data sharing. The benefits for the users of this design are primarily 1) the move of specialized hardware and software maintenance to a centralized facility, 2) rapid access to new tools and developments, requiring a minimum of local software installation, and 3) tools standardization and improved environments for data sharing.

The *Neuroinf project* is a *Technology Transfer Project* funded by the Norwegian High Performance Computing Consortium. The goal of the project is to build and distribute tools for exploitation of high performance computing capabilities, to the neuroscience community. We have both built new tools and improved existing ones by adding parallelization to computationally intensive subroutines. The goal is ultimately to be able to handle ever larger and more complex models. The package of software tools is called `m3d`.

The data and geometries that we are modelling in this project, originate from the problem of "localization in the brain". This task is difficult for several reasons, for instance, many regions in the brain can be defined ambiguously because they may share characteristics to some degree, and also because the regions actually do overlap.

One particular way of defining a region exemplifies this, namely a definition based on the density of some substance or other single-valued measure, e.g., a density or a probability. The first problem is to define the threshold value of the region, that is, which value should be used to distinguish between the outside and the inside of the region? Then, the problem may be the presence of only a small and perhaps sparse set of measures of this parameter. How shall we use the seemingly too few values to define a clear geometry? These are examples of some of the problems considered in the Neuroinf project, which will be discussed in the present chapter. For a more expanded introduction to this problem of localization in the brain, see for example Bjaalie [6].

We will start by describing the background and the problems posed with respect to surface reconstruction and modelling. We then proceed by discussing some possible solutions, ranging from our choice of implicit geometry representation, and methods for finding such representations, extracting geometry from scalar fields, to the postprocessing of geometry by the *scalar field editing*. We briefly discuss design and implementational details with respect to parallelization and employment of tools at users' locations.

This chapter complements the chapter "Surface Modelling from Biomedical Data" in Bjaalie et al. [8], in the sense that some of the main problems of the approach described therein are solved in another way in the current exposition. Further elaboration follows in Section 3.

2 Neuroanatomy and the Geometry Pipeline

In the field of neuroanatomy, researchers study issues such as the properties of nerve cells, or *neurons*, groups of such, the interplay of neuronal populations, distribution, and localization of functions. Modern imaging tools are only the last in a succession of tools, and do not reach the high level of resolution provided by classical microscopy techniques that are applied in this science. In return, they have other advantages, they can give complete volumes of

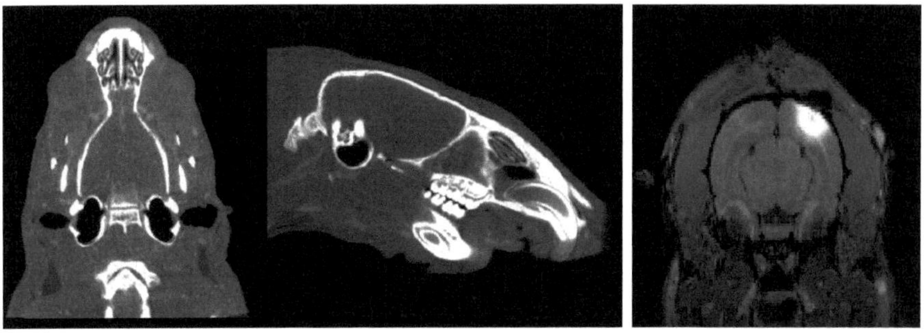

Fig. 1. From the left: Two raw images from stacks of CT scans from a rat head, acquired with a microCAT scanner. Courtesy of T. Hjørnevik, H. Qu, and J. G. Bjaalie. To the right: MR scan of high field strength through a rat head. The intensity in this scan reflects magnetic properties in the tissue and the concentration of a paramagnetic agent inserted in the brain for the purpose of labeling specific nerve cell tracts. Courtesy of T. B. Leergaard, J. G. Bjaalie, and A. Dale.

data without slicing the subject, and they can be applied to living tissue and beings without being destructive. Before embarking on the description and explanation of the currently implemented software tools, we want to briefly introduce some key concepts leading up to the point where we have actual data to process on the computer.

2.1 Background and Motivation

The brain is divided into multiple regions and areas. Within these, populations of neurons are present, and these populations occupy smaller or larger subvolumes of tissue. A *neuronal population* is defined by some characteristic that may vary over the total volume. Such a characteristic can be the category or type of neurons, identified via morphological measures or specific labeling techniques, or density. The kind of geometries that we are interested in modelling will be closed surfaces encapsulating brain regions and neuronal populations.

2.2 Imaging in Neuroanatomy

Traditionally, neuroanatomy researchers have modelled, processed and visualized 3D geometries of brain regions and more finely grained information "virtually", by inspecting series of images of 2D intersections taken at one angle, or by comparing insersections cut at different angles, see Figure 1. In this particular case, the entire shape of the brain of the rat is viewed inside the skeletal structures, that are the most prominent tissues in the CT scan, see Figure 1a and 1b, or viewed more in detail in the MR scan, Figure 1c.

The limitations of such images are obvious, it is difficult if not impossible to really grasp the geometry of complex structures from such data. The natural step beyond inspection of single images is 3D reconstruction of data for further processing and visualization on the desktop computer.

2.3 Geometric Modelling in Neuroanatomy

In addition to just defining geometries, it is often desirable to be able to manipulate these geometries, either because they are not believed to accurately model the different neuronal populations in question, or for the purpose of simplification, extracting primary features and eliminating details. For example, it may be difficult to formulate a scalar function that precisely discriminates regions or tissue elements identified by point coordinate sets that are already determined by the neuroscientist to be outliers. In such a case, the possibility of editing generated 3D models may be the most practical course of action. In other cases, it may be necessary to edit geometries to better expose occluded parts of the reconstructed region.

With these purposes in mind, we also need a way of defining, extracting and rendering geometry that lends itself easily to various kinds of editing. We will now elaborate on the view of the geometry as an interface between regions in 3D space on which a scalar function returns different values, depending on which side of the geometry we perform the evaluation. We will start by listing some methods of acquisition. The data so obtained may to different degrees dictate how we define our geometries.

2.4 Acquisition

There are a number of methods for acquiring images in neuroanatomy, and we shall see that they often lend themselves to particular ways of geometry reconstruction.

The main forms in which our acquired data appears are point clouds, sets of contours, or just stacks of images. Point clouds and contour lines may be extracted from microscopic sections, or from other imaging data including *computerized axial tomography* (CT) scans, *magnetic resonance imaging* (MRI), and *positron emission tomography* (PET) scans. A pair of CT scans and an MR scan, are shown in Figure 1.

Imaging Techniques

Many available imaging techniques, whether based on classical microscope systems, higher resolution microscopy such as confocal or multifoton microscopy, or lower resolution imaging such as MR, can be treated in a similar fashion. Both colour images and monochromatic images can be transformed to obtain a scalar function over the spatial domain that is depicted by the image. Occasionally, this scalar function can be applied more or less directly in our *implicit representation* approach, as outlined in Section 4.

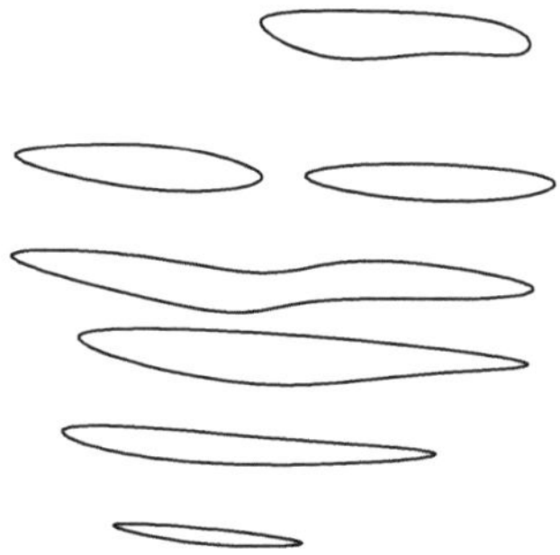

Fig. 2. Bifurcation example with stack of contours, one contour per plane at the top and the bottom, two in the middle. In this case we cannot simply interpolate the contours, for example with a tensor product spline surface. (Also called *lofting*.) Such an approach was used and discussed in [8], but then for contour stacks without such branches.

As a step in the processing of these images, any amount of *segmentation* can be performed, providing classification of regions in the images according to their various qualities. Such segmentation can be done both automatically and manually. The segmentation can be based on e.g., intensity values, colours, texture qualities like graininess, and density of certain structures. An important factor can also be the human knowledge component provided by an operator performing manual segmentation.

The output is typically a new image with more clearly defined regions or features, or a set of curves in the original image. The defined features may be clearly bounded regions, in which case a contour could be placed at the border of each such region, or in the other extreme, they may be just marked points. We have implemented modules to make use of exactly these two variations, stacks of contours and point clouds. (Note, however, the difference between segmentation and geometry construction.)

Segmented Images, Contour Stacks

Given a stack of contours, there are a number of ways to generate geometries from this. One *explicit* method, is to do *lofting*, using for instance splines. In such a scheme, all these contours are interpolated in the "stack-direction", i.e., transversal to each plane containing a contour. Such approaches were discussed in Bjaalie et al. [8]. For some sets of data this works out quite well, for others it is more difficult. An example of the latter is easily illustrated if a run of consecutive planes of the stack contains different numbers of contours, as shown in Figure 2.

In Section 4 we will see how we can benefit from converting such a stack of contours *back* to an implicit representation, where the contours are contained in a particular iso-surface of a scalar function on that part of $\mathbb{R}^3$ in which all the contours lie.

Point Clouds

Many processes generate 3D point clouds as output, e.g., use of microscopes, structured light scanners, and laser range scanning devices.

At SINTEF, a project group has developed an inexpensive scanner based on consumer technology, which is using structured light, see [9], or [23]. See also the "state-of-the-art-report" (STAR) from the AIM@SHAPE project, [1]. Very briefly, the technique is based on projection of light patterns using a commodity projector, and reading the projection of these patterns off of the object being scanned with a camera. These images can then be analysed and a matrix of distances to the object calculated, giving a 3D point cloud. Resolution is restricted by the resolution of the projector and the camera, and the speed of these devices. Typically, the speed with which the projector can switch from showing a lot of "minimal intensity pixels" to "maximal intensity pixels" and vice versa, is a limiting factor with respect to the speed of scanning. The scanning range will also be severely restricted by the light output by the projector.

A laser range scanner works differently, in that it sends out laser pulses that are reflected back. From the time delay between emission and reception of the reflection, the distance to the object can be determined.

More relevant for the medical application being discussed here, would be microscopy methods. These do not directly produce a point cloud, but rather a stack of images. As we will shortly discuss in Section 3, we can either produce a point cloud from this, or in other ways use the output more directly.

Point clouds can also be the output of more manual procedures in which an operator puts down markers on images.

Some point clouds lend themselves more easily to geometry reconstruction than others, for instance point clouds sampled on the surfaces of geometries, but even these cases can be difficult. We discuss various kinds of point clouds in Section 3 and then in more detail in Section 4.4.

3 Surface Reconstruction from Unorganized Data

Given some sort of unorganized data, the problem of reconstructing geometry is most of the time easy to state, like in these examples:

1. Given a point cloud, presumably sampled from the surface of a real object, reconstruct that surface in the form of an approximation preserving the features and topology of the object.
2. Given a point cloud in a plane, with regions with differing density of points, reconstruct a contour encapsulating the parts with a particular density.
3. Given a stack of contours, each embedded in its own plane, construct a surface through all contours.

4. Given a stack of CT scans or other images, construct a surface describing a predefined organ being imaged.

The reconstruction in some of these cases is more of a *segmentation* problem, as for instance when contours dividing regions in 2D images are made, while *surface reconstruction* comes into play when turning stacks of contours into 3D surfaces. Still, the stage from the segmented 2D image to a 1D contour in the (2D) plane is of course also a problem of geometry reconstruction, just simpler. The segmentation and reconstruction in example 2 and example 3 above, can be regarded as two steps in a segmentation/reconstruction like the one in example 4.

The term *unorganized data* refers to the fact that point clouds, contours etc. are not given on regular grids, with regular spacing etc. This typically means that more general methods, with more general possible geometries and topologies are needed. In some cases, data from different sources of a more organized kind (e.g., different image *modalities*) are combined to produce unorganized data, an example being a set of range scanner "images" of one object taken from different angles.

In all cases the more general approach is to use methods for unorganized data, where the sacrifice will typically be that one has to settle for less than optimal efficiency.

Regarding methods for surface reconstruction, there are many issues that can be elaborated upon, among them, what approximation space to find solutions in, approximation methods, which features to prioritize preservation of, etc.

3.1 Previous Work

In Bjaalie et al. [8], the focus was placed on constructing surfaces from point clouds in 3D space, where the surfaces were to enclose all the points. Furthermore the point clouds were organized in subclouds in stacked planes, typically acquired in some iterated imaging process. Their approach was to create first a closed boundary contour in each plane, and then a closed tensor product spline surface interpolating these contours. The use of a spline surface led to problems with the parameterization of each contour. In our current approach, this is not an issue, as we shall see.

In different research fields, different methods for the surface reconstruction problem have been devised, led by the particular needs in these fields. We can broadly group these approaches into three disciplines, *image processing*, *computational geometry* and *computer graphics*. Activities like medical imaging and surface reconstruction, computer vision, computer aided design etc., may find a foothold in any and all of these camps, to some degree.

In image processing, the problem is to determine a boundary around a region of the image with a particular characteristic, i.e., segmentation, and then construct a curve representation of this boundary. Perhaps the most common methods of doing this are level set methods, of which the *snakes algorithm* is the reference method. Briefly stated, the principle is to define a boundary in the image, together with an error functional, and then devise an iterative algorithm for modifying the boundary towards one with a smaller error. Typical problems with this approach are that the procedure is sensitive to the initial boundary, that there is often a large number of parameters that must be tuned for good performance, and computational efficiency. These methods are readily generalized to three dimensions. For further details, see for example Sethian [24], Osher et al. [20] and Osher et al. [21].

The computational geometry approach normally consists of finding a hull enclosing and linearly interpolating point sets in space. Methods may vary in how they do this, how efficiently they do this, and the resulting triangle meshes. What they tend to have in common, is that they interpolate the points, rather than approximate them. One major contribution is the Power Crust algorithm, see Amenta et al. [3]. The Power Crust interpolates a point cloud through a three-dimensional polyhedral solid described by an approximate *medial axis transform*. The medial axis transform links the *medial axis* and the associated radius function to the object. One could also mention the usage of *alpha shapes*, introduced by Edelsbrunner et al. [11]. Definitions and software are also discussed in Akkiraju et al. [2].

The tradition in computer graphics and computer aided geometric design leans more to approximation as opposed to interpolation. A very much cited reference in this domain is that of Hoppe et al. [13]. There, a surface approximating a point cloud is constructed, with its two main features being first that the surface is finally meshed out as an iso-surface using *marching cubes* or similar algorithms, and second that an estimated tangent plane orientation is used to define a *signed distance function.*

Overall, there are difficulties with respect to interpolation versus approximation, with normal orientation, with obtaining "nice polygons" (which may mean different things in different contexts) when marching out e.g., iso-surfaces, with the speed of reconstruction algorithms, and with robustness and automation.

In the current work, we have not set out to solve these problems, or even making use of the best solutions, instead we have chosen to focus on the following features:

- Interactive editing of geometries.
- Easy distribution, usage and maintenance of software.
- Distribution and usage of the software on comparatively weak clients, with the number crunching done on remote servers, possibly parallel computers, clusters, etc.

3.2 Implicit versus Explicit Representation and Methods

All the data in the different modalities listed so far, point clouds and contour stacks, fall in the class of *unorganized data.* The resulting geometry from the reconstruction process can be represented in one of two ways, *implicitly*, typically as the iso-surface of a function, or *explicitly* as a parametric surface. The reconstruction methods themselves can also be classified according to which kind of representation is used for the results.

Parametric Representation

A surface S is *parametrically represented* if every point on it can be defined as the value of a function of a number of independent variables, or *parameters*. Here, we only mention that such a parametric representation in general is not unique, more details about parametrically represented geometries can be found in Sections 3.2 and 3.3 of Andersen et al. [4]. *Parameterization of a surface* amounts to the finding, or defining, of such a function.

For some kinds of more organized data, a typical example being the point cloud produced by a laser range scanner, such a parametric surface is a very obvious choice. The surface scanned can be parameterized with a scalar function on a rectangular domain, the function in parameter point $f(\mathbf{x}_i)$ being the distance measured along the ray with parameter $\mathbf{x}_i$.

This is a special case of surface reconstruction since the problem is reduced to that of finding a "height function". In a *Geographic Information Systems/-Geographic Information Technology* (GIS/GIT) setting, this is often the case when heights are scanned from the air, such data was for example used to reconstruct roof geometries of buildings in Nygaard [19].

When combining several such "height maps" recorded from different angles, or combining acquired data of different modalities, this parametric approach may not be as practical, and one would perhaps be better off tackling the completely general surface reconstruction problem.

Implicit Representation

An alternative to the parametric representation is to represent the geometry *implicitly*. When doing this, we define a function

$$f : \mathbb{R}^d \to \mathbb{R}, \tag{1}$$

such that the iso-surface

$$S(\alpha) = \{x \in \mathbb{R}^d | f(x) = \alpha\} \tag{2}$$

is exactly the surface we want to produce from the given input data. This eliminates the problem of trying to find a way to parameterize it well, when there is no obvious such way. As mentioned above, Hoppe et al., [13], describes a method for reconstructing surfaces from point clouds, which has proven itself to be very robust. One of the features making it such a robust method, is the definition of a signed distance function, of which zero is a regular value, so that

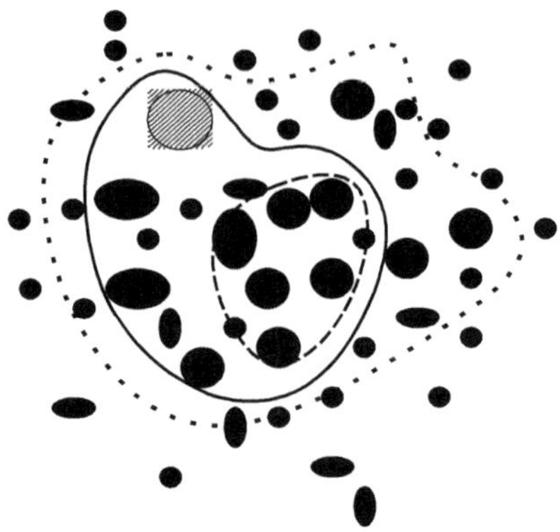

Fig. 3. Cell profiles of different kinds. The dashed line is a contour bounding a region in which the average size of cells is larger than some given value.

the implicit function theorem tells us that the zero set of this distance function is a smooth surface. This zero set, then, is what is used as the approximation of the unknown surface.

This implicitly assumes that the point cloud indeed consists of samples on, or close to, the unknown surface. In our present work, this is not the case. Granted, we also have such cases, but presently we want to focus on those for which this does not hold true. A typical scenario is the imaging, or even manual counting, of tissue with some varying characteristic, for instance cell profile size. In such an image, the segmentation may be the definition of regions with a particular density, as seen in Figure 3.

In this case, we do not have a point cloud sampled from a surface (or contour, in 2D) but rather a more general *scalar field* from which we want to extract an iso-surface, or iso-curve in 2D. For a situation like the one depicted in Figure 3, it is not completely clear what this scalar field should or could be. Note that there are of course an infinite number of scalar fields with a particular iso-surface or iso-curve, so there is some amount of leeway here.

Another significant advantage, in addition to not having to find or devise a way of finding a parameterization, is that of being easily able to represent geometries of arbitrary genus. In our context, it means for example that groups of cells of a particular kind, can form a number of closed geometries. One example could be a collection of disconnected neural tumors dispersed in a body.

As can be noted in Figure 3, the boundary of such a region may not be very well defined. To a certain extent, it may be easy for a neural reseracher to determine that both the dashed candidate and the dotted candidate are improper segmentations, but maybe not so easy to determine if the solid contour is the optimal. When we represent the contours implicitly, all three contours in Figure 3 can be iso-curves of the same scalar field, and part of a set of contours parameterized by the iso-curve parameter. This way, one can easily adjust the contour/segmentation by increasing or decreasing the iso-curve parameter. As the change in iso-curve parameter tends to zero, the operation on the curve tends to an *offset*.

Our Choice of Geometry Representation in the Neuroinf Project

Both representations, implicit and parametric, also have their disadvantages. For instance, a parametric surface is often easier to evaluate in given points, as its location in space simply is a function on the parameter domain. On the other hand, this representation does not go very well with surfaces that cannot be represented by a single such function on a simple domain. Here, the implicit methods have the upper hand.

Due to both the nature of our data, e.g., images with intensities, point clouds where the spatial point density is the feature of interest, etc., and to the nature of the acquisition methods, e.g., imaging methods, we have chosen to use an implicit representation in the Neuroinf project.

Finally, as we will see in Section 4.7, the implicit representation opens up for some convenient ways of geometry editing. Again, using Figure 3 as an example, editing out the effect of the semitransparent cell will turn out to be trivial with this editing scheme.

4 The Scalar Field

In most cases, the characteristic defining a neuronal population can be translated into a continuous scalar function over the volume in which it resides, where the value of the function in a point says something about the probability of that point belonging to the population. Then, the neuronal population can be defined as the part of the volume for which the function falls into some specified range. This scalar may be an intensity in an image, appropriate for coloured tissue volumes as well as electrical or chemical measures done on tissues. Densities can be correlated to growth properties of cells, and so on. Geometries in the form of iso-surfaces can then be extracted as suggested above.

Any vector-valued function must in some way be mapped into a scalar function in order to actually define a geometry, since the geometry will always be equivalent to the interface between regions of volume for which such a single-valued function takes on the values "inside" and "outside". This segmentation problem, i.e., finding some scalar field with the sought after iso-surface, can be solved with different methods, like e.g., level-set methods, as mentioned in Section 3.1. In the current work, the focus has not been on accurately finding well-defined surfaces, but rather on offering neuro researchers a variety of methods together with means for adjusting the results. Thus, we have composed a toolbox of simpler methods together with various filter operations, plus an editing capability.

4.1 Discretization

The *scalar field* is typically a map $f : \mathbb{R}^d \to \mathbb{R}$, and we need to discretize and store it. Among the main considerations are:

- **Accuracy and storage space** Just as the accuracy with which we can store particular scalar fields is very dependent on the dimension of the discrete approximation space, so are the storage requirements. Take for instance an original stack of images, maybe from a confocal microscope. If this consists of $512 = 2^9$ images of 512×512 pixels with a value between 0 and 255, which is not an unreasonable assumption, we may regard this as a (quantized) function in a space spanned by 2^{27} "box" functions. Just storing this requires 128 MB of memory.
  Assuming some smoothness of the data, and accepting some approximation/discretization error, we may approximate this with a combination of fewer but higher-order basis functions, e.g. a tri-variate cubic spline. Still, a tensor product of an n-dimensional approximation space leads to storage requirements of order $\mathcal{O}(n^d)$, with possibly inefficient space utilization. One way of reducing this is to use some form of compression e.g., using methods based on wavelets.
- **Evaluation** For the extraction of surfaces, a large number of function evaluations are needed, and for lighting computations necessary for visualization, we also need derivatives for the normals. Typically, the more compactly the function is represented, the more computationally intensive these evaluations are.
- **Modification** As we shall see in Section 4.7, we will edit geometry by modifying the scalar field, thus obtaining geometry modification as a form of "side effect". In order for this to work fast and smoothly, we need to be able to do local modifications to the scalar field quickly. Again, a simple and storage-inefficient representation will be simpler to modify.

Note that if we want to store a normal-field together with the scalar field in a case as outlined, the amount of data quadruples. Such a normal field is for instance useful during visualization. It may also be advantageous to store the numbers as single precision numbers rather than bytes, again giving a quadrupling of the data. Thus, the uncompressed 512^3 image turns into a massive 2 GB.

This exercise tells us that the compromise between simplicity and compactness becomes important when we need to loop through all the data. We do that both for evaluation and for modification of the data, e.g., during editing operations. For the model residing in computer memory, we have chosen the simplest of representations, by using uniformly spaced samples and tri-linear interpolation, amounting to *tri-variate linear splines on uniform knot vectors.* Additional compression can then be performed for tasks like file writing etc.

Depending on the usage, a size such as 512^3 may not be necessary. For most of our needs, which is mainly on-screen visualization, this size is an overkill. When using a 3D grid of n^3 uniform samples, one typically ends up with surfaces consisting of n^2 triangles. With the *per-pixel-lighting computations* possible on modern graphics hardware, it becomes almost impossible to see improvements for n larger than 128 on typical computer screens. See for instance Figure 4, where the grid is actually only 64^3.

4.2 Generating the Implicit Representation

We now want to look more into how we can produce the implicit representation of the geometry, that is, how to define the function f of (1) and (2). Some approaches, particularly for point clouds sampled from objects, were reviewed in Section 3.1. We will not discuss these cases further, but rather look into what other kinds of data we have experimented with.

In some special cases, S itself may be given, for instance as a spline surface, and the problem is just to find a function f (among the many possible solutions) satisfying (2). A good choice would then be a signed distance function if S is closed, and unsigned otherwise. Most of the time, however, S is not given explicitly, and using f to find S is our main interest. Note also that $d = 3$ corresponds to S being a surface, and that $d = 2$ corresponds to the reduced case where S is a curve in the plane. Both cases are interesting, for the latter may be applied to the special case where our input consists of several curves in different planes.

Our input data comes in mainly two forms, contour sets and point clouds, but exactly what surface they should represent may neither be clear nor well defined. Because of this, we have implemented a set of different methods in our application, and the user may choose the one which best fits the needs in each case.

4.3 Implicitization from Contour Sets

When we have contour sets as inputs, the situation is similar to the most frequently seen point cloud situation, in that we want to interpolate or approximate them, assuming that they are somehow sampled from the surface. We have restricted ourselves to closed contours, but note that we may still have "branching", i.e., a different number of closed contours in one plane than in the adjacent, see Figure 4.

We digress briefly to comment on the behaviour depicted in the bottom row of images in Figure 4. A distance field has been produced, and an interpolating surface extracted, such as displayed. We see that as the leftmost contour is translated to the right, the interpolating iso-surface undergoes a topological change, from genus zero to genus one. This is an effect of the definition of the scalar field, or distance field, used in this particular case, and the location of the contours relative to each other. If some other behaviour is wanted, one must change the definition of the scalar field, or alternatively, edit the surface using a technique like the one to be described in Section 4.7.

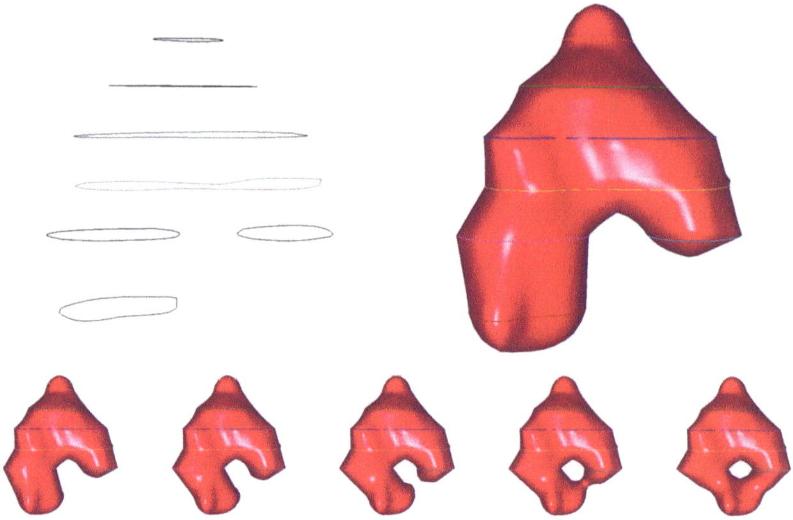

Fig. 4. To the left, we see that the two contours in the second lowermost plane are followed by a single contour in the plane above it. To the right, we see one possible geometry interpolating the given contours. Notice how the two branches passing through that second lowermost plane are not joined in the contour at the bottom, while they are above! In the bottom row, we see the effect of different horizontal positions of the single lowest contour. From left to right, the contour is slided rightward. See the text for additional comments.

We have followed these two approaches: First, we consider all contours lying in a common plane separately, one plane at a time, making a two-dimensional implicit representation followed by an extension to the whole three-dimensional volume. Second, we treat all contours simultaneously to find a global solution to the problem of finding the scalar field interpolating them.

Implicitizing the Contours Separately

In short, we find an implicit representation of each contour, in the plane in which it is embedded, and fill in the rest of the three-dimensional volume by interpolating these planes. This interpolation could be of any sophistication, but it will probably be wise to define the function f in the planes with contours in such a way that there is a "correspondence" between the iso-curves in different planes for the same "iso-value" α, so that this interpolation also makes sense.

For instance, in Figure 4, we have used a signed Euclidean distance, and linear interpolation between the planes. In addition, some smoothing filter operations have been applied. As a very concrete example, consider the simple case where we have two contours, each in the two planes $z=0$ and $z=1$. Say that we have the contours

$$\gamma_i = \{(x, y, z) \in \mathbb{R}^2 \times \{i\} \mid f_i(x, y) = 0\}, \quad i = 0, 1,$$

with functions f_i being e.g., distance functions for the two contours, respectively. We can then form a scalar threevariate function

$$f(x, y, z) = f_1(x, y)(1 - z) + f_2(x, y)z$$

that will have an iso-surface for $f = 0$ that interpolates the two contours.

Rather than defining what "correspondence" should mean, we give two examples to illustrate our point. In all cases we would of course like exactly the same iso-curve parameter to actually interpolate all contours in all planes, and thus, the corresponding iso-surface will be interpolating *all* contours.

Signed Approximate Distances

For a contour $\gamma_i \in \mathbb{R}^2$, we define $f_i : \mathbb{R}^2 \to \mathbb{R}$ such that

$$f_i(\mathbf{x}) = \begin{cases} \mathrm{dist}_i(\mathbf{x}, \gamma_i) & \text{if } \mathbf{x} \text{ is inside } \gamma_i; \\ -\mathrm{dist}_i(\mathbf{x}, \gamma_i) & \text{otherwise,} \end{cases}$$

for $\mathbf{x} \in \mathbb{R}^2$ where the distance $\mathrm{dist}_i(\mathbf{x}, \gamma_i)$ may be defined in different ways, but we need $\mathrm{dist}_i(\mathbf{x}, \gamma_i) = 0$ for $\mathbf{x} \in \gamma_i$, and $\mathrm{dist}_i(\mathbf{x}, \gamma_i) \geq 0$. We do not have to use the same distance function dist_i for all contours, but it is desirable. If for example the Euclidean distance is used, other iso-surfaces than the one interpolating the contours, i.e., $S = \{\mathbf{x} \in \mathbb{R}^3 \mid f(\mathbf{x}) = 0\}$, with f being an interpolate of all f_i, will be approximations of *offset surfaces* of S. (To the extent that these offset surfaces are sufficiently smooth and well-defined.) Given a surface $S \subset \mathbb{R}^3$ and offset distance α, the offset surface S_α is defined as $S_\alpha = \{\mathbf{x} \in \mathbb{R}^3 \mid \mathrm{d}(\mathbf{x}, S) = \alpha\}$, with $\mathrm{d}(\mathbf{x}, S)$ being the signed Euclidean distance function, i.e., the (signed) shortest distance between $\mathbf{x}$ and any point on S.

For the distance function, we have had success with simply measuring the four distances from a point $\mathbf{x} \in \mathbb{R}^2 \times \{\eta\}$ along the axes, to the nearest point on the contour in the plane $z=\eta$, and then choosing the smallest one of these. For the sign, we count the number of intersections between the contour and the ray from (x, y) to (∞, y), or in some other predetermined direction, an odd number gives positive sign, even number negative sign. The main contribution to the execution time comes from the ray-segment intersection test.

This last computation we speed up using the following algorithm. Say that a contour (or an approximation of a contour) is piecewise linear, we must then find all intersections between these straight line segments and the ray from (x, y) to (∞, y). We organize the line segments in a "triply sorted three-dimensional list". Say that we want to intersect the ray with all line segments

$((x_i, y_i), (x'_i, y'_i))_{i=1}^n$. By sorting all n segments with respect to $\min(y_i, y'_i)$, and then for each segment $((x_i, y_i), (x'_i, y'_i))$ in this list, we have another list with the remainder of the list, but now sorted with respect to $\max(y_i, y'_i)$, and finally, for each segment $((x_i, y_i), (x'_i, y'_i))$ in *this* list again, we have a last list with the remainder of the list, i.e., the "second list", but now sorted with respect to $\max(x_i, x'_i)$. Thus, to do the intersections, we do a (binary) search, and can immediately discard all segments fully above the ray, then a new search (in an even shorter list) results in the discarding of all segments fully below the ray, and finally, a third search in yet a shorter list will discard all segments fully to the left of the starting point of the ray. This leaves just a small number of segments to actually intersect the ray with. Typically, the number of segments to compute intersections for will be twice the number of crossings of the ray, so for most contours this will be of the order $\mathcal{O}(1)$.

This way of defining and computing f has been found to give such nice results that we use it as the default choice for contour sets, see Figure 4 for examples of the iso-surfaces generated from such sets of contours.

There are other ways of computing distance fields fast, done correctly one should be able to do it with complexity $\mathcal{O}(n)$, where n is the number of grid points. See for instance Sigg et al. [25], or Donnelly [10], in which even graphics card hardware is made use of to efficiently compute the Euclidean distances.

2D MBA in Each Plane

Another method which we have implemented and tested, is to use multilevel B-spline approximations (MBA) (also in combination with Multigrid (MG) methods) to approximate the contours in each plane. See for instance Lee et al. [16] and Hjelle [12] for descriptions of this approximation method. To use this method, we assign to contour $\gamma \subset \mathbb{R}^2$ some arbitrary "height", say $z = 0$, and sample the contour to get a set of three-dimensional vertices $(x_i, y_i, z_i = 0)_{i=1}^n$ in the plane. The algorithm takes such a set as input and produces a surface which approximates the points. If the specified space in which the approximation is to be found is large enough, we get interpolation, this is one of the features of this method. In order for the algorithm not to return the trivial surface consisting of the plane $z = 0$, we add some other points, for instance some points inside the contour, with height > 0, and some points outside, with height < 0.

Again, we interpolate each such plane, to get a global function f for the whole three-dimensional volume. One problem now is that other iso-surfaces than the one for $\alpha = 0$ may not make as much sense as we would like. For instance, if we choose α in such a way that the new iso-curve $\gamma_\alpha = \{(x, y) | f(x, y) = \alpha\}$ in the plane containing the original curve γ is chosen such that γ_α is a good approximation to an offset with distance d of γ, the surface S_γ is not necessarily a good approximation to the d-offset of the surface S_0 containing all the original curves.

This can be improved upon by taking more care in choosing the additional points when calling upon the MBA(+MG) algorithm. For instance, one could choose to approximate (even interpolate) points with an assigned "height" equal to a signed distance to the contour.

Global Approach

Just as we found the implicit representations of the contours one at a time, we may do all of them simultaneously. For the *signed* distances, we will have to determine insidedness for three-dimensional points inside closed triangulations instead of two-dimensional points inside contours, and likewise, the distance computation will have to be modified accordingly. Or, one may for instance use an adaption of Hoppe et al. [13], or others, i.e., estimating tangent plane orientations.

For the MBA(+MG) approach things will be different. Again, we go from two to three dimensions, with the increased complexity that involves, but one advantage is that we do not get the problem with "uncorrelated" two-dimensional implicitizations in the different planes. We have implemented this, and the results are indeed better than the plane-wise MBA(+MG) method, but the computational cost is so much higher that in all cases we prefer the "signed distance" method.

One advantage, for instance in the MBA(+MG) approximation case, is that we solve for all the "planes" simultaneously, so that we do not get a set of different implicitizations for each plane that has to somehow be combined into a three-dimensional implicitization afterward.

4.4 Implicitization from Point Clouds

Point clouds can be interpreted in different ways, most often perhaps as being sampled from some geometry. Then one may want to either interpolate the points, a subset of the points, or maybe approximate the points. In general, as for the contour-set case, the geometry S, i.e., the $(d-1)$-dimensional surface, should be the iso-surface for parameter $\alpha = 0$ of some function $f : \mathbb{R}^d \to \mathbb{R}$, depending on the given point cloud, such that $S(\alpha) = \{x \in \mathbb{R}^d | f(x) = \alpha\}$. Most of the point clouds experienced in our present context of neurology have not been of this kind, but rather samples from some unknown *density field*. This *density* could be a probability density, a mass density or in fact any other scalar-valued function. The kind of geometries we seek are then iso-surfaces in this scalar field. The difference from before is now that we do not have to approximate the geometry from the point cloud, then compute an implicit representation in the form of a scalar field from this, but rather that we have to find a scalar field from a set of samples.

Example 1: A researcher has obtained an image of some tissue coloured by insertion of a colouring agent in e.g., the bloodstream, sliced it, and put down points marking locations reached by the colouring agent. Clearly, there

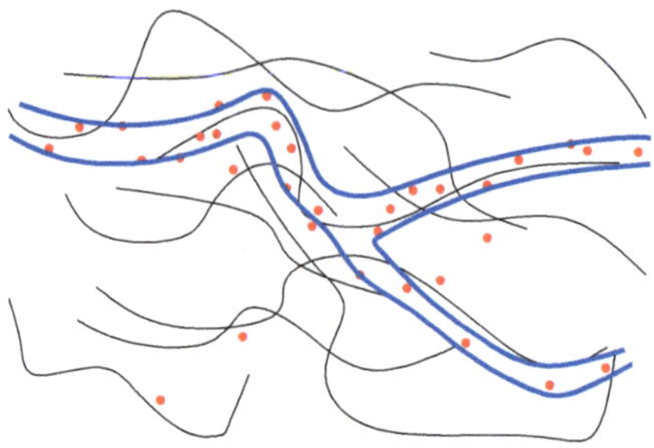

Fig. 5. Illustration of a possible density field. Suppose the thin, black lines belong to an image, with the red dots being put where some substance inserted in blood vessels has been detected. The sought-after blood vessel walls are outlined in blue. We can imagine a scalar field having this blue curve as an iso-curve in this 2D intersection.

is a larger probability of a small region with lots of markers being part of a blood vessel than a region without any markers at all; see Figure 5.

Example 2: In a stack of microscoped slices of tissue, all cell nuclei of a given kind are marked by a point, resulting in a stack of 2D point clouds, or a 3D point cloud, if regarded all simultaneously. If one is interested in the density of cells, the density of the point cloud might be directly applicable. To extract a geometry enclosing regions with a specific density of cells, one would then extract an iso-surface in this density field defined by the point cloud.

The remaining question is how to relate this notion of "density of points" to a scalar (density) function from which to extract the iso-surface. In both these cases the scalar function may be thought of as a probability density, and the determination of this function is the subject of the next subsection.

Numerical Estimation of the Probability Density Function (PDF)

If the point cloud defines some density, and we want iso-curves (iso-surfaces in 3D) as illustrated by the blue alternative in Figure 6, the problem to find f is one of *numerical pdf estimation*, or *non-parametric density estimation.* If the density was of a known parametric family, like e.g., of a normal distribution, the problem would be to estimate a finite number of unknown parameters, in the normal distribution case, mean and variance. If not even such a parametric family is known, we have an "infinite-parametric" estimation problem, hence *non-parametric density estimation.*

The d-dimensional empirical *cumulative distribution function* (cdf) is defined by

$$F^d(\mathbf{x}) = \frac{1}{n}\sum_{i=1}^{n} I(x_i^1 \leq x^1 \wedge \cdots \wedge x_i^d \leq x^d),$$

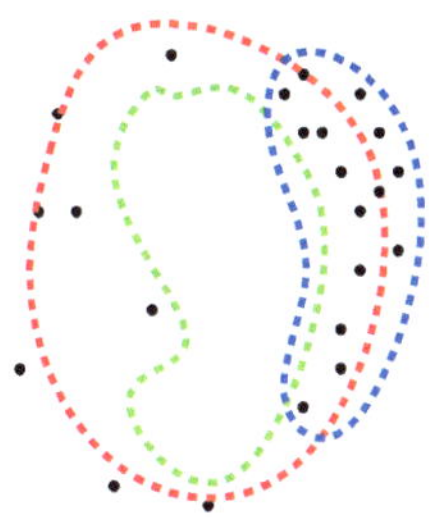

Fig. 6. Possible iso-curves representing geometries extracted from the same point cloud. In red, a possible iso-curve representing a sampled geometry. The curve is not interpolating the points, but approximating them. In green, a circuminscribed curve to the points, which also could be a valid geometry for this particular point cloud. Finally, in blue, the region with the highest density of black dots is enclosed by the iso-curve. This is the main interpretation for the neurological data analysed in our context.

where $\mathbf{x} = (x^1, \ldots, x^d) \in \mathbb{R}^d$, $\mathbf{x}_i = (x_i^1, \ldots, x_i^d) \in \mathbb{R}^d$ and $I(P)$ is an indicator function with value 1 if P is true, and with value 0 otherwise. The given points $\mathbf{x}_i$ are assumed drawn from the same unknown random distribution X. The density function is the derivative of the cdf, but it is not useful to just differentiate the empirical cdf F^d, since it clearly will be either zero or undefined in all points. Possible solutions include the approximation of the empirical cdf with a sufficiently smooth function for differentiation, for instance by using a spline. Most of the time just using the difference ratio

$$f^1(x) = \frac{F^1(x + h/2) - F^1(x - h/2)}{h} \tag{3}$$

(in 1D) approximating the derivative of the cdf, or the histogram with bin size h in more dimensions, has proven quite useful for us.

For the general and multivariate case (dimension d), assuming $\mathbf{x}_i$ to be sampled from some random variable $X_i = \mathbf{x}_i + h \cdot \mathbf{k}$ with h a small positive number and $\mathbf{k}$ a random variable with density function K, leads to the density function

$$f_i(\mathbf{x}) = \frac{1}{h^d} K(\frac{\mathbf{x} - \mathbf{x}_i}{h}),$$

for X_i by use of the multivariate change of variable formula. Averaging these gives the multivariate *kernel density estimator* $\hat{f}$,

$$\hat{f}(\mathbf{x}) = \frac{1}{nh^d} \sum_{i=1}^{n} K(\frac{\mathbf{x} - \mathbf{x}_i}{h}),$$

for X. This kernel density estimator is also known as the *Parzen window density estimator*, see Parzen [22], with *window width h*. With K the density function of a uniform random variable on $[-1/2, 1/2]$, the kernel density estimator coincides with the use of the numerical derivative (3), and is reminiscent of histogram binning with bin size h.

Many other kernels may be appropriate for a kernel density estimator. For example, Prakasa Rao [15] uses wavelets, Hwang et al. [14] compare certain radial basis functions to what they term an *exploratory projection pursuit technique*, Archambeau et al. [5] investigate *Finite Gaussian Mixture Models* and Zegers et al. [26] use neural networks.

A thorough testing or implementation of sophisticated estimation methods has not been within the scope of this project, and we have found rather naive approaches approximating and then differentiating the cumulative distribution function to work sufficiently well.

We have more or less heuristically chosen the window width h as small as possible for any given point cloud, but also experimented with this being a user adjustable parameter in our software. The number n of samples $\mathbf{x}_i$ will typically dictate h. Smaller n necessitates larger h for the combination of kernels to produce a smooth and plausible density function, or for the histogram binning to produce an appropriate range of possible numbers of samples in each bin.

In the case of histogram binning, we smooth the histogram by e.g., spline approximation or interpolation, to achieve a scalar field which is at least C^0. This simple method has given surprisingly good results, and is thus left as the default method of choice in the application. Note that there is little sense in using a method for scalar field generation which is much more sophisticated than the method of point cloud acquisition itself.

Other Alternative Scalar Functions

As the exact method by which the point cloud was obtained may not always be available to those making use of the point cloud for geometry modelling, it is advantageous to have a set of scalar function generation methods available. The following are some of the methods offered by the `m3d`-application to come out of the Neuroinf project. In these cases, we have used a three-dimensional regular grid of samples, with trilinear interpolation within the grid. These methods are offered without any rationale behind the design, as "experimental visualization tools" only.

1. For each node in the grid, the distance r to the closest vertex in the point cloud is found, and the node is given the value of the volume of a sphere of this radius, $4/3\pi r^3$.
2. Same as the previous one, except that not the closest vertex to each node is used, but rather the n-th closest.
3. For each point in the 3D grid, all vertices in the point cloud are looped through, and a contribution equal to the one from the radial basis function $\exp(-\text{const} \cdot r^2)$ centered at the vertex and with distance r from the node,

is added to that node, i.e., $f(\mathbf{x}) = \sum_{i=1}^{n} \exp(-\text{const} \cdot |\mathbf{x} - \mathbf{x}_i|^2)$ for point cloud $(\mathbf{x}_i)_{i=1}^{n}$. The rationale is that when the point cloud is dense, the scalar field is larger, and also that each vertex influences a larger region around itself. One problem with the straightforward implementation of this naive idea is that it is slow.

In addition to this, the user may set up a chain of "filters", or more precisely, a combination of predefined procedural transformations. If, for instance, the user wishes to make a scalar field out of the shortest distance from a given point in the grid to the nearest vertex, he can choose the first method listed above, and combine it with an $x^{1/3}$ transformation.

Other such transformations are: min, max and median filters, simple averaging, Gaussian filters, scalar transformations like $1/x$, x^2 and so on.

4.5 Surface Extraction

For the actual extraction and rendering of iso-surfaces, a recursive, multi-threaded "marching cubes-like" algorithm is used. The marching cubes algorithm is simple, and based on the assumption that a larger iso-surface can only intersect a small ("atomic") cell in a certain limited number of ways, see Lorensen et al. [17]. This relies on the assumption that the surface is reasonably well-behaved and smooth. For instance, splitting a three-dimensional volume into small rectangular boxes, sampling the scalar function in the corners of the box, there are only 2^8 ways the iso-surface can pass through this box. (Actually, the number is smaller when symmetries are accounted for.) This number results from the fact that at each corner the sampled scalar function must be either less than, or greater than, the iso-surface parameters, corresponding to each corner being either on the outside or the inside of the surface. From eight corners, we then get the 2^8 combinations. See Figure 7 for an example of the possible configurations.

As is shown in Figure 7c, not all cases are unambiguous. In the case shown, both the two dashed curves and the two solid curves are compatible with the configuration of the four corners. The only way to resolve this is to further subdivide the box, and consider scalar field values inside it. With respect to the case shown, if x is on the same side as the "+"-corners, the dashed curves are correct, otherwise the solid ones should be selected. Note that both alternatives result in closed iso-contours, although not necessarily the same number of such.

For each small rectangular box, a number of triangles are then generated. The intersections of the triangles' edges with the edges of the box are depending on the actual magnitude of the sampled function in the corners. Because the full volume is tiled with adjacent boxes sharing corners, triangles from neighbouring boxes will also be adjacent to each other. The topological information connecting triangles to each other does not come out easily but neither is this information needed for the visualization of the surfaces. On the plus

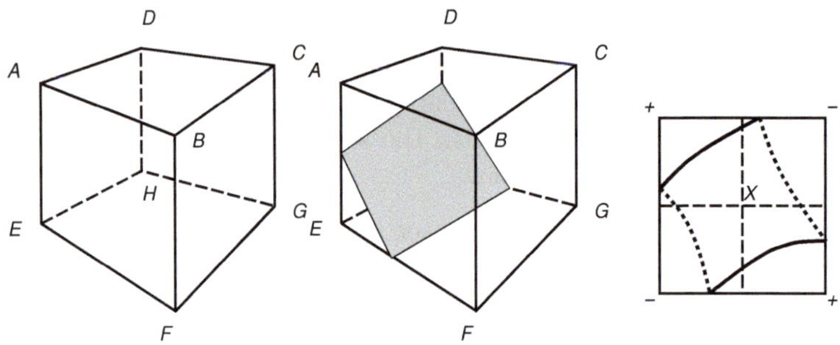

Fig. 7. From the left, we see a) A rectangular box with the eight corners labeled. b) A bilinear approximation of the iso-surface passing through the box, when the scalar function in E and H is less than the given iso-surface parameter, while the scalar function in the other corners is larger, or vice versa. c) Not all cases are unambiguous. The curve should pass twice through the box, but both the fat solid and the dotted alternatives are consistent with the corner values, where "+" and "-" indicate that they are on each side of the iso-curve parameter. Note that subdivision of the box (dashed straight lines) and determination of the value in X will most likely resolve the ambiguity in the 3D and 2D case, respectively.

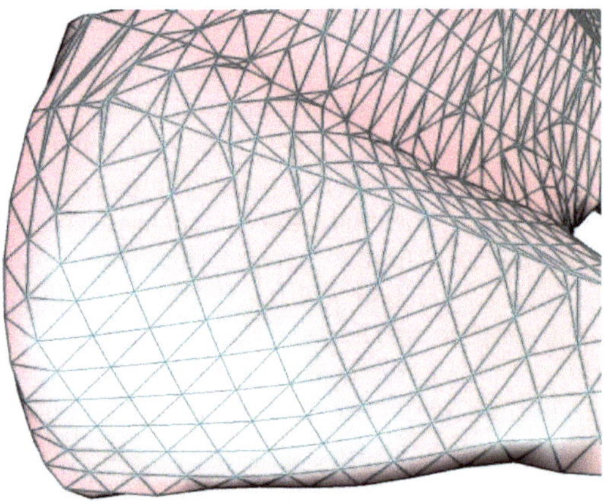

Fig. 8. To the upper left, we see the kind of thin triangles that may be produced by the marching cubes like algorithm. For visualization this is not important, and impossible to spot unless shown in such a "wireframe mode".

side of using this approach, we get triangles bounded in size (by the size of the rectangular boxes.) but on the other hand, we risk getting very thin triangles. Again, for visualization purposes, this is not an important drawback. See Figure 8 for an example of the kind of triangles that are produced.

4.6 Visualization Methods

In our application, the extracted geometry in the form of iso-surfaces is not used for any processing other than visualization and *picking* by the human operator. For this, we use the collection of triangles produced by running the marching cubes like algorithm on a 3D grid of samples of the scalar field. With a grid spacing small enough for transitions between these triangles to be unnoticeable on the screen, this still runs fast enough to 're-extract" large portions of the surfaces for every frame displayed. This makes the real-time editing to be discussed below in Section 4.7 possible.

To visualize these triangles, we have used `Java3D` and OpenGL. In our OpenGL-based client, we have also made use of modern *shader* programs to achieve per-pixel-lighting computations, which to an even higher degree makes transitions between triangles look nicer.

4.7 Editing

Since the extraction of iso-surfaces is quick enough to be (partially) done for every frame in an animation, we can treat the scalar field as our main representation for the geometry, and do editing operations directly on that. The side effects then manifest themselves immediately in the modified iso-surfaces extracted and visualized. Together with two other components, this comprises our editing system. The other two components are, 1) the *picking* on the current geometry and translation to the scalar field, and 2) the editing operation to be applied to the scalar field in the region picked.

Picking

Picking is simply the action of selecting parts of a 3D geometry visualized on the screen. When visualizing the 3D geometry we have a "world model" in which both the observer and the screen are placed, together with the 3D geometry. When we select a point on the screen with the mouse, we define a straight line connecting the observer and this pixel on the screen. Extending this line, and intersecting it with the 3D geometry, e.g., a set of iso-surfaces, we find a set of intersection points in 3D. The one closest to the observer is the *picked point*, and this is the part of the geometry we want to edit.

Editing Operations on the Scalar Field

Depending on what effect on the iso-surfaces we seek, there are many interesting operations we can do on scalar fields. For example, let $\phi_1 : \mathbb{R}^3 \to \mathbb{R}$ have the iso-surface $S_1 = \{x \in \mathbb{R}^3 | \phi_1(x) = 0\}$ and $\phi_2 : \mathbb{R}^3 \to \mathbb{R}$ have the iso-surface $S_2 = \{x \in \mathbb{R}^3 | \phi_2(x) = 0\}$. We might as well interpret the inside of S_i as a *solid* and call it solid S_i. Then we can use these solids together with Boolean operations like e.g., *union* and *intersection* to create new solids. An

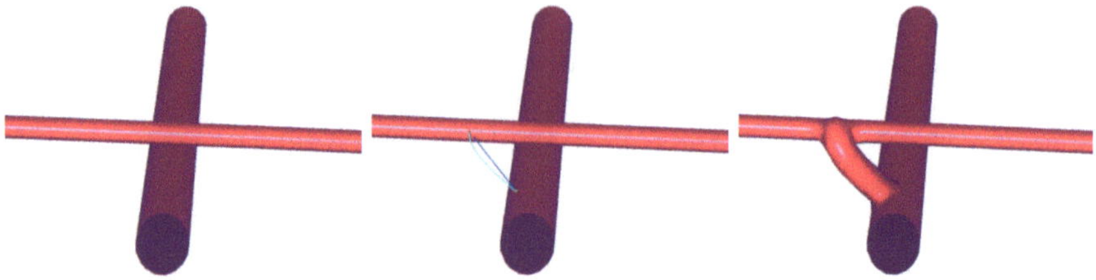

Fig. 9. From the left: a) Two straight non-intersecting cylindrical "pipes" are implicitly represented, so what we see is an iso-surface of some scalar field. b) In our application, we pick two points on the geometry, here one on each cylinder, the light blue line will be the spine of a connection between the two cylinders. c) A connection has been made, in the form of a scalar field of which we see the iso-surface, and a combined scalar field is made to contain the original plus the new geometry.

implicit representation of the solid $S_1 \cup S_2$ will then be $\phi_{1\cup 2} = \max(\phi_1, \phi_2)$, i.e.,

$$\phi_{1\cup 2}(\mathbf{x}) = \max(\phi_1(\mathbf{x}), \phi_2(\mathbf{x})),$$

and likewise, $\phi_{1\cap 2} = \min(\phi_1, \phi_2)$ will have the bounding surface of the intersection of solids S_1 and S_2 as iso-surface.

This way of combining solid shapes, here defined as the inside or outside regions of our iso-surfaces, through Boolean operators, is known as *constructive solid geometry*. This way, complex solids can be expressed as trees, with primitive solids as leafs and branches connected with operators.

We can apply such techniques locally to modify our geometries. In Figure 9 we show the result of an editing operation which attaches two cylindrical solids with a "hose-like" connection. This is simply done by picking two connection points on the original cylinders, and then the parameters for fitting a smooth offset of a spline curve attached perpendicularly to the cylinders are computed. Finally, the scalar field of the original cylinders is modified with the implicit representation of the spline-offset forming the "hose".

The most useful editing tool in what we may regard almost as a palette of "brushes" for 3D painting in the scalar field, is one which adds or subtracts "mass" locally around the picked point: We pick a point, and while holding down e.g., a mouse button, key, or similar, and moving the mouse, we add or subtract a mass proportional to the distance the mouse has moved from the picked point. This "(signed) addition" can take any shape, and have different influence radii, just like the brushes in an ordinary 2D paint program. For instance, given a plane iso-surface, and a "brush" with circular and finite influence radius which decays away from its center, the action will be to make a spherical bulge in the plane. On which side it appears as a bulge, depends on the scalar field, the "brush" function, and the "add/subtract-button" selected.

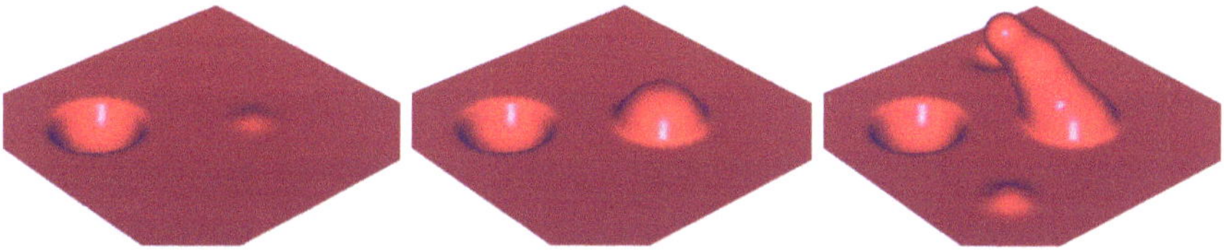

Fig. 10. A plane is modified with the 3D "scalar field paint brush". We see from the left, snapshots a) during the forming of the bulge to the right, after a dip has been made to the left, b) both the dip and the bulge are now made, c) additional "material" has been added to the bulge. Note that making this bulge re-join the surface itself in another spot, thus creating the topology of a torus, is not a problem.

See Figure 10, which hardly does justice to the effect of seeing the real-time changes to the geometry on-screen, while doing the editing.

Our main purpose for these painting tools is to correct, or otherwise enhance the neurological visualizations made from the contour sets and point clouds discussed so far. Such editing may be necessary when our (semi-)automatic methods for scalar field generation fail to produce the correct geometry in form of iso-surfaces, or when effects of outliers, etc. must be counteracted by the neuro-scientist. See Figure 11 for examples of using the tool.

5 High Performance Computing

One of the ideas behind the Neuroinf collaboration was the application of *high performance computing* and parallelization to perform the computationally intensive tasks anticipated. It became apparent through the course of the project that the tasks were going to be more interactive and not so computationally demanding as first believed, but we have nevertheless been able to make good use of parallel hardware and software to speed up parts of our pipeline.

5.1 Introduction

We chose to implement the front-end of our tool in Java/Java3D, for portability and easier installation and maintenance, but for the back-end we did not find this necessary. The back-end is implemented in C++. A schematic overview of the system is depicted in Figure 12.

The working principle of this setup assumes that the user on the client side will have a point cloud, that this is passed on to the server side which

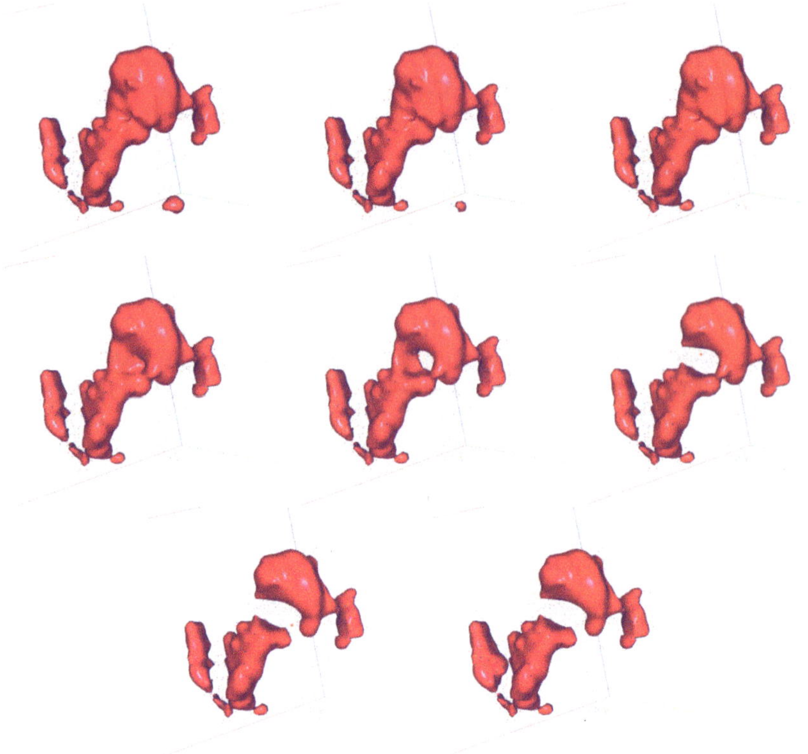

Fig. 11. Showing editing operations and per pixel shading. From left to right, top to bottom, a-h. a) A particular iso-surface of a point cloud-defined scalar field. Notice the small, closed surface in the lower right. b) We pick the small, closed surface, and adjust the scalar field in a local region, making the local iso-surface smaller, c) ...until it disappears. d) We perform the same operation in the middle of the large structure, e) ... in effect "digging a hole" in it. f) Digging some more, the genus jumps from one to zero for this surface... g) ...before it splits off into two closed surfaces. h) To the left, we see now that we have "added some mass" to the leftmost, local surface.

generates a scalar field and renders an iso-surface, which is then returned to the client for display. This architecture was chosen mainly for two reasons, it was assumed that the user would need very large scalar fields computed by time-consuming means from the point clouds, and secondly, it offered an easy way of storing the users' data centrally. The latter would then be beneficial for users wanting to access their data with web-browsers on computers in different locations.

In Figure 12, we see that scalar fields themselves may be sent also *from the client to the server*, this is because during editing operations, generating/-modifying the scalar field is faster on the client. At the same time, the server must be made aware of changes to the scalar field.

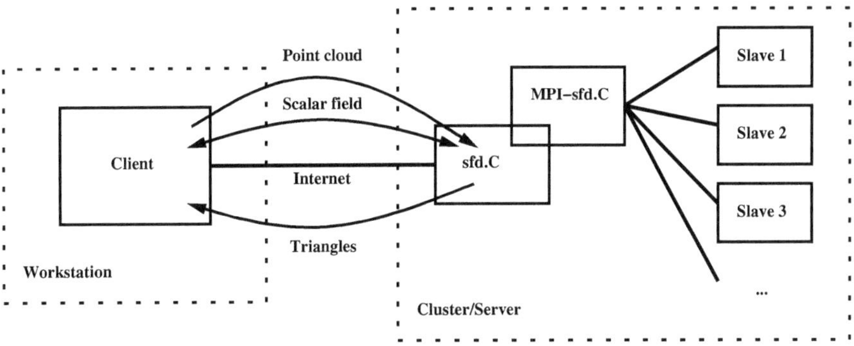

Fig. 12. Schematic overview of the client/server system.

5.2 Choice of Parallelization Technology

There are different aspects to consider, most importantly the amount of overhead connected with parallelizing the various tasks. We have parallelized in two directions.

First, a number of relatively small, but trivially parallelizable tasks have been implemented using *POSIX threads*. POSIX threads is a standardized threads application programming interface (API) offering what is also termed *lightweight processes*, which permits a workload being split onto distinct processors. Such a task is for example the application of filters to 3D grids of samples. This involves a lot of data (e.g., n^3 scalars, with $n = 256$, gives 128 MB when using double precision numbers) but relatively little CPU usage per data point. For this, the threads model with shared memory is suitable, and it scales very well on so-called *symmetric multi-processing* (SMP) machines.

Second, all the operations of extracting surfaces and generating scalar fields have been parallelized using the *message passing interface* (MPI) standard, so that the task can be distributed over a set of nodes in a cluster. For instance, if a number of point clouds are given, and scalar fields and triangles are to be produced independently for each of them, we can just distribute different point clouds to different nodes, and then collect the results after each node finishes the processing of its point cloud.

There are many alternative approaches that maybe would work even better, like using the *parallel virtual machine* (PVM) standard instead of MPI, the *open multi-processing* (OpenMP) API instead of POSIX threads directly, direct usage of Unix sockets for fast communication, etc. However, the cost of development and testing must also be considered.

5.3 Results

Our decision to implement the geometry modelling tools as an interactive application makes it necessary to harness high performance computing resources with low latency with respect to availability and interactivity. The server part of the application is parallelized at two levels, so that it should benefit from almost any parallel hardware setup.

Our experience is that the POSIX threads parallelization works very efficiently. Using this, we have chosen only to parallelize the "trivially parallelizable" tasks, that will scale perfectly in theory. Testing on dual CPU SMP machines, we have observed almost perfect scaling for this. Although this should scale equally well on more CPUs in principle, more overhead connected to cache misses, bus starvation and other hardware-oriented issues would have to be expected as the number of CPUs and threads increases.

The MPI-approach to distribution of work across many nodes in a cluster also works, but it is rather sensitive to the kind of data being processed. It will work better as the number and size of point clouds increase, but these numbers are most of the time too small to justify the overhead of the parallel execution, at least for most of our current test cases. It is possible that a better tailored interface for communication and work distribution could improve on this situation, at the cost of more expensive implementation.

Our conclusion with respect to parallelization is thus that SMP machines with shared memory offer the best benefit to cost of implementation ratio for our problem.

6 Application Development and Maintenance

A web client based on Java3D, with full 3D viewing capability and interactive manipulation was chosen as the tool for the user. This was motivated mainly by the ease of deployment and maintenance on the users' machines. For instance, this ensures full independence of user equipment, both hardware and software, as long as the user has access to a Java-enabled web browser and sufficient 3D viewing capability.

Since much of the computation is done on the central server, as described in Section 5, the client is relatively lightweighted. Also, some of the prerequsite software, like, e.g., Java3D, only has to be installed once on a user's machine. All of these installations can be done by the user, through the web browser. The effect of this is that both the client and the server modules can be maintained and updated almost without requiring involvement of the user.

The client has live communication with the centralized server, which may be a more powerful computer than single users have access to. Also, data may be stored on this server. We foresee that the application will be used by individual scientists that will make occasional use of the tool and will benefit from the platform independence and remote processing capability. A part of the tool is embedded in the FACCS application (Functional Anatomy of the Cerebro-Cerebellar System), available via The Rodent Brain Workbench (`www.rbwb.org`). One vulnerability of this system is the dependence on both the server side and the link (internet connection) connecting the client and

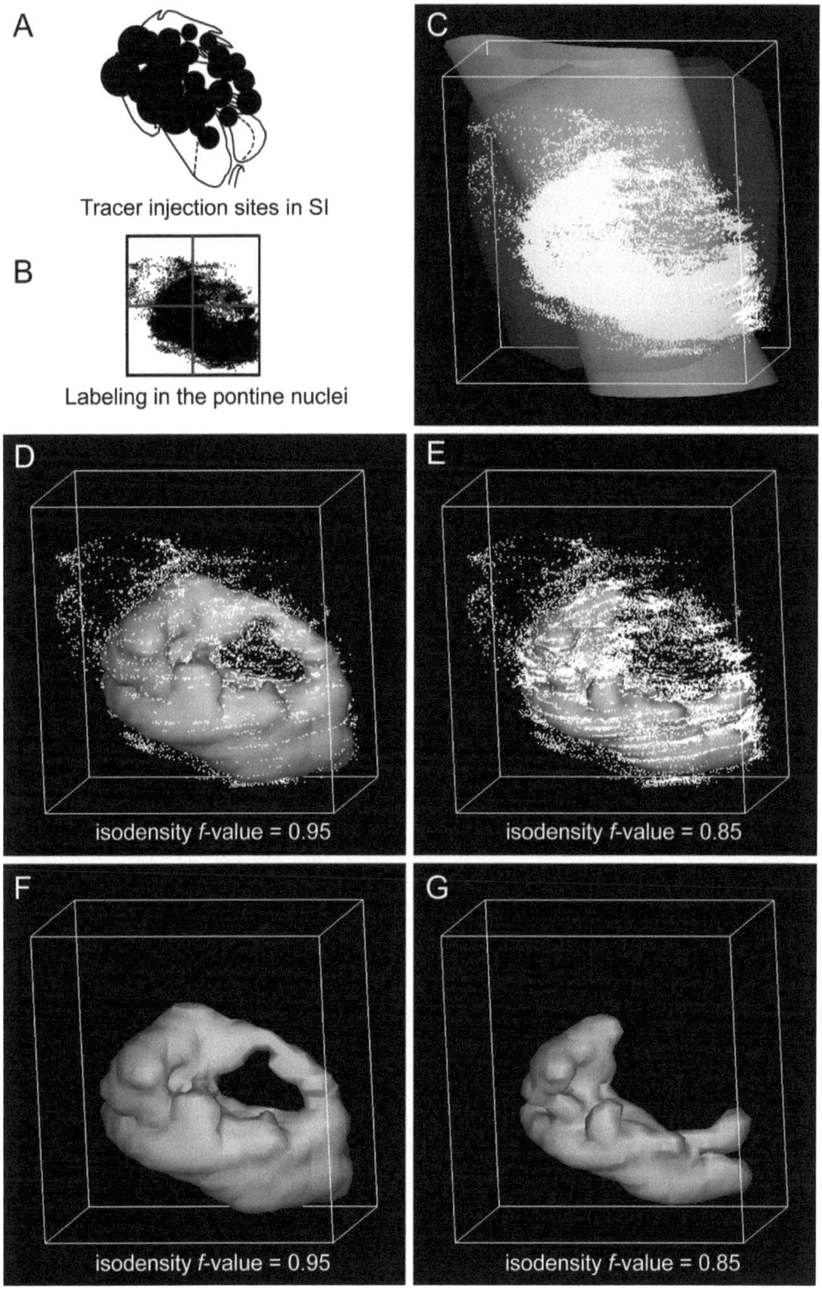

Fig. 13. Figures produced with the help of the application demonstrated in the current project. Reproduced from Moene et al. [7], with permission. See Section 6 for a discussion.

the server. This vulnerability may be reduced by extending the client with a local backup server. This is relatively easy to implement, since the client is already written in Java, with the server written in C++.

An example analysis based on the application is shown in Figure 13, adopted from Moene et al. [7]. In the figure, we see a typical kind of point cloud representing distributions of a defined set of neurons in a limited part of the brain, with corresponding surface geometries constructed to visualize the appropriate neurological structures. Most elements of the *graphical user interface* (GUI) are stripped from the images, except a coordinate system (bounding box) that helps to mediate the perspective and orientation as well as an exactly defined location, which is not readily seen from the geometries alone. We get an impression of how the surfaces and point clouds relate to each other, and the effect of adjusting choices of iso-surface parameter values (figures F and G in the lowermost row) for the same scalar field.

7 Conclusions

We have some early user experiences, resulting in users migrating to this new tool for the production of illustrations for research papers. The feedback so far has centered on the appreciation of the editing capabilities and the many degrees of freedom when extracting geometry from point clouds modelling density, as discussed in Section 4.4. These degrees of freedom include the set of methods available, resolution of "binning" for the numerical density estimation, arbitrary combinations of a number of filters and of course the choice of iso-surfaces indexed by an iso-surface parameter.

References

1. AIM@SHAPE-D5.1-group. Survey acquisition and reconstruction. Technical report, AIM@SHAPE, 2005.
2. N. Akkiraju, H. Edelsbrunner, M. Facello, P. Fu, E. P. Mücke, and C. Varela. Alpha shapes: Definition and software. In *Proceedings of the 1st International Computational Geometry Software Workshop*, pages 63–66, 1995.
3. N. Amenta, S. Choi, and R. K. Kolluri. The power crust. In *Proceedings of the sixth ACM symposium on Solid modeling and applications*, pages 249–266. ACM Press, 2001.
4. O. Andersen, G. Vasilakis. Building an ontology of CAD model information. In *this book*.
5. C. Archambeau and M. Verleysen. Fully nonparametric probability density function estimation with finite Gaussian mixture models, 2003.
6. J. G. Bjaalie. Localization in the brain: new solutions emerging. *Nature Rev. Neurosci.*, 3:322–325, April 2002.
7. I. A. Moene, S. Subramaniam, D. Darin, T. B. Leergaard, and J. G. Bjaalie. Towards a workbench for rodent brain image data: systems architecture and design. *Neuroinformatics*, in press, 2007.

8. J. G. Bjaalie, M. Dæhlen, and T. V. Stensby. Surface Modelling from Biomedical Data In *Numerical Methods and Software Tools in Industrial Mathematics*, pages 329–346. Birkhäuser, 1997.
9. F. Couweleers, Ø. Skotheim, and B. W. Tveiten. Optical measurement of weld toe geometry with structured light. In *Proceedings of 8th International Symposium on Measurement and Quality Control in Production (ISQMC)*. VDI Verlag GmbH, October 2004.
10. W. Donnelly. *GPUGems 2 : Programming Techniques for High-Performance Graphics and General-Purpose Computation*, chapter Per-Pixel Displacement Mapping with Distance Functions, page 130. Addison-Wesley, 2005.
11. H. Edelsbrunner and E. P. Mücke. Three-dimensional alpha shapes. *ACM Trans. Graph.*, 13(1):43–72, 1994.
12. Ø. Hjelle. Approximation of scattered data with multilevel B-splines. Technical report, SINTEF, 2001.
13. H. Hoppe, T. DeRose, T. Duchamp, J. McDonald, and W. Stuetzle. Surface reconstruction from unorganized points. In *Proceedings of the 19th annual conference on Computer graphics and interactive techniques*, pages 71–78. ACM Press, 1992.
14. J. Hwang, S. Lay, and A. Lippman. Nonparametric multivariate density estimation: a comparative study, 1994. *IEEE Trans. on Signal Processing*, 42(10):2795-2810, October 1994.
15. P. R. Indian. Nonparametric estimation of partial derivatives of a multivariate probability density by the method of wavelets. citeseer.ist.psu.edu/627077.html.
16. S. Lee, G. Wolberg, and S. Y. Shin. Scattered data interpolation with multilevel b-splines. *IEEE Transactions on Visualization and Computer Graphics*, 3(3):228–244, 1997.
17. W. E. Lorensen and H. E. Cline. Marching cubes: A high resolution 3D surface construction algorithm. In *SIGGRAPH '87: Proceedings of the 14th annual conference on Computer graphics and interactive techniques*, pages 163–169, New York, NY, USA, 1987. ACM Press.
18. Neuroinf project, 2001-2003. Web pages hosted by the NESYS group at the University of Oslo, `www.nesys.uio.no`.
19. J. O. Nygaard. Semiautomatic reconstruction of 3D buildings from map data and aerial laser scans. *J. Digit. Inform. Manag.*, 2(4):164–170, December 2004.
20. S. Osher and R. Fedkiw. *Level Set Methods and Dynamic Implicit Surfaces.* Number 153 in Applied Mathematical Sciences. Springer-Verlag, New York, 2003.
21. S. Osher and N. Paragios. *Geometric Level Set Methods in Imaging, Vision, and Graphics.* Springer-Verlag New York, Inc., 2003.
22. E. Parzen. On the estimation of a probability density function and mode. *Ann. Math. Stat.*, 33:1065–1076, 1962.
23. G. Sansoni, M. Carocci, and R. Rodella. Three-dimensional vision based on a combination of gray-code and phase-shift light projection: analysis and compensation of the systematic errors. *Applied Optics*, 38:6565–6573, November 1999.
24. J. A. Sethian. *Level set methods: Evolving interfaces in geometry, fluid mechanics, computer vision, and materials science.* Number 3 in Cambridge monographs on applied and computational mathematics. Cambridge University Press, Cambridge, U.K., 1996. 218 pages.

25. C. Sigg, R. Peikert, and M. Gross. Signed distance transform using graphics hardware. In R. Moorhead, G. Turk, and J. van Wijk, editors, *Proceedings of IEEE Visualization '03*. IEEE Computer Society Press, October 2003.
26. P. Zegers and J. G. Johnson. Density function estimation with multilayer perceptrons. Brazilian Symposium on Intelligent Automation (SBAI), September 2005.

An Introduction to General-Purpose Computing on Programmable Graphics Hardware

Tor Dokken, Trond Runar Hagen, and Jon Mikkelsen Hjelmervik

Summary. Using graphics hardware for general–purpose computations (GPGPU) has for selected applications shown a performance increase of more than one order of magnitude compared to traditional CPU implementations. The intent of this paper is to give an introduction to the use of graphics hardware as a computational resource. Understanding the architecture of graphics hardware is essential to comprehend GPGPU-programming. This paper first addresses the fixed functionality graphics pipeline, and then explains the architecture and programming model of programmable graphics hardware. As the CPU is instruction driven, while a graphics processing unit (GPU) is data stream driven, a good CPU algorithm is not necessarily well suited for GPU implementation. We will illustrate this with some commonly used GPU algorithms. The paper winds up with examples of GPGPU-research at SINTEF within simulation, visualization, image processing, and geometry processing.

1 Introduction

In this paper we will address how programmable graphics hardware and its impressive capability for processing floating-point numbers, can be used for scientific computations. The high performance of the graphics hardware is due to its parallel nature, where independent tasks are performed simultaneously. This leads us to a new programming paradigm for desktop computers, where parallel computation is necessary to take full advantage of the hardware.

Most modern PCs have programmable graphics processing units (GPUs). Such GPUs typically give a floating point computational power that is more than one order of magnitude higher compared to the CPU in a modern PC. In the coming decade, the computational power of GPUs is expected to grow considerably faster than the computational power of CPUs, because the GPU-architecture is more scalable.

While CPUs are instruction driven, GPUs are data-stream driven. This means that the GPU executes the same instruction sequence on large data sets. The instruction sequence to be executed is uploaded to the GPU, before the execution is triggered by a data-stream being assigned. The result of the computation can then be used for visualization, processed by a new instruction sequence, or read back to the CPU.

The use of parallel processing has traditionally been hampered by the high cost of specialized hardware. With the current introduction of clusters and more recently, multi-core CPUs, the cost of hardware for parallel processing is drastically reduced. In the coming years the increase in computational performance for CPUs is expected to be achieved by an increased number of computational cores in the CPU. This indicates that we are moving from a sequential programming era to parallel programming, even for standard PCs. Therefore many of the techniques that are used in GPU-based applications will be interesting also for future CPU-based implementations.

In this paper we address the challenges of General-Purpose Computations Using Graphics Hardware (GPGPU). In Section 2 we describe the evolution of graphics hardware followed by an overview of the graphics system in Section 3. In Section 4 we present an overview of the operations performed by graphical systems as defined in OpenGL [34]. An introduction to programming the GPU, with an example implementation, is presented in Section 5. As mentioned above, not all algorithms are well suited for GPU implementation. In Section 6 we therefore consider three problems that one is commonly faced with in scientific computing, and show how they can be solved using a GPU-implementation. In Section 7 we present the results of GPGPU-related research performed at SINTEF. For a more detailed overview of related works we refer to [28].

2 The Evolution of Commodity GPUs

At the start of the 1990s, high-performance 3D graphics hardware was fairly expensive, and consequently used predominantly for research and development, and by advanced industry. In general, 3D graphics was a tool for special applications, and the number of systems sold was fairly limited. During the late 1980s and first half of the 1990s, Silicon Graphics and other vendors launched on a regular basis new generations of 3D-systems with an improved price/performance ratio. Still, the prices were too high to allow broad-scale use of 3D graphics. A typical 3D workstation at the start of the 1990s used some proprietary UNIX-version and a RISC-type CPU from e.g., Digital Equipment, Sun, IBM, Hewlett-Packard, or Silicon Graphics. The computational performance of RISC CPUs were at that time considerably better than the less expensive PCs. However, the performance gap between PCs and UNIX-workstations disappeared during the 1990s, and at the same time the market for computer games on PCs grew rapidly, with a continuously increasing demand for 3D accelerated graphics.

With a large customer base and limited demands for backward compatibility, the computer game industry has been able to utilize new features in graphics hardware as they have been introduced. Thus, the combined innovation and development of graphics hardware and computer games have accelerated the evolution of low-price, high-performance graphics hardware.

The first 3D graphics hardware to be sold in large quantities for gaming purposes was introduced late in 1996. Even though the functionality was heavily reduced compared to graphics hardware made for workstations, this hardware allowed game developers to create real-time visual effects that had not been available in home computers until then. The first generation had only parts of the graphics pipeline implemented in hardware, and relied on the CPU to perform major parts of the computations. In late 1999, low-cost 3D graphics hardware that included the entire pipeline was introduced. As the performance increased, more and more features were implemented in hardware, allowing game developers to increase the complexity of their 3D graphics. The flexibility of the hardware increased, and in 2001 the first graphics hardware with programmable vertex processors was introduced. In Section 4.2 we describe the functionality of the different processors inside a GPU.

In 2002, GPUs with programmable fragment processors also became available, providing application developers full control over the calculations made to decide the color of each pixel. At this point also floating-point arithmetic for color calculations was introduced; although 32-bit precision was not available until 2003. Programmable vertex and fragment processors, combined with 32-bit floating-point precision opened up a wide range of possibilities, both for graphical and non-graphical applications.

Since GPUs are designed for real-time applications, new hardware features have not been added until they were usable with real-time performance. Similarly, an increase in performance is of little use unless it allows developers to produce better applications. Therefore the performance and the feature set of a GPU are dependent of each other, and we expect the flexibility of the processors to continue to increase in the future as the performance improves.

3 Hardware Overview

The reason why GPUs are well suited for general-purpose computations lies in the architecture. GPUs are stream processors, and are therefore designed to uniformly process large amounts of data. As a consequence of this, the memory on a graphics card is fast and the internal bandwidth is very high. In this section we will describe key differences between CPUs and GPUs, and why this makes the GPUs attractive for many different types of scientific algorithms.

3.1 Instruction-Based versus Stream-Based Processing

Data processing on a CPU is traditionally based on an instruction-driven model as illustrated in Figure 1. This model is called *Single Instruction, Single*

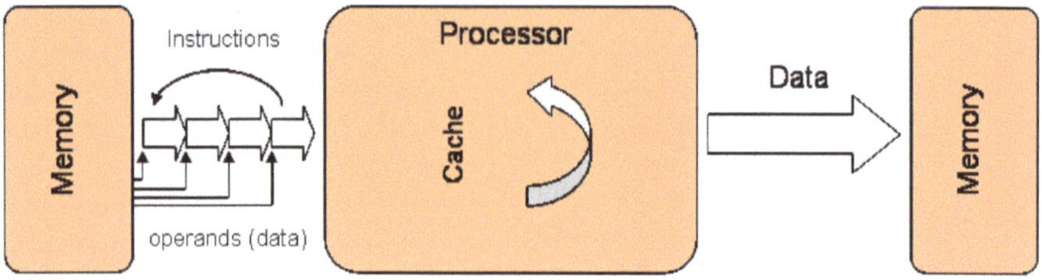

Fig. 1. Instruction-based processing.

Data (SISD) and corresponds to the von Neumann architecture. In this architecture, a single processor executes a single instruction stream to operate on data stored in the same memory as the instructions. The instructions needed in the execution of the program in turn refer to data, or to other instructions in the case of branching. The data needed for the execution of an instruction are loaded into the cache memory during the processing. The cache is a high-speed memory integrated on a chip (e.g., the CPU). If the data are not present in cache, they are loaded from the system memory over the relatively slow front-side bus, as described in Figure 3. This makes instruction-driven processing flexible but also inefficient when it comes to uniform operations on large blocks of data.

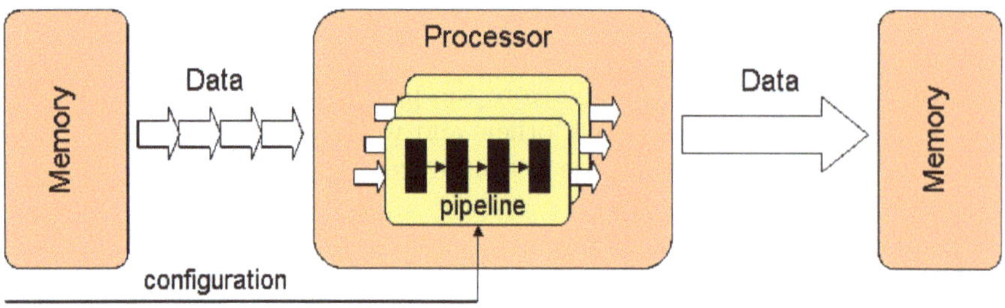

Fig. 2. Data stream processing.

GPUs are based on data-stream processing using a model called *Single Instruction, Multiple Data (SIMD)*. In this model, the processor is first configured by the instructions that will be executed, and then the data stream is processed, as illustrated in Figure 2. In other words, all data are processed by the same instructions until the processor is reconfigured. This model often leads to cache-friendly implementations on the GPU. The execution is parallelized by distributing the processing among several pipelines doing the same operations. The operations performed in these pipelines will be discussed in detail in Section 4.

3.2 Different Computing Paradigms

The computing paradigms for the CPU and the GPU are very different because the CPU is traditionally based on an instruction driven model, while the GPU is based on a stream processing model. As an example we may want to create a $m \times n$ matrix C by adding two $m \times n$ matrices A and B. On the CPU, the addition of the two matrices is done by a double for-loop, where we traverse all elements in the matrix and do the computations sequentially, e.g.,

```
// instruction stream
for(i=0; i<m; i++)
   for(j=0; j<n; j++)
      C[i][j] = A[i][j] + B[i][j];
```

In the stream-based computing model, we set up a data stream consisting of the two matrices A and B as input and matrix C as output. Then we create a computational kernel that takes one element from each data stream, adds them, and outputs the result. Finally the corresponding processing pipeline is 'executed'. In pseudo-code, this reads:

```
// data stream
setInputArrays (A, B);
setOutputArrays (C);
loadKernel ("matrix_sum_kernel");
execute ();
```

where the computational kernel simply corresponds to:

```
return (A[i][j]+B[i][j]);
```

As opposed to traditional instruction-driven algorithms, the nested for-loop and the call to the computational kernel never appear explicitly in the code. The for-loop is replaced by the mechanism that feeds the two input streams through the processing pipeline, and the computational kernel is called 'automatically' each time new elements from the two input streams arrives at one of the data-processing units.

On the GPU, our abstract processing pipeline is the graphics pipeline that renders graphics primitives to the screen. To add the matrices, we simply draw a rectangle with a resolution of $m \times n$ pixels, and set the color of each pixel in the rectangle equal to the sum the corresponding pixels in the two $m \times n$ input textures A and B. The for-loop is then called implicitly when the geometry is rendered.

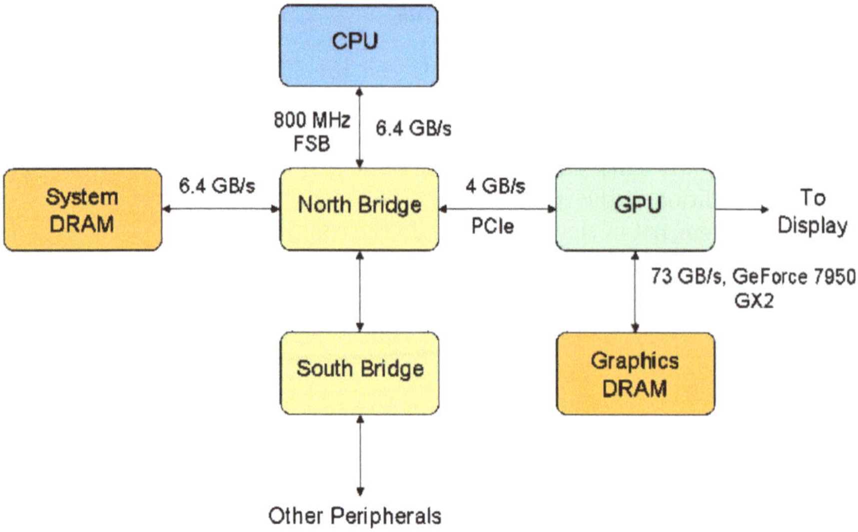

Fig. 3. The overall system architecture of a typical PC.

3.3 System Architecture

The overall system architecture of a PC can be illustrated as in Figure 3. The *North Bridge* and the *South Bridge* are the two main motherboard chips and together these are often referred to as the *chipset*. The North Bridge typically handles the communication between CPU, memory, graphics card, and the South Bridge. The South Bridge handles communication with the "slower" peripherals, like the network card, the hard-disk controller, and more. In some systems these controllers are included in the South Bridge. The front-side bus (FSB) is the term used to describe the data bus that carries all information passing from the CPU to other devices within the system.

In a computer system (as of autumn 2006) the communication between the CPU and the GPU is through a graphics connector called the *PCI Express*, which has a theoretical bandwidth of 4 GB/s simultaneously in each direction.

The internal memory bandwidth of the GPU is typically one order of magnitude higher compared to the CPU memory interface. For instance, the CPU memory interface is 6.4 GB/s with a 800 MHz FSB, whereas on a NVIDIA GeForce 7800 GTX 512 GPU the bandwidth is 55 GB/s. Algorithms that run on the GPU can take advantage of this higher bandwidth to achieve performance improvements, and this is one of the main reasons why GPUs are well suited for general-purpose computations.

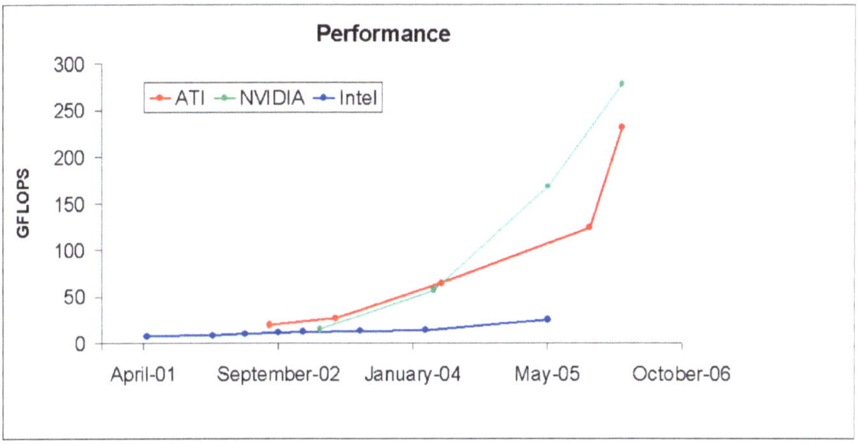

Fig. 4. Floating-point performance of commodity Intel CPUs versus commodity ATI and NVIDIA GPUs.

3.4 GPU Floating-Point Performance

Modern GPUs have a high number of pipelines performing the same type of operations, e.g., the new ATI chip with code name R600 is expected to have 64 pipelines. A higher number of pipelines gives the GPU a higher degree of parallelism. The GPU is designed to process 4-component floating-point vectors. Arithmetic operations that can be performed simultaneously for all four components, are therefore implemented efficiently in a GPU.

Implementations that take advantage of the GPU architecture can give very high floating-point performance. Figure 4 is based on data from GPU-Bench [12], and illustrates the performance of CPUs versus GPUs in recent years. The performance gap between CPUs and GPUs is expected to grow, making GPUs even more attractive as a computational resource in the future.

4 The Graphics Processing Pipeline

A GPU is designed to process geometric data (vertices) and convert these into colors (pixels) on the screen. This involves a number of operations, which will be described in this section. OpenGL [34] is the most commonly used platform-independent interface to graphics hardware, and we will therefore use OpenGL to describe the graphics system in this paper.

All operations performed in OpenGL are applied in a particular order, which can be described as the *graphics processing pipeline*. A key to understand how to use the GPU as a computational resource, is to understand each stage in the pipeline and how states set for these stages affect the final image rendered to the screen (or to an off-screen buffer). OpenGL is designed as a *state machine.* This means that all states/modes will remain unchanged until they are set to something else by the application. The state settings define the behavior of the pipeline and setting one state does not effect the others. Before looking at the *programmable pipeline* we will start by describing the *fixed-functionality pipeline.* It is important to understand the fixed-functionality pipeline because it heavily influences the programmable pipeline.

4.1 The Fixed Functionality Pipeline

Albeit all hardware vendors have their own specific implementations of OpenGL, the main stages in the pipeline are common for all implementations. Figure 5 describes a simplified view of the fixed functionality pipeline.

Before any operation is performed in the pipeline, we have to pass some geometry, in the form of vertices, to the graphics system. There are several ways to specify geometry data in OpenGL, and we refer to the OpenGL Programming Guide [34] for details.

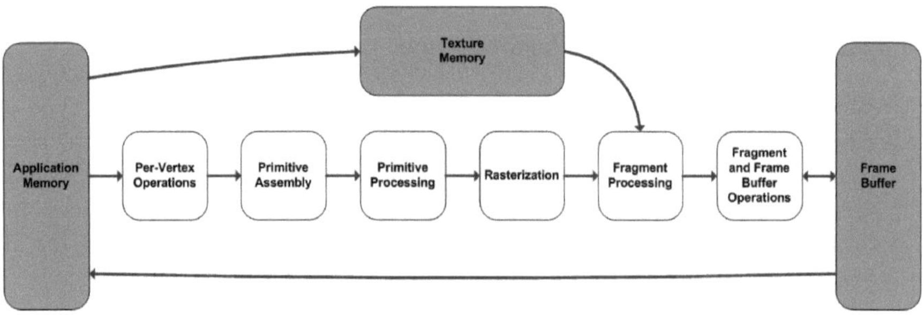

Fig. 5. A simplified view of the fixed functionality pipeline.

Per-Vertex Operations

The first stage in the pipeline is called *per-vertex operations*, also known as the *transformation and lighting* stage. The main operations at this point are transformation of vertex positions, normals, and texture coordinates, in addition to lighting calculations that are applied to modify the base color.

In OpenGL we use four component vectors to specify vertices, where the fourth-component is called the *homogeneous component* and is set to 1 in all standard rendering techniques. Vertex positions and surface normals are defined in a coordinate system called *object space*, see Figure 6. All three-dimensional objects can be defined in their own unique object space, so we need a common coordinate system that can contain a variety of objects. This coordinate system is called *world space*, and the transformation between object space and world space is called the *modeling transformation.* The next

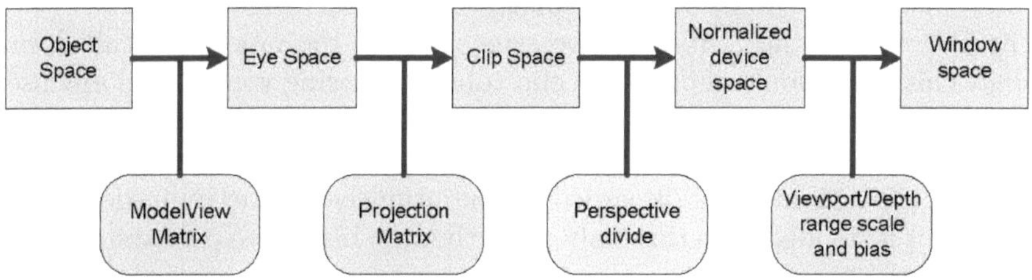

Fig. 6. Coordinate spaces and transformations in OpenGL.

step is to specify the viewing parameters, which can be combined in a matrix called the *viewing matrix*. When multiplied with this matrix, the coordinates are transformed from world space to *eye space*. In OpenGL, the modeling matrix and the viewing matrix are combined to a single matrix known as the *modelview matrix*. The matrices used in OpenGL are 4x4 matrices. Normally, light source positions are stored in eye space, which means that surface normals also need to be transformed to eye space for per-vertex lighting calculations to be performed correctly. This can be done by transforming surface normals defined in object space by the inverse transpose of the modelview matrix.

The next step is to create a viewing volume, which is the region that will be visible in the final image. This viewing volume is defined by specifying the projection parameters in a matrix called the *projection matrix*. The projection matrix is then used to transform objects in the viewing volume into a new space called *clip space*, which is suitable for geometry clipping, i.e., the removal of objects or parts of objects that are outside the viewing volume.

The process of transforming vertex coordinates from object space to clip space and determining all other attributes associated with each vertex, is done in the per-vertex operations stage. Figure 6 shows all coordinate spaces and transformations in OpenGL.

Primitive Assembly

After the per-vertex operations, the vertices are sent to a stage called *primitive assembly*. At this point, the vertex data are used to create primitives, i.e., points, lines, triangles, quadrilaterals and polygons. Each primitive is created from a set of vertices specified in the application, e.g., by using vertex arrays [34], and sent to the next stage of processing.

Primitive Processing

The first step in the *primitive processing* stage of the pipeline is called *clipping*. This operation compares the clip volume (viewing volume and any user-defined clipping planes) against each primitive. If the whole primitive is inside the clip volume, it is passed on for further processing; if the primitive is outside the clip volume it is discarded. If the primitive is partly inside the clip volume, it is divided such that only parts that are inside are passed for further processing.

Another operation called *culling* is also performed at this stage. This operation can cull (discard) all faces that face either towards or away from the camera.

After the per-vertex operations stage, all vertices are transformed to clip space. The next step is the transformation into the *normalized device space*. This operation divides each component of the clip-space coordinates by the homogeneous coordinate w, and is called *perspective divide*, because w is modified in the perspective transformation. This means that all visible primitives are transformed into a cubic region, where each component will range from $[-1, 1]$. The final transformation step is the transformation from the normalized device space to *window space*. The window space ranges from $[0, width - 1]$ in the x direction and $[0, height - 1]$ in the y direction. These transformations are described in the right part of Figure 6.

Rasterization

The *rasterization* is the process of decomposing graphics primitives into smaller elements corresponding to pixels in the frame buffer, hence rasterization of primitives occurs using window coordinates. The elements generated by the rasterization process are referred to as *fragments*. A fragment consists of a window coordinate, a depth value and other attributes like color, texture coordinates, etc. The values associated with each fragment are calculated by interpolation between vertex attributes for the primitive being processed. Figure 7 shows how fragments are created based on operations in the earlier stages of the pipeline.

Fragment Processing

A number of operations take place on fragments after the rasterization. This collection of operations is called the *fragment processing* stage. In this stage, there is one type of operations that is more important than the others, namely texturing operations.

Textures are used to specify the color of a fragment. This is done by creating one, two, or three-dimensional images containing the color details, and using the interpolated texture coordinates (from the rasterization stage) for texture lookup, see Figure 8. The process of attaching an image to geometry

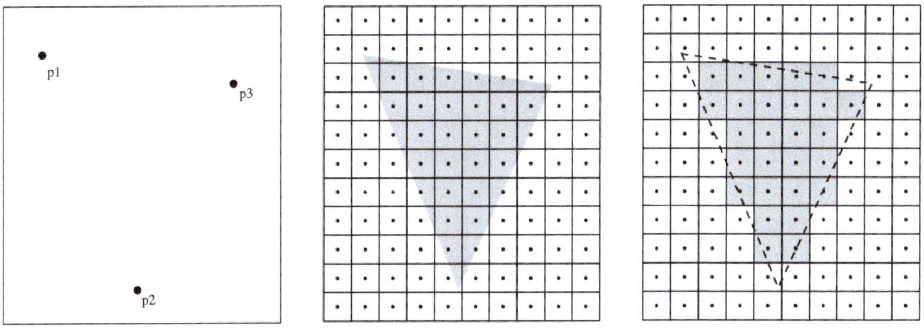

(a) Vertices transformed into clip space.

(b) Triangle created at the primitive assembly stage.

(c) Triangle after rasterization.

Fig. 7. Rasterization of a triangle.

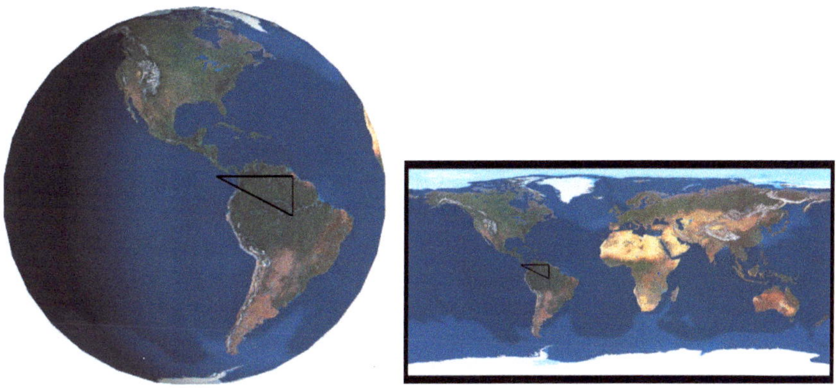

(a) Textured sphere.

(b) Texture illustrating the earth.

Fig. 8. Figure (a) shows a textured sphere where one triangle is marked with a black line. Figure (b) shows the texture used, and the corresponding triangle.

this way is often referred to as *texture mapping*. Texture mapping enables a vast variety of effects, and many extensions to OpenGL have been defined in this area. It is in this part of the pipeline most GPGPU calculations are performed.

Fragment and Frame-Buffer Operations

The last stage in the pipeline before the final pixels are stored in the frame buffer is called *fragment and frame-buffer operations*. Some of these operations are relatively simple and are normally implemented very efficiently in hardware, hence utilizing them for GPGPU purposes can be highly efficient. This includes operations like *alpha test*, *stencil test*, *depth test*, and *blending*.

The alpha test is used for accepting or rejecting a fragment based on its alpha (opacity) value. The stencil test compares a reference value with the value stored in the stencil buffer. Depending on the result of the test, the value in the stencil buffer is modified. There has to be a stencil buffer enabled for the stencil test to occur. The depth test is used to decide if the fragment should be drawn by comparing the depth (distance from the viewpoint) of the incoming fragment to the depth stored in the frame buffer. Blending is used to blend the color of the incoming fragment with the color stored in the frame buffer, based on current blending states.

4.2 The Programmable Pipeline

As described in Section 2, the complexity of the graphics hardware has grown rapidly in the last few years and as a consequence, the fixed functionality has been replaced with programmability. Currently, two stages of the pipeline are programmable, see Figure 9. The per-vertex operation stage has been replaced with a programmable *vertex processor* and the fragment processing stage has been replaced with a programmable *fragment processor*. All other parts of the pipeline remain unchanged.

A program intended to run on graphics hardware is called a shader. There are two types of shaders; *vertex shaders* and *fragment shaders* intended to run on the vertex processor and the fragment processor, respectively. When a shader is enabled, the fixed-functionality operations are replaced by the operations from a compiled shader source code. The programmability enables many new possibilities for high-quality rendering and general-purpose calculations.

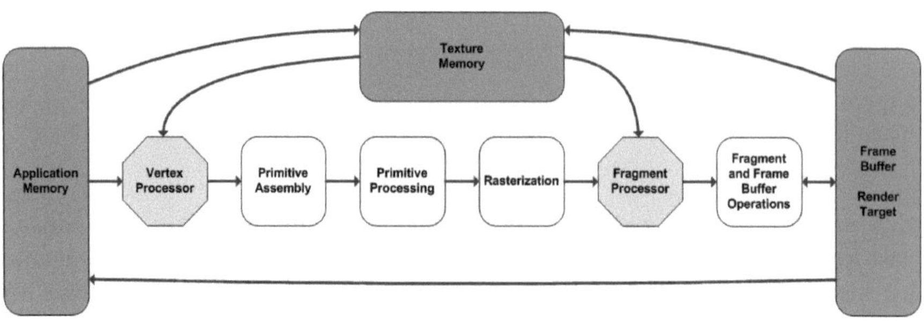

Fig. 9. A simplified view of the programmable pipeline. The programmable stages are highlighted in gray.

Vertex Processor

The vertex processor is designed to execute the same type of operations as in the per-vertex operation stage of the fixed-functionality pipeline; that is, transformation of vertex positions, normals, texture coordinates, and any other per-vertex operation the user specifies. Since the next stage in the pipeline operates on clip-space coordinates, the shader must include a coordinate transformation from local coordinate space to clip space. Modern GPUs also allow vertex shaders to fetch texture data, which are shared with the fragment processor.

Fragment Processor

The fragment processor operates on fragment values and is intended to do the same type of operations as in the fragment processing stage of the fixed-functionality pipeline. The attributes of a single fragment are optionally combined with texture data and written to one to four separate color buffers called *render targets*. However, the result may be discarded in the shader, and therefore not written to the render target. Normally, depth and stencil values are forwarded by the fragment processor to the next stage in the pipeline, but depth values can be replaced by a computed value from the fragment shader.

5 Programming the GPU

When programmable graphics hardware was introduced, it was an entirely new architecture and therefore the tools available were few and rather difficult to use. One of the difficulties was that at first there were no programming languages available, so the developers had to use an assembly-like interface. Fortunately, this has already changed and there now exist several high-level languages, namely Cg [11], HLSL [29] and the OpenGL Shading Language [31]. In this paper we focus on OpenGL Shading Language, because it is supported on a large variety of operating systems and hardware. In addition to compilers, there are also a number of tools and integrated development interfaces available, including Shader Designer, RenderMonkey and FX Composer. These may increase productivity both for application developers and artists because they allow rapid prototyping of shaders.

5.1 GPU as a Vector Unit

As described in Section 4.1, both vertex positions and colors are specified by four component vectors. Hence, four component vectors are the basic data type in GPU programming, and many functions on this data type are therefore implemented efficiently in hardware, including element-wise logarithmic and trigonometric functions. Since the result in one vector element is independent of the other elements, such functions can be evaluated in one clock cycle.

In addition to vectors, also matrices are important, as one of the most common tasks in vertex shaders is matrix-vector multiplication, viz.

```
vec4 x, b;  // Four-component vectors
mat4 M;     // 4x4 matrix
b=M*x;
```

Vector operations are not restricted to operate on the entire vector, but also include *swizzling*, which is the concept of extracting vector elements in arbitrary order, and *write mask*, which is to set only some of the elements and let the other elements be unchanged; e.g., the statements

```
vec4 a(1.0, 2.0, 3.0, 4.0);
vec4 b=a;      // b=(1.0, 2.0, 3.0, 4.0)
b.xyz=a.zyx;   // b=(3.0, 2.0, 1.0, 4.0)
b.wzyx=a;      // b=(4.0, 3.0, 2.0, 1.0)
```

The current generation of graphics hardware does not support integers at all, though this is likely to change in the near future in order to better support arrays and loops.

5.2 Flow Control

In procedural algorithms on CPUs, it is common to impose flow control through conditional statements (`if`, `while`, etc.). The availability of conditional statements may therefore seem obvious for the reader, but this was not the case for the first generation of fragment shaders. Even today, all the fragment pipelines execute the same path, and thus if an if/else structure is used, there are cases where the time required to execute the shader is the same as the time required for executing both branches. This overhead is avoided if fragments located near each other follow the same execution path, since the different pipelines are evaluating neighboring fragments. In general, details concerning branching are hardware-dependent and will change as new generations of hardware become available.

Data-dependent branching can increase the performance if parts of the shader can be omitted for large parts of the primitives being rendered, for example light calculations that can be omitted for points far from the light source. For other uses, dynamic branching in a fragment shader can be a challenge when it comes to performance; one should therefore consider if it is more efficient to use one of the following strategies:

Use arithmetic expressions. In many settings it is possible to rewrite the shader such that it uses arithmetic expressions instead of dynamic branching. This is especially advantageous when combined with the vector capabilities of the GPU (see e.g., the discussion in §4.4 of Hagen et al. [15] later in this book).

Z-culling. The depth test (also known as z-test) can be used to avoid the update of certain pixels. If one can determine in advance which pixels need to be updated, the depth buffer can be arranged so that pixels that should not be updated fail the depth test. This allows the GPU to do the depth test at an earlier stage than the execution of the fragment shader. Most modern GPUs perform coarse depth tests at such an early stage to avoid unnecessary computations, provided the fragment shader neither writes to the depth value nor discards fragments.

These tips are valid for the current generation of hardware and are likely to continue to be valid also in the future.

5.3 Execution Model for OpenGL Shaders

OpenGL is integrated into the operating system via a vendor specific OpenGL driver. The driver handles all communication between the CPU and the GPU. The compiler and linker for the OpenGL Shading Language is part of the OpenGL driver, as illustrated in Figure 10. This differs from HLSL and Cg, where the compiler translates the shader source code into assembly source code that can be sent to the underlying graphics API. OpenGL shaders must be compiled during application execution, which gives the hardware vendors more room for shader optimization and architectural innovation. On the other hand, this may result in a speed penalty when the application has to wait for shader compilation instead of providing precompiled code to the graphics API.

Shader source code is sent from the application to OpenGL as text strings using the OpenGL Shading Language API and loaded into a shader object in the driver. The next step is to compile the shader. If the compilation is successful, the compiled shader object is attached to a program object, which acts as a container for shader objects. The shader objects attached to the program object are then linked together and the resulting executable can be installed as part of the current OpenGL state.

5.4 OpenGL Shading Language

OpenGL Shading Language can be used for both vertex shaders and fragment shaders. The only differences between shaders written for the different processor types are the input and output. The fragment shader can either return the color value and the depth value of the fragment, or discard the fragment entirely. If the fragment shader calls the discard function, the fragment will not cause any changes to the frame buffer. The vertex shader on the other

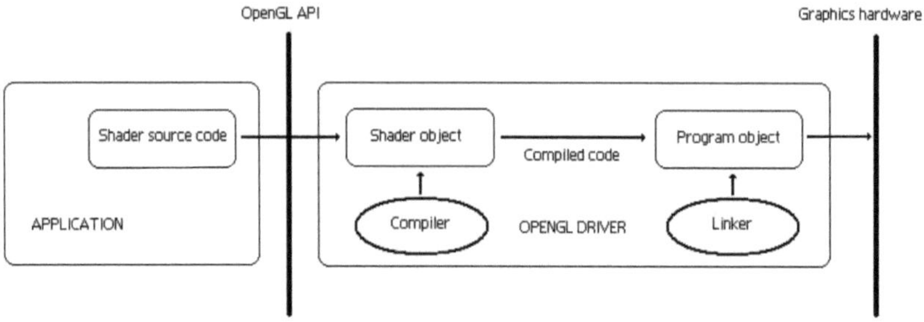

Fig. 10. Execution model for OpenGL shaders.

hand may not discard the vertex, but it may return any number of values in addition to the vertex position, provided that the fragment shader is able to read these values.

In this section we will give a brief overview of the OpenGL Shading Language (GLSL). For a more detailed description we refer to [31].

Basic Types

OpenGL Shading Language supports the basic data types shown in Table 1.

Type Qualifiers

The keyword `attribute` is a qualifier used to declare variables that are passed to a vertex shader from the application on a per-vertex basis. Hence, attributes are expected to change on every run of the vertex shader. The position and normal of a vertex are typical vertex attributes and can be declared in a vertex shader as follows:

```
attribute vec4 position;
attribute vec3 normal;
```

The `uniform` qualifier is used to declare a global variable whose values are the same for the whole primitive being processed. Contrary to the attribute qualifier, a uniform qualifier can be declared either in the vertex shader or in the fragment shader. An example of a declaration in a shader is:

```
uniform vec4 lightPosition;
```

`Varying` variables provide the interface between the vertex shader, fragment shader and the fixed functionality between them. Varying variables are set per vertex and interpolated, at the rasterization stage of the pipeline, over

Table 1. Basic data types in OpenGL Shading Language

void	for functions that do not return a value
bool	a conditional type, taking on values of true or false
int	a signed integer
float	a single floating-point scalar
vec2	a two component floating-point vector
vec3	a three component floating-point vector
vec4	a four component floating-point vector
bvec2	a two component boolean vector
bvec3	a three component boolean vector
bvec4	a four component boolean vector
ivec2	a two component integer vector
ivec3	a three component integer vector
ivec4	a four component integer vector
mat2	a 2x2 floating-point matrix
mat3	a 3x3 floating-point matrix
mat4	a 4x4 floating-point matrix
sampler1D	a handle for accessing a 1D texture
sampler2D	a handle for accessing a 2D texture
sampler3D	a handle for accessing a 3D texture
samplerCube	a handle for accessing a cube mapped texture
sampler1DShadow	a handle for accessing a 1D depth texture with comparison
sampler2DShadow	a handle for accessing a 2D depth texture with comparison

the primitive being processed. Varying variables must be declared both in the vertex shader and in the fragment shader (with the same name in both shaders), otherwise the linking will fail. An example of a declaration is:

varying vec4 textureCoordinate;

The `in`, `out`, and `inout` qualifiers are used for function parameters. Functions are called by value-return, which means that input arguments are copied into the function at call time and output arguments are copied back at the end of the function. The keyword `in` is used to denote a parameter that is copied into but not out of the function. `Out` is used to denote a parameter that is copied out but not in, and `inout` is used to denote that the parameter is copied both in and out.

Built-In Variables

Shaders communicate between fixed functionality in the pipeline through built-in variables, in addition to user defined variables. All built-in variables in the OpenGL Shading Language start with the prefix 'gl_' and are described in detail in [31]. We will in this section briefly describe some of the most important built-in variables.

The variable `gl_Position` is special for the vertex shaders, and is intended for writing the vertex coordinates in clip-space after transformation calculations in the vertex shader. For a vertex shader to be valid, this variable must be written to. The consequence of not writing to `gl_Position` is undefined.

There are three special output variables in the fragment language, namely `gl_FragColor`, `gl_FragData[]`, and `gl_FragDepth`. Writing to `gl_FragColor` or `gl_FragData[]` specifies the color that will be used in the following operations in the fixed-functionality pipeline. If a fragment shader does not write to any of these two variables the result is undefined. Writing to `gl_FragData[N]` specifies that the final result will end up in render target `N` after subsequent fixed functionality operations. The depth value of a fragment can be written to the variable `gl_Depth`. If `gl_Depth` is not written to, the fixed function depth value will be used in further operations. If a shader executes the `discard` keyword, the fragment is discarded, and the values in the fragment output variables will not be used.

Built-In Functions

OpenGL Shading Language defines a variety of built-in functions for scalar and vector operations. These functions should be used in shaders, rather than user-defined functions doing the same type of operations. Built-in functions are assumed to be optimal, hence they may have direct hardware support. The most common built-in functions in OpenGL Shading Language are listed below.

Most standard trigonometric functions like conversion between radians and degrees, sine, arcsine, etc., are supported in the language. All these functions operate component-wise:

```
radians degrees sin cos tan asin acos atan
```

The following exponential functions are defined in OpenGL Shading Language, and they all operate in a component-wise fashion:

```
pow exp log exp2 log2 sqrt inversesqrt
```

The common functions below all operate component-wise. These functions are common in most programming languages and often used in algorithms:

```
abs sign floor ceil fract mod min max clamp mix step
```

Geometric functions operate on vectors, and not component-wise as the functions described above. Most of these functions are common in vector libraries. The `ftransform` function is a special vertex shader function, used to calculate the transformations the same way as they would have been computed by the fixed-functionality pipeline.

```
length distance dot cross normalize ftransform
```

Texture lookup functions are used to read data from textures. All these functions have two arguments, `sampler` and texture coordinate. These arguments describe what texture to read from, and the coordinate where the data are stored in the texture. The texture coordinates are floating-point numbers and range from 0.0 to 1.0 in each direction.

```
texture1D texture2D texture3D textureCube
```

5.5 Application Programming Interfaces (APIs) for Shading Languages

Both OpenGL and DirectX are low-level APIs. Consequently, it is natural to use a high-level library as an interface rather than using OpenGL or DirectX directly. Graphical applications usually use a scene graph that handles all rendering-related tasks, including loading and activating shaders. Scene graphs are well suited for describing complex scenes and advanced rendering techniques. Scene graphs are less suitable for GPGPU applications, since they do not require complex scenes to be rendered. Instead, the shaders and the data flows in between are often complex in GPGPU applications. Consequently, there is a need for APIs specially designed for GPGPU applications.

Brook

Brook is a programming language developed for stream processors (see Section 3.1), and Brook for GPU [7] is a compiler for GPU programs. Developers who are more familiar with stream processing than computer graphics are the main users of this language. It extends ANSI C to allow data parallel operations, uses a preprocessor to generate C++ code, where the computational kernels are translated into fragment shaders, and includes its own compiler that generates shaders. Brook is also capable of generating code to be used on special stream processors.

Glift

Glift is a generic template library developed by Lefhon et al. [26]. Glift enables GPU programmers to separate algorithms from the definition of data structures, which simplifies the algorithm development and allows complex data structures to be used in GPU-based implementation.

Accelerator

Accelerator [36] is a system that simplifies GPGPU programming. The library translates operations on-the-fly to fragment shader code and API calls. The speed achieved by using Accelerator is typically within 50% of optimized handwritten fragment shader code.

Shallows

`Shallows` is a C++ library developed at SINTEF [20] that aims to simplify the development of GPGPU applications. All the libraries above aim to make GPU programming easier by generating the shaders, which may result in reduced performance. To avoid this potential performance loss, `shallows` requires that the application developer writes the shaders. Therefore, we have focused on developing a library that reduces the development time for the data flow in GPGPU applications. This approach requires that the users of `shallows` are familiar with the graphics pipeline, and have basic knowledge of OpenGL.

`Shallows` consists of intuitive C++ classes, that encapsulate the functions most commonly used in GPGPU applications. Both Cg and OpenGL shaders are supported in `shallows`. In addition to this, a number of utility functions for reading and writing textures are included.

5.6 Example: Simulation of Heat Conduction

Many algorithms are stencil updates that consist of a discrete set of cells and a computational kernel that is invoked for each cell to compute the new cell value, for example applications in physical simulation and image processing. A typical CPU implementation would consist of a construction that explicitly loops over all cells and calls the kernel for each of them. Since the programmable fragment and vertex processors have replaced their fixed-function counterparts, the shaders cannot be called explicitly, but are instead invoked by the graphics pipeline, as discussed briefly in Section 3.2

Here, we will present an implementation of a simple, yet illustrative example of general-purpose computing on a GPU, namely the simulation of heat conduction using the linear heat equation. Visually, this corresponds to repeatedly applying a Gaussian blur to an image. For this example we use C++ with the `shallows` programming library together with OpenGL Shading Language.

The 2D heat equation describes the transport of heat in a body:

$$u_t = u_{xx} + u_{yy}.$$

Using standard finite differences, this equation is easily discretized as

$$U_{i,j}^{n+1} = (1-4r)U_{i,j}^{n} + r\left(U_{i-1,j}^{n} + U_{i,j-1}^{n} + U_{i+1,j}^{n} + U_{i,j+1}^{n}\right), \qquad (1)$$

where h is the grid size, k is the length of the time step, and $r = k/h^2$. This discretization requires that $r \leq 1/4$ to guarantee stability, thus $k = h^2/4$ is the largest time step that provides a correct solutions. From (1) we see that the value of a grid cell can be computed by combining the values of the neighboring cells and the value of the cell itself at the previous time step, giving the stencil illustrated in Figure 11.

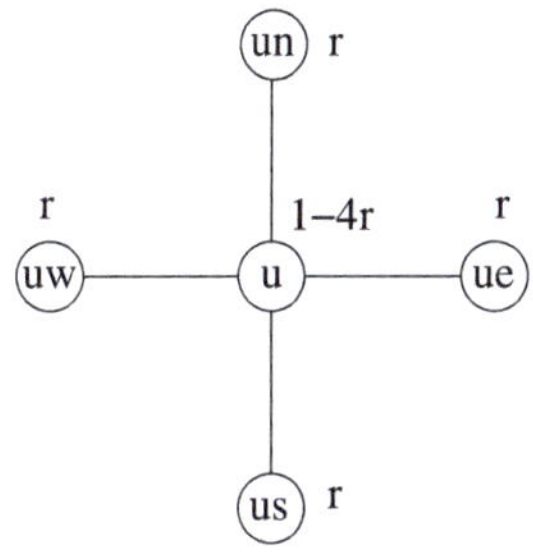

Fig. 11. The stencil used for simulating heat conduction.

(a) Original image. (b) Image after 100 iterations. (c) Image after 500 iterations.

Fig. 12. Repeated Gaussian blur applied to a 512x512 image.

Listing 1 contains the C++ source code used in the simulation of heat conduction. First, we allocate the render targets we use as computational domain. We require one off-screen frame buffer containing two render targets, one for holding the state at the current time step U^{n+1} and one for the previous time step U^n.

```
fb.reset(new OffScreenBuffer(BUFF_DIM, BUFF_DIM));
rt[0]=fb->createRenderTexture2D();
rt[1]=fb->createRenderTexture2D();
```

Throughout the simulation we "ping-pong" between these two buffers, so that we always read from the most recently updated render target. After allocating these render targets, we load the fragment shader given in Listing 2. This will make `shallows` compile and link the source code, viz:

```
heatProg.readFile("heat.shader");
```

Then, we set the output frame buffer:

```
heatProg.setFrameBuffer(fb);
```

Listing 1. Finite-difference solver for the heat equation using `shallows`.

```
shared_ptr<OffScreenBuffer> fb;
fb.reset(new OffScreenBuffer(BUFF_DIM, BUFF_DIM));

shared_ptr<RenderTexture2D> rt[2];
rt[0]=fb->createRenderTexture2D();
rt[1]=fb->createRenderTexture2D();

GLProgram heatProg;
heatProg.readFile("heat.shader");
heatProg.setFrameBuffer(fb);
heatProg.setParam1f("h", 1.0/BUFF_DIM);
heatProg.useNormalizedTexCoords();

float *p;
  /*
   * fill p with initial state
   */
fillRGBATexture(rt[1]->getTexture(), p);

/* Start the simulation */
for (int i=0; i<MAX_ITERATIONS; i++) {
  heatProg.setInputTexture("texture", rt[(i+1)%2]->getTexture());
  heatProg.setOutputTarget(0, rt[i%2]);
  heatProg.run();
 }
```

The fragment shader takes the distance between grid cells as a uniform parameter, which we set with:

```
heatProg.setParam1f("h", 1.0/BUFF_DIM);
```

`Shallows` can set the texture coordinates to be in any range. By calling the function `useNormalizedTexCoords`, we specify that they should be in the range $[0, 1]$. The rasterization will interpolate the texture coordinate, so that the texture coordinate sent to the fragment shader is $((i+0.5)*h, (j+0.5)*h)$ for cell u_{ij}. The final step before starting the simulation is to set up the initial state. This is performed by filling a floating point array with the initial state, and copying it to a texture, by the following function call:

```
fillRGBATexture(rt[1]->getTexture(), p);
```

Listing 2. Fragment and vertex shader for the heat equation written in GLSL.

```
[Vertex shader]

void main ()
{
  // Pass vertex position to fragment shader
  gl_Position = gl_Vertex;
  // Pass texture coordinate to fragment shader
  gl_TexCoord[0]=gl_MultiTexCoord0;
}

[Fragment shader]

uniform sampler2D texture; // Texture containing previous timestep
uniform float h; // Distance between grid cells (in the texture)

void main ()
{
  const float r = 0.25;

  vec4 u = texture2D(texture, gl_TexCoord[0].xy);
  vec4 ue = texture2D(texture, gl_TexCoord[0].xy + vec2(h, 0.0));
  vec4 uw = texture2D(texture, gl_TexCoord[0].xy - vec2(h, 0.0));
  vec4 un = texture2D(texture, gl_TexCoord[0].xy + vec2(0.0, h));
  vec4 us = texture2D(texture, gl_TexCoord[0].xy - vec2(0.0, h));

  vec4 result = (1.0-4.0*r)*u + r*(ue + uw + un + us);
  gl_FragColor = result;
}
```

At the beginning of each iteration of the simulation we set the most recently updated render target to be used as texture. The output target of the program is then set to be the other render target.

```
heatProg.setInputTexture("texture",
                         rt[(i+1)%2]->getTexture());
heatProg.setOutputTarget(0, rt[i%2]);
```

To execute the fragment shader, we call the `run` function in the `GLProgram` object, which renders a geometric primitive covering the viewport. This function assumes that the transformation matrix is an identity matrix. The vertex shader in Listing 2 therefore does not transform the vertex position, only passes it through the rasterization (for interpolation) into the fragment shader.

From (1) we get the stencil illustrated in Figure 11, and a fragment shader that implements this scheme is given in Listing 2. The shader first reads the cell value from the texture containing the previous time step, followed by reading the neighboring cell values. Note that the texture coordinates are floating point numbers in the range $[0, 1]$, and that the distance between two data cells therefore is h. The values read are combined according to (1), the result is returned as the new color of the fragment, by writing it to the variable `gl_FragColor`.

6 Common Numerical GPU-Algorithms

Traditional CPU-based, instruction-driven algorithms can seldom be applied directly on stream-based architectures. This makes it difficult to directly transfer an existing CPU algorithm to a GPU. Still, many algorithms for solving common tasks related to numerical computations have been successfully ported to the GPU. Here we give a short presentation of some of these algorithms, where the typical GPU implementation differs from a typical CPU implementation.

6.1 Sorting a Vector

Sorting a one-dimensional vector of data is a common and important task for a wide range of applications. Several methods for instance in computer graphics are dependent on efficient sorting to be able to handle complex scenes with an acceptable frame rate. It is therefore natural to investigate if it is feasible to implement sorting on the GPU. However, traditional sorting algorithms developed for the CPU are data-driven and can therefore not easily be adapted for implementation on a GPU.

The basic building block for sorting algorithms consists of reading two data elements, comparing, and swapping them if necessary. The runtime of a sorting algorithm can therefore be described by the number of comparisons. The most widely used sorting algorithm is called *quicksort* and spends $O(n \log n)$ comparisons on the average case, and $O(n^2)$ in the worst case; this makes the algorithm very fast for most input data. Quicksort is based on recursively splitting the vector in two, and sorting each sub-vector. Unfortunately, the size of each sub-vector is data dependent, which makes quicksort unsuited for GPU implementation.

Sorting networks form a family of sorting algorithms, where comparisons are performed in a fully predefined manner. These algorithms use a higher number of comparisons than data-dependent algorithms like quicksort, but they fit the architecture of the GPU better, since the sequence of comparisons is predefined. Furthermore, these methods are developed to allow many of the comparisons to be executed simultaneously, which is essential for the method to be applicable for GPU implementation. Figure 13 shows such a network, where eight data elements are sorted in six stages, i.e., the vector can be sorted in six time steps provided four comparisons can be made in parallel.

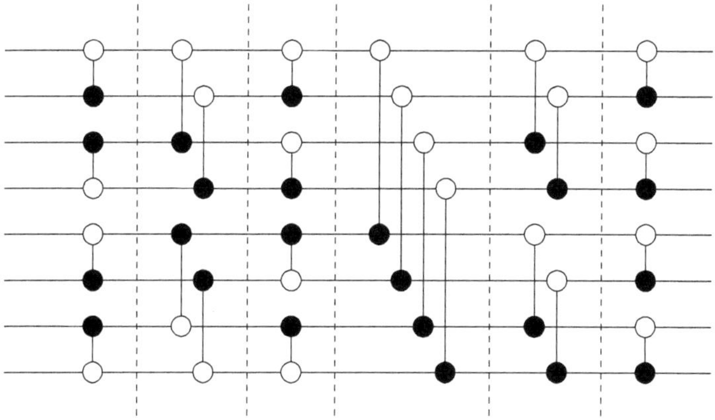

Fig. 13. Sorting network where eight values are sorted in six stages. The dashed lines separate the stages, and the black and white circles denote the destinations for maximum and minimum value respectively.

Bitonic sort [1] is such a sorting network, which has been successfully implemented on the Imagine stream processor by Kapasi et al. [24] and a GPU-based implementation has been carried out by Purcell et al. [30]. In the GPU implementation each of the stages is implemented as one rendering pass, with each pass rendering n pixels. The sorting is completed after $O((\log n)^2)$ such rendering passes. Even though $O(n(\log n)^2)$, the number of comparisons required, is much higher than quicksort uses on average ($O(n \log n)$), the GPU-based bitonic sort performs significantly better than CPU-based versions of quicksort.

Lesson 1 *The most efficient algorithm on a GPU is not necessarily the one that has the lowest operational count. The algorithm must fit the underlying architecture (here, an ability to do a large number of identical operations in parallel).*

6.2 Reduction Algorithms

Reduction algorithms, like summing the elements of a vector and finding the largest/smallest element, are common tasks that are trivial to implement on a sequential architecture, but often form a bottleneck when implementing an algorithm in parallel. The GPU implementation described here is similar to implementations that you may find for other parallel architectures.

Finding the largest element in a texture can be solved by using a multi-pass algorithm that reduces the size of the render target for each iteration, as illustrated in Figure 14. This example shows four passes, where the size of the render targets is one-fourth of the size of the input textures. The maximum of four neighboring values in the input texture is calculated in a fragment

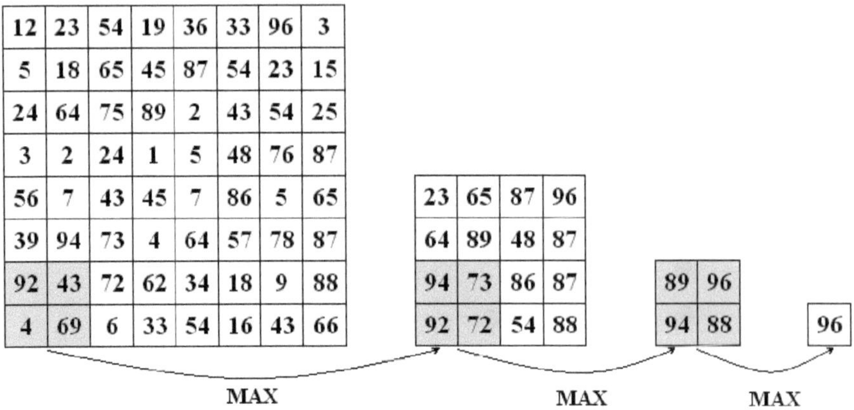

Fig. 14. Finding the maximum value using multiple passes.

shader, and stored in a render target that is used as input texture in the next pass.

It is possible to increase the number of values we want to compare in the fragment shader, which can decrease the number of passes needed. In some cases, it is faster to read back the result to the CPU before we have reduced the texture size to one single pixel and let the CPU do the final calculations. The same type of algorithm can be used to compute inner products, various norms of vectors, and other gather-type operations.

Lesson 2 *When searching for an algorithm that is suited for GPU implementation, it is often helpful to study algorithms written for other parallel architectures, e.g., clusters.*

6.3 Linear Solvers

The conjugate gradient method is a well known and popular iterative algorithm for solving linear systems of equations. Its popularity is not only due to the efficiency of the algorithm, but also because the method can be implemented without knowledge of the data structure used to store the linear system. The only requirements are that the following operations can be performed:

Matrix-vector product: $\mathbf{b} = \mathbf{A}\mathbf{x}$.
Inner product: $a =< \mathbf{x}, \mathbf{x} >= \sum_i x_i^2$.
Scalar multiplication and vector addition (saxpy): $\mathbf{z} = \mathbf{x} + s\mathbf{y}$

Provided these functions, the conjugated gradient method can be implemented without knowledge of the underlying data structure representing the linear system. This method requires that the matrix is symmetric positive definite. However, by adding $\mathbf{A}^T\mathbf{x}$ to the list of operations, one can also implement the bi-conjugated gradient method, which can be applied to non-symmetric matrices.

There are two main challenges when adapting this algorithm for GPU-based implementations; the first is which data structure to use to achieve efficient matrix-vector multiplication; the second is how to calculate the inner product. The latter challenge has already been discussed in Section 6.2, we will therefore concentrate on the matrix-vector multiplication.

The natural and naïve representation of a vector is to use a 1D texture and use the ordering of vector elements as the ordering of pixels. However this representation induces a restriction to the size of a vector, as the maximal number of elements is lower for 1D textures than for 2D textures. Furthermore, rendering 1D textures is generally slower than using 2D textures. Similarly, the obvious representation for a matrix is to use a 2D texture and use the natural ordering of the elements. The restriction on the sizes of textures prevents this from being usable for large matrices.

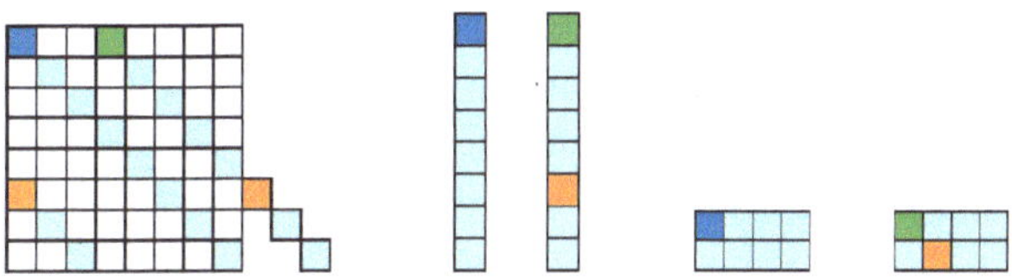

Fig. 15. Representing a matrix as a set of diagonal matrices packed into 2D textures.

Krüger and Westermann [25] proposed an alternative representation, which does not have these drawbacks. In their work, a matrix is represented as a set of vectors. A full matrix is represented as a set of column vectors, while a banded matrix is represented as a set of diagonal vectors. Each vector is stored in a 2D texture, which makes it possible to solve systems of equations involving banded matrices. Since the diagonal vectors are not of equal length, each vector (except the main diagonal) consists of two diagonals concatenated as shown in Figure 15. The actual layout of a vector in a 2D texture is not important for the implementation of the conjugated gradient method as long as the same scheme is used for all vectors. However, for performance purposes it is important that four-component textures are used.

The matrix-vector product is performed as a set of component-wise vector-vector products, adding the result to the current value of the vector element. One such component-wise multiplication and addition constitutes one rendering pass, and the complete matrix-vector product is calculated after performing one such rendering pass for each diagonal vector in the matrix. For banded matrices the algorithm requires one rendering pass per band in the matrix and is therefore very efficient.

A matrix-vector product can be performed by collecting the non-zero matrix elements on each row, multiplying each of them with the corresponding vector element. Bolz et al. [2] had this property in mind when developing a representation for general sparse matrices. The non-zero elements of each row are stored sequentially in a texture. Each element consists of the matrix value and the column number, the latter pointing to the corresponding vector element. With this structure, a fragment program can loop over the elements in a certain row, given the starting location of the matrix elements and the number of non-zero elements. Both these numbers can be stored in a texture. The unstructured memory access used in this algorithm provides the main challenge for obtaining high performance, together with the branching imposed by the varying number of elements per row. The branching can be avoided by grouping together all rows that have the same number of elements and then calculating the results for each group separately. This leaves the memory access pattern as the main challenge, which is also the case for the CPU-based implementations of sparse matrix solvers. Single precision may not be sufficient for many numerical applications, for example numerical simulations. Göddeke et. al. [14] have proposed a mixed precision defect-correction approach to handle the lack of double precision capabilities in todays GPUs.

Lesson 3 *In some cases the data structures used for CPU implementations are unsuited for GPU implementations. The use of untraditional data structures may be necessary to develop efficient algorithms for GPU implementations.*

7 SINTEF Contributions

Researchers at SINTEF have studied GPU-based implementations of algorithms both for visualization and general-purpose computation. Here we will provide a brief overview of some of our results, together with some related work. For a more complete overview of related work and applications that have been successfully adapted for GPU-based implementation, we refer to Owens et al. [28].

7.1 Numerical Solutions of PDEs

Mathematical models used for describing the behavior of natural and man-made systems are typically based on partial differential equations (PDEs) that typically give the rate of change both in time and/or space of physical quantities, e.g., conserved quantities like mass, momentum, and energy. Numerical solution of systems of PDEs is useful to reproduce past and predict future behavior of the system to broaden our understanding of the underlying mechanisms.

Simulations based on numerical solution of PDEs are of widespread use in science and engineering, in particular within various branches of the natural sciences. In almost all applications, there is an ever increasing demand for faster, larger, and more accurate and reliable computations. It is therefore not surprising that numerical algorithms for solving PDEs were among the first non-graphical applications considered on GPU hardware. An example of such a GPU-based implementation was given in Section 5.6 for the heat equation.

Most GPU implementations have focused on applications in image processing or fluid dynamics, and in particular the incompressible Navier–Stokes equations. Here the PDE-driven simulations involve equations that are time-dependent, where the initial situation is known and a PDE-solver is used to determine how the system progresses in time. Such problems are solved by an algorithm, which based on the current (and previous) time step(s) calculates the solution at the next time step. In general, it is common to use an implicit or semi-implicit temporal discretization, which means that one must solve a linear system for each time step. The earliest GPU-based implementations of implicit schemes were carried out before the programmable pipeline was available, e.g., by Rumpf and Strzodka [32]. Bolz et al. [2] and Krüger and Westermann [25] later introduced more advanced methods that take advantage of the programmable pipeline and use linear solvers as described in Section 6.3.

The GPU architecture, however, is much more suited for applications that are based on an explicit temporal discretization, for which the value of each cell can be computed independently, and hence in parallel. Rather than trying to surmount the problems of solving large sparse linear systems in parallel on the GPU, as considered by the majority of other GPGPU researchers focusing on scientific computing, our group at SINTEF have instead looked for other applications in which most state-of-the-art methods are explicit. That is, we have tried to find problems for which parallel processing on GPUs has a particular potential for providing high speedup and bringing simulations from batch to interactive mode.

Numerical solution of (systems of) hyperbolic conservation laws is one such area, where most modern high-resolution schemes are based on explicit temporal discretizations and as such are (almost) embarrassingly parallel. In addition, for systems of equations, the basic unknown in each cell is a vector, which means that implementations may benefit from the simultaneous processing of vectors available in GPUs. Moreover, modern high-resolution schemes have a high number of arithmetic operations per memory (texture) fetch due to complex spatial reconstruction schemes, nonlinear limiter functions, etc. Altogether, this makes high-resolution schemes for hyperbolic conservation laws a prime candidate for GPGPU implementation.

In our work, we have primarily focused on solving the shallow-water equations describing nonlinear (water) waves and the Euler equations for the dynamics of an ideal gas. Our results show that the GPU algorithms run between 15 and 30 times faster than similar CPU algorithms. A detailed description of our GPU-implementation can be found later in this book [15]. In [16] we present a GPU-based implementation of the shallow water equations with

Fig. 16. Simulation of a dambreak over an artificially constructed terrain.

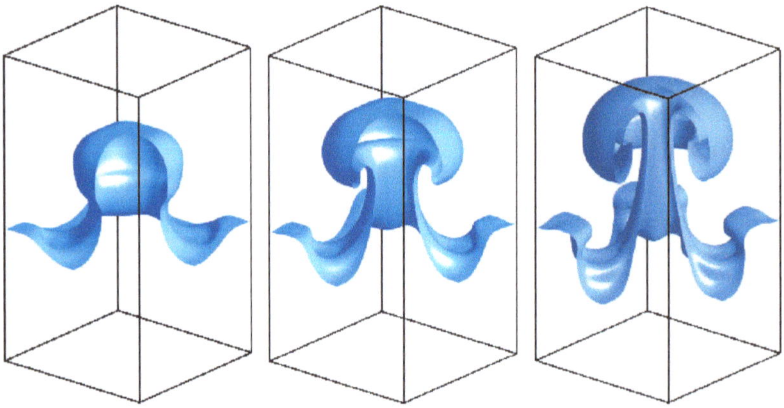

Fig. 17. The Rayleigh–Taylor instability at times $t = 0.5$, 0.6, and 0.7.

dry states and a variable bottom topography. This application has real-time performance for both simulation and visualization. Two time steps of the simulation are shown in Figure 16. In [17] we describe how to solve both 2D and 3D Euler equations on GPUs, and discuss how these types of problems can be solved on multi-GPU systems. This is used in a simulation of a Rayleigh–Taylor instability, see Figure 17. This instability arises when a layer of heavier fluid is placed on top of a lighter fluid and the heavier fluid is accelerated downwards by gravity. Currently, one of our master students (Martin Sætra), is developing a solver of hyperbolic conservation laws that can be used both on multi-GPU systems and clusters of PCs with programmable graphics hardware.

7.2 Visualization of Smooth Surfaces

CAD models are usually represented by NURBS surfaces. Similarly subdivision surfaces are popular in the entertainment industry. Still the GPU does not support rendering of either surface type. Instead, the surfaces must be tessellated (converted into sets of triangles) by the application before being sent to the graphics system. Applications using dynamic objects and those allowing extreme close-ups of static objects require evaluation of the surfaces at runtime.

Bolz and Schröder [3] demonstrated that the GPU can efficiently evaluate a subdivision surface that can be represented as a net of Bézier patches. The evaluation can be performed either in a vertex shader to position the vertex, in a fragment shader to calculate per-pixel normals (and thus lighting), or in a fragment shader calculating the position. The latter will allow the results to be used as vertex positions at a later rendering pass.

Guthe et al. [13] let the CPU determine the level of detail required to ensure that rendering errors are less than one pixel when rendering bi-cubic B-spline surfaces. The algorithm also handles trimmed surfaces while maintaining the error within one pixel. In [19] we presented a method that uses the rasterization to determine the parameter values where the surface is to be sampled, and connect the samples to form triangles. This is an extension of the method proposed by Kanai and Yasui [23], where the samples are not connected but rather drawn as points. Figure 18 shows a rendered image using our method.

An alternative to render a tessellated version of the surface is *ray-tracing*, where rays are traced from the eye back through the scene. Ray-tracing of real algebraic surfaces requires substantial geometry processing, as a large number of intersections between straight lines and the zero set of a tri-variate polynomial $q(x, y, z) = 0$ of total degree $m \geq 1$ have to be calculated. Loop and Blinn [27] address the visualization on GPUs of real algebraic surfaces of degrees up to four, while later in this book [33] we address an approach to visualize algebraic surfaces of any degree. However, in practice, the highest degree we can achieve right now is five. This restriction is due to limitations in the number of temporary variables in current GPUs. The next generation of GPUs will have a much higher number of temporary variables, which will allow our approach to also work for much higher polynomial degrees.

Modern GPUs have made it possible to use techniques such as *normal mapping* and *per-pixel lighting* to increase the visual realism of polygonal meshes. These techniques work well in most cases, but cannot hide the piecewise linear silhouette of a coarse triangular mesh. In previous work Dyken and Reimers [9] present a solution to this problem by calculating a continuous *silhouetteness* for every edge in the mesh, see Figure 19. In cooperation with our research group [10] this method was extended by calculating the *silhouetteness* on the GPU, as well as incorporating effective adaptive tessellations of silhouette triangles by extending Boubekeur and Schlick [5].

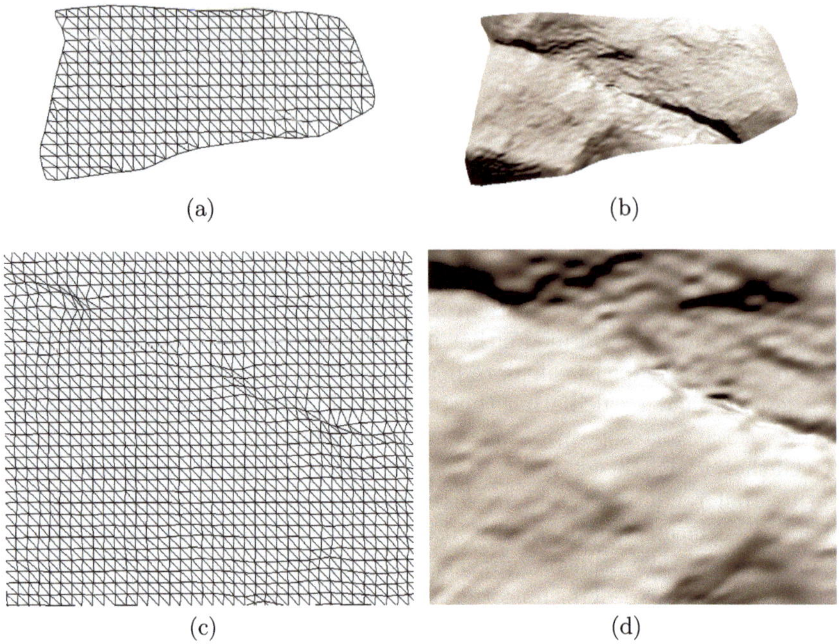

Fig. 18. Screen-space tessellation of a cubic spline surface; (a) and (b) show the entire surface, (c) and (d) are close-ups.

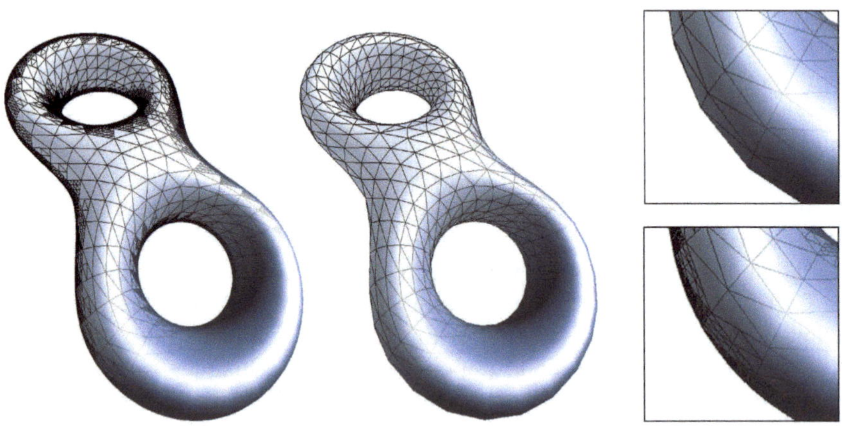

Fig. 19. Visualization with silhouette refinement to the left, and uniform tessellation to the right.

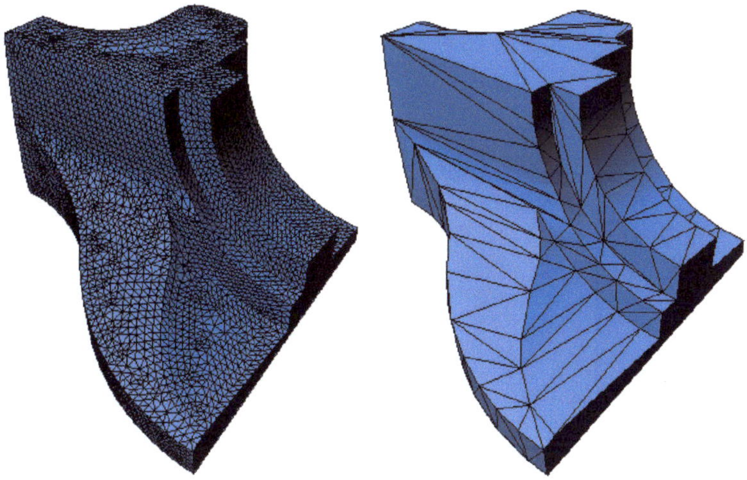

Fig. 20. Original 3D model to the left, and a simplified version to the right.

7.3 Geometry Processing

Intersecting geometric objects is a time-consuming task, that is essential in CAD systems. Often, time is spent to determine in which regions of the surfaces the intersections may lie. In another paper in this book [8], we address how to accelerate CAD-type intersection algorithms, including a brief overview of a GPU-based algorithm. A more detailed GPU-implementation is presented in [6], where we describe how the GPU can be used to identify potential intersection regions on spline surfaces. The paper also extends the algorithm to detect potential self-intersecting regions.

Preparing a CAD model for FEM analysis can be a time-consuming task, where mesh simplification plays an important role. It is important that the simplified model has the same properties as the original model, and that it is within a given error tolerance. Botsch et al. [4] proposed a GPU-based method for checking if a candidate triangle is within the given tolerance. Their method requires that a signed distance field is generated in the CPU, and uses the GPU to sample the signed distance field. In [21], we proposed a framework for GPU-based simplification, where vertex removal is performed for all vertices in an independent set. The GPU is used both to order the vertices, and to test if the re-meshed version is valid. Since the vertices that are removed are not neighbors, all calculations required to determine which vertices to remove, as well as the re-meshing, may be performed in parallel. This allows us to use the GPU to perform the computations. The performance boost provided by the GPU allows more complex criteria for validity testing to be used with acceptable performance. Figure 20 shows a 3D-model before and after our GPU-based simplification was performed.

Fig. 21. Image before inpainting to the left, and the reconstructed image after the GPU-based inpainting to the right.

7.4 Other GPU Activities

In addition to the activities addressed above, scientists and PhD students at SINTEF are in cooperation with partners developing GPU algorithms for a wide variety of applications. We will here give a short description of a selection of the works.

Inpainting

The problem of inpainting is to fill in image information in an area that may be damaged or simply missing, based upon the image information available outside this region. In our GPU implementation of inpainting, we have focused on a recent PDE-based approach introduced by Tai, Osher, and Holm [35] using the so-called TV-Stokes equations. The GPU implementation has been performed in collaboration with Talal Rahman who is currently working as a post doc in the GPU project.

The inpainting algorithm is divided into two steps. The first step is to find a set of tangential vectors to the level curves of the image in the missing region. These tangential vectors are found such that they have the smallest total variation norm over the missing region. The second step is to find a surface that fits the tangential vectors in the blank region. Each of these two steps are performed by solving a time dependent problem using explicit finite differences schemes on a staggered grid. Figure 21 shows an example of an inpainted image.

Marine Acoustics

In collaboration with researchers from NTNU in Trondheim, we have developed a GPU-based ray-tracing algorithm for underwater sound fields [18]. This work is based on a model called PlaneRay, developed by Hovem et al. [22]. Numerical modeling of the acoustic propagation in the oceans is an important issue in underwater acoustics and is required in many applications to assess the performance of acoustic equipments such as echo sounders, sonars, and communications systems.

In the PlaneRay algorithm the rays are spread out from a single point source. The sound speed in the ocean varies with the oceanic condition, in particular with the temperature and the salinity of the waters. Since the sound speed is not constant, the sound rays follow curved paths rather than straight ones. The PlaneRay model is based on an iterative algorithm for calculating the ray paths through the water. On the GPU all rays are calculated in parallel and each ray corresponds to a fragment in a texture. The processing of a ray propagation step only depends on the two previous steps and on the sound speed at the current water depth. This makes the algorithm embarrassingly parallel and the algorithm is implemented on the GPU by "ping-ponging" between three render targets. Figure 22 illustrates the acoustic waves calculated by the GPU-based algorithm. These calculations can be very time-consuming, and because of the parallel nature of these algorithms they are well suited for GPU-implementations. Our results show a speedup between one and two orders-of-magnitude compared to a CPU implementation, depending on how many rays are calculated simultaneously.

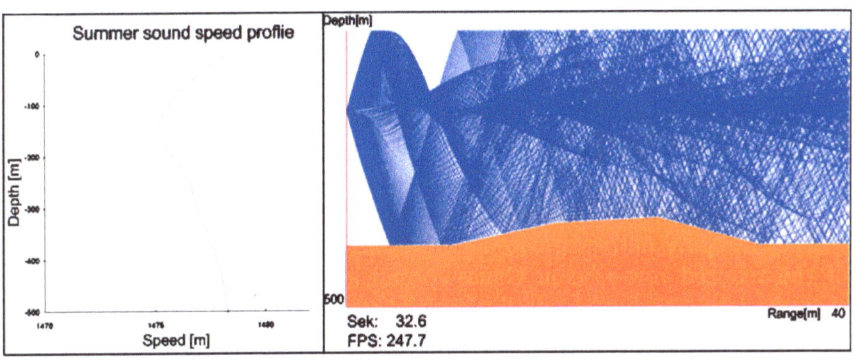

Fig. 22. Sound field using a summer profile of the sound speed and with the source at a depth of 150 meters.

A Matlab Interface to the GPU

In his masters thesis, André R. Brodtkorb is developing an interface to GPU programming in Matlab to enable scientists to take advantage of the speed boost of GPUs for standard operations on dense matrices. The left plot in Figure 23 reports the efficiency of a dense matrix-matrix multiplication implemented on the GPU and interfaced from Matlab compared to the highly tuned ATLAS routines otherwise used by Matlab. In addition to speed up the computation by a factor two, the GPU implementation may offload the CPU and thereby increase the efficiency of a computation.

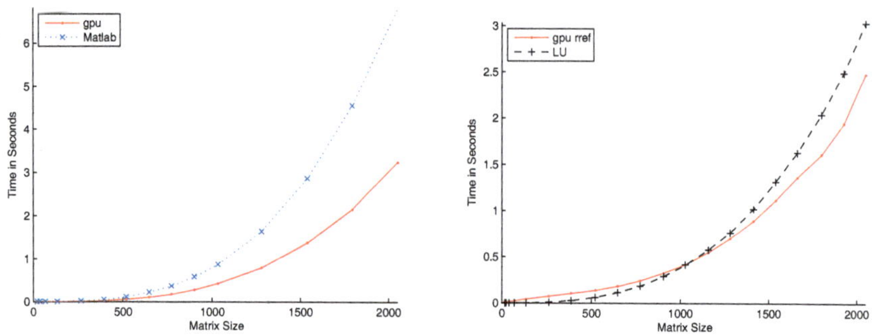

(a) (b)

Fig. 23. (a) Dense matrix-matrix multiplication on the GPU executed from Matlab, showing speed gains of about 2×. **(b)** Computation of the reduced row echelon form using Gauss-Jordan elimination on the GPU from Matlab, compared to LU factorization.

The right plot of Figure 23 reports a similar comparison for the inversion of a full matrix; i.e., solving $Ax = b$ for a dense matrix A and a known right-hand side b. In Matlab, this is typically done by calling `A\b`, which invokes a standard Gauss elimination `mldivide(A,b)`. The Matlab routine was significantly outperformed by our straightforward GPU implementation of the reduced row-echelon Gauss-Jordan algorithm, leading us to believe that the former is not optimally implemented. Figure 23 therefore reports a comparison with the ATLAS implementation of LU factorization instead, an algorithm that is far less memory and computationally demanding. Still, our GPU implementation was faster for matrix sizes of 1200 and larger.

8 Concluding Remarks

In recent years, GPGPU has evolved into a popular research field. Even though not all applications are suited for GPU-based implementations, an impressively wide range of algorithms have already been successfully adapted for GPU-based implementations. However, GPGPU is not yet in widespread use for mainstream applications, due to a somewhat uncommon programming model, and the lack of standardized software libraries, good debugging tools and integrated development environments. There is a lot of progress regarding these issues, but the field needs to mature more before we can expect GPGPU to reach its potential.

Graphics hardware is evolving fast both with respect to performance and functionality. Each generation of new GPUs has improved the programmability in terms of maximal shader length, flow control, and instruction set. This development is likely to continue, and will ease the implementation of GPU-algorithms.

Recent trends in processor design show that the best way to increase performance is by increasing parallelism. This has made standard multi-core processors popular, as well as more specialized processors like Cell processors. We believe that many of the algorithms developed in GPGPU related research can be implemented also for such architectures. In some sense the algorithms for stream processing and GPGPU are providing a glimpse into the coming programming paradigm.

Acknowledgement. This research was funded in part by the Research Council of Norway under grant no. 158911/I30.

References

1. K. E. Batcher. Sorting networks and their applications. In *AFIPS Spring Joint Computing Conference*, pages 307–314, 1968.
2. J. Bolz, I. Farmer, E. Grinspun, and P. Schröder. Sparse matrix solvers on the GPU: conjugate gradients and multigrid. *ACM Trans. Graph.*, 22(3):917–924, 2003.
3. J. Bolz and P. Schröder. Evaluation of subdivision surfaces on programmable graphics hardware. Submitted for publication, 2003.
4. M. Botsch, D. Bommes, C. Vogel, and L. Kobbelt. GPU-based tolerance volumes for mesh processing. In *Proceedings of the 12th Pacific Conference on Computer Graphics and Applications (PG '04)*, pages 237–243, Washington, DC, USA, 2004. IEEE Computer Society.
5. T. Boubekeur and C. Schlick. Generic mesh refinement on GPU. In *HWWS '05: Proceedings of the ACM SIGGRAPH/EUROGRAPHICS conference on Graphics hardware*, pages 99–104, New York, NY, USA, 2005. ACM Press.
6. S. Briseid, T. Dokken, T. R. Hagen, and J. O. Nygaard. Spline surface intersections optimized for GPUs. In V.N. Alexandrov, G.D. van Albada, P.M.A. Sloot, and J. Dongarra, editors, *Computational Science – ICCS 2006: 6th International Conference, Reading, UK, May 28-31, 2006, Proceedings, Part IV*, volume 3994 of *Lecture Notes in Computer Science (LNCS)*, pages 204–211. Springer, 2006.

7. I. Buck, T. Foley, D. Horn, J. Sugerman, K. Fatahalian, M. Houston, and P. Hanrahan. Brook for GPUs: stream computing on graphics hardware. *ACM Trans. Graph.*, 23(3):777–786, 2004.
8. T. Dokken and V. Skytt. Intersection algorithms and CAGD. In *this book.*
9. C. Dyken and M. Reimers. Real-time linear silhouette enhancement. In M. Dæhlen, K. Mørken, and L. Schumaker, editors, *Mathematical Methods for Curves and Surfaces: Tromsø 2004*, pages 135–144. Nashboro Press, 2004.
10. C. Dyken, J. S. Seland, and M. Reimers. Real time silhouette enhancement using graphics hardware. To appear in Computer Graphics Forum.
11. R. Fernando and M. J. Kilgard. *The Cg Tutorial: The Definitive Guide to Programmable Real-Time Graphics.* Addison–Wesley Longman Publishing Co., Inc., 2003.
12. http://graphics.stanford.edu/projects/gpubench.
13. M. Guthe, Á. Balázs, and R. Klein. GPU-based trimming and tessellation of NURBS and T-Spline surfaces. *ACM Trans. Graph.*, 24(3):1016–1023, 2005.
14. D. Göddeke, R. Strzodka, and S. Turek. Accelerating Double Precision FEM Simulations with GPUs In *Proceedings of ASIM 2005 - 18th Symposium on Simulation Technique*, pages 139–144, 2005.
15. T. R. Hagen, M. O. Henriksen, J. M. Hjelmervik, and K.-A. Lie. Using the graphics processor as a high-performance computational engine for solving systems of hyperbolic conservation laws. In *this book.*
16. T. R. Hagen, J. M. Hjelmervik, K.-A. Lie, J. R. Natvig, and M. O. Henriksen. Visual simulation of shallow-water waves. *Simulation Practice and Theory. Special Issue on Programmable Graphics Hardware*, 13(9):716–726, 2005.
17. T. R. Hagen, K.-A. Lie, and J. R. Natvig. Solving the Euler equations on graphical processing units. In V.N. Alexandrov, G.D. van Albada, P.M.A. Sloot, and J. Dongarra, editors, *Computational Science – ICCS 2006: 6th International Conference, Reading, UK, May 28-31, 2006, Proceedings, Part IV*, volume 3994 of *Lecture Notes in Computer Science (LNCS)*, pages 220–227. Springer Verlag, 2006.
18. O. Haugehåtveit, J. M. Hovem, and T. R. Hagen. Calculation of underwater sound fields on GPU. Preprint, 2006.
19. J. M. Hjelmervik and T. R. Hagen. GPU-based screen space tessellation. In M. Dæhlen, K. Mørken, and L. L.Schumaker, editors, *Mathematical Methods for Curves and Surfaces: Tromsø 2004*, pages 213–221. Nashboro Press, 2005.
20. J. M. Hjelmervik, T. R. Hagen, and J. Seland. Shallows, 2005. http://shallows.sourceforge.net.
21. J. M. Hjelmervik and J.-C. Léon. GPU-accelerated shape simplification for mechanical-based applications. Submitted for publication, 2006.
22. J. M. Hovem, H. Dong, and X. Li. A forward model for geoacoustic inversion based on ray tracing and plane-wave refection coefficients. In *Proc. of the 2nd Workshop on Acoustic Inversion methods and Experiments for Assessment of the Shallow Water Environment, Iscia, Italy, 28-30 June*, 2004.
23. T. Kanai and Y. Yasui. Surface quality assessment of subdivision surfaces on programmable graphics hardware. In *Proc. International Conference on Shape Modeling and Applications 2004*, pages 129–136, Los Alamitos, CA, 2004. IEEE CS Press.
24. U. J. Kapasi, W. J. Dally, S. Rixner, P. R. Mattson, J. D. Owens, and B. Khailany. Efficient conditional operations for data-parallel architectures.

In *MICRO 33: Proceedings of the 33rd annual ACM/IEEE international symposium on Microarchitecture*, pages 159–170, New York, NY, USA, 2000. ACM Press.
25. J. Krüger and R. Westermann. Linear algebra operators for GPU implementation of numerical algorithms. *ACM Trans. Graph.*, 22(3):908–916, 2003.
26. A. E. Lefohn, S. Sengupta, J. Kniss, R. Strzodka, and J. D. Owens. Glift: Generic, efficient, random-access GPU data structures. *ACM Transactions on Graphics*, 25(1):60–99, Jan. 2006.
27. C. Loop and J. Blinn. Real-time GPU rendering of piecewise algebraic surfaces. *ACM Transactions on Graphics*, 25(3), 2006.
28. J. D. Owens, D. Luebke, N. Govindaraju, M. Harris, J. Krüger, A. E. Lefohn, and T. J. Purcell. A survey of general-purpose computation on graphics hardware. In *Eurographics 2005, State of the Art Reports*, pages 21–51, Aug. 2005.
29. C. Peeper and J. L. Mitchell. Introduction to the DirectX 9 high level shading language. In W. Engel, editor, *ShaderX2: Introduction and Tutorials with DirectX 9.* Wordware, Plano, Texas, 2003.
30. T. J. Purcell, C. Donner, M. Cammarano, H. W. Jensen, and P. Hanrahan. Photon mapping on programmable graphics hardware. In *Proceedings of the ACM SIGGRAPH/EUROGRAPHICS Conference on Graphics Hardware*, pages 41–50. Eurographics Association, 2003.
31. R. J. Rost. *OpenGL Shading Language.* Addison–Wesley Longman Publishing Co., Inc., 2004.
32. M. Rumpf and R. Strzodka. Using graphics cards for quantized FEM computations. In *Proceedings of IASTED Visualization, Imaging and Image Processing Conference*, pages 193–202. 2001.
33. J. Seland and T. Dokken. Real-time algebraic surface visualization. In *this book.*
34. D. Shreiner, M. Woo, J. Neider, and T. Davis. *OpenGL Programming Guide: The Official Guide to Learning OpenGL, Version 2.* Addison–Wesley Longman Publishing Co., Inc., 2005.
35. X.-C. Tai, S. Osher, and R. Holm. Image inpainting using a TV-Stokes equation. In X.-C. Tai, K.-A. Lie, T. F. Chan, and S. Osher, editors, *Image Processing Based on Partial Differential Equations*, pages 3–22. Springer, 2007.
36. D. Tarditi, S. Puri, and J. Oglesby. Accelerator: simplified programming of graphics processing units for general-purpose uses via data-parallelism. Technical report, Microsoft Research, December 2005.

Real-Time Algebraic Surface Visualization

Johan Simon Seland and Tor Dokken

Summary. We demonstrate a ray tracing type technique for rendering algebraic surfaces using programmable graphics hardware (GPUs). Our approach allows for real-time exploration and manipulation of arbitrary real algebraic surfaces, with no pre-processing step, except that of a possible change of polynomial basis.

The algorithm is based on the blossoming principle of trivariate Bernstein-Bézier functions over a tetrahedron. By computing the blossom of the function describing the surface with respect to each ray, we obtain the coefficients of a univariate Bernstein polynomial, describing the surface's value along each ray. We then use Bézier subdivision to find the first root of the curve along each ray to display the surface. These computations are performed in parallel for all rays and executed on a GPU.

Key words: GPU, algebraic surface, ray tracing, root finding, blossoming

1 Introduction

Visualization of 3D shapes is a computationally intensive task and modern GPUs have been designed with lots of computational horsepower to improve performance and quality of 3D shape visualization. However, GPUs were designed to only process shapes represented as collections of discrete polygons. Consequently all other representations have to be converted to polygons, a process known as *tessellation*, in order to be processed by GPUs. Conversion of shapes represented in formats other than polygons will often give satisfactory visualization quality. However, the tessellation process can easily miss details and consequently provide false information.

The qualitatively best shape visualization is currently achieved by *ray tracing*; for each pixel in the image plane, straight lines from the center of projection through the pixel are followed until the shape is intersected. In these points shading is calculated and possibly combined with shading information calculated from refracted and reflected continuations of the line. This process

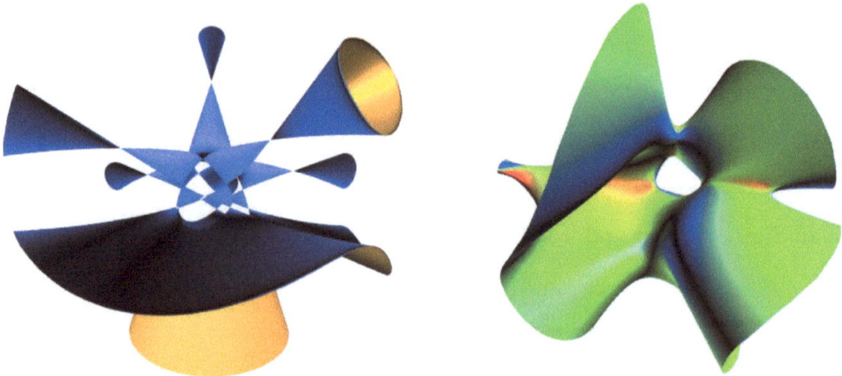

Fig. 1. A quintic surface with 31-double points, also known as a Dervish (left), and a randomly generated degree 4 surface with curvature visualization (right). Both surfaces above are rendered at interactive frame rates. We emphasize that no tessellation has taken place.

is computationally intensive and time consuming, and in general it is not possible to use ray tracing to achieve interactive frame rates.

For a function $f : \mathbb{R}^3 \to \mathbb{R}$, an implicit surface can be defined by the level set of the equation $f(x,y,z) = c$, where $x,y,z,c \in \mathbb{R}$. By reordering the terms, we can, without loss of generality, just consider the zero-set of the function, e.g., $f(x,y,z) - c = 0$. For this work, all functions and surfaces are considered to be real-valued. If the function f is an algebraic polynomial, the resulting surface is called an *algebraic surface*. Such surfaces can easily describe intricate, smooth shapes with varying topology, and also allow for the interpolation and blending of such surfaces. Their polynomial nature also makes it easy to find analytic directional derivatives, normals and curvature. The standard example of an algebraic surface is that of a sphere with radius r centered at the origin:

$$x^2 + y^2 + z^2 - r^2 = 0.$$

This paper proposes a method for high quality, high performance, ray tracing type rendering of algebraic surfaces using GPUs. The approach guarantees that the topology of the visualized surface is correct within pixel accuracy. The current prototype can be used for real-time visualization of algebraic surfaces of total degree five, the limitation being the current number of registers in each fragment processor and not the floating point capabilities of the GPU.

The approach aims at algebraic surfaces of degree 3 and higher, as conversion of these to high quality tessellations is a very challenging task. Such conversion requires detailed knowledge of the topology of the algebraic surface. For quadratic algebraic surfaces the traditional classification into spheres, cylinders, cones, ellipsoids, hyperboloids, as well as the degenerate cases aids the process of generating tessellations. Cubic algebraic surfaces can be classified into 45 classes [19], but for higher degrees no such useful classification exists. However, our method is applicable also for quadratic and cubic algebraic surfaces.

This paper is organized as follows: We begin with an overview of related work in Section 2. Thereafter we give a short, informal description of our method, focusing on how the GPU pipeline can be used for ray tracing. In Section 4 we describe details of our GPU implementation, as well as performance benchmarks of our method compared to state-of-the-art CPU ray tracing for algebraic surfaces. To avoid adding too much mathematical detail to Sections 3 and 4, we present the basic ideas first, using forward references to equations, which are presented in Section 5 together with more mathematical background. Finally we conclude and present some thoughts on future work in Section 6.

2 Related Work

Using the GPU for non-graphical simulations is an emerging field, called GPGPU (General-Purpose Computing on Graphics Processing Units). A survey of recent activity is given by Owens et al. [30]. The paper by Dokken, Hagen, and Hjelmervik [8] in this book details activity at SINTEF and programming techniques for GPUs.

Since programming GPUs is rather complex, several systems have been developed to use the CPU for automatic generation of shader code. A group at the University of Waterloo has developed Sh [26], which is a metaprogramming language that uses the syntax of C++ for the development of shader programs for different generations of GPUs. A group at Stanford University has developed Brook for GPUs [3], which is designed for using the GPU as a general stream processor.

Ray tracing on GPUs has seen a flurry of activity in recent years. A CPU simulation of GPU ray tracing was implemented by Purcell et al. [31] when programmable graphics hardware was starting to appear. Later the master theses by Karlsson and Ljungsted [21] and Christen [6] describe concrete implementations running on production hardware. Thrane and Simonsen [40] compare acceleration structures for GPU based ray tracing, and conclude that bounding volume hierarchies (BVH) is the best technique. Carr et al. [5] accelerate the BVH approach by storing it on the GPU as a *geometry image* [17].

Displacement mapping on GPUs has also been implemented by using ray tracing like techniques. Hirche et al. [18] displace a triangle mesh along the normal direction by replacing base mesh triangles with triangles that cover the area on the screen that would be affected by the displaced base mesh. Donnelly [10] uses sphere tracing to effectively implement bounded displacement mapping using fragment shaders.

There has also been a few recent publications about dedicated hardware for ray tracing: Carr et al. [4] describe the necessary modifications to existing GPU designs to support ray tracing. Schmittler et al. [35] describe a dedicated FPGA chip. However none of these designs are yet in commercial production. Using dedicated graphics hardware for ray tracing is however not a new idea, Olano et al. [28, 29] used the PixelFlow system developed at the University of North Carolina, to render spherical primitives in the mid nineties.

GPU assisted rendering of implicit surfaces (metaballs) has been demonstrated by Uralsky [41], using the geometry shader which will be made available on next generation (DX10) GPUs. However, his approach uses the GPU to tessellate the surface to discrete geometry in the classical way, by using a marching-cubes like algorithm, whereas our method avoids the tessellation step completely. Rendering of implicit curves using GPUs has been demonstrated by Loop and Blinn [22]. Their approach was based on rendering the inside of a curve defined over a triangle for use in font rendering. Our work can be seen as an extension of their approach to surfaces. During the later stages of our research, Loop and Blinn [23] published their extension to surfaces, which is very similar to our approach. However, they use analytic root finding algorithms, thereby limiting the maximum degree of surfaces they can visualize to general quartic surfaces, as it is not possible to analytically solve general equations of degree 5 and higher.

One existing comparable software package is also worth mentioning, namely the package Surfex [20], which uses the (CPU-based) ray tracer SURF [12] to visualize algebraic surfaces. We will provide some run-time comparisons with the SURF ray tracer in Section 4.5.

3 Overview of the Algorithm

Our algorithm is based on the classical ray tracing approach, where rays pass from the eye, and we wish to find the location of each ray's first intersection with the surface. For a given algebraic surface $f(x, y, z) = 0$ of total degree d, we are only interested in intersections inside a view volume V. Hence, it is natural to only consider the segment of each ray inside V. Let $\mathbf{f}$ and $\mathbf{b}$ represent the front and back coordinate of the ray as it enters and leaves V. The segment is then given by the parametrization $\mathbf{p}(t) = (1-t)\mathbf{f}+t\mathbf{b}$, $t \in [0, 1]$. We combine this parametrization with the equation of the algebraic surface, yielding a univariate polynomial $g(t) = f(\mathbf{p}(t))$ of degree d. Now the problem of finding the closest intersection to the eye point, can be reformulated to finding the smallest $t \in [0, 1]$ such that $g(t) = 0$.

Our algorithm is run fully on a GPU, and heavily exploits the floating point power and programmable capabilities of modern GPUs. Although we use the GPU to render graphics, our approach does not use the GPU pipeline in the

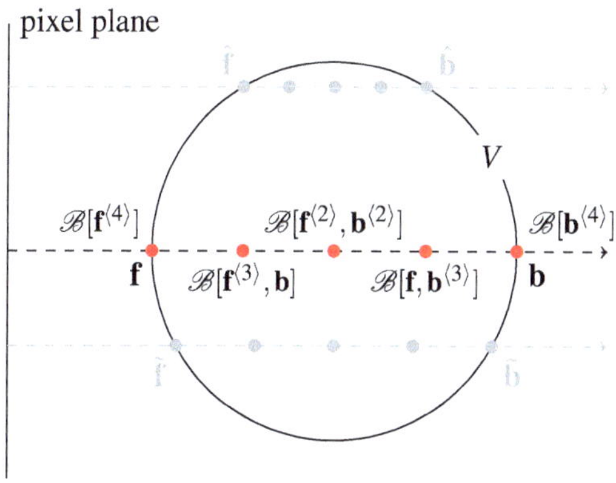

Fig. 2. For each ray segment inside the view volume V, we find the front ($\mathbf{f}$) and back ($\mathbf{b}$) vectors in barycentric coordinates. Then we use blossoming to find the coefficients of the univariate Bernstein-Bézier function along each ray. Since each ray corresponds to a pixel, and there is no interdependency between rays we can calculate all rays in parallel on a GPU.

classical way. In fact, our use of the graphics pipeline is more closely related to GPGPU techniques than to classical computer graphics. In the following, we give a short summary of the rendering pipeline of a GPU, highlighting just the stages relevant for our method.

- To initiate rendering, the CPU sends geometric primitives to the GPU as a set of triangles.
- Each vertex is thereafter processed in the *vertex processor*. The vertex processor may change the position of a vertex and also set up other variables that will be interpolated across the triangle (e.g., normals, colors etc.).
- Thereafter the *rasterizer*-stage generates a set of fragments (meta-pixels) from the associated geometry.
- The final stage is the *fragment processor*, which calculates the final color and depth of the pixel. The fragment processor cannot change the position of the pixel, but it may discard the pixel if it is not of interest.

Both the vertex and fragment processors can be programmed by the use of *shading languages*, and such programs are called *shaders*. High end GPUs are highly parallel and typically have 6-8 *vertex pipelines* and 16-48 *fragment pipelines* operating in parallel.

Our use of the GPU can then be summarized as follows

1. Issue *spanning geometry* to the GPU. The spanning geometry defines our view volume V, and initiates rendering.
2. Let the GPU rasterize the spanning geometry into a set of fragments, yielding the coordinates $\mathbf{f}$ and $\mathbf{b}$ for each ray.
3. Find the univariate polynomial, $g(t)$, representing the value of the algebraic surface along each ray.

4. Find the smallest t such that $g(t) = 0$. If no zeros are found, we discard the fragment.
5. For the surviving fragments, calculate the final depth and color by using information from the algebraic surface.

Steps 3 to 5 above are executed in the fragment processor of the GPU, and require the activation of a special shader program, which is outlined in Listing 1, and discussed further in Section 4.

As GPUs are limited to single precision floating point arithmetic, extreme care has to be taken with respect to representation formats and algorithmic components to guarantee the quality of the ray tracing. Our approach uses Bernstein bases both for representing the (trivariate) algebraic surfaces, see Section 5.1, and for representing the (univariate) polynomials describing the intersection of the ray segments of interest with the algebraic surface, see Section 5.2. The choice of the Bernstein bases is based on its numerical superiority compared to other polynomial bases for algorithms run in single precision arithmetic. It can be shown that the Bernstein basis is optimal both for univariate and multivariate polynomials, see[14, 24]. Some more details and numerical examples are given in the paper by Dokken and Skytt [9] in this book.

Another advantage is the possibility of representing coordinates in *barycentric coordinates* (see Section 5). This allows us to use a technique know as *blossoming* (see Section 5.3) to find an expression for the surface values along each ray (Item 3 above). Blossoming can be seen as a generalization of the *de Casteljau algorithm*, which is used for evaluating Bézier functions. Both of these algorithms are numerically very stable and effective. Barycentric coordinates over a domain in $\mathbb{R}^3$ are represented as 4-component vectors, which fit neatly on GPUs, as they are designed to operate on 4-component vectors.

The de Casteljau algorithm is however not optimal for evaluating a polynomial, as it uses $\mathcal{O}(d^2)$ floating point operations, whereas Horner's scheme just uses $\mathcal{O}(d)$ operations for an evaluation. However, numerical errors are much more likely to occur using Horner's scheme which might lead to visual artifacts.

4 GPU Implementation

We will now examine in detail how we have implemented our algorithm on programmable GPUs, using the OpenGL Shading Language. Other shading languages such as Cg or HLSL are equally well suited, since there are only minor differences between them.

Some more terminology from shader programming is necessary to describe our method. Shader programs are written in a C-like programming language, and compiled at run-time by the graphics driver. The host CPU may set some variables of the shader program at run-time. These are shared by all

Listing 1. The main steps of our fragment shader. This code is executed in parallel for all rays, and run completely on a GPU.

```
uniform int degree;

void main() {
  vec4 f = barycentric_front ();
  vec4 b = barycentric_back();
  float c[degree+1]; // Control points
  for(int i = 0; i < degree; ++i)
      c[i] = blossom(f, b, i);
  if (number_of_sign_changes(c) == 0)
      discard;
  float t = find_leftmost_zero(c);
  vec4 p = (1-t)*f + t*b;
  gl_FragColor = shade_and_light(p);
  gl_FragDepth = to_cartesian(p).z;
}
```

the pipelines of the GPU and are called *uniform variables*, since they have a uniform value over each primitive. Communication between the vertex and fragment shader programs takes the form of *varying variables*, which are interpolated (and thus varying) across geometric primitives. Uniform variables can only be read by the shader program and can be seen as global constants. The vertex processor may only write varying variables, and the fragment processor can only read varying variables. There can also be local variables for each vertex/fragment which the shader program can read and write at will. However their number is limited and there exist no stack or heap memories, so recursion and pointers are not available. For an in-depth overview of GPU programming see any book on shading languages, e.g., Rost [34] or Fernando and Kilgard [15].

This section uses forward references to equations that will be detailed in Section 5, as all mathematical prerequisites are collected there.

4.1 Issue Spanning Geometry

To initiate our algorithm, we must send geometry encompassing the barycentric domain to the GPU. Theoretically any convex shape will suffice, but in practice shapes for which it is trivial to calculate the barycentric front and back coordinates for each ray are best suited.

The easiest spanning geometry of all is the tetrahedron, where the barycentric coordinates can be encoded as the color property of each vertex. Linear interpolation of color across each face will then yield the correct coordinate.

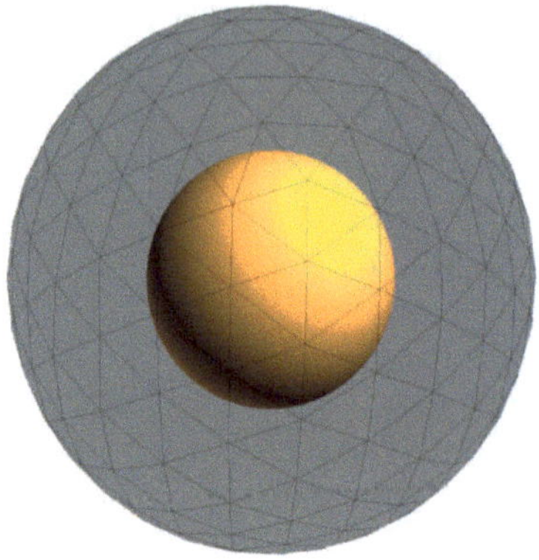

Fig. 3. The view volume (black edges) is all that is sent to the GPU, and the rasterization of the view volume geometry initiates our ray tracing type rendering.

However, visualizing an algebraic surface inside a tetrahedron clips the surface in an unnatural way, and gives a poor impression of the surface to a human observer. Instead, we have found the sphere to be a good spanning geometry. By issuing a moderately tessellated sphere and linearly interpolating the analytic normal at each vertex over each face, and thereafter using (3) to get the geometric normal (scaled by the radius) yields a good approximation to the barycentric coordinate at a pixel. Figure 3 above illustrates how the spanning geometry corresponds to the view volume.

To calculate the back coordinate $\mathbf{b}$, several methods can be used. We usually render the same bounding geometry, using front face culling in a separate render pass, and store the value in a texture. For simple shapes, like a sphere, it is possible to use standard intersection equations to calculate $\mathbf{b}$. It is also possible to set up tetrahedrons such that both coordinates may be calculated in the same pass, as demonstrated by Weiler et al. [42] or Wylie et al. [43].

4.2 Multivariate Blossoming on GPUs

The most computationally demanding subtask of the visualization is the calculation of the coefficients of the curve representing the value of f along each ray. We find these coefficients by blossoming and form an algorithm based on (9).

If recursion had been supported by the GPU, we could have implemented (9) directly. However, current generation GPUs do not support recursion, so for a given degree d, we use the CPU to unroll the recursion and end up with a long expression (of shader source code) representing the blossom, and use uniform variables to represent the coefficients of the surface.

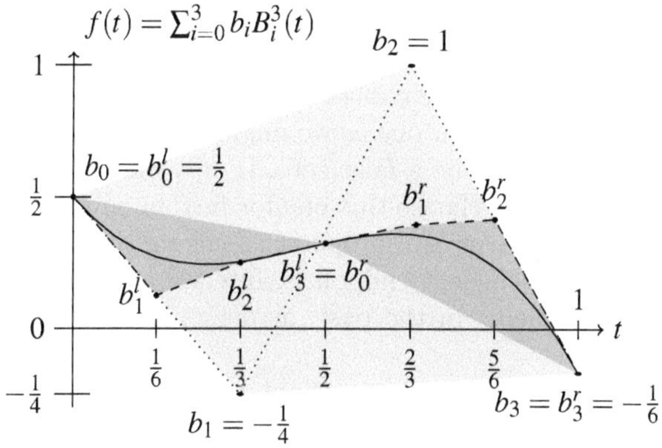

Fig. 4. A cubic Bézier curve with its original control polygon (dotted) and the control polygon after one iteration of Bézier-subdivision (dashes). The results of the subdivision are two new sets of control points, each covering half of the interval of the original control polygon. Observe that the dotted control polygon changes sign more often than $f(t)$. This false crossing vanishes after one iteration of subdivision. Also note that the curve is completely contained within the convex hull of the control points.

Since the blossom is recalculated for every frame, we can easily modify the surface by altering the coefficients, allowing for real-time shape manipulation.

As the degree of the surface increases, the complexity of the generated shader source code increases dramatically, and we have triggered several bugs in the shader compilers while developing this algorithm. Higher degrees also require more registers for intermediate calculations, and this is the limiting factor on the degree of surfaces we are able to visualize.

DeRose et al. [7] describe optimizations to the recursive approach for calculating the blossom. Relevant to our method is to reuse partial blossom calculations, as we calculate the control points. We implemented this on the GPU, but the added bookkeeping actually made our algorithm slower, even if we were able to reduce the amount of floating point computations, demonstrating the performance penalty of executing branching code on GPUs.

4.3 Polynomial Root Finding on GPUs

Root finding for polynomials in Bernstein form has been explored in detail by Spencer [39]. Recent work by Mørken and Reimers [27] demonstrates a totally convergent root finding algorithm with quadratic convergence based on knot insertion of B-splines. Unfortunately, modern root finders require data structures such as linked lists that are not yet suitable for GPUs.

Because of its simple data structure, we decided to use a heuristic method based on recursive subdivision of the coefficients, as described by Schneider [36]. The idea is to recursively subdivide the polynomial in half by using the de Casteljau algorithm (see Section 5.2). At each level of recursion we count the number of sign changes in the coefficients, and if both sets of coefficients contain sign changes we choose to continue to subdivide the leftmost of them,while storing the other one as we might need to unwind the recursion if the first one turns out to be a false zero. If only one set of coefficients has sign changes, we of course choose this one for further subdivision. This effectively uses the Bézier convex hull property to ensure that a zero is within an interval given by the coefficient endpoints. Interval arithmetic has been used successfully for ray tracing in the past, as demonstrated by Snyder[38] and Duff [11].

Rockwood et al. [33] present a variant of the above, where the Bézier curve is subdivided at the leftmost sign change of the control polygon. While marginally faster than Schneider's method, it has some problems near singularities when executed on a GPU, leading to visual artifacts.

Since it could happen that there is a false zero (as in Figure 4), we must be able to unwind the recursion and choose another interval to subdivide. On a regular CPU this would be trivial, but as the GPU lacks a stack we store the other potential coefficients and the recursion level, and when we detect a false zero we unroll to the last set of potential coefficients. The number of possible false zeros is bounded by the degree of the surface, so we know in advance how many sets of coefficients we might need to store. We terminate the recursion when we can decide that there is no root (no sign changes of the coefficients), or when we reach a fixed level of subdivisions.

When implementing CPU-based root finders, it is usually effective to have early termination criteria, such as the "flatness" test described by Schneider. Since such tests lead to more complex branching structures, we have found that such tests actually hamper performance on current generation GPUs.

Since we only need to find the roots of univariate polynomials, where the derivatives can easily be computed, one might believe that Newton-like methods would perform extremely well. However, we are plagued with the classical problem of such methods; a good initial value must be found in order for the method to converge. Using the control polygon's first intersection as the initial value for a naive Newton iteration yields correct zeros for the majority of cases, and more than triples the FPS quoted in Table 1. However, the relatively few incorrect zeros yield glaring visual artifacts and a wrong topology of the surface. We believe that significant speed increases are possible by using hybrid Newton-subdivision methods when dynamic branching of GPUs improves.

The convergence of the coefficients of a Bézier curve is $\mathcal{O}(h^2)$, where h is the interval length. If the original Bézier curve was defined over the interval $[0,1]$, ten levels of subdivision will give an interval length of $2^{-10} = \frac{1}{1024}$, with the maximal distance between the control polygon and the curve reduced by a factor of $2(\frac{1}{1024})^2 \approx 0.9536\,7 \times 10^{-6}$. This is approximately the accuracy of numbers in single precision floating point format used by the GPU. Consequently the control polygon will be near indistinguishable from the curve. A typical displayed image is currently around 1000×1000 pixels, thus ten levels

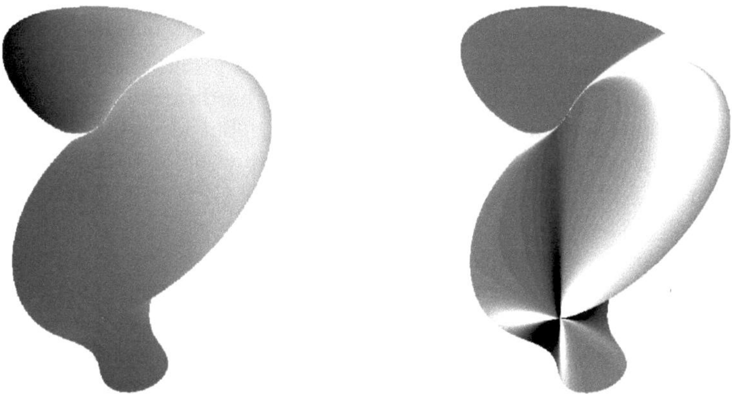

Fig. 5. A quartic surface where all first order derivatives vanish, leading to a breakdown in the lighting equation (which is based on calculating the normal by taking the cross product of derivatives). This breakdown is clearly visible in the left image, where the shape looks flat and uninteresting. The right image shows the curvature field, clearly depicting the singularity which is invisible in the first image.

of subdivision will give the same resolution orthogonal to the image plane as in the image plane.

4.4 Curvature and Lighting Calculations

When the intersection has been found by the root finder, we know its position t within the univariate parameter domain. The resulting position in the trivariate domain can then easily be found by calculating $\mathbf{u} = (1-t)\mathbf{f} + t\mathbf{b}$, where $\mathbf{f}$ and $\mathbf{b}$ are the front and back vectors, respectively. Furthermore we can use the blossom (see (11)) to calculate the directional derivatives and obtain the (barycentric) normal vector of the surface. Various per-pixel lighting models based on the normal (or higher order derivatives) can then be trivially implemented. This generally works very well for most surfaces, however, when all the directional derivatives vanish the equations break down, as illustrated in Figure 5.

Calculating the mean curvature of an implicit surface is possible, although the formulas become quite complex as they involve all second derivatives of the surface. However, the blossom can be used to find all these terms, and repeated use of (12) gives us the terms necessary to calculate the curvature, as illustrated by the right surface in Figure 1. The performance impact of this curvature calculation is negligible in comparison to regular shading.

4.5 Performance

We have implemented our algorithm using C++ and the OpenGL Shading Language using Nvidia 6000 and 7000-series hardware. Current ATI drivers have problems compiling our shader-programs, but we believe that once their drivers improve, our algorithm will be equally well (or better) suited for execution on ATI hardware, as present day ATI GPUs are better at dynamic branching.

For speed comparisons, we have compared rendering speeds of the same surface with the SURF ray tracer and our GPU implementation. We have benchmarked using surfaces with the highest known number of ordinary double points (singularities) for a given degree[1], thus stressing the root finder. We have also tested with random surfaces, where are coefficients are in the range $(-1, 1)$. All benchmarks have been executed on an AMD X2 4400 CPU with a Nvidia Geforce 7800 GT GPU, running Windows XP and Nvidia driver version 91.31. Moreover, both ray tracers were set up to use a single light source, the same resolution (966×892, the viewport of our prototype implementation when running in full-screen) and no anti-aliasing. The rendered surfaces also fill approximately the same number of pixels and the surfaces are oriented in the same direction. We emphasize that there is a strict correspondence between the resolution and the frame rate, as performance is directly affected by the number of pixels rendered. Since Loop and Blinn [23] do not provide any table of performance, a direct comparison is not possible. However, they use a tensor product formulation of the blossom, which has high computational complexity ($\mathcal{O}(d^4)$), but leads to very effective shader source code (a series of nested multiply-and-add (MAD) operations). The recursive approach has computational complexity $\mathcal{O}(d^3)$, but is not as GPU friendly as the tensor formulation. For lower degrees it is apparent that using the tensor product formulation is more effective, the crossing point is probably around degree 5, for which [23] is no longer applicable as they use an analytic root finder, which only exist in general for degree 4 and lower.

SURF is specially designed for rendering algebraic surfaces, and is significantly faster for such surfaces than the better known ray tracer POV-Ray. We have chosen to only benchmark against the fastest CPU solution available.

Table 1 summarizes our findings. We observe that the GPU based approach is significantly faster than SURF, but the difference decreases as the degree increases. The quintic Dervish surface (illustrated in Figure 1 (left)) is just 2.7 times faster on the GPU, while our algorithm is an order of magnitude faster for quartic surfaces. Since the computational load increases with each degree, one should expect the difference to be larger. However, the problem of a highly parallel GPU, is that the pixel pipelines must be synchronized, thus making early termination and adaptivity in the root finder almost useless. If one pixel requires recursion to the bottom, all pixels in the pipeline must wait until it is finished. Since higher degrees require more computations, this penalizes higher degrees more than the lower ones, as more computations must

Table 1. Speed comparisons of the CPU based ray tracer SURF and our GPU implementation. Resolution is 966×892 for all benchmarks. The quoted GPU model is Nvidia 7800 GT. The named surfaces are the surfaces with the highest known number of singularities for their degree.

Surface	Degree	SURF (FPS)	GPU (FPS)	Speedup	Dual GPU (FPS)	Speedup
Sphere	2	2.6	60.0	23.1	108.0	41.5
Cayley	3	1.3	22.1	16.9	38.0	29.2
random	3	1.3	27.6	21.2	45.0	34.6
Kummer	4	1.1	9.2	8.3	18.2	16.5
random	4	1.0	12.2	12.2	22.3	22.3
Dervish	5	1.2	3.2	2.7	6.4	5.3
random	5	0.9	3.9	4.3	7.6	8.4

be executed while the rest of the pixels are waiting. Future GPUs are expected to improve at dynamic branching, and we expect huge speed increases when more adaptive root finding algorithms become available.

SURF is also able to exploit sparse polynomials, which our GPU approach fails to do. This is especially apparent for random surfaces, where all coefficients are non-zero. SURF is actually slower on a random quartic surface than on the relatively sparse quintic Dervish surface. In general, for our GPU approach, we observe little difference in rendering speed between surfaces of the same degree, but random surfaces are in general a little bit faster than the ones quoted in Table 1, as the surfaces in the table are chosen for their complexity. While sparse polynomials are important for studying algebraic surfaces, industrial use of algebraic surfaces will most often address surfaces with most coefficients non-zero. Consequently for industrial users the timings for sparse polynomials have little interest, while they are relevant within algebraic surfaces research.

As there is no interdependency between render passes or neighboring pixels, our algorithms are very well suited for execution on multi-GPU based systems, such as ATI CrossFire or Nvidia SLI. We effectively double the performance when using two Nvidia 7800 GT cards in *Split Frame Rendering* (SFR) mode.

In conclusion we emphasize that our GPU approach is still more than twice as fast as SURF for a quintic surface. For lower degrees the GPU based approach is almost an order of magnitude faster than the CPU based approach. In addition the CPU is left free to do other tasks while rendering. However, to effectively render models represented by a large number of piecewise algebraic surfaces, one would need to render a large number of primitives, probably including efficient culling strategies, and rendering speed must be increased further for this to be feasible in real time. The current rendering speeds for quartic and quintic surfaces can however hardly be called real time, but it is still possible to interactively alter and explore the surface, and future GPUs are expected to increase speeds even further.

5 Bernstein-Bézier Techniques

This section presents the necessary mathematical background material to understand our method. The material on blossoming is quite abstract, and comes from spline and approximation theory. However, a deep understanding of this material is not crucial to understand our method. The interested reader may consult Ramshaw [32] or Farin [13] for good and thorough introductions to the material. Statements are given without proofs, as they can be found in the above texts.

For compactness, we use standard multi-index notation. Thus for tuples $\mathbf{j} = (j_1, \ldots, j_n)$, we let $\mathbf{j} \pm \mathbf{k} = (j_1 \pm k_1, \ldots, j_n \pm k_n)$, $|\mathbf{j}| = \sum_{i=1}^{n} j_i$, $\mathbf{j}! = \prod_{i=1}^{n} j_i!$ and $\mathbf{x}^{\mathbf{j}} = \prod_{i=1}^{n} x_i^{j_i}$. In the following, $\mathbf{x}$ will always denote a point in Cartesian coordinates, while $\boldsymbol{\beta}$ denotes a point in barycentric coordinates.

5.1 The Bernstein Basis

For the space of real polynomials of total degree d,

$$P_{d,s} := P_d(\mathbb{R}^s) = \left\{ p(\mathbf{x}) = \sum_{|\mathbf{j}| \le d} c_{\mathbf{j}} \mathbf{x}^{\mathbf{j}} : c_{\mathbf{j}} \in \mathbb{R} \right\}, \tag{1}$$

it is well known that they can be expressed in the Bernstein-Bézier (BB) basis, which is preferable for use in computations since it is more numerically stable.

The Bernstein basis functions $B_{\mathbf{i}}^d$ of degree d can be expressed as:

$$B_{\mathbf{i}}^d(\boldsymbol{\beta}) = \frac{d!}{\mathbf{i}!} \boldsymbol{\beta}^{\mathbf{i}}, \quad |\mathbf{i}| = d, \quad |\boldsymbol{\beta}| = 1. \tag{2}$$

Here, $\boldsymbol{\beta} = \boldsymbol{\beta}(\mathbf{x}) = (\beta_1(\mathbf{x}), \ldots, \beta_{s+1}(\mathbf{x}))$ denotes the barycentric coordinate of a point $\mathbf{x} \in \mathbb{R}^s$ with respect to a set of base points, $(\mathbf{v}_1, \ldots, \mathbf{v}_{s+1}) \in \mathbb{R}^s$, which form a non-degenerate simplex $\Sigma_s = \mathrm{conv}(\mathbf{v}_1, \ldots, \mathbf{v}_{s+1})$. For instance, the 0-simplex is a point, the 1-simplex a line segment, the 2-simplex is a triangle and the 3-simplex a tetrahedron.

Conversion between barycentric and Cartesian coordinates (and vice versa) is uniquely given by the relation:

$$\boldsymbol{\beta} = \boldsymbol{\beta}(\mathbf{x}) = A^{-1} \begin{pmatrix} \mathbf{x} \\ 1 \end{pmatrix}, \quad A = \begin{pmatrix} \mathbf{v}_1 & \ldots & \mathbf{v}_{s+1} \\ 1 & \ldots & 1 \end{pmatrix}. \tag{3}$$

The matrix A above does not depend on a particular point, and can thus be calculated (along with its inverse) once for a given domain, Σ_s.

The barycentric coordinates form a partition of unity, $\sum_{i=1}^{s+1} \beta_i = 1$, so even if we have increased the number of variables by one, we do not increase the dimension of the polynomial space. For a point in Σ_s, its barycentric coordinates are non-negative, which ensures that all computations are done by true convex combinations, yielding high numerical stability.

For illustration purposes, we expand the above notation for the trivariate case and use the standard unit simplex ($\mathbf{v}_1 = (1,0,0)^T$, $\mathbf{v}_2 = (0,1,0)^T$, $\mathbf{v}_3 = (0,0,1)^T$ and $\mathbf{v}_4 = (0,0,0)^T$). This gives the barycentric coordinate $\boldsymbol{\beta} = \big(\beta_1 = x,\ \beta_2 = y,\ \beta_3 = z,\ \beta_4 = w = (1-x-y-z)\big)$.

$$\begin{aligned} B^d_{i,j,k,\ell}(\boldsymbol{\beta}(x,y,z)) = B^d_{i,j,k,\ell}(x,y,z,w) &= \frac{d!}{i!j!k!\ell!} x^i y^j z^k w^\ell, \\ \ell = d-i-j-k. & \end{aligned} \tag{4}$$

For an algebraic polynomial p of total degree d, we now have two alternative formulations, using either the power or the Bernstein basis:

$$p(\mathbf{x}) = \sum_{|j| \le d} c_{\mathrm{j}} \mathbf{x}^j = \sum_{|\mathrm{i}|=d} b_{\mathrm{i}} B_{\mathrm{i}}(\boldsymbol{\beta}(\mathbf{x})) = f(\boldsymbol{\beta}), \quad \mathbf{x} \in \Sigma_s, \tag{5}$$

where b_{i} are called the BB-coefficients (Bernstein-Bézier) of the function. In the following we let $f(\boldsymbol{\beta})$ be any polynomial in Bernstein form.

We also summarize some key properties of the Bernstein basis:

Lemma 1. *On the unit simplex, the following hold:*

1. $B^d_{\mathrm{i}}(\boldsymbol{\beta}(\mathbf{x})) \ge 0$ *for all* $\mathbf{x} \in \Sigma_s$.
2. $\sum_{|\mathrm{i}|=d} B^d_{\mathrm{i}}(\boldsymbol{\beta}(\mathbf{x})) = 1$ *for all* $\mathbf{x} \in \mathbb{R}^s$.
3. $\frac{d!}{(d-|\mathbf{r}|)!}\mathbf{x}^{\mathbf{r}} = \sum_{|\mathrm{i}|=d} \frac{\hat{\mathbf{i}}}{(\hat{\mathbf{i}}-\mathbf{r})!} B^d_{\mathrm{i}}(\boldsymbol{\beta}(\mathbf{x})), \quad |\mathbf{r}| \le d,\ \hat{\mathbf{i}} = (i_1, \ldots, i_s).$

For a proof see [25]. The above lemma provides us with the foundations for a change of basis algorithm between the power and Bernstein bases.

5.2 Polynomials in Bernstein Form

Given an algebraic polynomial in Bernstein-Bézier form, the de Casteljau algorithm provides a fast and numerically stable way of evaluating the function. Let b_{i} denote the coefficients of an algebraic function in BB-form and compute repeated convex combinations of the base points by setting:

$$\begin{aligned} b^r_{\mathrm{i}}(\boldsymbol{\beta}) = \sum_{j=1}^{s+1} \beta_j b^{r-1}_{\mathrm{i}+\mathbf{e}_j}(\boldsymbol{\beta}), \quad b^0_{\mathrm{i}}(\boldsymbol{\beta}) = b_{\mathrm{i}}, \\ \mathbf{e}_1 = (1,0,\ldots,0),\ \ldots,\ \mathbf{e}_{s+1} = (0,\ldots,1). \end{aligned} \tag{6}$$

Then $b^d_{\mathbf{0}}(\boldsymbol{\beta})$ is the value of the polynomial at the parameter value $\boldsymbol{\beta}$ in a non-degenerate simplex. The de Casteljau algorithm can be seen as a pyramid-like

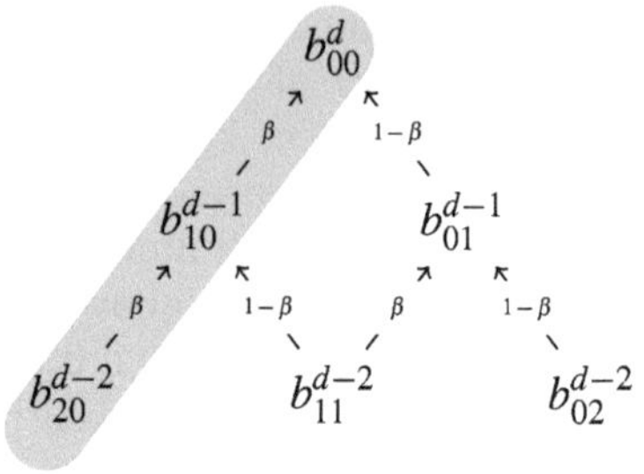

Fig. 6. The de Casteljau algorithm for a univariate BB-polynomial. Each value is formed by a convex combination of the values on a lower level. At the lowest level we find the BB-coefficients of the polynomial. The shaded area indicates the coefficients for the subdivided polynomial in the interval $[0, \beta]$.

structure, where each level is a convex combination of $s+1$ values at the level below, starting with the coefficients. This process is illustrated for a univariate polynomial in Figure 6.

The de Casteljau algorithm also admits a *subdivision formula* for univariate BB-polynomials, and hence allows us to find a formula for a polynomial in BB-form over a sub-interval of the original polynomial. To find the coefficients of a BB-polynomial over the interval $[0, \beta]$, we carry out (6) and store the leftmost intermediate value at each iteration:

$$c_{\mathbf{i}} = b_{\mathbf{0}}^{j}(\beta). \tag{7}$$

Conversely, the rightmost values give the coefficients for the BB-polynomial over the interval $[\beta, 1]$. For multivariate polynomials, blossoming (see Section 5.3 below), can be used to find the coefficients over a subset of the domain.

If we use the BB-coefficients of a univariate BB-polynomial f to form the points $\mathbf{b}_{\mathbf{i}} = \mathbf{b}_{i,j} = (j/d, b_{\mathbf{i}})$ we find the *control points* of a Bézier parametrization of f, often called a Bézier curve. The polygon formed by the control points is called the control polygon of the curve.

Such a parametrization has a number of useful properties:

Lemma 2.

1. *Bézier curves have no more sign changes than their control polygons. This is called the* variation diminishing property.
2. *The control points generated by repeated subdivision of the control polygon converge to the original Bézier curve.*
3. *The entire Bézier curve is contained in the convex hull of the control points.*

The above properties are used to form the root finding algorithm described in Section 4.3.

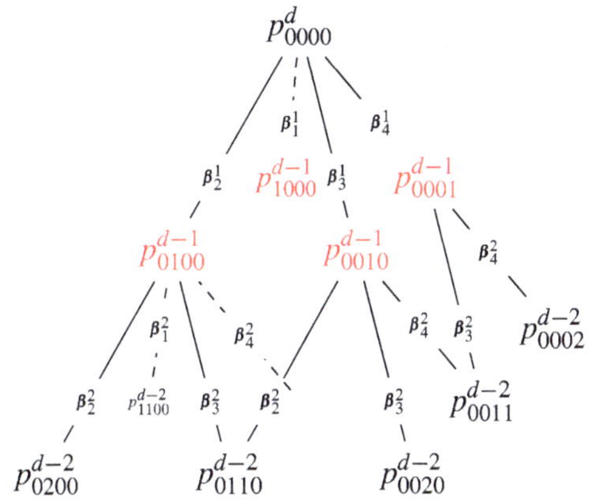

Fig. 7. The pyramidal structure of trivariate blossoming. Each value is formed as a combination of four values at the previous level, using $\boldsymbol{\beta}_i$ as weights. (Some nodes and weights have been removed for clarity.)

5.3 The Blossom

If we modify the de Casteljau algorithm by introducing a sequence of parameter points, $(\boldsymbol{\beta}_1, \ldots, \boldsymbol{\beta}_d)$ and use a distinct point at each level of recursion, we arrive at the blossom of the polynomial, which we denote as $\mathcal{B}$. An intermediate value of the blossom can be expressed as

$$p^r_{\mathbf{i}} = \sum_{j=1}^{s+1} \beta_{r_j} p^{r-1}_{\mathbf{i}+\mathbf{e}_j}, \quad p^0_{\mathbf{i}} = b_{\mathbf{i}}, \tag{8}$$

and the blossom itself is given by

$$\mathcal{B}(f)[\boldsymbol{\beta}_1, \ldots, \boldsymbol{\beta}_d] = p^d_{\mathbf{0}}. \tag{9}$$

An illustration of this process for a trivariate polynomial is given in Figure 7.

The blossom has traditionally been used as an analytic tool for Bézier and spline curves and surfaces. In our context we are interested in using it as a computational algorithm to calculate coefficients and derivatives of an algebraic surface. Some of the properties of the blossom are summarized below:

Lemma 3. *Every BB-polynomial f of degree d, has an associated unique functional, called the blossom, which we will denote as $\mathcal{B}$.*

The blossom has the following properties:

1. *It is symmetric in its arguments; $\mathcal{B}(f)[\boldsymbol{\beta}_1, \ldots, \boldsymbol{\beta}_d] = \mathcal{B}(f)[\pi(\boldsymbol{\beta}_1, \ldots, \boldsymbol{\beta}_d)]$, (here $\pi(\cdot)$ means any permutation of its arguments).*
2. *It is multi-affine in each of its variables; $\mathcal{B}(f)[s\boldsymbol{\alpha}+t\boldsymbol{\beta}, \ldots] = s\mathcal{B}(f)[\boldsymbol{\alpha}, \ldots] + t\mathcal{B}(f)[\boldsymbol{\beta}, \ldots]$, $s, t \in \mathbb{R}$.*
3. *It has the diagonal property; $\mathcal{B}(f)[\boldsymbol{\beta}, \ldots, \boldsymbol{\beta}] = f(\boldsymbol{\beta})$.*

It is common to define the blossom as a functional with the above properties. For details see [32, 13].

We use notation in the style of Farin and let repetition of arguments be denoted $\mathcal{B}(f)[\boldsymbol{\beta},\ldots,\boldsymbol{\beta},\ldots] = \mathcal{B}(f)[\boldsymbol{\beta}^{\langle n\rangle},\ldots]$ if the argument $\boldsymbol{\beta}$ is repeated n times.

The key to finding a curve representing the value of f along the line segment $\mathbf{xy}$ is a surprising property of the blossom, which we summarize in the following lemma:

Lemma 4. *Given two points $\boldsymbol{\alpha} = \boldsymbol{\beta}(\mathbf{x})$ and $\boldsymbol{\gamma} = \boldsymbol{\beta}(\mathbf{y})$, $\mathbf{x},\mathbf{y} \in \Sigma_s$ and a multivariate BB-polynomial $f(\boldsymbol{\beta})$. Then the straight line segment from $\mathbf{x}$ to $\mathbf{y}$ is mapped to a univariate BB-polynomial of degree d, and its coefficients are given by repeated blossoming of the endpoints,*

$$\mathcal{B}(f)[\boldsymbol{\alpha}^{\langle d\rangle}], \mathcal{B}(f)[\boldsymbol{\alpha}^{\langle d-1\rangle},\boldsymbol{\gamma}],\ldots,\mathcal{B}(f)[\boldsymbol{\gamma}^{\langle d\rangle}].$$

For a proof, see [13, 37, 16]. This property then leads to an algorithm for finding the coefficients $(c_{d,0},\ldots,c_{0,d})$ of the univariate polynomial along the ray:

$$c_{d-i,i} = \mathcal{B}(f)[\boldsymbol{\alpha}^{\langle d-i\rangle},\boldsymbol{\gamma}^{\langle i\rangle}], \quad i \in (0,\ldots,d), \tag{10}$$

as illustrated in Figure 2.

For a point $\boldsymbol{\beta}$ in the domain, we may also use the blossom to find derivatives with respect to the (barycentric) direction vector $\mathbf{d}$,

$$D_{\mathbf{d}^r} f(\boldsymbol{\beta}) = \frac{d!}{(d-r)!}\mathcal{B}(f)[\mathbf{d}^{\langle r\rangle},\boldsymbol{\beta}^{\langle d-r\rangle}], \quad r \le d. \tag{11}$$

In the same manner we can also calculate mixed directional derivatives. If $\mathbf{d}_1, \mathbf{d}_2$ represent two distinct vectors their mixed directional derivatives are:

$$D^{r,s}_{\mathbf{d}_1,\mathbf{d}_2} f(\boldsymbol{\beta}) = \frac{d!}{(d-r-s)!}\mathcal{B}(f)[\mathbf{d}_1^{\langle r\rangle},\mathbf{d}_2^{\langle s\rangle},\boldsymbol{\beta}^{\langle d-r-s\rangle}], \quad r+s \le d. \tag{12}$$

To summarize, the blossom provides us with the algorithms needed to convert a trivariate polynomial into a set of univariate polynomials with respect to each ray. Furthermore it allows us to find all derivatives of the surface at an arbitrary point.

6 Conclusion and Further Work

Our work presents a real-time rendering strategy for algebraic surfaces up to total degree five. However, as GPU technologies advance we believe it can be improved further. The two most obvious issues, and thus prime candidates for further work, are to increase rendering speed and the degree of surfaces we can visualize.

With regards to speed, we believe that better root finders than our current implementation can dramatically increase the performance. As GPUs get better at dynamic branching, we believe that hybrid methods will be very well suited for use on GPUs. Higher speeds will also allow a large number of primitives to be rendered simultaneously, which is necessary for displaying piecewise algebraic surfaces.

The limiting factor for the maximum degree of surface visualization is the number of registers available in each fragment pipeline. Our Nvidia 7800 GPU has 32 four-wide temporary registers, in comparison next generation GPUs (DX10) are expected to have 4096 registers[2], which will allow us to visualize surfaces of much higher total degree using our current approach.

Using the depth buffer and blending functionality of GPUs to do constructive solid geometry with algebraic surfaces, as well as integration with mesh based ray tracing techniques is also possible. Mode complex lighting models, including reflection, refraction and self-shadows could be possible, but these aspects will also rely on more registers and better dynamic branching.

With regards to applications, we believe that CAD-systems could benefit significantly from modeling shapes by using general algebraic surfaces. However, as algebraic surfaces often have a non-trivial topology, it has until now been necessary to analyze and establish the topology of the algebraic surface before tessellation and visualization can take place. The approach of this paper shows that such a topology analysis is not necessary and consequently allows efficient visualization of general algebraic surfaces in CAD-systems.

Acknowledgement. This work was supported by contract number 158911/I30 of the Research Council of Norway.

References

1. `http://mathworld.wolfram.com/OrdinaryDoublePoint.html`. Checked as of September 2006.
2. D. Blythe. The Direct3D 10 system. *ACM Trans. Graph.*, 25(3):724–734, 2006.
3. BrookGPU. `http://graphics.stanford.edu/projects/brookgpu/`.
4. N. A. Carr, J. D. Hall, and J. C. Hart. The ray engine. In *HWWS '02: Proceedings of the ACM SIGGRAPH/EUROGRAPHICS conference on Graphics hardware*, pages 37–46, Aire-la-Ville, Switzerland, Switzerland, 2002. Eurographics Association.
5. N. A. Carr, J. Hoberock, K. Crane, and J. C. Hart. Fast GPU ray tracing of dynamic meshes using geometry images. In *GI '06: Proceedings of the 2006 conference on Graphics interface*, pages 203–209, Toronto, Ont., Canada, 2006. Canadian Information Processing Society.
6. M. Christen. Ray tracing on GPU. Master's thesis, University of Applied Sciences, Basel, 2005. `http://www.clockworkcoders.com/oglsl/rt/`.
7. T. D. DeRose, R. N. Goldman, H. Hagen, and S. Mann. Functional composition algorithms via blossoming. *ACM Trans. Graph.*, 12(2):113–135, 1993.

8. T. Dokken, T. R. Hagen, and J. M. Hjelmervik. An introduction to general-purpose computing on programmable graphics hardware. In *this book.*
9. T. Dokken and V. Skytt. Intersection algorithms for CAGD. In *this book.*
10. W. Donnelly. *GPU Gems 2*, chapter Per-Pixel Displacement Mapping with Distance Functions, pages 123–136. Addison-Wesley, 2005.
11. T. Duff. Interval arithmetic recursive subdivision for implicit functions and constructive solid geometry. In *SIGGRAPH '92: Proceedings of the 19th annual conference on Computer graphics and interactive techniques*, pages 131–138, New York, NY, USA, 1992. ACM Press.
12. S. Endraß. SURF 1.0.5. `http://surf.sourceforge.net`.
13. G. Farin. *Curves and surfaces for CAGD: a practical guide.* Morgan Kaufmann Publishers Inc., San Francisco, CA, USA, 2002.
14. R. T. Farouki and T. N. T. Goodman. On the optimal stability of Bernstein basis. *Mathematics of Computation*, 65:1553–1556, 1996.
15. R. Fernando and M. J. Kilgard. *The Cg Tutorial: The Definitive Guide to Programmable Real-Time Graphics.* Addison-Wesley Longman Publishing Co., Inc., Boston, MA, USA, 2003.
16. R. Goldman. Using degenerate Bézier triangles and tetrahedra to subdivide Bézier curves. *Computer Aided Design*, 14(6):307–311, November 1982.
17. X. Gu, S. J. Gortler, and H. Hoppe. Geometry images. In *SIGGRAPH '02: Proceedings of the 29th annual conference on Computer graphics and interactive techniques*, pages 355–361, New York, NY, USA, 2002. ACM Press.
18. J. Hirche, A. Ehlert, S. Guthe, and M. Doggett. Hardware accelerated per-pixel displacement mapping. In *GI '04: Proceedings of the 2004 conference on Graphics interface*, pages 153–158, School of Computer Science, University of Waterloo, Waterloo, Ontario, Canada, 2004. Canadian Human-Computer Communications Society.
19. S. Holzer and O. Labs. *Algebraic Geometry and Geometric Modelling*, chapter Illustrating the classification of real cubic surfaces, pages 119–134. Mathematics and Visualization. Springer-Verlag, 2006.
20. S. Holzer and O. Labs. SURFEX 0.89. Technical report, University of Mainz, University of Saarbrücken, 2006. `http://www.surfex.AlgebraicSurface.net`.
21. F. Karlsson and C. J. Ljungstedt. Ray tracing fully implemented on programmable graphics hardware. Master's thesis, Department of Computer Engineering, Chalmers University of Technology, Göteborg, 2004.
22. C. Loop and J. Blinn. Resolution independent curve rendering using programmable graphics hardware. *ACM Trans. Graph.*, 24(3):1000–1009, 2005.
23. C. Loop and J. Blinn. Real-time GPU rendering of piecewise algebraic surfaces. *ACM Trans. Graph.*, 25(3):664–670, 2006.
24. T. Lyche and J. M. Pẽna. Optimally stable multivariate bases. *Advances in Computational Mathematics*, 20:149–159, 2004.
25. T. Lyche and K. Scherer. On the p-norm condition number of the multivariate triangular Bernstein basis. *J. Comput. Appl. Math.*, 119(1-2):259–273, 2000.
26. M. D. McCool, Z. Qin, and T. S. Popa. Shader metaprogramming. In *SIGGRAPH/Eurographics Graphics Hardware Workshop*, pages 57–68, September 2002. revised.
27. K. Mørken and M. Reimers. An unconditionally convergent method for computing zeros of splines and polynomials. *Mathematics of Computation*, To appear.

28. M. Olano. *A Programmable Pipeline for Graphics Hardware.* PhD thesis, Department of Computer Science, University of North Carolina, Chapel Hill, http://www.cs.unc.edu/ olano/papers/dissertation/, April 1998.
29. M. Olano, A. Lastra, and J. Leech. Procedural primitives in a high performance, hardware accelerated, Z-Buffer renderer. Technical Report Tr97–040, UNC Computer Science Technical Report, 1997.
30. J. D. Owens, D. Luebke, N. Govindaraju, M. Harris, J. Krüger, A. E. Lefohn, and T. J. Purcell. A survey of general-purpose computation on graphics hardware. In *Eurographics 2005, State of the Art Reports*, pages 21–51, August 2005.
31. T. J. Purcell, I. Buck, W. R. Mark, and P. Hanrahan. Ray tracing on programmable graphics hardware. In *SIGGRAPH '02: Proceedings of the 29th annual conference on Computer graphics and interactive techniques*, pages 703–712, New York, NY, USA, 2002. ACM Press.
32. L. Ramshaw. Blossoming: A connect-the-dots approach to splines. Technical report, Digital Systems Research, 1987.
33. A. Rockwood, K. Heaton, and T. Davis. Real-time rendering of trimmed surfaces. *Computer Graphics*, 23(3):107–116, July 1989.
34. R. J. Rost. *OpenGL(R) Shading Language.* Addison Wesley Longman Publishing Co., Inc., 2004.
35. J. Schmittler, S. Woop, D. Wagner, W. J. Paul, and P. Slusallek. Realtime ray tracing of dynamic scenes on an FPGA chip. In *HWWS '04: Proceedings of the ACM SIGGRAPH/EUROGRAPHICS conference on Graphics hardware*, pages 95–106, New York, NY, USA, 2004. ACM Press.
36. P. J. Schneider. *Graphics gems*, chapter A Bézier curve-based root-finder, pages 408–415. Academic Press Professional, Inc., San Diego, CA, USA, 1990.
37. H. P. Seidel. *A general subdivision theorem for Bézier triangles*, pages 573–581. Academic Press Professional, Inc., San Diego, CA, USA, 1989.
38. J. M. Snyder. Interval analysis for computer graphics. In *SIGGRAPH '92: Proceedings of the 19th annual conference on Computer graphics and interactive techniques*, pages 121–130, New York, NY, USA, 1992. ACM Press.
39. M. R. Spencer. *Polynomial real root finding in Bernstein form.* PhD thesis, Brigham Young University, Provo, UT, USA, 1994.
40. N. Thrane and L. O. Simonsen. A comparison of acceleration structures for GPU assisted ray tracing. Master's thesis, Department of Computer Science, University of Aarhus, August 2005.
41. Y. Uralsky. Practical metaballs and implicit surfaces. Game Developers Conference 2006. `http://developer.nvidia.com`.
42. M. Weiler, M. Kraus, and T. Ertl. Hardware-based view-independent cell projection. In *VVS '02: Proceedings of the 2002 IEEE symposium on Volume visualization and graphics*, pages 13–22, Piscataway, NJ, USA, 2002. IEEE Press.
43. B. Wylie, K. Moreland, L. A. Fisk, and P. Crossno. Tetrahedral projection using vertex shaders. In *VVS '02: Proceedings of the 2002 IEEE symposium on Volume visualization and graphics*, pages 7–12, Piscataway, NJ, USA, 2002. IEEE Press.

Part II

Numerical Simulation

Overview

Computer simulation is the rapidly evolving *third way* in science that complements classical experiments and theoretical models in the study of natural, man-made, and abstract systems and processes. Today, computer simulations are used to provide valuable insights in virtually all fields of science and engineering. A particular type of computer simulations that is widely used in engineering and the natural sciences is what we here refer to as *numerical simulation*, i.e., the use of computers to solve differential equations describing particular phenomena in natural and man-made processes. Numerical simulation is also the branch of applied mathematics and computer science that is concerned with the correctness, robustness, and efficiency of simulation algorithms and their computer implementation.

Through the history of SINTEF Applied Mathematics, our activity on numerical simulation has had two branches, one in Oslo and one in Trondheim. Whereas the activity in Oslo has sprung from a scientific tradition within applied mathematics, our Trondheim activity is more rooted in engineering disciplines and traditions.

In Oslo, the earliest activities within numerical simulation among people who were later to become part of SINTEF Applied Mathematics, started in the late 1960s and were concerned with wave analysis. In the beginning, the primary interest was in creating holographic computer displays, but later the research focused more on water waves. Among the highlights of this activity during the last three decades we mention the construction in 1985 of a prototype ocean-wave power plant at Toftestallen in Øygarden, outside Bergen, using the novel and innovative tapered channel principle, and the development in 1994–2002 of the UNDA computer simulator for calculations of fully nonlinear and dispersive waves around marine structures based on a novel spline-collocation method for potential flow. The development of UNDA was funded by a consortium of Norwegian oil companies (Norsk Hydro, Statoil, Norske Conoco, Saga Petroleum, and Norske Shell) and is a good example of fundamentally new mathematical and numerical algorithms being invented to answer current industry needs.

After the turn of the century, our activities within wave modelling started to focus more on using nonlinear wave models (higher-order modified nonlinear Schrödinger equations) as described in the paper by Trulsen. The aim of this activity was to describe various aspects of nonlinear ocean waves such as deterministic wave forecasting and statistical analysis of extreme waves (sometimes called freak or rogue waves) based on deterministic wave simulation. Potential benefits include better understanding of their occurrence, properties and statistics, including the possibility for predicting hazardous wave conditions that might affect marine operations. Due to the lack of a sufficient level of industry funding, the activity on wave modelling has been relocated from SINTEF to the University of Oslo. For more information on past and current activity, see `http://www.math.uio.no/~karstent/waves/`.

In Trondheim, an important milestone in the prehistory of SINTEF Applied Mathematics was the initiation of the national supercomputing project in late 1986 and the subsequent installation of the first Norwegian supercomputer, a Cray X-MP/24. Since then, our researchers have been concerned with high-performance computing, in the beginning on specialised hardware like the Cray X-MP and its predecessors, and later on cluster solutions based on commodity hardware. In early 2003, researchers in Oslo spotted an emerging and exotic field called GPGPU (general-purpose computing using graphics hardware), in which researchers around the world were trying to exploit the unrivalled ability of modern graphics hardware to process floating-point data in scientific computing. Soon we were convinced that this was going to be important in the future, and we therefore initiated a strategic project within this field with funding from the Research Council of Norway. To investigate the capabilities of graphics cards, we choose to study the implementation of high-resolution schemes for hyperbolic conservation laws, which are particularly well suited for the parallel computing paradigm of data-based stream processing. The paper by Hagen et al. gives a thorough introduction to the idea and the results we obtained, which are in fact amazing. By moving computations from the CPU to the GPU, we typically observe a speedup of at least one order of magnitude. Given the current trend with multi-core computers and emerging multi-GPU solutions, we believe that harnessing the combined computing power of instruction-driven CPUs and data-stream processors like GPUs *will* be the predominant way of scientific and technical computing in the future.

The Department of Numerical Simulation, the predecessor of the current simulation group in Oslo, was formed in the mid 1990s. Before the turn of the century, its researchers worked mainly on mathematical and numerical analysis of flow models—in particular models for flow in porous media—as part of two large strategic research programs. The main theme in these projects was the development of object-oriented numerical software tools, a new and groundbreaking idea in the early 1990s, which in SINTEF's case led to the development of Diffpack (`http://www.diffpack.com`) in collaboration with the University of Oslo. This work has been documented in a previous book by SINTEF Applied Mathematics, "Numerical Methods and Software Tools in Industrial Mathematics", Birkhäuser, 1997, and in the popular textbook "Computational Partial Differential Equations - Numerical Methods and Diffpack Programming" by Hans Petter Langtangen on Springer Verlag. These early strategic research programs had a strong focus on educating doctoral and master students and laid the foundations of what are now flourishing research groups at Simula Research Laboratory (`http://www.simula.no`).

The simulation group in Oslo still has a focus on educating students, as is clearly reflected in the paper by Aarnes, Gimse, and Lie. The paper gives an introduction to flow modelling in petroleum reservoirs, where the emphasis has been put on presenting simple, yet efficient Matlab codes that can later be a starting point for (advanced) students in their studies and research. The material has been used by several of our master and doctoral students and will hopefully prove to be useful to others as well.

Given Norway's role as a major international exporter of oil, it is natural that research related to the recovery of petroleum resources from the North Sea has been an important activity at SINTEF Applied Mathematics throughout the past fifteen years. The focus of our geometric modelling group has been on geological modelling and in particular on surface reconstruction of horizons and faults from large sets of geological data for customers like Roxar and former Technoguide (now Schlumberger Petrel). The simulation group, on the other hand, focuses more on discretisation of models for fluid flow and development of solvers with very high efficiency. On the industry side, the group has had a long-term alliance with the FrontSim-group within Schlumberger, dating back to the former company Technical Software Consultants, with whom we have developed streamline methods for three-phase and compositional flow. Similarly, in industry projects for Statoil, our researchers have studied flow in pore networks, models for nuclear magnetic resonance in core samples, and aspects of upscaling small-scale geological models.

In 2004, our activities took a new direction, when we were awarded a strategic institute program on multiscale methods for reservoir simulation from the Research Council of Norway; see `http://www.math.sintef.no/GeoScale` for a description of the project. Here, this activity is represented through the paper by Aarnes, Kippe, Lie, and Rustad, which describes problems in modelling and simulation related to the large discrepancies in physical scales in geological models. A thorough discussion of various upscaling methods is given, leading up to our current research on multiscale simulation methods. From being purely academic research ten years ago, multiscale techniques are now starting to migrate into industry. Our group has made several important contributions to this end; for instance, the extension of the methodology to the complex grid types used by the petroleum industry.

Starting from our basic research on multiscale methodologies in the strategic GeoScale project, we have during the last year built up a portfolio of industry projects aiming at the development of highly efficient solvers for direct flow simulation of industry-standard geomodels of highly heterogeneous and fractured porous media. At the time of writing (October 2006), the focus of our research is on extending the multiscale methodology to complex grids with nonmatching faces used, for instance, to model faults. Another application of our multiscale methodologies is within history matching (i.e., adjustment of the geological reservoir description to reflect observed production history), where these methods can be used to enable fast matching of models with multimillion cells.

In the final paper in the second part of the book, Utnes gives an introduction to stratified geophysical flow over variable topography. Simulating fine-scale processes like e.g., extreme wind and turbulence in mountain areas, is a main challenge within weather prediction. Researchers in our simulation group in Trondheim have since around 1990 worked on predicting local flow in mountainous terrain in an effort to increase aviation safety. This work has been motivated by the belief that turbulence-windshear has been the major cause of at least 11 fatal accidents in Norwegian aviation during the last 25 years. High-resolution non-hydrostatic models enable a much better description of the vertical motion in geophysical flows in steep terrain and in small-scale dynamic weather systems, e.g., as seen around a number of Norwegian airports. SINTEF has developed a high-resolution non-hydrostatic finite-element code denoted SIMRA. This code is today used in daily operation by the Norwegian Meteorological Institute in a nested system for local wind prediction at a number of airports in Norway in an effort to increase aviation safety by avoiding the most hazardous flying conditions; the paper reports an example of use for the Trondheim Airport at Værnes.

Knut–Andreas Lie
Chief Scientist
Editor, Numerical Simulation

Weakly Nonlinear Sea Surface Waves — Freak Waves and Deterministic Forecasting

Karsten Trulsen

1 Introduction

The material contained here is to a large extent motivated by the so-called Draupner "New Year Wave", an extreme wave event that was recorded at the Draupner E platform in the central North Sea on January 1st 1995 [4, 5]. This location has an essentially uniform depth of 70 m. The platform is of jacket type and is not expected to modify the wave field in any significant way. The platform had been given a foundation of a novel type, and for this reason was instrumented with a large number of sensors measuring environmental data, structural and foundation response. We are particularly interested in measurements taken by a down looking laser-based wave sensor, recording surface elevation at a speed of 2.1333 Hz during 20 minutes of every hour. The full 20 minute time series recorded starting at 1520 GMT is shown in Figure 1 and a close-up of the extreme wave event is shown in Figure 2. To remove any doubt that the measurements are of good quality, Figure 3 shows an even finer close-up with the individual measurements indicated. It is clear that the extreme wave is not an isolated erroneous measurement. The minimum distance between the sensor and the water surface was 7.4 m.

The significant wave height, defined as four times the standard deviation of the surface elevation, was $H_s = 11.9$ m while the maximum wave height was 25.6 m with a crest height of 18.5 m above the mean water level. Haver [4] states that the wave itself was not beyond design parameters for the platform, but basic engineering approaches did not suggest that such a large wave should occur in a sea state with such a small value of H_s. Some damage was reported to equipment on a temporary deck. The wave has often been referred to as a freak or rogue wave, although there is lacking consensus exactly what this means, this is further discussed in section 2.

During the past decade, a considerable effort has been undertaken by many people to understand the Draupner wave, and rogue waves in general. Statoil should be commended for their policy of releasing this wave data to the public, thus igniting an avalanche of exciting research. The work reported here

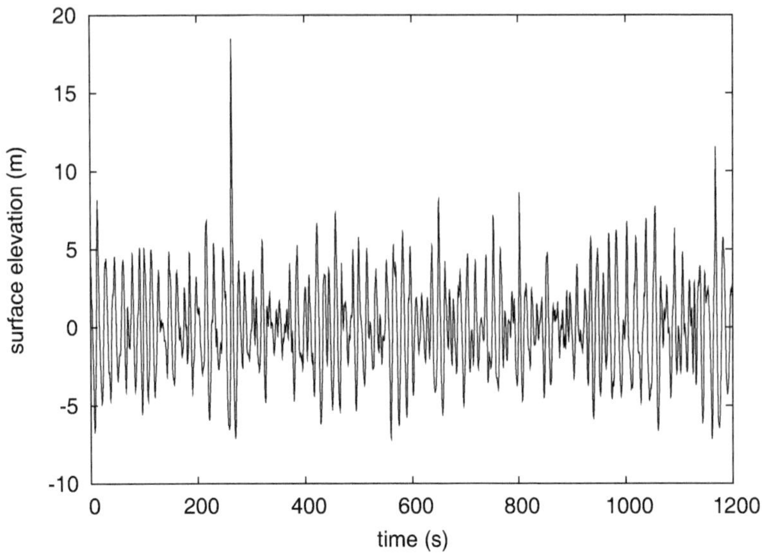

Fig. 1. Draupner 20 minute time series starting at 1520 GMT.

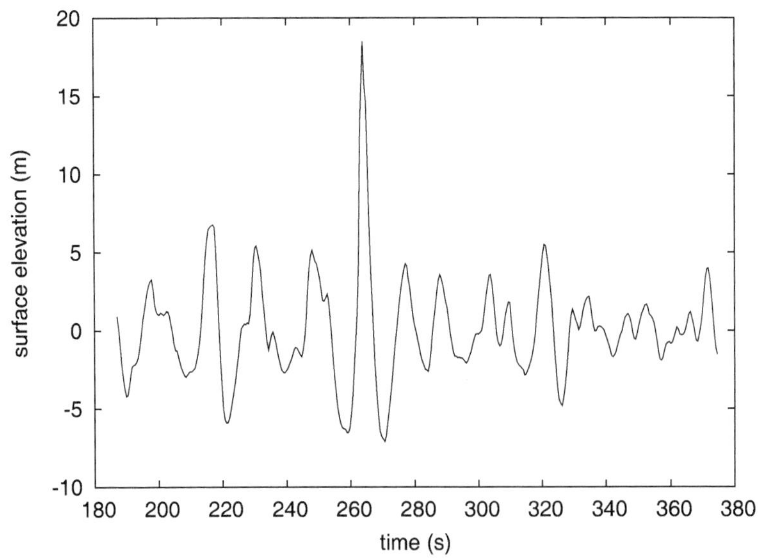

Fig. 2. Extract of Draupner time series close to the extreme wave.

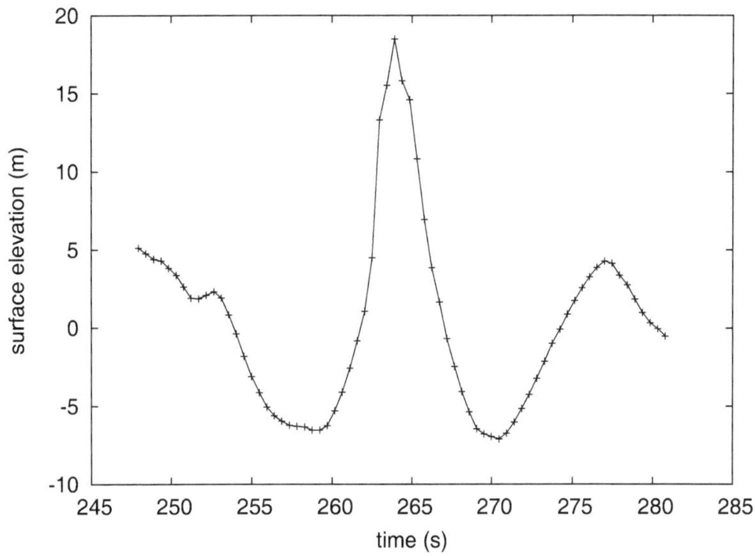

Fig. 3. Extract of Draupner time series close to the extreme wave, with discrete measurements indicated.

has been carried out by the motivation to understand the wave conditions under which the Draupner wave occurred, describe simplified models suitable to model these wave conditions, and explore the possibility to describe the spatiotemporal evolution of these waves with a view to deterministically forecast freak waves.

As a general introduction to ocean waves relevant for section 2, the books by Kinsman [6], Ochi [8] and Goda [3]. The perturbation analysis in section 3 is strongly motivated by Mei [7].

2 Empirical Description of the Draupner "New Year Wave"

We define the mean water level as the arithmetic mean position during the $N = 2560$ discrete measurements of duration 20 minutes. Figures 1, 2 and 3 show the surface elevation relative to the mean water level, $\zeta_n \equiv \zeta(t_n)$, where $t_n = n\Delta t$ are the discrete times. Thus the time average is by definition

$$\frac{1}{N}\sum_{n=0}^{N-1}\zeta_n = 0.$$

The wave amplitude is defined as the vertical distance between the mean water level and a crest. The wave height is defined as the vertical distance between a trough and a neighboring crest. Given a discrete time series, we do not necessarily have measurements of the exact crests and troughs. More robust estimates for the overall amplitude and wave height are achieved using the standard deviation of the surface elevation. The standard deviation, or root-mean-square, of the discrete measurements can be computed as

$$\sigma = \sqrt{\overline{\zeta^2}} = \sqrt{\frac{1}{N}\sum_{n=0}^{N-1} \zeta_n^2} = 2.98\text{m}.$$

The *significant wave height* is defined as four times the standard deviation

$$H_s = 4\sigma = 11.9\text{m}. \tag{1}$$

We define the *characteristic amplitude* as

$$\bar{a} = \sqrt{2}\sigma = 4.2\text{m}. \tag{2}$$

To compute the wave height, it is first necessary to define what constitutes a crest and a trough. In the following we define a crest to be the largest positive surface displacement above the mean water level between a zero-up-crossing and a subsequent zero-down-crossing, while a trough is defined to be the largest negative surface displacement below the mean water level between a zero-down-crossing and a subsequent zero-up-crossing. It is also necessary to distinguish between wave heights defined by zero-up-crossing or zero-down-crossing. The given time series has 106 down-crossing wave heights and 105 up-crossing wave heights.

If we assume that the discrete measurements correspond to the actual crests and troughs, then the maximum crest height (wave amplitude) is 18.5 m, the maximum down-crossing wave height is 25.6 m and the maximum up-crossing wave height is 25.0 m.

After all the individual wave heights are found, we can sort them in decreasing order. For any positive number α, the average of the $1/\alpha$ highest wave heights is denoted as $H_{1/\alpha}$. Particularly common is the average of the 1/3 highest wave heights; the down-crossing $H_{1/3}$ is 11.6 m and the up-crossing $H_{1/3}$ is 11.4 m. These values are almost identical to H_s. It can be shown that for the limiting case of a narrow-banded and Gaussian sea, the wave height has a Rayleigh distribution, and $H_{1/3} = 4.0\sigma$, see [8]. The present sea state is only approximately narrow-banded and Gaussian.

The maximum crest height of 18.5 m is equal to 6.2 standard deviations. The maximum wave height of 25.6 m is equal to 8.6 standard deviations, or equivalently, 2.1 significant wave heights. One of the most common definitions of a freak wave is that the maximum wave height is greater than twice the significant wave height. The Draupner "New Year Wave" qualifies to be called freak according to this definition, however, the crest height certainly appears more "freak" than the wave height.

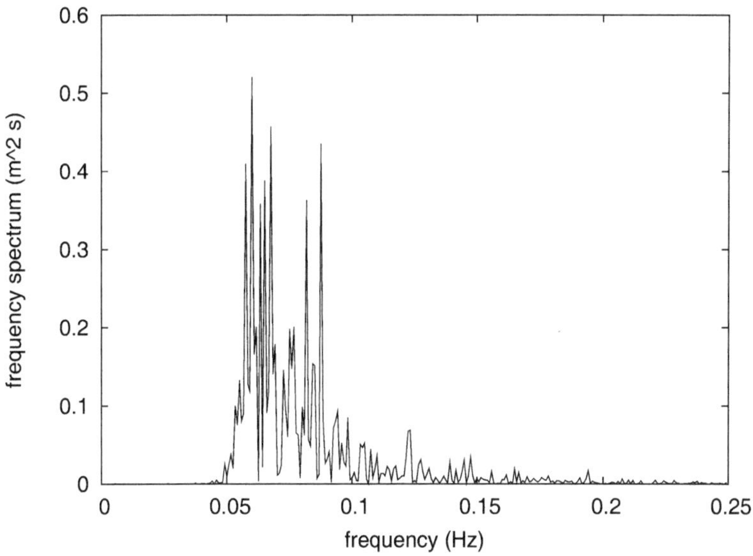

Fig. 4. Frequency spectrum estimated from $2|\hat{\zeta}(\omega)|^2$ without smoothing, linear axes.

The angular frequency spectrum $S(\omega)$ can be estimated by the square magnitude of the Fourier transform of the time series $2|\hat{\zeta}(\omega)|^2$. Recall that the angular frequency is $\omega = 2\pi f$ where f is the frequency. We employ the discrete Fourier transform of the time series

$$\hat{\zeta}_j = \frac{1}{N} \sum_{n=0}^{N-1} \zeta_n e^{i\omega_j t_n}, \tag{3}$$

and use only non-negative frequencies $\omega_j = 2\pi j/T$ for $j = 0, 1, \ldots, N/2$ for the angular frequency spectrum. Here T is the duration of the time series with N discrete measurements. Figure 4 shows the estimated spectrum with linear axes, while Figure 5 shows the same with logarithmic axes, in both cases without any smoothing. Figure 5 suggests how ω^{-5} and ω^{-4} power laws fit the high-frequency tail of the observed spectrum; the first law is often used for engineering design spectra [3, 8] while the second law is often found to be a better model near the spectral peak [2].

Based on the Fourier transform of the time series for surface elevation, we may estimate a characteristic angular frequency as the expected value of the spectral distribution

$$\omega_c = 2\pi f_c = \frac{2\pi}{T_c} = \frac{\sum_j |\omega_j||\hat{\zeta}_j|^2}{\sum_j |\hat{\zeta}_j|^2} = 0.52 \text{ s}^{-1} \tag{4}$$

corresponding to a characteristic frequency of $f_c = 0.083$ Hz and a characteristic period of $T_c = 12$ s.

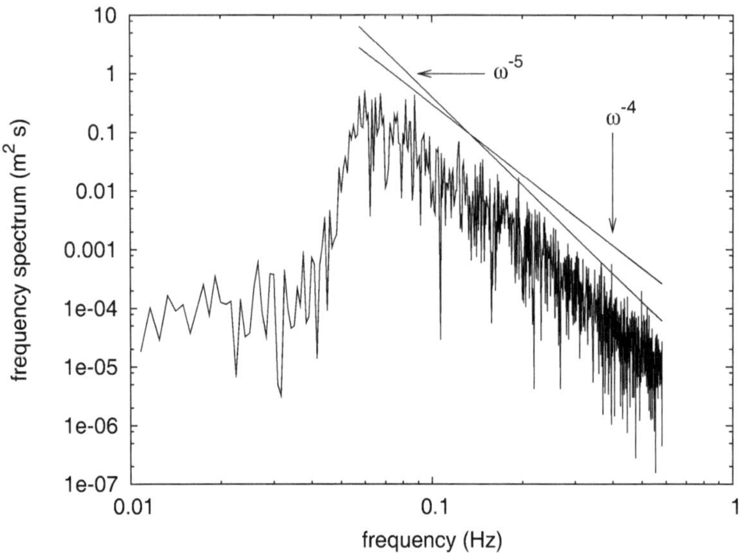

Fig. 5. Frequency spectrum estimated from $2|\hat{\zeta}(\omega)|^2$ without smoothing, logarithmic axes. Two possible power laws, ω^{-4} and ω^{-5}, are indicated for the high frequency tail.

The characteristic wavenumber $k_c = 2\pi/\lambda_c$, where λ_c is the characteristic wavelength, can be estimated based on the assumption that the waves are linear to leading order. Then we simply solve the linear dispersion relation

$$\omega^2 = gk \tanh kh \tag{5}$$

were $g = 9.81 \text{ m/s}^2$ is the acceleration of gravity and $h = 70$ m is the relevant depth. We find the characteristic wavenumber $k_c = 0.0289 \text{ m}^{-1}$, the characteristic wave length $\lambda_c = 217$ m and the characteristic non-dimensional depth $k_c h = 2.0$. As far as estimating the characteristic wavenumber is concerned, no great error is done using infinite depth in which case the dispersion relation becomes

$$\omega^2 = gk \tag{6}$$

and we get $k_c = 0.0279 \text{ m}^{-1}$ and $\lambda_c = 225$ m.

Knowing the characteristic wavenumber and amplitude, we compute the characteristic steepness $\epsilon = k_c \bar{a} = 0.12$, which is a measure of the strength of nonlinearity.

It will be useful to have some measure of the characteristic width $\Delta\omega$ of the angular frequency spectrum around the characteristic angular frequency ω_c. We define the dimensionless bandwidth to be the ratio between the deviation and the mean

$$\delta_\omega = \frac{\Delta\omega}{\omega_c}. \tag{7}$$

The most simple-minded approach to estimate $\Delta\omega$ is to stare at Figure 4. This exercise may lead to the conclusion that $\Delta\omega/2\pi \approx 0.025$ Hz, and thus $\delta_\omega \approx 0.32$.

Some other approaches to compute $\Delta\omega$, for example computing the standard deviation of the frequency spectrum, typically yield unreasonably large and less useful values of $\Delta\omega$. This is because the high-frequency tail of the spectrum is largely due to nonlinearly bound waves (waves that are not free, i.e. waves that do not satisfy the linear dispersion relation) and measurement noise. Our estimate for bandwidth will only be applied to the derivation of the simplified models to be described in the following. The most relevant measure of spectral width should then only account for linear free waves, and therefore the simple-minded approach turns out to be quite recommendable.

In conclusion of these empirical considerations, the important observations are given by three non-dimensional numbers: The characteristic steepness is $\epsilon = k_c\bar{a} \approx 0.12$, the characteristic bandwidth is $\delta_\omega = \Delta\omega/\omega_c \approx 0.32$ and the characteristic non-dimensional depth is $k_c h \approx 2.0$. People who work in the field of wave hydrodynamics will commonly say that that these waves are weakly nonlinear, have moderately small bandwidth and are essentially on deep water. Notice also that the bandwidth is slightly greater than the steepness $\delta_\omega > \epsilon$. In the next section we show how to take advantage of these non-dimensional numbers to construct simplified mathematical models that can describe this wave field.

Here the reader may likely object: How can we claim that the unusually extreme wave event in Figures 1–3 is weakly nonlinear? Our notion of weak nonlinearity is in the sense that the instantaneous dependence on the steepness is weak for the overall wave field. We do not claim that long-term accumulated effects are weakly dependent on the steepness. We also do not claim that the Draupner wave crest by itself is weakly nonlinear. An accurate description of such an extreme wave crest may indeed have to be strongly nonlinear. However, since the background wave field can be considered weakly nonlinear, we anticipate that important information about such waves is contained in a weakly nonlinear description, such as when and where they occur, their approximate magnitude, velocity and acceleration fields, etc.

It is instructive to see how the sea state of the Draupner wave compares to typical sea states in the northern North Sea. Figure 6 contains approximately 70 000 data points, each representing a 20 minute wave record from the northern North Sea. T_p denotes the wave period corresponding to the frequency at the spectral peak. From Figure 5 we deduce the value $T_p \approx 15$ s for the Draupner time series, greater than the estimated value of T_c (in general we typical find $T_p \gtrsim T_c$). It appears that the sea state of the Draupner "New Year Wave" may be considered slightly unusual due to its large H_s, but the sea state is definitely not unusual due to its overall steepness.

The above discussion reveals the need to distinguish two questions: How unusual is the sea state in which the Draupner "New Year Wave" occurred, and how unusual is the Draupner "New Year Wave" within the sea state in

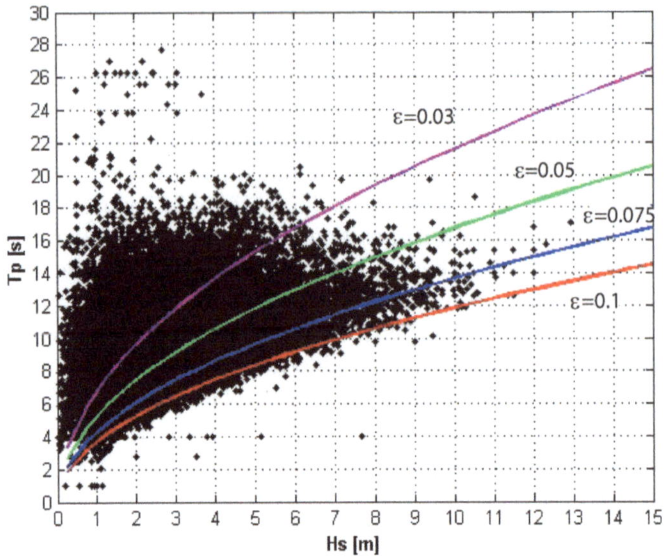

Fig. 6. Scatter diagram for H_s and T_p from the northern North Sea. Pooled data 1973–2001, from the platforms Brent, Statfjord, Gullfax and Troll. Curves of constant steepness are also shown, using a definition of steepness $\epsilon = \sqrt{2}\pi^2 H_s/(gT_p^2)$ slightly different from that used in the text above since $T_p \gtrapprox T_c$. The figure was prepared by K. Johannessen, Statoil.

which it occurred? Partial answers to the second question have been given in our recent work using deterministic evolution equations similar to those in the next section for Monte–Carlo simulations, see [9]. However, in the following we limit the discussion to weakly nonlinear waves from a purely deterministic point of view.

3 Weakly Nonlinear Models for Sea Surface Waves

We desire a simplified model capable of describing individual waves in a sea state as that shown in Figure 1. Suitable simplified models are achieved by taking advantage of the smallness of the steepness ϵ, i.e. weakly nonlinear models, and by taking advantage of the moderately small bandwidth δ, and by assuming that the appropriate depth is sufficiently large that the characteristic wave can be estimated by the deep-water dispersion relation (6).

Assuming that the flow is inviscid and irrotational, there exists a velocity potential $\phi(\mathbf{x}, t)$ such that the velocity field is $\mathbf{u}(\mathbf{x}, t) = \nabla\phi(\mathbf{x}, t)$. Here $\mathbf{x} = (x, y, z)$ is the spatial position vector and t is time. The water is confined between the bottom at depth $z = -h$ and the free surface at $z = \zeta(x, y, t)$. If the water is considered incompressible then the conservation of mass requires

$$\nabla^2\phi = 0 \qquad -h < z < \zeta. \tag{8}$$

At the free surface the kinematic and dynamic free-surface conditions are

$$\frac{\partial\zeta}{\partial t} + \nabla\phi\cdot\nabla\zeta = \frac{\partial\phi}{\partial z} \qquad z = \zeta, \tag{9}$$

$$\frac{\partial\phi}{\partial t} + g\zeta + \frac{1}{2}(\nabla\phi)^2 = 0 \qquad z = \zeta, \tag{10}$$

where g is the acceleration of gravity. The kinematic condition at the bottom is

$$\frac{\partial\phi}{\partial z} = 0 \qquad z = -h. \tag{11}$$

We assume asymptotic expansions in terms of higher harmonics of the characteristic wave

$$\phi = \bar{\phi} + \mathrm{Re}\left\{A\mathrm{e}^{\mathrm{i}\theta} + A_2\mathrm{e}^{2\mathrm{i}\theta} + A_3\mathrm{e}^{3\mathrm{i}\theta} + \cdots\right\} \tag{12}$$

$$\zeta = \bar{\zeta} + \mathrm{Re}\left\{B\mathrm{e}^{\mathrm{i}\theta} + B_2\mathrm{e}^{2\mathrm{i}\theta} + B_3\mathrm{e}^{3\mathrm{i}\theta} + \cdots\right\} \tag{13}$$

The phase $\theta = k_c x - \omega_c t$ describes the principal direction and evolution of the wave field, where k_c and ω_c are the characteristic values for the wavenumber and angular frequency, see (4) and (5). Slow modulation is anticipated on the spatial scales δx, δy and the temporal scale δt, where δ is the non-dimensional bandwidth, see (7). The magnitude of first harmonic complex amplitudes A and B is proportional to the steepness ϵ; these first harmonic terms represent free waves, i.e. waves with angular frequencies and wave vectors that are related by the linear dispersion relation. The higher harmonics A_n and B_n turn out to have magnitudes proportional to ϵ^n. The zeroth harmonic terms, i.e. the induced flow $\bar{\phi}$ and surface displacement $\bar{\zeta}$, represent the mean slow response due to the waves; on deep water their magnitudes are proportional to ϵ^2 and $\epsilon^2\delta$, respectively. We have now set up the framework for a multiple-scales perturbation analysis with two small parameters ϵ and δ, see [7]. In the following we suffice by only summarizing the results.

The spatial evolution of the first harmonic complex amplitude of the wave field, *i.e.* the evolution in the horizontal x-direction of a wave field given in terms of time t, is given by a nonlinear Schrödinger equation. In the following we consider the limiting case that the waves are perfectly long-crested, *i.e.* there is no dependence on y.

Nonlinear Schrödinger equations for water waves can be given either in terms of A (velocity potential) or B (surface displacement) as the fundamental unknown. The resulting equations for A and B are similar, but not completely equal. There is no commonly accepted standard that A is used for velocity potential and B is used for surface elevation, this can sometimes cause some confusion. In the following we express all equations in terms of B (surface elevation) as the fundamental unknown since these equations are more natural to use for the applications we have in mind.

In [14] it was shown how the nonlinear Schrödinger equations can be extended to include exact linear dispersion by introducing a pseudo-differential operator for the linear part. This extension is particularly useful when the bandwidth is greater than the steepness, $\delta > \epsilon$, which is the typical case for sea surface waves. For the spatial evolution of long-crested waves, exact linear dispersion can be expressed in a remarkably simple form. The linear Schrödinger equation with exact linear dispersion takes the form

$$\frac{\partial B}{\partial x} + \frac{2k_c}{\omega_c}\frac{\partial B}{\partial t} + \frac{\mathrm{i}k_c}{\omega_c^2}\frac{\partial^2 B}{\partial t^2} = 0. \tag{14}$$

In the following we limit the perturbation analysis to $\delta = \epsilon$, however, it should be kept in mind that the linear part for long-crested waves is already given exactly in (14) for arbitrarily broad bandwidth.

If leading nonlinear effects are included we get the classical nonlinear Schrödinger (NLS) equation up to order ϵ^3

$$\frac{\partial B}{\partial x} + \frac{2k_c}{\omega_c}\frac{\partial B}{\partial t} + \frac{\mathrm{i}k_c}{\omega_c^2}\frac{\partial^2 B}{\partial t^2} + \mathrm{i}k_c^3|B|^2B = 0. \tag{15}$$

The appropriate reconstruction of the water surface is achieved with (13) using only the terms with the first harmonic B and the second harmonic B_2, where

$$B_2 = \frac{k_c}{2}B^2. \tag{16}$$

Including nonlinear and dispersive effects of the slightly higher order ϵ^4, we get the modified nonlinear Schrödinger (MNLS) equation, also known as the Dysthe equation [1]

$$\begin{aligned}\frac{\partial B}{\partial x} &+ \frac{2k_c}{\omega_c}\frac{\partial B}{\partial t} + \frac{\mathrm{i}k_c}{\omega_c^2}\frac{\partial^2 B}{\partial t^2} + \mathrm{i}k_c^3|B|^2B \\ &- \frac{8k_c^3}{\omega_c}|B|^2\frac{\partial B}{\partial t} - \frac{2k_c^3}{\omega_c}B^2\frac{\partial B^*}{\partial t} - \frac{4\mathrm{i}k_c^3}{\omega_c^2}\frac{\partial \bar{\phi}}{\partial t}B = 0 \quad \text{at} \quad z = 0,\end{aligned} \tag{17}$$

$$\frac{\partial \bar{\phi}}{\partial z} = -k_c\frac{\partial}{\partial t}|B|^2 \quad \text{at} \quad z = 0, \tag{18}$$

$$\frac{4k_c^2}{\omega_c^2}\frac{\partial^2 \bar{\phi}}{\partial t^2} + \frac{\partial^2 \bar{\phi}}{\partial y^2} + \frac{\partial^2 \bar{\phi}}{\partial z^2} = 0 \quad \text{for} \quad -h < z < 0, \tag{19}$$

$$\frac{\partial \bar{\phi}}{\partial z} = 0 \quad \text{at} \quad z = -h. \tag{20}$$

Now the reconstruction of the surface displacement involves all four terms of (13), with the following formulas

$$\bar{\zeta} = -\frac{k_c}{\omega_c^2}\frac{\partial \bar{\phi}}{\partial t}, \qquad B_2 = \frac{k_c}{2}B^2 + \frac{ik_c}{\omega_c}B\frac{\partial B}{\partial t}, \qquad B_3 = \frac{3k_c^2}{8}B^3. \tag{21}$$

Similar equations for short-crested waves were derived in [1] and in [14]. In [2] we showed that the MNLS equation in two horizontal dimensions (for short-crested waves) predicts the commonly observed ω^{-4} power law for the high-frequency tail near the peak of the spectrum, suggested in Figure 5.

4 Deterministic Wave Forecasting — Validation with Laboratory Experiments

One application of the equations developed in the previous section is deterministic forecasting. The evolution equations (17)–(20) together with the reconstruction formulas (13) and (21) contain in principle all information about how individual waves evolve. In the following we give an example of deterministic forecasting of laboratory waves.

4.1 Error Criterion

Assuming that prediction of amplitude and phase are equally important, we introduce the error criterion

$$e(x) = \sqrt{\frac{\sum_i(\zeta_{\text{sim}}(x,t_i) - \zeta_{\text{exp}}(x,t_i))^2}{\sum_i(\zeta_{\text{exp}}(x,t_i))^2}} \tag{22}$$

where $\zeta_{\text{sim}}(x,t_i)$ is the simulated surface elevation and $\zeta_{\text{exp}}(x,t_i)$ is the measured surface elevation.

4.2 Application to Bichromatic Waves

The experiment was carried out in the 270 m long and 10.5 m wide towing tank at Marintek [10, 11, 12, 15]. Nine probes were located along the centerline of the tank, while one probe was located off the centerline. A sketch of the arrangement of wave probes is shown in Figure 7.

Measurements at probe 1 are used for initialization, and detailed comparisons between simulated and measured waves are presented here for probes 2, 4, 5, 7 and 10. The probes were located at distances 9.3 m, 40 m, 80 m, 120 m, 160 m and 200 m from the wave maker, respectively. The sampling period was 0.05 s. In order to get rid of seiching in the experimental data, the measured time series were high-pass filtered at about 0.08 Hz.

The depth of the wave tank is 10 m for the first 80 m closest to the wave maker, and 5 m elsewhere. However, the wave periods and heights employed are so small that the waves are essentially on deep water. We have set the depth to 10 m in the simulations. While the non-uniform depth is practically

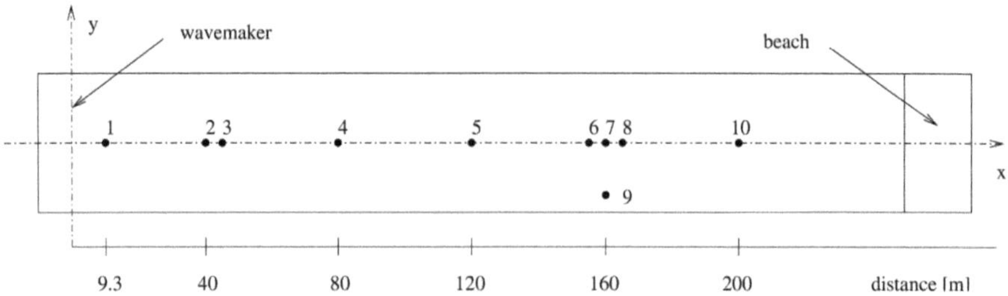

Fig. 7. Sketch of towing tank with wave probes.

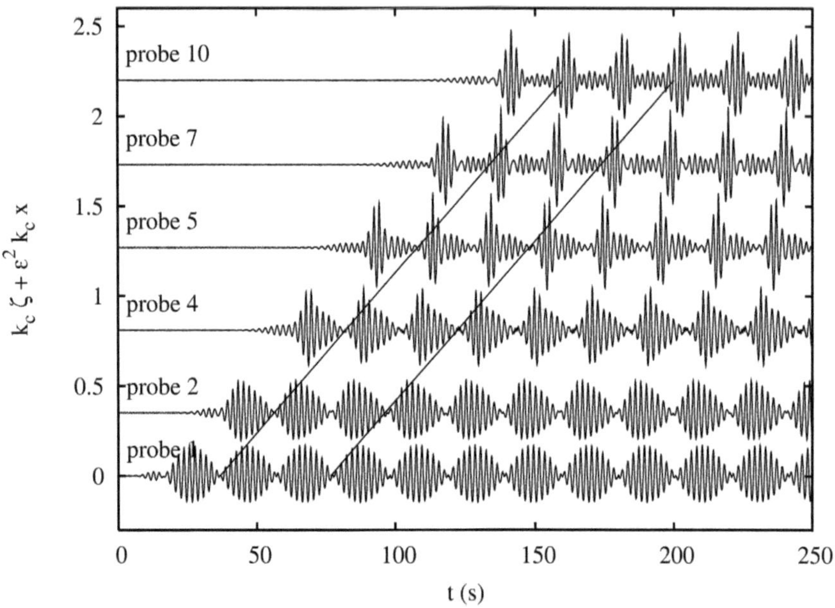

Fig. 8. Bichromatic waves. Time series from experiment. The two slanted lines show the moving time window used for validation of the simulations, the time window moves with the group velocity.

invisible at the spatiotemporal scales of individual waves, it may be visible for the zeroth harmonic $\bar{\phi}$ and $\bar{\zeta}$ that depend on longer spatiotemporal scales. The effect of non-uniform depth becomes more important as the peak period increases, however an extended validation study using a large collection of experimental data suggests that the best agreement between simulation and experiment is obtained for longer peak periods. Hence there is reason to believe that the neglect of non-uniform depth is not a principal source of error in the present case.

The bichromatic waves were generated with nominal periods 1.9 s and 2.1 s, both with nominal wave heights 0.16 m. The steepness of the portion of

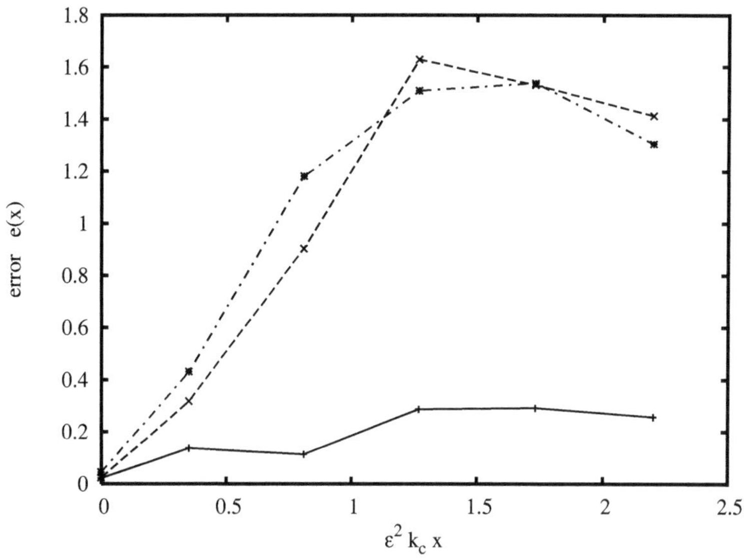

Fig. 9. Normalized error $e(x)$ of simulations as a function of normalized propagation distance at selected probes. The linear simulation (dashed-dotted curve) and the NLS simulation (dashed curve) give rather similar results, while the MNLS simulation (solid curve) is better.

the time series at probe 1 used for initialization is estimated to be $\epsilon = 0.11$. We use a time domain of length 204.8 s. The simulations are initialized by the relevant part of the experimental time series at probe 1. Comparisons for validation are made with a time interval of length 40 s, starting at shifted times corresponding to a time window moving with the linear group velocity. Figure 8 shows all six time series at the selected wave probes. The slanted pair of lines indicate the moving time window used for validation, the time window moves with the group velocity.

Simulations were carried out with the linear equation (14) using linear reconstruction of the surface elevation, with the NLS equation (15) using second order reconstruction (16), and with the MNLS equation (17)–(20) using third-order reconstruction of the surface elevation (21). In all cases the Fourier transform of the measured waves used for initialization was bandpass filtered around the characteristic angular frequency ω_c; almost all energy contained in the interval $0 < \omega < 2\omega_c$ was used to initialize B.

The normalized errors $e(x)$ of the three simulations, computed within the moving time window, are shown in Figure 9. The linear simulation (dashed-dotted curve) and the NLS simulation (dashed curve) are rather similar in quality, while the MNLS simulation (solid curve) is much better. It is clear that the NLS equation is not a significant improvement over the linear theory,

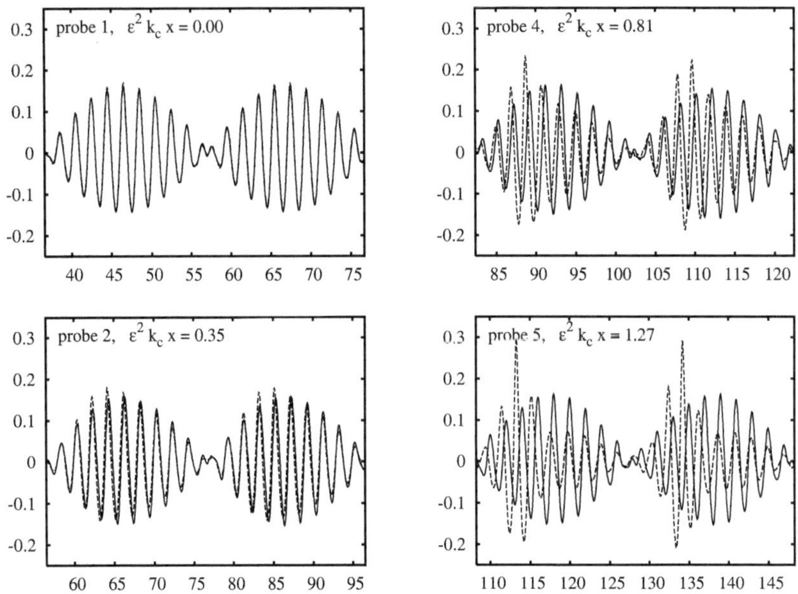

Fig. 10. Bichromatic waves: Linear simulation (solid curves) and experimental measurements (dotted curves) at four probes. Horizontal axis is time in seconds. Vertical axis is normalized surface displacement $k_c\zeta$.

at least not for deterministic forecasting of bichromatic waves, but the MNLS equation is clearly a major improvement.

The exact details of how the simulated waves compare with the experimental measurements within the moving time window is seen for the linear simulation in Figure 10, for the NLS simulation in Figure 11 and for the MNLS simulation in Figure 12. In all three cases the groups maintain their total length.

Linear evolution underpredicts both the phase and group velocities observed in the experiment. Linear evolution also does not account for the change in shape of wave groups.

NLS evolution accounts for nonlinear increase in phase velocity and therefore yields good agreement with the observed phase velocity in the experiment. The NLS evolution does not account for the nonlinear increase in group velocity that is seen to occur. It does account for nonlinear increase in amplitude of the groups, but does not capture the asymmetric forward-leaning evolution seen in the experiments.

MNLS evolution accounts for nonlinear increase in both the phase and group velocities, in good agreement with the experimental observations. The nonlinear increase in amplitude of the groups and the asymmetric forward-leaning evolution of the groups seen in the experiments are also captured.

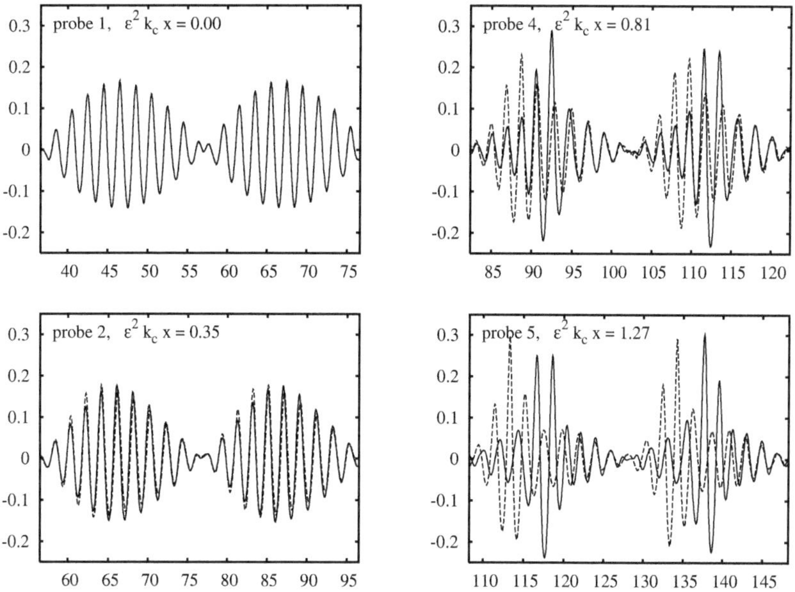

Fig. 11. Bichromatic waves: NLS simulation simulation (solid curves) and experimental measurements (dotted curves) at four probes. Horizontal axis is time in seconds. Vertical axis is normalized surface displacement $k_c\zeta$.

There is still some disagreement between the evolution predicted by the MNLS equation and the experiment. Possible reasons include inaccurate initialization from measurements, the presence of transversal modulation, free waves that are not narrow-banded, and accumulated error in the spatial evolution due to the truncation of the perturbation analysis leading to the MNLS equation.

5 Case Study of a Freak Wave

We have taken the measured time series in Figure 1 as initial condition for simulations forward and backward in space according to the MNLS equation (17)–(20), see [13]. This simulation most likely has little to do with reality since we are neglecting the fact that the waves were not perfectly long-crested, however there is some empirical evidence that the loads on the platform at least were unidirectional, see [5].

The predicted time histories at 50 meter intervals upstream and downstream are shown in Figure 13. At 500 m upstream of Draupner E there appears to be a large wave group about one minute before the freak wave hits the platform. This wave group appears to split up into a short and large leading group that runs away from a longer and smaller trailing group that

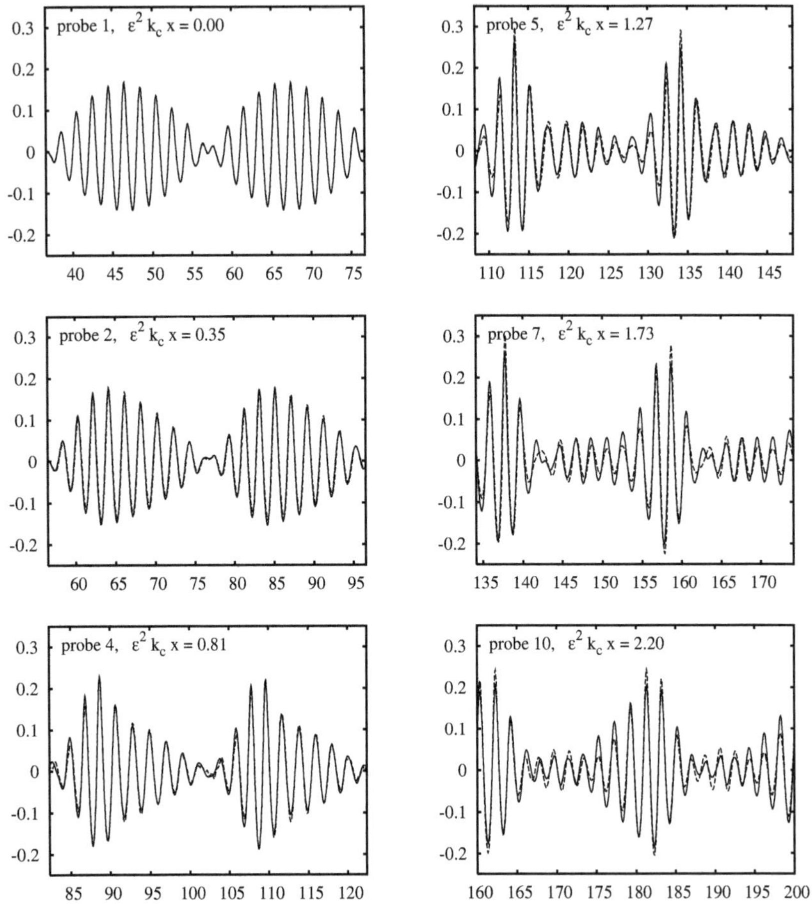

Fig. 12. Bichromatic waves: MNLS simulation (solid curves) and experimental measurements (dotted curves) at six probes. Horizontal axis is time in seconds. Vertical axis is normalized surface displacement $k_c\zeta$.

becomes rather diffuse as it approaches the platform. The freak wave that hits the platform is in the middle of the short leading group. After the impact with the platform, this wave group broadens, becomes less steep, and slows down slightly. A large trough (a "hole" in the ocean) probably occurred slightly upstream of the platform.

As far as the occurrence of the Draupner "New Year wave" is concerned, few certain conclusions can be drawn since we do not have detailed information about directional wave distribution, i.e. details about short-crestedness. We may conclude that if the waves had been long-crested, then the extreme wave likely did not occur suddenly and unexpectedly. An imaginary observer on the platform, looking north, would probably have seen large waves (a wall

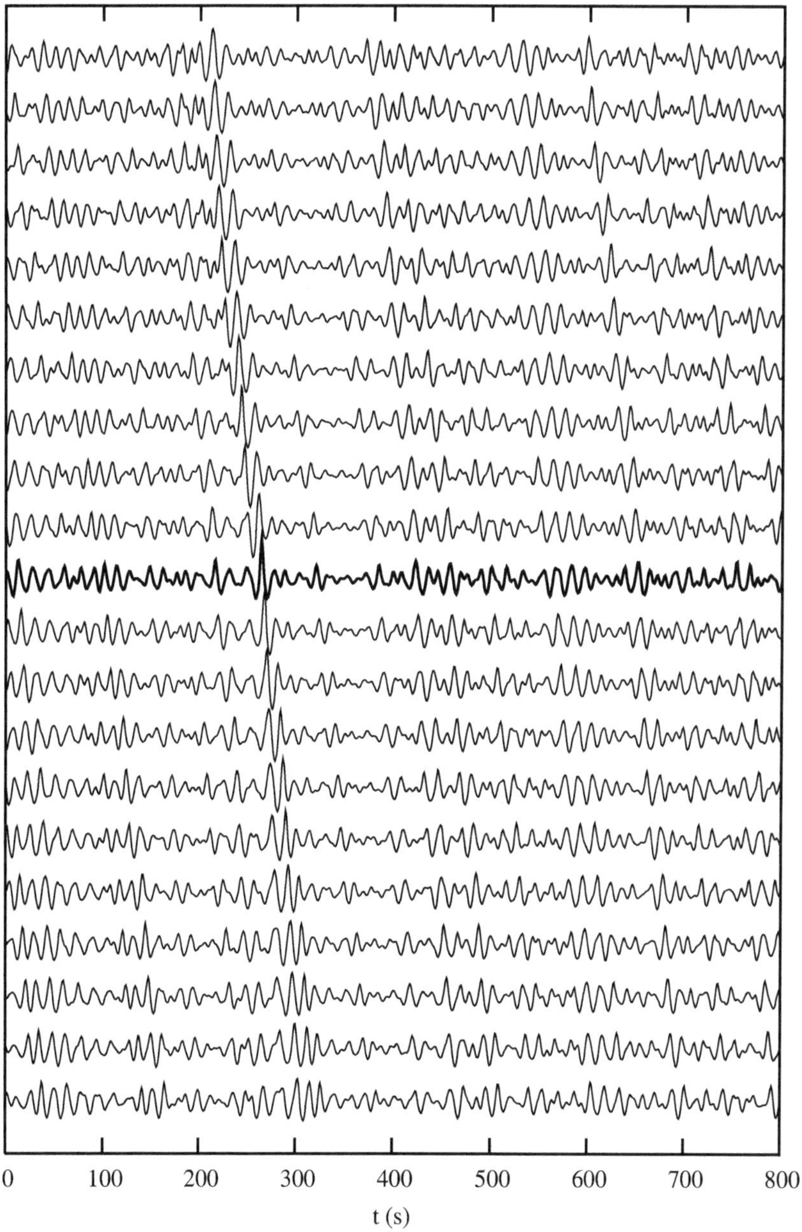

Fig. 13. Spatiotemporal evolution of the Draupner wave field, hindcasted with the MNLS spatial evolution equation. The simulation is presented from upstream (top) to downstream (bottom) as time series at equidistant intervals of 50 meters. The measured time series used for initialization of the hindcast is in bold. The zero lines for hindcasted time series are shifted vertically in comparison with the measured time series.

of water) approaching the platform for at least a minute or so, including an apparent hole in the ocean. This description is indeed similar to stories told by seafarers who survived ship founderings due to large waves, however, such stories have often not been believed in the past.

Acknowledgement. A significant part of the work reported here was done while the author was at SINTEF Applied Mathematics. The analytical and numerical development was supported by the Research Council of Norway (139177/431, 109328/410), the European Union (EVG1–CT–2001–00051, MAS3–CT96–5016), Norsk Hydro and Statoil. Validation studies were funded by Norsk Hydro. Field data was made available by Statoil. Laboratory data was made available by Carl Trygve Stansberg at Marintek. The Marintek towing tank experiment was funded by the Research Council of Norway. This paper improved much thanks to the thorough reviews by two referees.

References

1. K. B. Dysthe. Note on a modification to the nonlinear Schrödinger equation for application to deep water waves. *Proc. R. Soc. Lond. A*, 369:105–114, 1979.
2. K. B. Dysthe, K. Trulsen, H. E. Krogstad, and H. Socquet-Juglard. Evolution of a narrow band spectrum of random surface gravity waves. *J. Fluid Mech.*, 478:1–10, 2003.
3. Y. Goda. *Random seas and design of maritime structures.* World Scientific, 2000.
4. S. Haver. A possible freak wave event measured at the Draupner jacket Januar 1 1995. In *Rogue Waves 2004*, pages 1–8, 2004. http://www.ifremer.fr/web-com/stw2004/rw/fullpapers/walk_on_haver.pdf
5. D. Karunakaran, M. Bærheim, and B. J. Leira. Measured and simulated dynamic response of a jacket platform. In C. Guedes-Soares, M. Arai, A. Naess, and N. Shetty, editors, *Proceedings of the 16th International Conference on Offshore Mechanics and Arctic Engineering*, volume II, pages 157–164. ASME, 1997.
6. B. Kinsman. *Wind waves.* Dover, 1984.
7. C. C. Mei. *The applied dynamics of ocean surface waves.* World Scientific, 1989. 740 pp.
8. M. K. Ochi. *Ocean waves.* Cambridge, 1998.
9. H. Socquet-Juglard, K. Dysthe, K. Trulsen, H. E. Krogstad, and J. Liu. Probability distributions of surface gravity waves during spectral changes. *J. Fluid Mech.*, 542:195–216, 2005.
10. C. T. Stansberg. Propagation-dependent spatial variations observed in wavetrains generated in a long wave tank. data report. Technical Report MT49 A93–0176, Marintek, 1993.
11. C. T. Stansberg. Spatially developing instabilities observed in experimental bichromatic wave trains. In A. J. Grass, editor, *26th IAHR Congress (HYDRA 2000)*, volume 3, pages 180–185. Thomas Telford, 1995.
12. C. T. Stansberg. On the nonlinear behaviour of ocean wave groups. In *Proc. Ocean Wave Measurement and Analysis, Fifth International Symposium WAVES 1997*, volume 2, pages 1227–1241, 1998.

13. K. Trulsen. Simulating the spatial evolution of a measured time series of a freak wave. In M. Olagnon and G. Athanassoulis, editors, *Rogue Waves 2000*, pages 265–273. Ifremer, 2001.
14. K. Trulsen, I. Kliakhandler, K. B. Dysthe, and M. G. Velarde. On weakly nonlinear modulation of waves on deep water. *Phys. Fluids*, 12:2432–2437, 2000.
15. K. Trulsen and C. T. Stansberg. Spatial evolution of water surface waves: Numerical simulation and experiment of bichromatic waves. In *Proc. 11th International Offshore and Polar Engineering Conference*, volume 3, pages 71–77, 2001.

How to Solve Systems of Conservation Laws Numerically Using the Graphics Processor as a High-Performance Computational Engine

Trond Runar Hagen, Martin O. Henriksen, Jon M. Hjelmervik, and Knut–Andreas Lie

Summary. The paper has two main themes: The first theme is to give the reader an introduction to modern methods for systems of conservation laws. To this end, we start by introducing two classical schemes, the Lax–Friedrichs scheme and the Lax–Wendroff scheme. Using a simple example, we show how these two schemes fail to give accurate approximations to solutions containing discontinuities. We then introduce a general class of semi-discrete finite-volume schemes that are designed to produce accurate resolution of both smooth and nonsmooth parts of the solution. Using this special class we wish to introduce the reader to the basic principles used to design modern high-resolution schemes. As examples of systems of conservation laws, we consider the shallow-water equations for water waves and the Euler equations for the dynamics of an ideal gas.

The second theme in the paper is how programmable graphics processor units (GPUs or graphics cards) can be used to efficiently compute numerical solutions of these systems. In contrast to instruction driven micro-processors (CPUs), GPUs subscribe to the data-stream-based computing paradigm and have been optimised for high throughput of large data streams. Most modern numerical methods for hyperbolic conservation laws are explicit schemes defined over a grid, in which the unknowns at each grid point or in each grid cell can be updated independently of the others. Therefore such methods are particularly attractive for implementation using data-stream-based processing.

1 Introduction

Evolution of conserved quantities such as mass, momentum and energy is one of the fundamental physical principles used to build mathematical models in the natural sciences. The resulting models often appear as systems of quasi-linear hyperbolic partial differential equations (PDEs) in divergence form

$$Q_t + \nabla \cdot F(Q) = 0, \tag{1}$$

where $Q \in \mathbb{R}^m$ is the set of conserved quantities and F is a (nonlinear) flux function. This class of equations is commonly referred to as 'hyperbolic conservation laws' and govern a broad spectrum of physical phenomena in acoustics, astrophysics, cosmology, combustion theory, elastodynamics, geophysics, multiphase flow, and nonlinear material science, to name a few [9, 28, 54, 8, 33, 50, 15]. The simplest example is the linear transport equation,

$$q_t + aq_x = 0,$$

which describes the passive advection of a conserved scalar quantity q. However, the most studied case is the nonlinear Euler equations that describe the dynamics of a compressible and inviscid fluid, typically an ideal gas.

In scientific and industrial applications, the evolution of conserved quantities is often not the only modelling principle. Real-life models therefore tend to be more complex than the quasilinear equation (1). Chemical reactions, viscous forces, spatial inhomogeneities, phase transitions, etc, give rise to other terms and lead to more general evolution models of the form

$$A(Q)_t + \nabla \cdot F(x,t,Q) = S(x,t,Q) + \nabla \cdot \big(D(Q,x,t)\nabla Q\big).$$

Similarly, the evolution equation (1) is often coupled with other models and side conditions. As an example, we refer to the discussion of reservoir simulation elsewhere in this book [1, 2], where injection of water into an oil reservoir is described by a system consisting of an elliptic equation for the fluid pressure and a hyperbolic equation for the transport of fluid phases. Here we will mainly focus on simplified models of the form (1) and discuss some of the difficulties involved in solving such models numerically.

The solutions of nonlinear hyperbolic conservation laws exhibit very singular behaviour and admit various kinds of discontinuous and nonlinear waves, such as shocks, rarefactions, propagating phase boundaries, fluid and material interfaces, detonation waves, etc. Understanding these nonlinear phenomena has been a constant challenge to mathematicians, and research on the subject has strongly influenced developments in modern applied mathematics. In a similar way, computing numerically the nonlinear evolution of possibly discontinuous waves has proved to be notoriously difficult. Two seminal papers are the papers by Lax [29] from 1954, Godunov [16] from 1959, and by Lax and Wendroff [30] from 1960. Classical high-order discretisation schemes tend to generate unphysical oscillations that corrupt the solution in the vicinity of discontinuities, while low-order schemes introduce excessive numerical diffusion. To overcome these problems, Harten [19] introduced so-called high-resolution methods in 1983. These methods are typically based upon a finite-volume formation and have been developed and applied successfully in many disciplines [33, 50, 15]. High-resolution methods are designed to capture discontinuous waves accurately, while retaining high-order accuracy on smooth parts of the solution. In this paper we present a particular class of high-resolution methods and thereby try to give the reader some insight into a fascinating research field.

High-resolution methods can also be applied to more complex flows than those described by (1), in which case they need to be combined with additional numerical techniques. We will not go into this topic here, but rather refer the interested reader to e.g., the proceedings from the conference series "Finite Volumes for Complex Applications" [6, 53, 21].

High-resolution methods are generally computationally expensive; they are often based on explicit temporal discretisation, and severe restrictions on the time-steps are necessary to guarantee stability. On the other hand, the methods have a natural parallelism since each cell in the grid can be processed independently of its neighbours, due to the explicit time discretisation. High-resolution methods are therefore ideal candidates for distributed and parallel computing. The computational domain can be split into overlapping sub-domains of even size. The overlaps produce the boundary-values needed by the spatial discretisation stencils. Each processor can now update its sub-domain independently; the only need for communication is at the end of the time-step, where updated values must be transmitted to the neighbouring sub-domains. To achieve optimal load-balancing (and linear speedup), special care must be taken when splitting into sub-domains to reduce communication costs.

In this paper we will take a different and somewhat unorthodox approach to parallel computing; we will move the computation from the CPU and exploit the computer's graphics processor unit (GPU)—commonly referred to as the 'graphics card' in everyday language—as a parallel computing device. This can give us a speedup of more than one order of magnitude. Our motivation is that modern graphics cards have an unrivalled ability to process floating-point data. Most of the transistors in CPUs are consumed by the L2 cache, which is the fast computer memory residing between the CPU and the main memory, aimed at providing faster CPU access to instructions and data in memory. The majority of transistors in a GPU, on the other hand, is devoted to processing floating points to render (or rasterize) graphical primitives (points, lines, triangles, and quads) into frame buffers. The commodity GPUs we will use in this study (NVIDIA GeForce 6800 and 7800) have 16 or 24 parallel pipelines for processing fragments of graphical primitives, and each pipeline executes operations on vectors of length four with a clock frequency up to 500 MHz. This gives an *observed* peak performance of 54 and 165 Gflops, respectively, on synthetic tests, which is one order of magnitude larger than the *theoretical* 15 Gflops performance of an Intel Pentium 4 CPU. How we actually exploit this amazing processing capability for the numerical solution of partial differential equations is the second topic of the paper. A more thorough introduction to using graphics cards for general purpose computing is given elsewhere in this book [10] or in a paper by Rumpf and Strzodka [46]. Currently, the best source for information about this thriving field is the state-of-the-art survey paper by Owens et al. [42] and the GPGPU website `http://www.gpgpu.org/`.

The aim of the paper is two-fold: we wish to give the reader a taste of modern methods for hyperbolic conservation laws and an introduction to the use of graphics cards as a high-performance computational engine for partial differential equations. The paper is therefore organised as follows: Section 2 gives an introduction to hyperbolic conservation laws. In Section 3 we introduce two classical schemes and discuss how to implement them on the GPU. Then in Section 4 we introduce a class of high-resolution schemes based upon a semi-discretisation of the conservation law (1), and show some examples of implementations on GPUs. In Sections 3.5 and 5 we give numerical examples and discuss the speedup observed when moving computations from the CPU to the GPU. Finally, Section 6 contains a discussion of the applicability of GPU-based computing for more challenging problems.

2 Hyperbolic Conservation Laws

Systems of hyperbolic conservation laws arise in many physical models describing the evolution of a set of conserved quantities $Q \in \mathbb{R}^m$. A conservation law states that the rate of change of a quantity within a given domain Ω equals the flux over the boundaries $\partial\Omega$. When advective transport or wave motion are the most important phenomena, and dissipative, second-order effects can be neglected, the fluxes only depend on time, space and the conserved quantities. In the following we assume that $\Omega \in \mathbb{R}^2$ with outer normal $n = (n_x, n_y)$ and that the flux has two components (F, G), where each component is a function of Q, x, y, and t. Thus, the conservation law in integral form reads

$$\frac{d}{dt}\iint_\Omega Q\,dxdy + \int_{\partial\Omega} (F,G)\cdot(n_x,n_y)\,ds = 0. \tag{2}$$

Using the divergence theorem and requiring the integrand to be identically zero, the system of conservation laws can be written in differential form as

$$\partial_t Q + \partial_x F(Q,x,y,t) + \partial_y G(Q,x,y,t) = 0. \tag{3}$$

In the following, we will for simplicity restrict our attention to cases where the flux function only depends on the conserved quantity Q. When the Jacobi matrix $\partial_Q(F,G)\cdot n$ has only real eigenvalues and a complete set of real eigenvectors for all unit vectors n, the system is said to be hyperbolic, and thus belongs to the class of 'hyperbolic conservation laws'. These equations have several features that in general make them hard to analyse and solve numerically. For instance, a hyperbolic system may form discontinuous solutions even from smooth initial data and one therefore considers weak solutions of (3) that satisfy the equation in the sense of distributions; that is, $Q(x,y,t)$ is a weak solution of (3) if

$$\iiint\limits_{\mathbb{R}^2\times\mathbb{R}^+} \Big(Q\partial_t\phi + F(Q)\partial_x\phi + G(Q)\partial_y\phi\Big)\,dxdydt = 0. \tag{4}$$

for any smooth test function $\phi(x,y,t) \in \mathbb{R}^m$ with compact support on $\mathbb{R}^2 \times \mathbb{R}^+$. Weak solutions are not necessarily unique, which means that the conservation law in general must be equipped with certain side-conditions to pick the physically correct solution. These conditions may come in the form of entropy inequalities, here in weak integral form

$$\iiint_{\mathbb{R}^2\times\mathbb{R}^+} \Big(\eta(Q)\phi_t + \psi(Q)\phi_x + \varphi(Q)\phi_y\Big)\, dxdydt \\ + \iint_{\mathbb{R}^2} \phi(x,y,0)\eta(Q(x,y,0))\, dxdy \geq 0,$$

where $\eta(Q)$ denotes the entropy, (ψ, φ) the associated pair of entropy fluxes, and ϕ is a positive test function. The concept of entropy inequalities comes from the second law of thermodynamics, which requires that the entropy is nondecreasing. Generally, entropy is conserved when the function Q is smooth and increases over discontinuities. For scalar equations it is common to use a special family of entropy functions and fluxes due to Kružkov [24],

$$\eta_k(q) = |q - k|, \qquad \begin{cases} \psi_k(q) = \operatorname{sign}(q-k)[F(q) - F(k)], \\ \varphi_k(q) = \operatorname{sign}(q-k)[G(q) - G(k)], \end{cases}$$

where k is any real number.

A particular feature of hyperbolic systems is that all disturbances have a finite speed of propagation. For simplicity, let us consider a linear system in one-dimension,

$$U_t + AU_x = 0$$

with initial data $U(x,0) = U_0(x)$. Hyperbolicity means that the constant coefficient matrix A is diagonalisable with real eigenvalues λ_i, i.e., we can write A as $R\Lambda R^{-1}$, where $\Lambda = \operatorname{diag}(\lambda_1, \ldots, \lambda_m)$, and each column r_i of R is a right eigenvector of A. Premultiplying the conservation law by R^{-1} and defining $V = R^{-1}U$, we end up with m independent scalar equations

$$\partial_t V + \Lambda \partial_x V = 0.$$

Here V is called the vector of *characteristic variables*, and the decomposition $R\Lambda R^{-1}$ is called a *characteristic decomposition.*

Each component v_i of V satisfies a linear transport equation $(v_i)_t + \lambda_i(v_i)_x = 0$, with initial data $v_i^0(x)$ given from $V^0 = R^{-1}U_0$. Each linear transport equation is solved by $v_i(x,t) = v_i^0(x - \lambda_i t)$, and hence we have that

$$U(x,t) = \sum_{i=0}^{m} v_i^0(x - \lambda_i t) r_i.$$

This equation states that the solution consists of a superposition of m simple waves that are advected independently of each other along so-called characteristics. The propagation speed of each simple wave is given by the corresponding eigenvalue. In the nonlinear case, we can write $Q_t + A(Q)Q_x = 0$, where $A(Q)$ denotes the Jacobi matrix dF/dQ. However, since the eigenvalues and eigenvectors of $A(Q)$ now will depend on the unknown vector Q, we cannot repeat the analysis from the linear case and write the solution in a simple closed form. Still, the eigenvalues (and eigenvectors) give valuable information of the local behaviour of the solution, and we will return to them later.

Before we introduce numerical methods for conservation laws, we will introduce two canonical models. The interested reader can find a lot more background material in classical books like [9, 28, 54, 8] or in one of the recent textbooks numerical methods for conservation laws[15, 22, 33, 50] that introduce both theory and modern numerical methods. A large number of recent papers on mathematical analysis and numerical methods for hyperbolic conservation laws can be found on the *Conservation Laws Preprint Server* in Trondheim, `http://www.math.ntnu.no/conservation/`. Excellent images of a large number of flow phenomena can be found in Van Dyke's *Album of Fluid Motion* [11] or by visiting e.g., the *Gallery of Fluid Mechanics*, `http://www.galleryoffluidmechanics.com/`.

2.1 The Shallow-Water Equations

In its simplest form, the shallow-water equations read

$$Q_t + F(Q)_x = \begin{bmatrix} h \\ hu \end{bmatrix}_t + \begin{bmatrix} hu \\ hu^2 + \frac{1}{2}gh^2 \end{bmatrix}_x = \begin{bmatrix} 0 \\ 0 \end{bmatrix}. \qquad (5)$$

The two equations describe conservation of mass and momentum. This system of equations has applications to a wide range of phenomena, including water waves in shallow waters, (granular) avalanches, dense gases, mixtures of materials, and atmospheric flow. If the application is water waves, h denotes water depth, u the depth-averaged velocity, and g is the acceleration of gravity.

The shallow-water equations have the following eigenvalues

$$\lambda(Q) = u \pm \sqrt{gh}. \qquad (6)$$

and possess only nonlinear waves in two families corresponding to the slow and the fast eigenvalue. These nonlinear waves may be continuous transitions called rarefactions or propagating discontinuities called shocks. A rarefaction wave satisfies a differential equation on the form

$$\frac{dF(Q)}{dQ}Q'(\xi) = \lambda(Q)Q'(\xi), \qquad \xi = \xi(x,t) = (x - x_0)/(t - t_0),$$

whereas a shock between right and left values Q_L and Q_R satisfies an algebraic equation of the form

$$(Q_L - Q_R)s = F(Q_L) - F(Q_R),$$

where s is the shock speed. (This algebraic equation is called the Rankine–Hugoniot condition, see [31].)

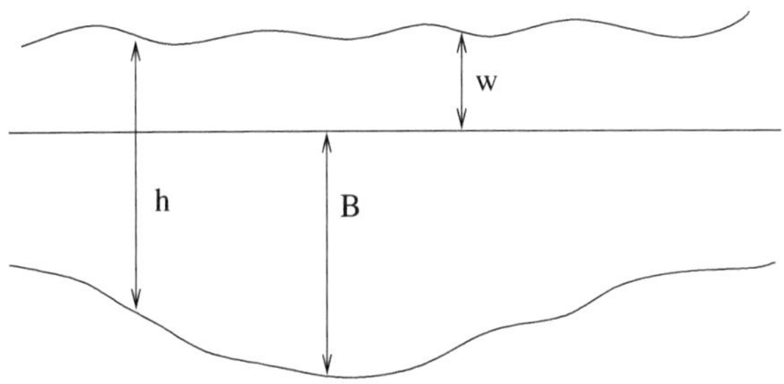

Fig. 1. The quantities involved in the shallow-water equations.

The shallow-water equations form a depth-integrated model for free surface flow of a fluid under the influence of gravity, see Figure 1. In the model, it is assumed that the height of the top surface above the bottom is sufficiently small compared with some characteristic length of motion. Moreover, vertical motion within the fluid is ignored, and we have hydrostatic balance in the vertical direction. For water waves over a bottom topography (bathymetry) $B(x, y)$ in two spatial dimensions, this gives rise to the following inhomogeneous system

$$\begin{bmatrix} h \\ hu \\ hv \end{bmatrix}_t + \begin{bmatrix} hu \\ hu^2 + \frac{1}{2}gh^2 \\ huv \end{bmatrix}_x + \begin{bmatrix} hv \\ huv \\ hv^2 + \frac{1}{2}gh^2 \end{bmatrix}_y = \begin{bmatrix} 0 \\ -ghB_x \\ -ghB_y \end{bmatrix}. \tag{7}$$

The right-hand side of (7) models the influence of a variable bathymetry. Figure 1 illustrates the situation and the quantities involved. The function $B = B(x, y)$ describes the bathymetry, while $w = B(x, y, t)$ models the water surface elevation, and we have that $h = w - B$. For brevity, we will refer to this system as

$$Q_t + F(Q)_x + G(Q)_y = H(Q, Z).$$

The system has the following eigenvalues,

$$\lambda(Q, n) = (u, v) \cdot n \pm \sqrt{gh},\ (u, v) \cdot n$$

where a linear shear wave with characteristic wave-speed v has been introduced in addition to the two nonlinear waves described above. For simplicity, we will henceforth assume a flat bathymetry, for which $H(Q, Z) \equiv 0$. Example 6 contains a short discussion of variable bathymetry.

2.2 The Euler Equations

Our second example of a hyperbolic system is given by the Euler equations for an ideal, polytropic gas. This system is probably *the* canonical example of a hyperbolic system of conservation laws that describes conservation of mass, momentum (in each spatial direction) and energy. In one spatial dimension the equations read

$$Q_t + F(Q)_x = \begin{bmatrix} \rho \\ \rho u \\ E \end{bmatrix}_t + \begin{bmatrix} \rho u \\ \rho u^2 + p \\ u(E+p) \end{bmatrix}_x = 0. \tag{8}$$

and similarly in two dimensions

$$\begin{bmatrix} \rho \\ \rho u \\ \rho v \\ E \end{bmatrix}_t + \begin{bmatrix} \rho u \\ \rho u^2 + p \\ \rho uv \\ u(E+p) \end{bmatrix}_x + \begin{bmatrix} \rho v \\ \rho uv \\ \rho v^2 + p \\ v(E+p) \end{bmatrix}_y = 0. \tag{9}$$

Here ρ denotes density, u and v velocity in x- and y- directions, p pressure, and E is the total energy (kinetic plus internal energy) given by

$$E = \tfrac{1}{2}\rho u^2 + p/(\gamma - 1), \qquad \text{or} \qquad E = \tfrac{1}{2}\rho(u^2 + v^2) + p/(\gamma - 1).$$

The gas constant γ depends on the species of the gas, and is, for instance, equal to 1.4 for air. The eigenvalues of (8) are

$$\lambda(Q, n) = (u, v) \cdot n \pm \sqrt{\gamma p/\rho},\ (u, v) \cdot n,$$

giving three waves: a slow nonlinear wave, a linear contact wave, and a fast nonlinear wave. In two space dimensions $\lambda(Q, n) = (u, v) \cdot n$ is a double eigenvalue corresponding both to a contact wave and a shear wave. A contact wave is characterised by a jump in density together with no change in pressure and velocity, i.e., ρ is discontinuous and p and $(u, v) \cdot n$ are constant, whereas a shear wave means a discontinuity in the tangential velocity $(u, v) \cdot n^{\perp}$.

3 Two Classical Finite Volume Schemes

A standard approach to numerical schemes for PDEs is to look for a *pointwise* approximation and replace temporal and spatial derivatives by finite differences. Alternatively, we can divide the domain Ω into a set of finite sub-domains Ω_i and then use the integral form of the equation to compute approximations to the *cell average* of Q over each grid cell Ω_i. In one spatial dimension, the cell-average is given by

$$Q_i(t) = \frac{1}{\Delta x_i} \int_{x_{i-1/2}}^{x_{i+1/2}} Q(x, t)\, dx, \tag{10}$$

and similarly in two dimensions

$$Q_i(t) = \frac{1}{|\Omega_i|} \iint_{\Omega_i} Q(x,y,t)\, dxdy. \tag{11}$$

The corresponding schemes are called *finite-volume schemes.*

To derive equations for the evolution of cell averages $Q_i(t)$ in one spatial dimension, we impose the conservation law on integral form (2) on the grid cell $\Omega_i = [x_{i-1/2}, x_{i+1/2}]$, which then reads

$$\frac{d}{dt} \int_{x_{i-1/2}}^{x_{i+1/2}} Q(x,t)\, dx = F\big(Q(x_{i-1/2},t)\big) - F\big(Q(x_{i+1/2},t)\big) \tag{12}$$

Using the integral form of the equation rather than the differential form is quite natural, as it ensures that also the discrete approximations satisfy a conservation property.

From (12) we readily obtain a set of ordinary differential equations for the cell averages $Q_i(t)$,

$$\frac{d}{dt} Q_i(t) = -\frac{1}{\Delta x_i}\big[F_{i+1/2}(t) - F_{i-1/2}(t)\big], \tag{13}$$

where the fluxes across the cell-boundaries are given by

$$F_{i\pm 1/2}(t) = F\big(Q(x_{i\pm 1/2},t)\big). \tag{14}$$

To derive schemes for the two-dimensional shallow-water and Euler equations we will follow a similar approach. We will assume for simplicity that the finite volumes are given by the grid cells in a uniform Cartesian grid, such that

$$\Omega_{ij} = \big\{\, (x,y) \in \big[(i-1/2)\Delta x, (i+1/2)\Delta x\big] \times \big[(j-1/2)\Delta y, (j+1/2)\Delta y\big] \,\big\}.$$

To simplify the presentation, we henceforth assume that $\Delta x = \Delta y$. Then the semi-discrete approximation to (2) reads

$$\begin{aligned} \frac{d}{dt} Q_{ij}(t) = L_{ij}(Q(t)) = &-\frac{1}{\Delta x}\Big[F_{i+1/2,j}(t) - F_{i-1/2,j}(t)\Big] \\ &-\frac{1}{\Delta y}\Big[G_{i,j+1/2}(t) - G_{i,j-1/2}(t)\Big], \end{aligned} \tag{15}$$

where $F_{i\pm 1/2,j}$ and $G_{i,j\pm 1/2}$ are numerical approximations to the fluxes over the cell edges

$$\begin{aligned} F_{i\pm 1/2,j}(t) &\approx \frac{1}{\Delta y} \int_{y_{j-1/2}}^{y_{j+1/2}} F\big(Q(x_{i\pm 1/2},y,t)\big)\, dy, \\ G_{i,j\pm 1/2}(t) &\approx \frac{1}{\Delta x} \int_{x_{i-1/2}}^{x_{i+1/2}} G\big(Q(x,y_{j\pm 1/2},t)\big)\, dx. \end{aligned} \tag{16}$$

Equation (15) is a set of ordinary differential equations (ODEs) for the time-evolution of the cell-average values Q_{ij}.

Modern high-resolution schemes are often derived using a finite-volume framework, as we will see in Section 4. To give the reader a gentle introduction to the concepts behind high-resolution schemes we start by deriving two classical schemes, the Lax–Friedrichs [29] and the Lax–Wendroff schemes [30], using the semi-discrete finite-volume framework. The exact same schemes can be derived using a finite-difference framework, but then the interpretation of the unknown discrete quantities would be different. The distinction between finite-difference and finite-volume schemes is not always clear in the literature, and many authors tend to use the term 'finite-volume scheme' to denote conservative schemes derived using finite differences. By conservative we mean a scheme that can be written on conservative form (in one spatial dimension):

$$Q_i^{n+1} = Q_i^n - r(F_{i+1/2} - F_{i-1/2}), \tag{17}$$

where $Q_i^n = Q_i(n\Delta t)$ and $r = \Delta t/\Delta x$. The importance of the conservative form was established through the famous Lax–Wendroff theorem [30], which states that if a sequence of approximations computed by a consistent and *conservative* numerical method converges to some limit, then this limit is a weak solution of the conservation law.

To further motivate this form, we can integrate (12) from t to $t+\Delta t$ giving

$$\begin{aligned} Q_i(t+\Delta t) &= Q_i(t) \\ &\quad - \frac{1}{\Delta x_i}\Big[\int_t^{t+1\Delta t} F(Q(x_{i+1/2},t))\,dt - \int_t^{t+1\Delta t} F(Q(x_{i-1/2},t))\,dt\Big]. \end{aligned}$$

By summing (17) over all cells (for $i = -M : N$) we find that the cell averages satisfy a discrete conservation principle

$$\sum_{i=-M}^{N} Q_i^{n+1}\Delta x = \sum_{i=-M}^{N} Q_i^n \Delta x - \Delta t(F_{N+1/2} - F_{-M-1/2})$$

In other words, our discrete approximations mimic the conservation principle of the unknown function Q since the sum $\sum_i Q_i \Delta x$ approximates the integral of Q.

3.1 The Lax–Friedrichs Scheme

We will now derive our first numerical scheme. For simplicity, let us first consider one spatial dimension only. We start by introducing a time-step Δt and discretise the ODE (13) by the forward Euler method. Next, we must evaluate the flux-integrals (14) along the cell edges. We now make the assumption that $Q(x, n\Delta t)$ is piecewise constant and equals the cell average $Q_i^n = Q_i(n\Delta t)$ inside each grid cell. Then, since all perturbations in a hyperbolic system travel with a finite wave-speed, the problem of estimating the solution at the cell interfaces can be recast as a set of locally defined initial-value problems on the form

$$Q_t + F(Q)_x = 0, \qquad Q(x,0) = \begin{cases} Q_L, & x < 0, \\ Q_R, & x > 0. \end{cases} \tag{18}$$

This is called a Riemann problem and is a fundamental building block in many high-resolution schemes. We will come back to the Riemann problem in Section 4.1. For now, we take a more simple approach to avoid solving the Riemann problem. For the Lax–Friedrichs scheme we approximate the flux function at each cell-boundary by the average of the fluxes evaluated immediately to the left and right of the boundary. Since we have assumed the function to be piecewise constant, this means that we will evaluate the flux using the cell averages in the adjacent cells,

$$F(Q(x_{i+1/2}, n\Delta t)) = \frac{1}{2}\Big(F(Q_i^n) + F(Q_{i+1}^n)\Big), \quad \text{etc.}$$

This approximation is commonly referred to as a *centred* approximation.

Summing up, our first scheme reads

$$Q_i^{n+1} = Q_i^n - \frac{1}{2} r \Big[F(Q_{i+1}^n) - F(Q_{i-1}^n)\Big].$$

Unfortunately, this scheme is notoriously unstable. To obtain a stable scheme, we can add the artificial diffusion term $(\Delta x^2/\Delta t)\partial_x^2 Q$. Notice, however, that by doing this, it is no longer possible to go back to the semi-discrete form, since the artificial diffusion terms blow up in the limit $\Delta t \to 0$. If we discretise the artificial diffusion term by a standard central difference, the new scheme reads

$$Q_i^{n+1} = \frac{1}{2}\Big(Q_{i+1}^n + Q_{i-1}^n\Big) - \frac{1}{2} r \Big[F(Q_{i+1}^n) - F(Q_{i-1}^n)\Big]. \tag{19}$$

This is the classical first-order Lax–Friedrichs scheme. The scheme is stable provided the following condition is fulfilled

$$r \max_{i,k} |\lambda_k^F(Q_i)| \leq 1, \tag{20}$$

where λ_k^F are the eigenvalues of the Jacobian matrix of the flux function F. The condition is called the CFL condition and simply states that the domain of dependence for the exact solution, which is bounded by the characteristics of the smallest and largest eigenvalues, should be contained in the domain of dependence for the discrete equation.

A two-dimensional scheme can be derived analogously. To approximate the integrals in (16) one can generally use any numerical quadrature rule. However, the midpoint rule is sufficient for the first-order Lax–Friedrichs scheme. Thus we need to know the point values $Q(x_{i\pm1/2}, y_j, n\Delta t)$ and $Q(x_i, y_{j\pm1/2}, n\Delta t)$. We assume that $Q(x, y, n\Delta t)$ is piecewise constant and equal to the cell average within each grid cell. The values of $Q(x, y, n\Delta t)$ at the midpoints of the edges can then be estimated by simple averaging of the one-sided values, like in the one dimensional case. Altogether this gives the following scheme

$$\begin{aligned} Q_{ij}^{n+1} =& \frac{1}{4}\Big(Q_{i+1,j}^n + Q_{i-1,j}^n + Q_{i,j+1}^n + Q_{i,j-1}^n\Big) \\ &- \frac{1}{2}r\Big[F(Q_{i+1,j}^n) - F(Q_{i-1,j}^n)\Big] - \frac{1}{2}r\Big[G(Q_{i,j+1}^n) - G(Q_{i,j-1}^n)\Big], \end{aligned} \tag{21}$$

which is stable under a CFL restriction of 1/2.

The Lax–Friedrichs scheme is also commonly derived using finite differences as the spatial discretisation, in which case Q_{ij}^n denotes the point values at $(i\Delta x, j\Delta y, n\Delta t)$. The scheme is very robust. This is largely due to the added numerical dissipation, which tends to smear out discontinuities. A formal Taylor analysis of a single time step shows that the truncation error of the scheme is of order two in the discretisation parameters. Rather than truncation errors, we are interested in the error at a fixed time (using an increasing number of time steps), and must divide with Δt, giving an error of order one. For smooth solutions the error measured in, e.g., the L^∞ or L^1-norm therefore decreases linearly as the discretisation parameters are refined, and hence we call the scheme *first-order*.

A more fundamental derivation of the Lax–Friedrichs scheme (21) is possible if one introduces a *staggered* grid (see Figure 9). To this end, let Q_{i+1}^n be the cell average over $[x_i, x_{i+2}]$, and Q_i^{n+1} be the cell average over the staggered cell $[x_{i-1}, x_{i+1}]$. Then (19) is the *exact* evolution of the piecewise constant data from time t^n to t^{n+1} followed by an averaging onto the staggered grid. This averaging introduces the additional diffusion discussed above. The two-dimensional scheme (21) can be derived analogously.

3.2 The Lax–Wendroff Scheme

To achieve formal second-order accuracy, we replace the forward Euler method used for the integration of (13) by the midpoint method, that is,

$$Q_i^{n+1} = Q_i^n - r\Big[F_{i+1/2}^{n+1/2} - F_{i-1/2}^{n+1/2}\Big]. \tag{22}$$

To complete the scheme, we must specify the fluxes $F_{i\pm1/2}^{n+1/2}$, which again depend on the values at the midpoints at time $t^{n+1/2} = (n+\frac{1}{2})\Delta t$. One can show that these values can be *predicted* with acceptable accuracy using the Lax–Friedrichs scheme on a grid with grid-spacing $\frac{1}{2}\Delta x$,

$$Q_{i+1/2}^{n+1/2} = \frac{1}{2}\Big(Q_{i+1}^n + Q_i^n\Big) - \frac{1}{2}r\Big[F_{i+1}^n - F_i^n\Big]. \tag{23}$$

Altogether this gives a second-order, predictor-corrector scheme called the Lax–Wendroff scheme [30]. Like the Lax–Friedrichs scheme, the scheme in

(22)–(23) can be interpreted both as a finite difference and as a finite volume scheme, and the scheme is stable under a CFL condition of 1. For extensions of the Lax–Wendroff method to two spatial dimensions, see [44, 15, 33]. To get second-order accuracy, one must include cross-terms, meaning that the scheme uses nine points. A recent version of the two-dimensional Lax–Wendroff scheme is presented in [37].

We have now introduced two classical schemes, a first-order and a second-order scheme. Before we describe how to implement these schemes on the GPU, we will illustrate their approximation qualities.

Example 1. Consider a linear advection equation on the unit interval with periodic boundary data

$$u_t + u_x = 0, \qquad u(x,0) = u_0(x), \qquad u(0,t) = u(1,t).$$

As initial data $u_0(x)$ we choose a combination of a smooth squared sine wave and a double step function,

$$u(x,0) = \sin^2\Big(\frac{x-0.1}{0.3}\pi\Big)\chi_{[0.1,0.4]}(x) + \chi_{[0.6,0.9]}(x).$$

Figure 2 shows approximate solutions after ten periods ($t = 10$) computed by Lax–Friedrichs and Lax–Wendroff on a grid with 100 cells for $\Delta t = 0.95\Delta x$. Both schemes clearly give unacceptable resolution of the solution profile. The first-order Lax–Friedrichs scheme smears both the smooth and the discontinuous part of the advected profile. The second-order Lax–Wendroff scheme preserves the smooth profile quite accurately, but introduces spurious oscillations at the two discontinuities.

Consider next the nonlinear Burgers' equation,

$$u_t + \big(\tfrac{1}{2}u^2\big)_x = 0, \qquad u(x,0) = u_0(x), \qquad u(0,t) = u(1.5,t),$$

with the same initial data as above. Burgers' equation can serve as a simple model for the nonlinear momentum equation in the shallow-water and the Euler equations. Figure 3 shows approximate solutions at time $t = 1.0$ computed by Lax–Friedrichs and Lax–Wendroff on a grid with 150 cells for $\Delta t = 0.95\Delta x$. As in Figure 2, Lax–Friedrichs smooths the nonsmooth parts of the solution (the two shocks and the kink at $x = 0.1$), whereas Lax–Wendroff creates spurious oscillations at the two shocks. On the other hand, the overall approximation qualities of the schemes are now better due to self-sharpening mechanisms inherent in the nonlinear equation.

To improve the resolution one can use a so-called composite scheme, as introduced by Liska and Wendroff [37]. This scheme is probably the simplest possible high-resolution scheme and consists of e.g., three steps of Lax–Wendroff followed by a smoothing Lax–Friedrichs step. The idea behind this scheme is that the numerical dissipation in the first-order Lax–Friedrichs scheme should dampen the spurious oscillations created by the second-order Lax–Wendroff scheme. In Figure 4 we have recomputed the results from Figures 2 and 3 using the composite scheme.

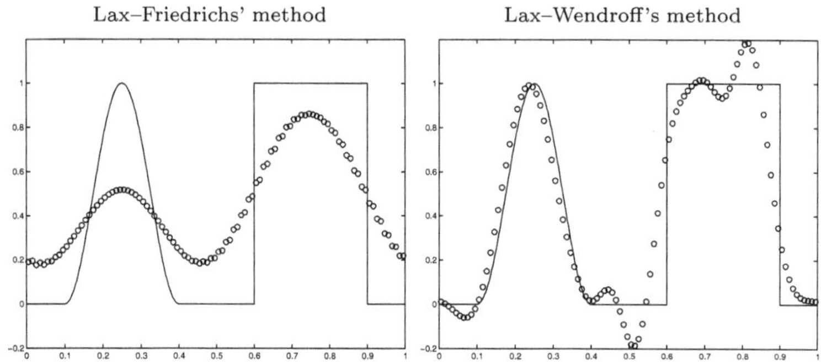

Fig. 2. Approximate solutions at time $t = 10.0$ for the linear advection equation computed by the Lax–Friedrichs and the Lax–Wendroff schemes.

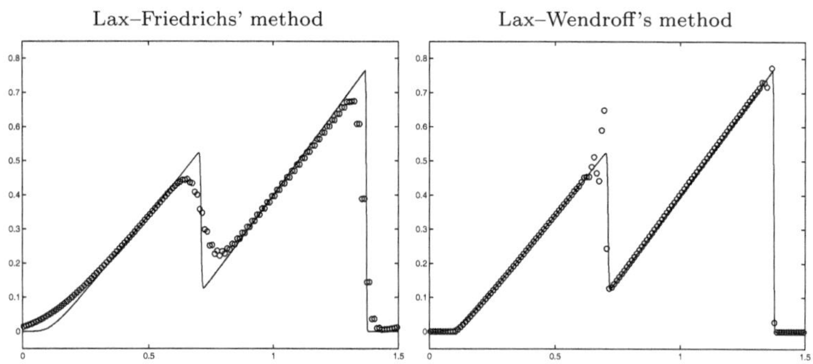

Fig. 3. Approximate solutions at time $t = 1.0$ for Burgers' equation computed by the Lax–Friedrichs and the Lax–Wendroff schemes.

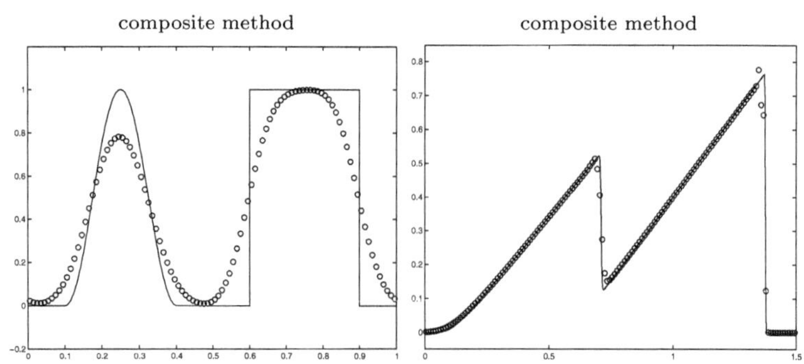

Fig. 4. Approximate solution at time $t = 10.0$ for the linear advection equation (left) and at time $t = 1.0$ for Burgers' equation (right) computed by a composite scheme.

The previous example demonstrated the potential shortcomings of classical schemes, and such schemes are seldom used in practice. On the other hand, they have a fairly simple structure and are therefore ideal starting points for discussing how to implement numerical methods on the GPU. In the next subsection, we therefore describe how to implement the Lax–Friedrichs scheme on the GPU.

3.3 Implementation on the GPU

Before we start discussing the actual implementation the Lax–Friedrichs scheme, let us look briefly into the design of a graphics card and its associated programming model.

GPUs operate in a way that is similar to a shared-memory parallel computer of the single-instruction-multiple-data (SIMD) type. In a SIMD computer, a single control unit dispatches instructions to a large number of processing nodes, and each node executes the same instruction on its own local data. SIMD computers are obviously particularly efficient when the same set of operations can be applied to all data items. If the execution involves conditional branches, there may be a performance loss since processing nodes in the worst case (depending on the underlying hardware) must execute both branches and thereafter pick the correct result. In a certain sense, explicit high-resolution schemes are good candidates for a SIMD implementation, since the update of each grid-cell involves a fixed set of given arithmetic operations and seldom introduces conditional branches in the instruction stream.

Let us now look at how computations are dispatched in a GPU. The main objective of the GPU is to render geometrical primitives (points, lines, triangles, quads), possibly combined with one or more textures (image data), as discrete pixels in a frame buffer (screen or off-screen buffer). When a primitive is rendered, the rasterizer samples the primitive at points corresponding to the pixels in the frame buffer. Per vertex attributes such as texture coordinates are set by the application for each vertex and the rasterizer calculates an interpolated value of these attributes for each point. These bundles of values then contain all the information necessary for calculating the colour of each point, and are called *fragments*. The processing pipeline is illustrated in Figure 5. We see that whereas a CPU is instruction-driven, the programming model of a GPU is data-driven. This means that individual data elements of the data-stream can be pre-fetched from memory before they are processed, thus avoiding the memory latency that tends to hamper CPUs.

Graphics pipelines contain two programmable parts, the *vertex processor* and the *fragment processor*. The vertex processor operates on each vertex without knowledge of surrounding vertices or the primitive to which it belongs. The fragment processor works in the same manner, it operates on each fragment in isolation. Both processors use several parallel pipelines, and thus work on several vertices/fragments simultaneously. To implement our numerical schemes (or a graphical algorithm) we must specify the operations in the

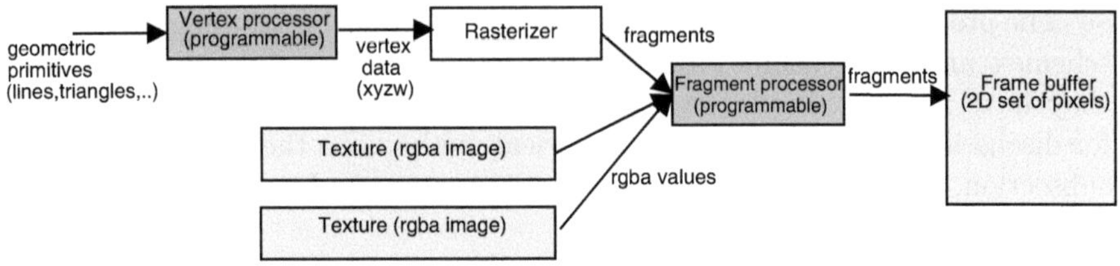

Fig. 5. A schematic of the graphics pipeline.

vertex and fragment processors. To this end, we make one or more *shaders*; these are codes written in a high-level, often C-like, language like Cg or Open GL Shading Language (GLSL), and are read and compiled by our graphical application into programs that are run by the vertex and fragment processors. Consult the references [14] and [45] for descriptions of these languages. The compiled shaders are executed for each vertex/fragment in parallel.

Our data model will be as follows: The computational domain is represented as an off-screen buffer. A geometric primitive is rendered such that a set of $N_x \times N_y$ fragments is generated which covers the entire domain. In this model, we never explicitly loop over all cells in the grid. Instead, processing of each single grid-cell is invoked implicitly by the rasterizer, which interpolates between the vertices and generates individual data for each fragment. Input field-values are realised in terms of 2D textures of size $N_x \times N_y$. Between iterations, the off-screen buffer is converted into a texture, and another off-screen buffer is used as computational domain. To solve the conservation law (3), the fragment processor loads the necessary data and performs the update of cell averages, according to (21) for the Lax–Friedrichs scheme. In this setting, the vertex processor is only used to set texture coordinates that are subsequently passed to the fragment processor. The fragment processor is then able to load the data associated with the current cell and its neighbours from the texture memory.

In the following subsection we will discuss the implementation of fragment shaders in detail for the Lax–Friedrichs scheme. However, to start the computations, it is necessary to set up the data stream to be processed and to configure the graphics pipeline. This is done in a program running on the CPU. Writing such a program requires a certain familiarity with the graphics processing jargon, and we do not go into this issue—the interested reader is referred to [46, 13, 43].

Shaders for Lax–Friedrichs

Listing 1 shows the vertex and fragment shader needed to run the Lax–Friedrichs scheme (21) for the two-dimensional shallow-water equations (7) with constant bottom topography $B \equiv 0$. In the listing, all entities given in boldface are keywords and all entities starting with `gl_` are built-in variables, constants, or attributes.

Listing 1. Fragment and vertex shader for the Lax–Friedrichs scheme written in GLSL.

```
[Vertex shader]

varying vec4 texXcoord;
varying vec4 texYcoord;
uniform vec2 dXY;

void main(void)
{
  texXcoord=gl_MultiTexCoord0.yxxx+vec4(0.0,0.0,-1.0,1.0)*dXY.x; // j, i, i-1, j+1
  texYcoord=gl_MultiTexCoord0.xyyy+vec4(0.0,0.0,-1.0,1.0)*dXY.y; // i, j, j-1, j+1

  gl_Position  = gl_ModelViewProjectionMatrix * gl_Vertex;
}


[Fragment shader]

varying vec4 texXcoord;
varying vec4 texYcoord;

uniform sampler2D QTex;

uniform float r;
uniform float halfG;

vec4 fflux(in vec4 Q)
{
  float u=Q.y/Q.x;
  return vec4( Q.y, (Q.y*u + halfG*Q.x*Q.x), Q.z*u, 0.0 );
}

vec4 gflux(in vec4 Q)
{
  float v=Q.z/Q.x;
  return vec4( Q.z, Q.y*v, (Q.z*v + halfG*Q.x*Q.x), 0.0 );
}

void main(void)
{
  vec4 QE = texture2D(QTex, texXcoord.wx);
  vec4 QW = texture2D(QTex, texXcoord.zx);
  vec4 QN = texture2D(QTex, texYcoord.xw);
  vec4 QS = texture2D(QTex, texYcoord.xz);

  gl_FragColor = 0.25*(QE + QW + QN + QS)
    - 0.5*r*(fflux(QE) - fflux(QW))
    - 0.5*r*(gflux(QN) - gflux(QS));
}
```

The vertex shader is executed once for all vertices in the geometry. Originally, the geometry is defined in a local coordinate system called object-space or model coordinates. The last line of the vertex shader transforms the vertices from object-space to the coordinate system of the camera (view coordinates, or eye-space). The two coordinate systems will generally be different. During the rendering phase only points inside the camera view (i.e., inside our computational domain) are rendered. In the shader, the model coordinates are given by `gl_Vertex` and the projected view coordinates are returned by `gl_Position`. The coordinate transformation is an affine transformation given by a global matrix called `gl_ModelViewProjectionMatrix`. (Since the camera does not move, this transformation is strictly speaking not necessary to perform for every time step and could have been done once and for all on the CPU).

The first two lines of the vertex shader compute the texture coordinates. These coordinates will be used by the fragment shader to fetch values in the current cell and its four nearest neighbours from one or more texture. The keywords `vec2` and `vec4` declare vector variables with 2 or 4 elements respectively. The construction `varying vec4 texXcoord` means that the four-component vector `texXcoord` will be passed from the vertex shader to the fragment shader. This happens as follows: The vertex shader defines one vector-value `V` at each vertex (the four corners of our computational rectangle), and the rasterizer then interpolates bilinearly between these values to define corresponding values for all fragments within the camera view. In contrast, `uniform vec2 dXY` means that `dXY` is constant for each primitive; its value must be set by the application program running on the CPU. Notice how we can access individual members of a vector as `V.x`, `V.y`, `V.z` and `V.w`, or several elements e.g., `V.xy` or `V.wzyx`. The last example demonstrates 'swizzling'; permuting the components of the vector.

We now move on to the fragment shader. Here we see the definitions of the texture coordinates `texXcoord` and `texYcoord`, used to pass coordinates from the vertex processor. `QTex` is a texture handle; think of it as a pointer to an array. The constants `r` and `halfG` represent $r = \Delta t/\Delta x$ and $\frac{1}{2}g$, respectively. To fetch the cell-averages from positions in the texture corresponding to the neighbouring cells, we use the function `texture2D`. The new cell-average is returned as the final fragment colour by writing to the variable `gl_FragColor`.

To ensure a stable scheme, we must ensure that the time step satisfies a CFL condition of the form (20). This means that we have to gather the eigenvalues from all points in the grid to determine the maximum absolute value. In parallel computing, such a gather operation is usually quite expensive and represents a bottleneck in the processing. By rethinking the gather problem as a graphics problem, we can actually use hardware supported operations on the GPU to determine the maximal characteristic speed. Recall that the characteristic speed of each wave is given by the corresponding eigenvalue. To this end we pick a suitably sized array, say 16×16, and decompose the computational domain $N_x \times N_y$ into rectangular sub-domains (quads) so that each sub-domain is associated with a 16×16 patch of the texture containing the Q-values. All quads are moved so that they coincide in world-coordinates, and hence we get $N_x N_y/16^2$ overlapping quads. By running the fragment shader in Listing 2 they are rendered to the *depth buffer*. The purpose of the depth buffer is to determine whether a pixel is behind or in front of another. When a fragment is rendered to the depth buffer, its depth value is therefore compared with the depth value that has already been stored for the current position,

Listing 2. Fragment shader written in GLSL for determining the maximum characteristic speed using the depth buffer.

```
uniform sampler2D QnTex;
uniform float scale;
uniform float G;

void main(void)
{
  vec4 Q = texture2D(QnTex, gl_TexCoord[0].xy);
  Q.yz /= Q.x;
  float c = sqrt( G*Q.x );
  float r = max(abs(Q.y) + c, abs(texQ.z) + c);
  r *= scale;

  gl_FragDepth = r;
}
```

and the maximum value is stored. This process is called *depth test.* In our setting this means that when all quads have been rendered, we have a 16×16 depth buffer, where each value represents the maximum over $N_x N_y / 16^2$ values. The content of the 16×16 depth buffer is read back to the CPU, and the global maximum is determined. Since the depth test only accepts values in the interval $[0, 1]$, we include an appropriate scaling parameter `scale` passed in from the CPU.

3.4 Boundary Conditions

In our discussions above we have tacitly assumed that every grid cell is surrounded by neighbouring cells. In practice, however, all computations are performed on some bounded domain and one therefore needs to specify boundary conditions. On a CPU, an easy way to impose boundary conditions is to use *ghost cells*; that is, to extend the domain by extra grid cells along the boundaries, whose values are set at the beginning of each time step (in the ODE solver). See [33] for a thorough discussion of different boundary conditions and their implementation in terms of ghost cells.

There are many types of boundary conditions:

Outflow (or absorbing) boundary conditions simply let waves pass out of the domain without creating any reflections. The simplest approach is to extrapolate grid values from inside the domain in each spatial direction.

Reflective boundary conditions are used to model walls or to reduce computational domains by symmetry. They are realised by extrapolating grid values from inside the domain in each spatial direction and reversing the sign of the velocity component normal to the boundary.

Periodic boundary conditions are realised by copying values from grid cells inside the domain at the opposite boundary.

Inflow boundary conditions are used to model a given flow into the domain through certain parts of the boundary. These conditions are generally more difficult to implement. However, there are many cases for which inflow can be modelled by assigning fixed values to the ghost cells.

In this paper we will only work with outflow boundary conditions on rectangular domains. The simplest approach is to use a zeroth order extrapolation, which means that all ghost cells are assigned the value of the closest point inside the domain. Then outflow boundaries can be realised on the GPU simply by specifying that our textures are *clamped*, which means that whenever we try to access a value outside the domain, a value from the boundary is used. For higher-order extrapolation, we generally need to introduce an extra rendering pass using special shaders that render the correct ghost-cell values to extended textures.

3.5 Comparison of GPU versus CPU

We will now present a few numerical examples that compare the efficiency of our GPU implementation with a straightforward single-threaded CPU implementation of the Lax–Friedrichs scheme. The CPU simulations were run on a Dell Precision 670 which is a EM64T architecture with a dual Intel Xeon 2.8 GHz processor with 2.0 GB memory running Linux (Fedora Core 2). The programs were written in C and compiled using `icc -O3 -ipo -xP` (version 8.1). The GPU simulations were performed on two graphics cards from consecutive generations of the NVIDIA GeForce series: (i) the GeForce 6800 Ultra card released in April 2004 with the NVIDIA Forceware version 71.80 beta drivers, and (ii) the GeForce 7800 GTX card released in June 2005 with the NVIDIA Forceware version 77.72 drivers. The GeForce 6800 Ultra has 16 parallel pipelines, each capable of processing vectors of length 4 simultaneously. On the GeForce 7800 GTX, the number of pipelines is increased to 24.

In the following we will for simplicity make comparisons on $N \times N$ grids, but this is no prerequisite for the GPU implementation. Rectangular $N \times M$ grids work equally well. Although the current capabilities of our GPUs allow for problems with sizes up to 4096×4096, we have limited the size of our problems to 1024×1024 (due to high computational times on the CPU).

Example 2 (Circular dambreak). Let us first consider a simple dambreak problem with rotational symmetry for the shallow-water equations; that is, we consider two constant states separated by a circle

$$h(x,y,0) = \begin{cases} 1.0, & \sqrt{x^2+y^2} \le 0.3, \\ 0.1, & \text{otherwise}, \end{cases}$$
$$u(x,y,0) = v(x,y,0) = 0.$$

We assume absorbing boundary conditions. The solution consists of an expanding circular shock wave. Within the shock there is a rarefaction wave

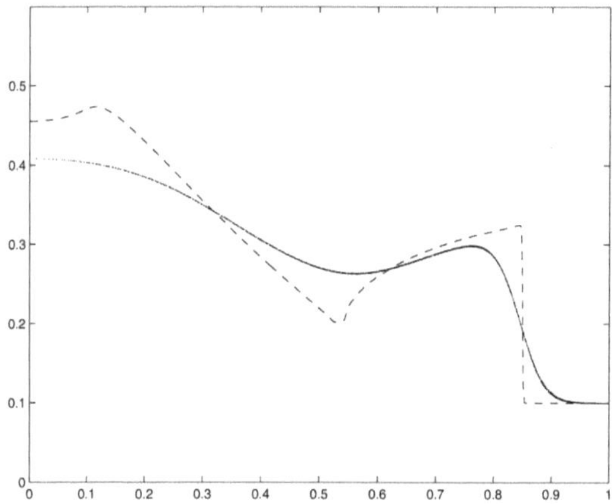

Fig. 6. Scatter plot of the circular dambreak problem at $t = 0.5$ computed by Lax–Friedrichs scheme on a 128×128 grid. The dashed line is a fine-grid reference solution.

transporting water from the original deep region out to the shock. In the numerical computations the domain is $[-1.0, 1.0] \times [-1.0, 1.0]$ with absorbing boundary conditions. A reference solution can be obtained by solving the reduced inhomogeneous system

$$\begin{bmatrix} h \\ hu_r \end{bmatrix}_t + \begin{bmatrix} hu \\ hu_r^2 + \frac{1}{2}gh^2 \end{bmatrix}_r = -\frac{1}{r}\begin{bmatrix} hu_r \\ hu_r^2 \end{bmatrix},$$

in which $u_r = \sqrt{u^2 + v^2}$ denotes the radial velocity. This system is nothing but the one-dimensional shallow-water equations (5) with a geometric source term.

Figure 6 shows a scatter plot of the water height at time $t = 0.5$ computed by the Lax–Friedrichs. In the scatter plot, the cell averages of a given quantity in all cells are plotted against the radial distance from the origin to the cell centre. Scatter plots are especially suited for illustrating grid-orientation effects for radially symmetric problems. If the approximate solution has perfect symmetry, the scatter plot will appear as a single line, whereas any symmetry deviations will show up as point clouds. The first-order Lax–Friedrichs scheme clearly smears both the leading shock and the tail of the imploding rarefaction wave. This behaviour is similar to what we observed previously in Figures 2 and 3 in Example 1 for the linear advection equation and the nonlinear Burgers' equation.

Table 1 reports a comparison of average runtime per time step for GPU versus CPU implementations of the Lax–Friedrichs scheme. Here, however, we see a that the 7800 GTX card with 24 pipelines is about two to three times faster than the 6800 Ultra card with 16 pipelines. The reason is as follows: For the simple Lax–Friedrichs scheme, the number of (arithmetic) operations

Table 1. Runtime per time step in seconds and speedup factor for the CPU versus the GPU implementation of Lax–Friedrichs scheme for the circular dambreak problem run on a grid with $N \times N$ grid cells, for the NVIDIA GeForce 6800 Ultra and GeForce 7800 GTX graphics cards.

N	CPU	6800	speedup	7800	speedup
128	2.22e-3	7.68e-4	2.9	2.33e-4	9.5
256	9.09e-3	1.24e-3	7.3	4.59e-4	19.8
512	3.71e-2	3.82e-3	9.7	1.47e-3	25.2
1024	1.48e-1	1.55e-2	9.5	5.54e-3	26.7

performed per fragment in each rendering pass (i.e., the number of operations per grid point) is quite low compared with the number of texture fetches. We therefore cannot expect to fully utilise the computational potential of the GPU. In particular for the 128×128 grid, the total number of fragments processed per rendering pass on the GPU is low compared with the costs of switching between different rendering buffers and establishing each pipeline. The cost of these context switches will therefore dominate the runtime on the 6800 Ultra card, resulting in a very modest speedup of about 3–4. For the 7800 GTX, the cost of context switches is greatly reduced due to improvements in hardware and/or drivers, thereby giving a much higher speedup factor.

Example 3 (Shock-bubble interaction). In this example we consider the interaction of a planar shock in air with a circular region of low density.

The setup is as follows (see Figure 7): A circle with radius 0.2 is centred at $(0.4, 0.5)$. The gas is initially at rest and has unit density and pressure. Inside the circle the density is 0.1. The incoming shock wave starts at $x = 0$ and propagates in the positive x-direction. The pressure behind the shock is 10, giving a 2.95 Mach shock. The domain is $[0, 1.6] \times [0, 1]$ with symmetry about the x axis. Figure 7 shows the development of the shock-bubble interaction in terms of (emulated) Schlieren images. Schlieren imaging is a photo-optical technique for studying the distribution of density gradients within a transparent medium. Here we have imitated this technique by depicting $(1 - |\nabla\rho|/\max|\nabla\rho|)^p$ for $p = 15$ as a greymap.

Figure 8 shows the result of a grid refinement study for Lax–Friedrichs. Whereas the scheme captures the leading shock on all grids, the weaker waves in the deformed bubble are only resolved satisfactorily on the two finest grids.

To study the efficiency of a GPU implementation versus a CPU implementation of the Lax–Friedrichs scheme, we reduced the computational domain to $[0, 1] \times [0, 1]$ and the final simulation time to $t = 0.2$. Runtime per time step and speedup factors are reported in Table 2. Comparing with the circular dambreak case in Table 1, we now observe that whereas the runtime per time step on the CPU increases by a factor between 1.3–1.4, the runtime on the GPU increases at most by a factor 1.1. The resulting increase in speedup factor is thus slightly less than 4/3. This can be explained as follows: Since the

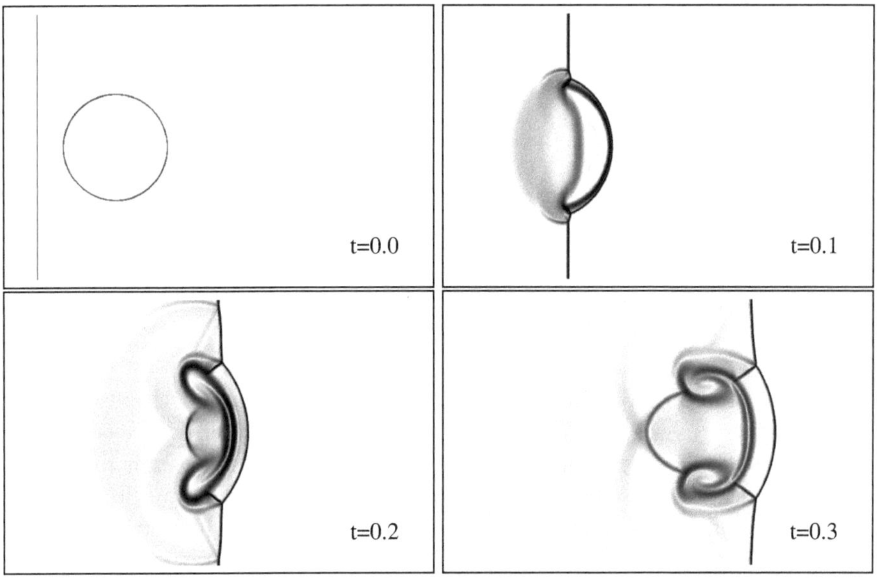

Fig. 7. Emulated Schlieren images of a shock-bubble interaction. The approximate solution is computed by the composite scheme [37] on a 1280×800 grid.

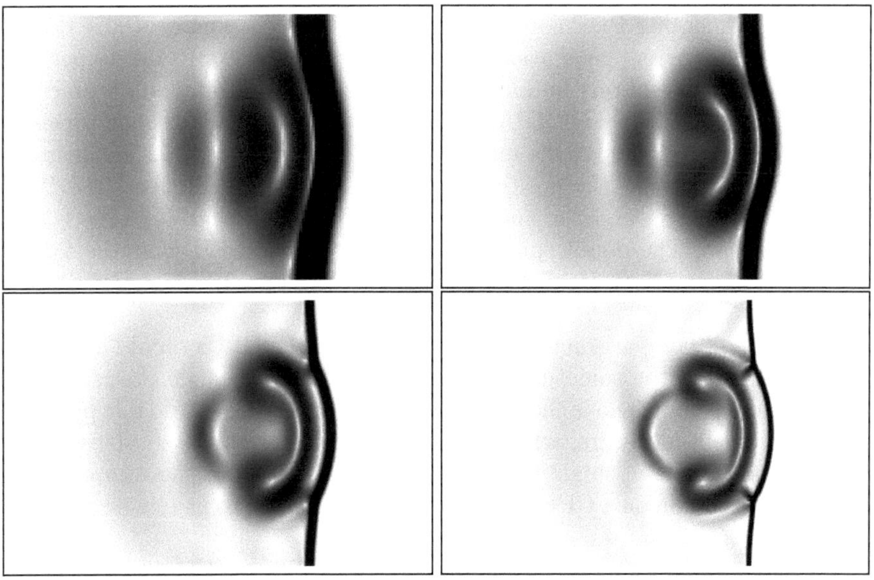

Fig. 8. Grid refinement study of the shock-bubble case at time $t = 0.3$ computed with the Lax–Friedrichs scheme for $\Delta x = \Delta y = 1/100$, $1/200$, $1/400$, and $1/800$.

Table 2. Runtime per time step in seconds and speedup factor for the CPU versus the GPU implementation of Lax–Friedrichs for the shock-bubble problem run on a grid with $N \times N$ grid cells, for the NVIDIA GeForce 6800 Ultra and GeForce 7800 GTX graphics cards.

N	CPU	6800	speedup	7800	speedup
128	3.11e-3	7.46e-4	4.2	2.50e-4	12.4
256	1.23e-2	1.33e-3	9.3	5.56e-4	22.1
512	4.93e-2	4.19e-3	11.8	1.81e-3	27.2
1024	2.02e-1	1.69e-2	12.0	6.77e-3	29.8

shallow-water equations have only three components, the GPU only exploits 3/4 of its processing capabilities in each vector operation, whereas the Euler equations have four component and can exploit the full vector processing capability. However, since the flux evaluation for the Euler equations is a bit more costly in terms of vector operations, and since not all operations in the shaders are vector operations of length four, the increase in speedup factor is expected to be slightly less than 4/3.

3.6 Floating Point Precision

When implementing PDE solvers on the CPU it is customary to use double precision. The current generations of GPUs, however, are only able to perform floating point operations in single precision (32 bits), but double precision has been announced by NVIDIA to appear in late 2007. The single precision numbers of the GPU have the format 's23e8', i.e., there are 6 decimals in the mantissa. As a result, the positive range of representable numbers is $[1.175 \cdot 10^{-38}, 3.403 \cdot 10^{38}]$, and the smallest number ϵ_s, such that $1 + \epsilon_s - 1 > 0$, is $1.192 \cdot 10^{-7}$. The double precision numbers of our CPU have the format 's52e11', i.e., mantissa of 15 decimals, range of $[2.225 \cdot 10^{-308}, 1.798 \cdot 10^{308}]$, and $\epsilon_d = 2.220 \cdot 10^{-16}$.

Ideally, the comparison between the CPU and the GPUs should have been performed using the same precision. Since CPUs offer single precision, this may seem to be an obvious choice. However, on current CPUs, double precision is supported in hardware, whereas single precision is emulated. This means that computations in single precision may be *slower* than the computations in double precision on a CPU. We therefore decided not to use single precision on the CPU.

The question is now if using single precision will effect the quality of the computed result. For the cases considered herein, we have computed all results on the CPU using both double and single precision. In all cases, the maximum difference of any cell average is of the same order as ϵ_s. Our conclusion based on these *quantitative studies* is that the GPU's lack of double precision does not deteriorate the computations for the examples considered in Sections 3.5 and

5. Please note that the methods considered in this paper are stable, meaning they cannot break down due to rounding errors.

4 High-Resolution Schemes

Classical finite difference schemes rely on the computation of point values. In the previous section we saw how discontinuities lead to problems like oscillations or excessive smearing for two basic schemes of this type. For pedagogical reasons we presented the *derivation* of the two schemes in a semi-discrete, finite-volume setting, but this does not affect (in)abilities of the schemes in the presence of discontinuities. Modern high-resolution schemes of the Godunov type aim at delivering high-order resolution of all parts of the solution, and in particular avoiding the creation of spurious oscillations at discontinuities. There are several approaches to high-resolution schemes. Composite schemes form one such approach, see [37]. In the next sections we will present another approach based on geometric considerations.

4.1 The REA Algorithm

Let us now recapitulate how we constructed our approximate schemes in Section 3. Abstracted, our approach can be seen to consist of three basic parts:

1. Starting from the known cell-averages Q_{ij}^n in each grid cell, we *reconstruct* a piecewise polynomial function $\widehat{Q}(x, y, t^n)$ defined for all (x, y). For a first-order scheme, we let $\widehat{Q}(x, y, t^n)$ be constant over each cell, i.e.,

$$\widehat{Q}(x, y, t^n) = Q_{ij}^n, \qquad (x, y) \in [x_{i-1/2}, x_{i+1/2}] \times [y_{j-1/2}, y_{j+1/2}].$$

   In a second-order scheme we use a piecewise bilinear reconstruction, and in a third order scheme a piecewise biquadratic reconstruction, and so on. The purpose of this reconstruction is to go from the discrete set of cell-averages we use to *represent* the unknown solution and back to a globally defined function that is to be evolved in time. In these reconstructions special care must be taken to avoid introducing spurious oscillations into the approximation, as is done by classical higher-order schemes like e.g., Lax–Wendroff.
2. In the next step we *evolve* the differential equation, exactly or approximately, using the reconstructed function $\widehat{Q}(x, y, t^n)$ as initial data.
3. Finally, we *average* the evolved solution $\widehat{Q}(x, y, t^{n+1})$ onto the grid again to give a new representation of the solution at time t^{n+1} in terms of cell averages Q_{ij}^{n+1}.

This three-step algorithm is often referred to as the REA algorithm, from the words reconstruct-evolve-average.

The Riemann Problem

To evolve and average the solution in Steps 2 and 3 of the REA algorithm, we generally use a semi-discrete evolution equation like (15). We thus need to know the fluxes out of each finite-volume cell. In Section 3.1 we saw how the problem of evaluating the flux over the cell boundaries in one spatial dimension could be recast to solve a set of local Riemann problems of the form

$$Q_t + F(Q)_x = 0, \qquad Q(x,0) = \begin{cases} Q_L, & x < 0, \\ Q_R, & x > 0. \end{cases}$$

The Riemann problem has a similarity solution of the form $Q(x,t) = V(x/t)$. The similarity solution consists of constant states separated by simple waves that are either continuous (rarefactions) or discontinuous (shocks, contacts, shear waves, etc) and is therefore often referred to as the Riemann fan. Thus, to compute the flux over the cell edge, all we need to know is $V(0)$. However, obtaining either the value $V(0)$ or a good approximation generally requires detailed knowledge of the underlying hyperbolic system. The algorithm used to compute $V(0)$ is called a Riemann solver. Riemann solvers come in a lot of flavours. Complete solvers, whether they are exact or approximate, use detailed knowledge of the wave structure and are able to resolve all features in the Riemann fan. Exact Riemann solvers are available for many systems, including the Euler and the shallow-water equations, but are seldom used in practice since they tend to be relatively complicated and expensive to compute. Incomplete solvers, like the central-upwind or HLL solver presented in Section 4.5, lump all intermediate waves and only resolve the leading waves in the Riemann fan, thereby giving only a rough estimate of the centre state. A significant amount of the research on high-resolution schemes has therefore been devoted to developing (approximate) numerical Riemann solvers that are fast and accurate. The Riemann problem is thoroughly discussed in the literature, the interested reader might start with [31] and then move on to [50, 51] for a thorough discussion of different Riemann solvers for the Euler and the shallow-water equations.

4.2 Upwind and Centred Godunov Schemes

The REA algorithm above is a general approach for constructing the so-called Godunov schemes that have ruled the ground in conservation laws during the last decades.

Godunov schemes generally come in two flavors: upwind or centred schemes. To illustrate their difference, we will make a slight detour into fully discrete schemes. For simplicity, we consider only the one-dimensional scalar case. Instead of the cell average (10) we introduce a sliding average

$$\bar{Q}(x,t) = \frac{1}{\Delta}x \int_{x-\Delta x/2}^{x+\Delta x/2} Q(\xi,t)\, d\xi.$$

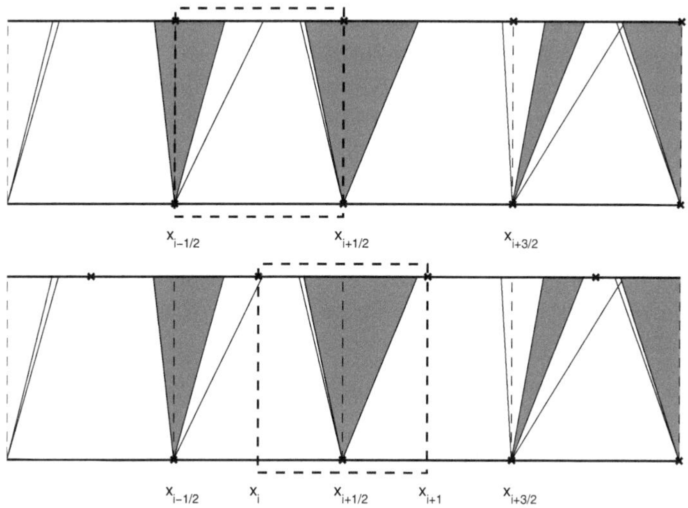

Fig. 9. Sliding average for upwind schemes (top) and centred schemes (bottom). The dashed boxes show the integration volume in the (x,t) plane. The shaded triangles and solid lines emanating from $x_{i-1/2}$, $x_{i+1/2}$ and $x_{i+3/2}$ illustrate the self-similar Riemann fans, whereas the dashed lines indicate the cell edges at time t^n.

Analogously to (13), we can derive an evolution equation for the sliding average

$$\begin{aligned}\bar{Q}(x,t+\Delta t) &= \bar{Q}(x,t) \\ &\quad - \frac{1}{\Delta x}\int_t^{t+\Delta t}\left[F\left(Q(x+\tfrac{\Delta x}{2},\tau)\right) - F\left(Q(x-\tfrac{\Delta x}{2},\tau)\right)\right] d\tau. \qquad (24)\end{aligned}$$

Within this setting, we obtain an upwind scheme if we choose $x = x_i$ in (24) and a centred scheme for $x = x_{i+1/2}$. Figure 9 illustrates the fundamental difference between these two approaches. In the upwind scheme, the flux is integrated over the points $x_{i\pm1/2}$, where the reconstruction is generally discontinuous. To integrate along the cell interface in time we will therefore generally have to find the self-similar solution of the corresponding Riemann problem (18) along the space-time ray $(x - x_{i\pm1/2})/t = 0$. To this end, one uses either exact or approximate Riemann solvers, as discussed above.

Centred schemes, on the other hand, use a sliding average defined over a *staggered* grid cell $[x_i, x_{i+1}]$, cf. the dashed box in the lower half of Figure 9. Thus the flux is integrated at the points x_i and x_{i+1}, where the solution is continuous if Δt satisfies a CFL condition of $1/2$. This means that no explicit wave propagation information for the Riemann fans is required to integrate the flux, since in this case the Riemann fan is inside confinement of the associated staggered cells. Since they use less information about the specific conservation law being solved, centred schemes tend to be less accurate than upwind schemes. On the other hand, they are much simpler to implement and work pretty well as black-box solvers.

The first staggered high-resolution scheme based upon central differences was introduced by Nessyahu and Tadmor [40], who extended the staggered Lax–Friedrichs scheme (19) discussed in Section 3.1 to second order. Similarly, the scheme has later been extended to higher order and higher dimensions by e.g., Arminjon et al. [3], Russo et al. [7, 34], and Tadmor et al. [38, 23]. Joint with S. Noelle, the fourth author (Lie) has contributed to the research on staggered schemes, see [35, 36]. A complete overview of staggered high-resolution methods is outside the scope of the current presentation, instead we refer the reader to Tadmor's *Central Station*

`http://www.cscamm.umd.edu/~tadmor/centralstation/`

which has links to virtually all papers written on high-resolution central schemes, and in particular, on staggered schemes.

Non-staggered centred schemes can be obtained if the new solution at time $t + \Delta t$ is first reconstructed over the staggered cell and then averaged over the original cell to define a new cell-average at time $t + \Delta t$. In certain cases, semi-discrete schemes can then be obtained in the limit $\Delta t \to 0$. We do not present derivations here; the interested reader is referred to e.g., [26, 25].

4.3 Semi-Discrete High-Resolution Schemes

Let us now return to the semi-discretised equation (15). Before we can use this equation for computing, we must answer two questions: how do we solve the ODEs evolving the cell-average values Q_{ij}, and how do we compute the edge fluxes (16)? The ODEs are usually solved by an explicit predictor-corrector scheme of Runge–Kutta type. In each stage of the Runge–Kutta method the right-hand side of (15) is computed by evaluating the flux integrals according to some standard quadrature rule. To apply a quadrature rule, two more points need to be clarified: how to compute point-values of Q from the cell-averages, and how to evaluate the integrand of the edge fluxes at the integration points. Listing 3 shows a general algorithm based on (15).

Different schemes are distinguished by the three following points: the reconstruction used to compute point-values, the discretisation and evaluation of flux integrals, and the ODE-solver for (15). In the following we will describe all these parts in more detail. We start with the ODE-solver, move on to the flux evaluation, and in the next section consider the reconstruction. We return to the flux evaluation in Section 4.5.

Runge–Kutta ODE Solver

To retain high-order accuracy in time without creating spurious oscillations, it is customary to use so-called TVD Runge–Kutta methods [47] as the ODE-solver. These methods employ a convex combination of forward Euler steps

Listing 3. General algorithmic form of a semi-discrete, high-resolution scheme, written in quasi-C code

```
Q[N_RK]=make_initial_data();

// Main loop
for (t=0; t<T; t+=dt) {
  Q[0] = Q[N_RK];
  dt = compute_Dt_from_CFL(Q[0]);

  // Runge−Kutta steps
  for (n=1; n<=N_RK; n++) {
    Q[n−1] = set_boundary_conditions(Q[n−1]);
    Qp     = reconstruct_point_values(Q[n−1]);
    [F,G]  = evaluate_edge_fluxes(Qp);
    Q[n]   = compute_RK_step(n, dt, Q, F, G);
  }
}
```

to advance the solution in time. They are especially designed to maintain the TVD property (30), i.e., ensure that the solution is total variation diminishing, see [31]. The second-order method (RK2) reads:

$$\begin{aligned} Q_{ij}^{(1)} &= Q_{ij}^n + \Delta t L_{ij}(Q^n), \\ Q_{ij}^{n+1} &= \frac{1}{2}Q_{ij}^n + \frac{1}{2}\big[Q_{ij}^{(1)} + \Delta t L_{ij}(Q^{(1)})\big], \end{aligned}$$

and similarly for the third-order method (RK3)

$$\begin{aligned} Q_{ij}^{(1)} &= Q_{ij}^n + \Delta t L_{ij}(Q^n), \\ Q_{ij}^{(2)} &= \frac{3}{4}Q_{ij}^n + \frac{1}{4}\big[Q_{ij}^{(1)} + \Delta t L_{ij}(Q^{(1)})\big], \\ Q_{ij}^{n+1} &= \frac{1}{3}Q_{ij}^n + \frac{2}{3}\big[Q_{ij}^{(2)} + \Delta t L_{ij}(Q^{(2)})\big]. \end{aligned}$$

Listing 4 shows the fragment shader implementing one step of the Runge–Kutta solver. The different steps are realised by choosing different values for the parameter c; $c = (0, 1)$ for the first, and $c = (\frac{1}{2}, \frac{1}{2})$ for the second step of RK2. As in the previous shaders, we use handles of type `uniform sampler2D` to sample texture data, and return the computed values in `gl_FragColor`.

Quadrature Rules

To evaluate the edge fluxes (16), we use a standard quadrature rule. For a second-order scheme the midpoint rule suffices, i.e.,

Listing 4. Fragment shader for the Runge–Kutta solver

```
varying vec4 texXcoord;
varying vec4 texYcoord;

uniform sampler2D QnTex;
uniform sampler2D FnHalfTex;
uniform sampler2D GnHalfTex;

uniform vec2 c;
uniform vec2 dXYf;
uniform float dT;

void main(void)
{
  vec4 FE = texture2D(FnHalfTex, texXcoord.yx);          // F_{i+1/2,j}
  vec4 FW = texture2D(FnHalfTex, texXcoord.zx);          // F_{i-1/2,j}
  vec4 GN = texture2D(GnHalfTex, texYcoord.xy);          // G_{i,j+1/2}
  vec4 GS = texture2D(GnHalfTex, texYcoord.xz);          // G_{i,j-1/2}

  vec4 Lij = -((FE-FW)/dXYf.x + (GN-GS)/dXYf.y);        // Right-hand side

  vec4 Q = texture2D(QnTex, texXcoord.yx);               // Q_{ij}^{(k)}
  vec4 QFirst = texture2D(QnFirstTex, texXcoord.yx);     // Q_{ij}^{n}

  gl_FragColor = c.x*QFirst + c.y*(Q + dT*Lij);          // Q_{ij}^{(k+1)}
}
```

$$\int_{-1/2}^{1/2} f(x)\,dx = f(0). \tag{25}$$

For a third-order scheme, we can use Simpson's rule

$$\int_{-1/2}^{1/2} f(x)\,dx = \frac{1}{6}\Big[f(-\tfrac{1}{2}) + 4f(0) + f(\tfrac{1}{2})\Big], \tag{26}$$

or, as we will do in the following, use the fourth order Gauss quadrature rule

$$\int_{-1/2}^{1/2} f(x)\,dx = \frac{1}{2}\Big[f(\tfrac{-1}{2\sqrt{3}}) + f(\tfrac{1}{2\sqrt{3}})\Big]. \tag{27}$$

Hence, the edge fluxes (16) become

$$\begin{aligned} F_{i\pm 1/2,j}(t) &= \frac{1}{2}\Big[F\big(Q(x_{i\pm 1/2}, y_{j+\alpha}, t)\big) + F\big(Q(x_{i\pm 1/2}, y_{j-\alpha}, t)\big)\Big],\\ G_{i,j\pm 1/2}(t) &= \frac{1}{2}\Big[G\big(Q(x_{i+\alpha}, y_{j\pm 1/2}, t)\big) + G\big(Q(x_{i-\alpha}, y_{j\pm 1/2}, t)\big)\Big], \end{aligned} \tag{28}$$

where $x_{i\pm\alpha}$ and $y_{j\pm\alpha}$ denote the integration points of the Gauss quadrature rule.

At this point it is probably clear to the reader why we need the reconstruction from Step 1 in Section 4.1. Namely, whereas our semi-discrete scheme (15) evolves cell averages, the flux quadrature requires *point values* at the Gaussian integration points $(x_{i\pm 1/2}, y_{j\pm\alpha})$ and $(x_{i\pm\alpha}, y_{j\pm 1/2})$. The second step in the computation of fluxes therefore amounts to obtaining these point values through a piecewise polynomial reconstruction.

4.4 Piecewise Polynomial Reconstruction

The major challenge in developing high-order reconstructions is to cope with nonsmooth and discontinuous data. We cannot expect to maintain high-order accuracy at a discontinuity. Instead, the aim is to minimise the creation of spurious oscillations. In this section we introduce a simple piecewise linear reconstruction that prevents spurious oscillations, but retains second-order spatial accuracy away from discontinuities. This will be achieved by introducing a so-called limiter function that compares left and right-hand slopes and picks a nonlinear average in such a way that the total variation of the reconstructed function is no greater than that of the underlying cell averages.

The reconstruction will be introduced for a scalar equation, and we tacitly assume that it can be extended in a component-wise manner to nonlinear systems of equations.

Bilinear Reconstruction

For a second-order scheme we can use a bilinear reconstruction $\widehat{Q}(x,y,t^n)$ given by

$$\widehat{Q}_{ij}(x,y,t^n) = Q^n_{ij} + s^x_{ij}(x - x_i) + s^y_{ij}(y - y_j). \tag{29}$$

Here the slopes s^x_{ij} and s^y_{ij} can be estimated from the-cell average values of the neighbouring cells. For simplicity, let us consider the one-dimensional case. Here there are three obvious candidate stencils for estimating the slope:

$$s^-_i = \frac{Q^n_i - Q^n_{i-1}}{\Delta x}, \qquad s^+_i = \frac{Q^n_{i+1} - Q^n_i}{\Delta x}, \qquad s^c_i = \frac{Q^n_{i+1} - Q^n_{i-1}}{2\Delta x}.$$

Which stencil we should choose will depend on the local behaviour of the underlying function, or in our case, on the three cell averages. As an example, assume that $Q^n_k = 1$ for $k \leq i$ and $Q^n_k = 0$ for $k > i$. In this case, $s^-_i = 0$, $s^+_i = -1/\Delta x$, and $s^c_i = -1/2\Delta x$. Hence, $\widehat{Q}(x)$ will remain monotonic if we choose $s^x_i = s^-_i$, whereas a new maximum will be introduced for s^+_i and s^c_i. Solutions of scalar conservation laws are bounded by their initial data in the sup-norm, preserve monotonicity and have diminishing total variation. The last point may be expressed as

$$\mathrm{TV}(Q(\cdot,\tau)) \leq \mathrm{TV}(Q(\cdot,t)), \qquad \text{for } t \leq \tau, \tag{30}$$

where the total variation functional is defined by

$$\mathrm{TV}(v) = \limsup_{h\to 0} \frac{1}{h}\int |v(x) - v(x-h)|\, dx.$$

We want our discrete solution to mimic this behaviour. Therefore we need to put some "intelligence" into the stencil computing the linear slopes. The key to obtaining this intelligence lies in the introduction of a nonlinear averaging function Φ capable of choosing the slope. Given Φ, we simply let

$$\Delta x\, s_{ij}^x = \Phi\big(Q_{ij}^n - Q_{i-1,j}^n, Q_{i+1,j}^n - Q_{ij}^n\big).$$

This function is called a *limiter*, and is applied independently in each spatial direction. By limiting the size of the slopes in the reconstruction, this function introduces a nonlinearity in the scheme that ensures that solutions are nonoscillatory and total variation diminishing (TVD). An example of a robust limiter is the minmod limiter

$$\mathrm{MM}(a,b) = \tfrac{1}{2}\big(\mathrm{sgn}(a) + \mathrm{sgn}(b)\big)\min(|a|,|b|) \tag{31}$$

$$= \begin{cases} 0, & \text{if } ab \le 0, \\ a, & \text{if } |a| < |b| \text{ and } ab > 0, \\ b, & \text{if } |b| < |a| \text{ and } ab > 0, \end{cases} \tag{32}$$

which picks the least slope and reduces to zero at extrema in the data. For systems of conservation laws, solution are not necessarily bounded in sup-norm or have diminishing total variation. Still, it is customary to apply the same design principle as for scalar equations. In our experiments, we have also used the modified minmod limiter and the superbee limiter

$$\mathrm{MM}_\theta(a,b) = \begin{cases} 0, & \text{if } ab \le 0, \\ \theta a, & \text{if } (2\theta - 1)ab < b^2, \\ \tfrac{1}{2}(a+b), & \text{if } ab < (2\theta-1)b^2, \\ \theta b, & \text{otherwise}, \end{cases}$$

$$\mathrm{SB}_\theta(a,b) = \begin{cases} 0, & \text{if } ab \le 0, \\ \theta a, & \text{if } \theta ab < b^2, \\ b, & \text{if } ab < b^2, \\ a, & \text{if } ab < \theta b^2 \\ \theta b, & \text{otherwise}. \end{cases}$$

Away from extrema, the minmod always chooses the one-sided slope with least magnitude, whereas the other two limiters tend to choose steeper reconstruction, thus adding less numerical viscosity into the scheme. For other families of limiter, see e.g., [36, 50].

Listing 5 shows the fragment shader implementing the bilinear reconstruction. This shader needs to return two vectors corresponding to $\Delta x \partial_x Q$ and $\Delta y \partial_y Q$. At the time of implementation, GLSL was not capable of simultaneously writing to two textures due to problems with our drivers, we therefore chose to implemented the shader in Cg using so-called *multiple render targets*. In our implementation, we write to two textures using the new structure

Listing 5. Fragment shader in Cg for the bilinear reconstruction

```
float4 minmod(in float4 a, in float4 b)    // minmod function
{
  float4 res = min(abs(a), abs(b));
  return res*(sign(a)+sign(b))*0.5;
}

struct f2b {                               // Return two fragment buffers
  float4 color0  : COLOR;                  // d_x Q
  float4 color1  : COLOR1;                 // d_y Q
};

struct v2f_input {                         // Current stencil
  float4 pos         : POSITION;           // Position
  float4 texXcoord;                        // Offsets in x
  float4 texYcoord;                        // Offsets in y
};

f2b main(v2f_input IN, uniform sampler2D QnTex, uniform float2 dXYf)
{
  f2b OUT;

  float4 Q  = tex2D(QnTex, IN.texXcoord.yx);          // Q_{ij}
  float4 QE = tex2D(QnTex, IN.texXcoord.wx);          // Q_{i+1,j}
  float4 QW = tex2D(QnTex, IN.texXcoord.zx);          // Q_{i-1,j}
  OUT.color0 = minmod(Q-QW, QE-Q);                    // d_x Q_{ij}

  float4 QN = tex2D(QnTex, IN.texYcoord.xw);          // Q_{i,j+1}
  float4 QS = tex2D(QnTex, IN.texYcoord.xz);          // Q_{i,j-1}
  OUT.color1 = minmod(Q-QS, QN-Q);                    // d_y Q_{ij}

  return OUT;
}
```

`f2b`. Moreover, to avoid introducing conditional branches that can reduce the computational efficiency, we have used the non-conditional version (31) rather than the conditional version (32) of the minmod-limiter.

Higher-order accuracy can be obtained by choosing a higher-order reconstruction. Two higher-order reconstructions are presented in Appendix A; the CWENO reconstruction in Appendix A.1 is a truly multidimensional reconstruction giving third order spatial accuracy, whereas the dimension-by-dimension WENO reconstruction in Appendix A.2 is designed to give fifth order accuracy at Gaussian integration points.

4.5 Numerical Flux

In the previous section we presented a reconstruction approach for obtaining one-sided point values at the integration points of the flux quadrature (28). All that now remains to define a fully discrete scheme is to describe how to use these point-values to evaluate the integrand itself. In the following we will introduce a few fluxes that can be used in combination with our high-resolution semi-discrete schemes.

Centred Fluxes

In Sections 3.1 and 3.2 we introduced two classical centred fluxes, the Lax–Friedrichs flux:

$$F^{LF} = \frac{1}{2}\big[F(Q_L) + F(Q_R)\big] - \frac{1}{2}\frac{\Delta}{x}\Delta t\big[Q_R - Q_L\big],$$

and the Lax–Wendroff flux

$$\begin{aligned} F^{LW} &= F(Q^*), \\ Q^* &= \frac{1}{2}\big[Q_L + Q_R\big] - \frac{1}{2}\frac{\Delta}{t}\Delta x\big[F(Q_R) - F(Q_L)\big]. \end{aligned}$$

By taking the arithmetic average of these two fluxes, one obtains the FORCE flux introduced by Toro [50]

$$F^{FORCE} = \frac{1}{4}\big[F(Q_L) + 2F(Q^*) + F(Q_R)\big] - \frac{1}{4}\frac{\Delta}{x}\Delta t\big[Q_R - Q_L\big], \tag{33}$$

In the above formulae Q_L and Q_R denote the point values extrapolated from left and right at the Gaussian integration points. If used directly with the cell averaged values (and with forward Euler as time integrator), the FORCE flux gives an oscillatory second-order method.

A Central-Upwind Flux

So far, wave propagation information has only been used to ensure stability through a CFL condition. Let us now incorporate information about the largest and smallest eigenvalues into the flux evaluation, while retaining the centred approach. This will lead to so-called central-upwind fluxes, as introduced by Kurganov, Noelle, and Petrova [25]. To this end, we define

$$\begin{aligned} a^+ &= \max_{Q\in\{Q_L,Q_R\}} \big(\lambda_m(Q), 0\big), \\ a^- &= \min_{Q\in\{Q_L,Q_R\}} \big(\lambda_1(Q), 0\big), \end{aligned}$$

where λ_1 is the smallest and λ_m the largest eigenvalues of the Jacobian matrix of the flux F. If the system is not genuinely nonlinear or linearly degenerate, the minimum and maximum is taken over the curve in phase space that connects Q_L and Q_R. The values (a^-, a^+) are estimates of how far the Riemann fan resulting from the discontinuity (Q_L, Q_R) extends in the negative and positive direction. Using this information, the evolving solution can be divided into two unions of local domains: one covering the local Riemann fans, where the solution is possibly discontinuous and one covering the area in-between, where the solution is smooth, see Figure 10. By evolving and averaging separately in the two sets, the following flux function can be derived [25]

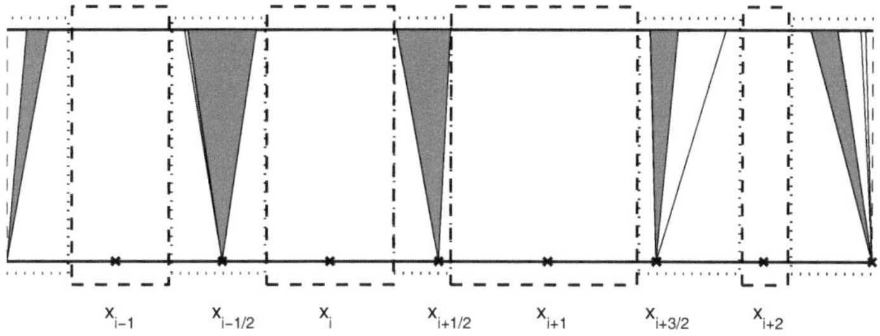

Fig. 10. Spatial averaging in the REA algorithm for the central-upwind scheme.

$$F^{CUW} = \frac{a^+F(Q_L) - a^-F(Q_R)}{a^+ - a^-} + \frac{a^+a^-}{a^+ - a^-}\Big(Q_R - Q_L\Big). \tag{34}$$

The name "central-upwind" comes from the fact that for monotone flux functions, the flux reduces to the standard upwind method. For example, for scalar equations with $F'(Q) \geq 0$, a^- is zero and $F^{CUW} = F(Q_L)$. Moreover, the first-order version of the scheme (i.e., for which $Q_L = Q_i$ and $Q_R = Q_{i+1}$) is the semi-discrete version of the famous HLL upwind flux [20]. Finally, if $a^+ = -a^- = \Delta x/\Delta t$, then F^{CUW} reduces to the Lax–Friedrichs flux F^{LF}.

Listing 6 shows the fragment shader implementing the computation of the edge-flux (16) using a fourth-order Gauss quadrature (27), bilinear reconstruction, and central-upwind flux. In the implementation, we compute estimates of the wave-speeds a^+ and a^- at the midpoint of each edge and use them to evaluate the flux function at the two Gaussian integration points. Notice also the use of the built-in inner-product `dot` and the weighted average `mix`.

A Black-Box Upwind Flux

For completeness, we also include a 'black-box' upwind flux, which is based upon resolving the local Riemann problems *numerically.* In the Multi-Stage (MUSTA) approach introduced by Toro [52], local Riemann problems are solved numerically a few time steps using a predictor-corrector scheme with a centred flux. This 'opens' the Riemann fan, and the centre state $V(\xi = 0)$ can be picked out and used to evaluate the true flux over the interface. The MUSTA algorithm starts by setting $Q_L^{(1)} = Q_L$ and $Q_R^{(1)} = Q_R$, and then iterates the following steps 3–4 times:

1. Flux evaluation: set $F_L^{(\ell)} = F(Q_L^{(\ell)})$ and $F_R^{(\ell)} = F(Q_R^{(\ell)})$ and compute $F_M^{(\ell)}$ as the FORCE flux (33).
2. Open Riemann fan

$$Q_L^{(\ell+1)} = Q_L^{(\ell)} - \frac{\Delta}{t}\Delta x\Big(F_M^{(\ell)} - F_L^{(\ell)}\Big), Q_R^{(\ell+1)} = Q_R^{(\ell)} - \frac{\Delta}{t}\Delta x\Big(F_R^{(\ell)} - F_M^{(\ell)}\Big),$$

This approach has the advantage that it is general and requires no knowledge of the wave structure of a specific system.

Listing 6. Fragment shader for computing the edge-flux in the x-direction using central-upwind flux for the two-dimensional Euler equations.

```
varying vec4 texXcoord;
uniform sampler2D QnTex;
uniform sampler2D SxTex;
uniform sampler2D SyTex;
uniform float gamma;

float pressure(in vec4 Q)
{
  return (gamma-1.0)*(Q.w - 0.5*dot(Q.yz,Q.yz)/Q.x);
}

vec4 fflux(in vec4 Q)
{
  float u = Q.y/Q.x;
  float p = (gamma-1.0)*(Q.w - 0.5*dot(Q.yz,Q.yz)/Q.x);
  return vec4(Q.y, (Q.y*u)+p, Q.z*u, u*(Q.w+p) );
}

void main(void)
{
  // Reconstruction in (i,j)
  vec4 Q  = texture2D(QnTex, texXcoord.yx);
  vec4 Sx = texture2D(SxTex, texXcoord.yx);
  vec4 Sy = texture2D(SyTex, texXcoord.yx);
  vec4 QL = Q + Sx*0.5;
  vec4 QLp = Q + Sx*0.5 + Sy*0.2886751346;
  vec4 QLm = Q + Sx*0.5 - Sy*0.2886751346;

  // Reconstruction in (i+1,j)
  vec4 Q1 = texture2D(QnTex, texXcoord.wx);
  vec4 Sxp = texture2D(SxTex, texXcoord.wx);
  vec4 Syp = texture2D(SyTex, texXcoord.wx);
  vec4 QR = Q1 - Sxp*0.5;
  vec4 QRp = Q1 - Sxp*0.5 + Syp*0.2886751346;
  vec4 QRm = Q1 - Sxp*0.5 - Syp*0.2886751346;

  // Calculate ap and am
  float c, ap, am;
  c = sqrt(gamma*QL.x*pressure(QL));
  ap = max((QL.y + c)/QL.x, 0.0);
  am = min((QL.y - c)/QL.x, 0.0);
  c  = sqrt(gamma*QR.x*pressure(QR));
  ap = max((QR.y + c)/QR.x, ap);
  am = min((QR.y - c)/QR.x, am);

  // Central-upwind flux
  vec4 Fp = ((ap*fflux(QLp) - am*fflux(QRp)) + (ap*am)*(QRp - QLp))/(ap - am);
  vec4 Fm = ((ap*fflux(QLm) - am*fflux(QRm)) + (ap*am)*(QRm - QLm))/(ap - am)

  gl_FragColor = mix(Fp, Fm, 0.5);
}
```

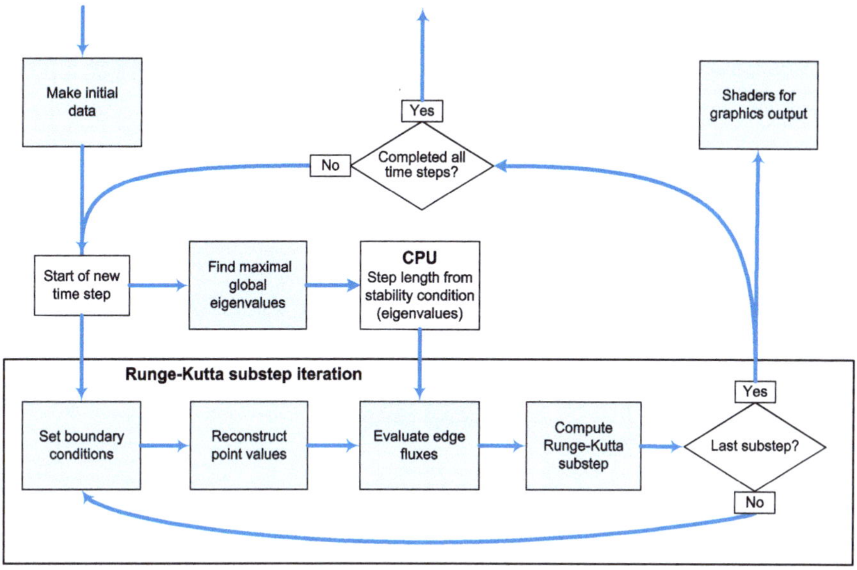

Fig. 11. Flow-chart for the GPU implementation of a semi-discrete, high-resolution scheme. White boxes refer to operations on the CPU and shaded boxes to operations on the GPU.

4.6 Putting It All Together

In the above sections we have presented the components needed to implement a high-resolution scheme on the GPU. Let us now see how they can be put together to form a full solver. Figure 11 shows a flow-chart for the implementation. In the flow-chart, all the logic and setup is performed on the CPU, whereas all time-consuming operations are performed in parallel on the GPU. The only exception is the determination of the time step, which is performed by postprocessing results from the depth buffer (see Listing 2). The three essential shaders are the reconstruction given in Listing 5, the evaluation of edge fluxes in Listing 6, and the Runge–Kutta steps in Listing 4. To make a complete scheme, we also need two shaders to set initial and boundary data, respectively. These shaders are problem specific and must be provided for each individual test case.

5 Numerical Examples

We now present a few numerical examples to highlight a few features of the high-resolution schemes and assess the efficiency of a GPU versus a CPU implementation. As in Section 3.5, all CPU simulations were run on a Dell Precision 670 with a dual Intel Xeon 2.8 GHz processor and the GPU simulation were performed on two NVIDIA GeForce graphics card, a 6800 Ultra

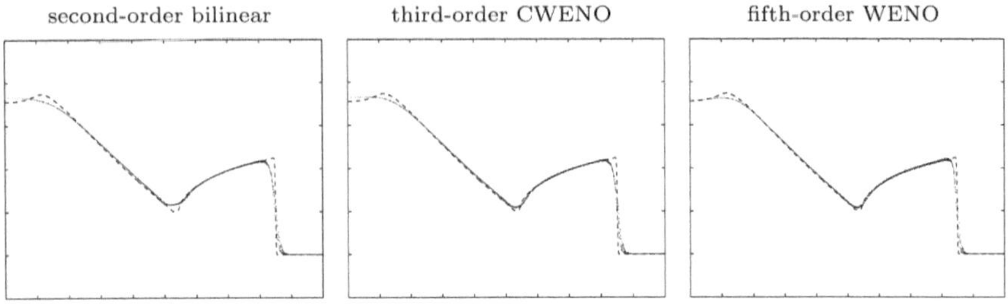

Fig. 12. Scatter plot of the circular dambreak problem at $t = 0.5$ computed on a 128×128 grid with reconstructions: bilinear with minmod limiter, CWENO, and WENO. The dashed line is a fine-grid reference solution.

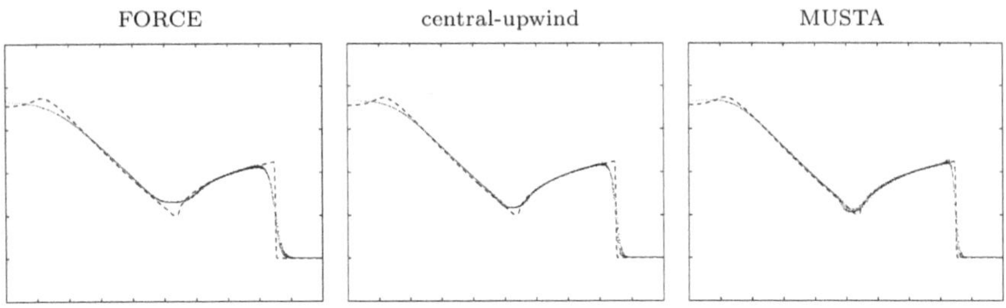

Fig. 13. Scatter plot of the circular dambreak problem at $t = 0.5$ computed on a 128×128 with bilinear reconstruction and FORCE, central-upwind, and MUSTA fluxes. The dashed line is a fine-grid reference solution.

and a 7800 GTX. (See also our two journal papers on shallow-water waves [17] and the 3D Euler equations [18] for supplementary numerical results.)

Example 4 (Circular dambreak). In the first example, we revisit the circular dambreak problem from Example 2. Figure 12 shows approximate solutions computed using three different reconstructions. (To isolate the effects of the reconstruction, all three computations used third-order Runge–Kutta, central-upwind flux, and fourth order Gauss quadrature). Starting with the bilinear reconstruction, we see that this scheme gives superior resolution of both the leading shock and the rarefaction wave compared with the Lax–Friedrichs scheme in Figure 6. By increasing the order of the reconstruction from bilinear to the third-order CWENO, we improve the local minimum at $r \approx 0.5$. Finally, by introducing the fifth-order WENO reconstruction, we also obtain a satisfactory resolution of the constant state near the origin (and a slight improvement near the shock).

Figure 13 shows a comparison of the three different fluxes for the bilinear scheme. The FORCE flux is a centred flux that does not incorporate any information of the local wave propagation. As a result, the corresponding scheme smears the leading shock and both ends of the rarefaction wave (at

Table 3. Runtime per time step in seconds and speedup factor for the CPU versus the GPU implementation of bilinear interpolation with modified minmod limiter for the dambreak problem run on a grid with $N \times N$ grid cells, for the NVIDIA GeForce 6800 Ultra and GeForce 7800 GTX graphics cards. The upper part uses second-order and the bottom part third-order Runge–Kutta time stepping.

N	CPU	6800	speedup	7800	speedup
128	3.06e-2	3.78e-3	8.1	1.27e-3	24.2
256	1.22e-1	8.43e-2	14.5	4.19e-3	29.1
512	4.86e-1	3.18e-2	15.3	1.68e-2	28.9
1024	2.05e-0	1.43e-1	14.3	6.83e-2	30.0
128	4.56e-2	5.58e-3	8.2	1.90e-3	23.9
256	1.83e-1	1.24e-2	14.8	6.23e-3	29.4
512	7.33e-1	4.69e-2	15.6	2.51e-2	29.2
1024	3.09e-0	2.15e-1	14.3	1.04e-1	29.7

$r \approx 0.5$ and $r \approx 0.15$, respectively). The central-upwind flux incorporates local wave propagation information in the form of two-sided estimates of the local wave speeds and therefore gives sharper resolution of both the shock and the rarefaction ends. Finally, the MUSTA flux is an iterative solver for the local Riemann problem that gives very accurate approximation of the Godunov edge-flux. The corresponding scheme therefore gives good resolution of both the leading shock and the rarefaction ends.

In Table 3 we have rerun the experiments from Table 1 in Example 2, now using the high-resolution scheme with central-upwind flux and bilinear reconstruction (and modified minmod limiter with $\theta = 1.3$). Compared with the simple Lax–Friedrichs scheme, the bilinear scheme involves a larger number of arithmetic operations per grid cell. This means that the costs of data fetch and data processing perfectly balance, giving a considerably higher speedup then for Lax–Friedrichs.

Example 5 (Shock-bubble). In the second example, we revisit the shock-bubble interaction from Example 3. We will now consider two different simulation methods using the second-order bilinear reconstruction from Section 4.4 and the third-order CWENO reconstruction from Appendix A.1. Apart from this, both methods use the second Runge–Kutta method, fourth-order Gauss quadrature, and central-upwind flux.

Figure 14 illustrates the effect of choosing different reconstructions. In the bilinear reconstruction, we have used the minmod limiter (31), the modified minmod-θ limiter (with $\theta = 1.3$) and the superbee limiter (with $\theta = 1.5$). The two latter limiters tend to choose steeper slopes in the local reconstructions, thereby introducing less numerical viscosity in the approximate solution. The interface making up the walls of the collapsing bubble is instable and tends to break up for the least dissipative superbee limiter, whereas the minmod limiter has sufficient numerical dissipation to suppress the instability. We notice in

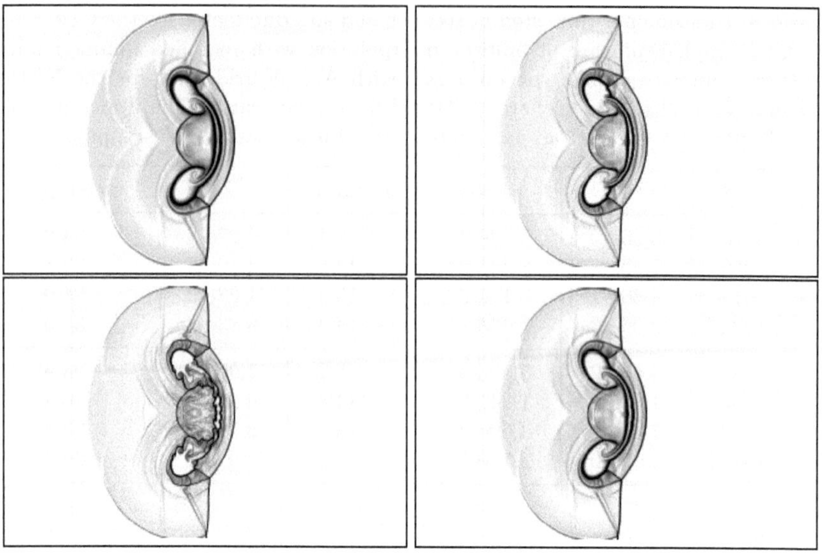

Fig. 14. Approximate solution for $\Delta x = 1/800$ of the shock-bubble problem at time $t = 0.2$ computed with the bilinear scheme for using three different limiters: minmod (top-left), modified minmod (top-right), and superbee (bottom-left) and by the CWENO scheme (bottom-right).

particular that although the CWENO reconstruction has higher order, it also contains more numerical dissipation than the bilinear reconstruction with the superbee limiter and therefore suppresses the breaking of the bubble interface.

In Table 4 we have rerun the experiments from Table 2 in Example 3, now using the high-resolution scheme with bilinear reconstruction (and modified minmod limiter with $\theta = 1.3$). Comparing Tables 3 and 4, we observe that the runtime for the GPU simulations only increases a few percent when going from the shallow-water to the Euler equations, indicating that for the bilinear scheme the runtime is strongly dominated by vector operations that are perfectly parallel. The resulting speedup factors of order 20 and 40 for GeForce 6800 and 7800, respectively, are in fact amazing, since the GPU implementations have not involved any low-level optimisation apart from the obvious use of vector operations whenever appropriate. (Using the widespread `gcc` compiler with full optimisation rather than `icc` to compile the CPU code gave approximately 50% higher runtime on the CPU, and hence speedup factors of magnitude up to 70 for the GeForce 7800 card!)

Table 5 shows corresponding runtimes and speedup factors for the third-order CWENO reconstruction. As seen from the discussion in Appendix A.1, we cannot expect to retain the good speedup observed for the bilinear scheme. Still, a speedup of 8–9 for GeForce 6800 and 23–24 for GeForce 7800 is indeed impressive.

Table 4. Runtime per time step in seconds and speedup factor for the CPU versus the GPU implementation of bilinear interpolation with modified minmod limiter for the shock-bubble problem run on a grid with $N \times N$ grid cells, for the NVIDIA GeForce 6800 Ultra and GeForce 7800 GTX graphics cards. The upper part uses second-order and the bottom part third-order Runge–Kutta time stepping.

N	CPU	6800	speedup	7800	speedup
128	4.37e-2	3.70e-3	11.8	1.38e-3	31.7
256	1.74e-1	8.69e-3	20.0	4.37e-3	39.8
512	6.90e-1	3.32e-2	20.8	1.72e-2	40.1
1024	2.95e-0	1.48e-1	19.9	7.62e-2	38.7
128	6.27e-2	5.22e-3	12.0	1.97e-3	31.9
256	2.49e-1	1.28e-2	19.5	6.44e-3	38.6
512	9.89e-1	4.94e-2	20.0	2.56e-2	38.6
1024	4.24e-0	2.20e-1	19.3	1.13e-1	37.5

Table 5. Runtime per time step in seconds and speedup factor for the CPU versus the GPU implementation of CWENO reconstruction for the shock-bubble problem run on a grid with $N \times N$ grid cells, for the NVIDIA GeForce 6800 Ultra and GeForce 7800 GTX graphics cards. The upper part uses second-order and the bottom part third-order Runge–Kutta time stepping.

N	CPU	6800	speedup	7800	speedup
128	1.05e-1	1.22e-2	8.6	4.60e-3	22.8
256	4.20e-1	4.99e-2	8.4	1.74e-2	24.2
512	1.67e-0	1.78e-1	9.4	6.86e-2	24.3
1024	6.67e-0	7.14e-1	9.3	2.99e-1	22.3
128	1.58e-1	1.77e-2	8.9	6.80e-3	23.2
256	6.26e-1	6.78e-2	9.2	2.59e-2	24.2
512	2.49e-0	2.66e-1	9.4	1.02e-1	24.3
1024	9.98e-0	1.06e-0	9.4	4.45e-1	22.4

Example 6. In the last example, we will consider the full shallow-water equations (7) with a variable bathymetry. A particular difficulty with this system of balance laws (conservation laws with source terms are often called balance laws) is that it admits steady-state or solutions in which the (topographical) source terms are exactly balanced by nonzero flux gradients,

$$\left[hu^2 + \tfrac{1}{2}gh^2\right]_x + \left[huv\right]_y = -ghB_x,$$

and similarly in the y-direction. Capturing this delicate balance without introducing spurious waves of small amplitude is a real challenge for any numerical scheme. Of particular interest is the so-called *lake-at-rest* problem, in which $hu = hv = 0$ and $w = h + B =$ constant.

For high-resolution schemes a lot of research has been devoted to develop so-called well-balanced schemes, which are capable of accurately resolving steady-state and near-steady-state solutions; see e.g., [32, 5, 27, 4, 41] and references therein. In [17] we discussed a GPU-implementation of the well-balanced approach from [27]. The key points in this approach is: (i) to reconstruct the surface elevation $w = h + B$ rather than the water depth h as a piecewise polynomial function, and (ii) to use a special quadrature rule for the source term S. For the second component of the source term, the quadrature rule reads

$$\begin{aligned} S_{ij}^{(2)} &= -\frac{1}{|\Omega_{ij}|}\int_{y_{j-1/2}}^{y_{j+1/2}}\int_{x_{i-1/2}}^{x_{i+1/2}} g(w-B)_x\,dxdy \\ &\approx -\frac{g}{2\Delta x}\big(h^L_{i+1/2,j-\alpha}+h^R_{i-1/2,j-\alpha}\big)\big(B_{i+1/2,j-\alpha}-B_{i-1/2,j-\alpha}\big) \\ &\quad -\frac{g}{2\Delta x}\big(h^L_{i+1/2,j+\alpha}+h^R_{i-1/2,j+\alpha}\big)\big(B_{i+1/2,j+\alpha}-B_{i-1/2,j+\alpha}\big), \end{aligned} \tag{35}$$

This source term is then included in the semi-discrete evolution equation for the cell-averages

$$\frac{d}{dt}Q_{ij} = -\frac{F_{i+1/2,j}-F_{i-1/2,j}}{\Delta x} - \frac{G_{i,j+1/2}-G_{i,j-1/2}}{\Delta y} + S_{ij}, \tag{36}$$

where the flux-terms and the ODE are discretised as discussed above.

We will now use this scheme to simulate a dambreak in a mountainous terrain. Figure 15 shows four snapshots from the GPU simulation. Initially, the water in the lake is at rest in the upper lake (steady state). At time zero, the dam breaks and water starts to flood down into the neighbouring valley, where it creates strong flood waves that wash up on the hillsides. As time increases, the simulation *slowly* approaches a new steady state with equal water height in the two valleys.

The presence of dry-bed zones (i.e., areas with $h = 0$) poses extra challenges for the simulation, since these areas must be given special treatment to avoid (nonphysical) negative water heights; see [17].

6 Concluding Remarks

In this paper we have tried to give the reader an introduction to two exciting research fields: numerical solution of hyperbolic conservation laws and general purpose computation using graphics hardware. Research on high-resolution methods for hyperbolic conservation laws is a mature field, evolved during the last three decades, and both its mathematical and numerical aspects are supported by a large body of publications. When it comes to the use of graphics hardware as a computational resource, the situation is the opposite. Although there exist some early papers on the use of fixed-pipeline graphics

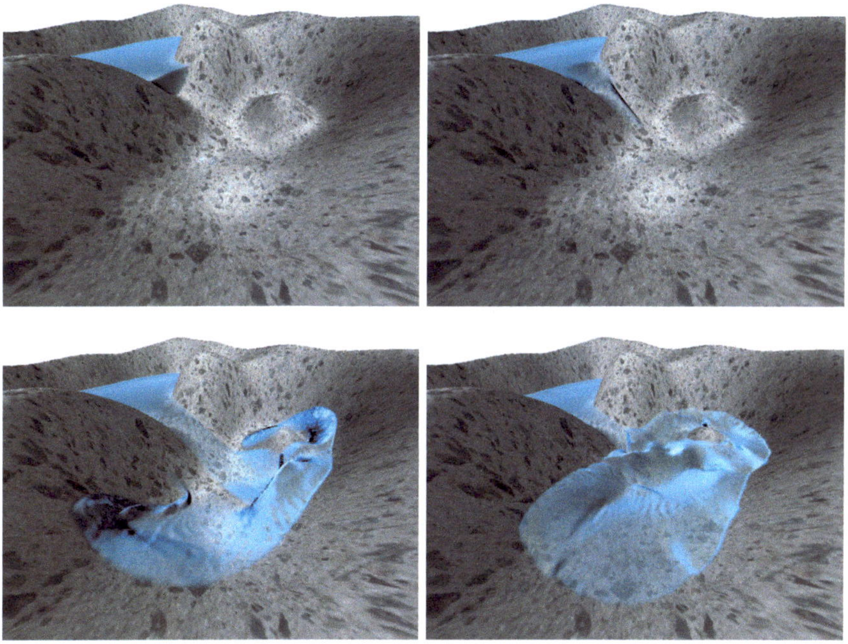

Fig. 15. Snapshots of a dambreak simulation in an artificially constructed terrain.

cards for non-graphical purposes, the interest in this field has exploded since the first GPUs with fully programmable fragment processors were introduced in 2003. Since then, the use of graphics cards for general purpose computing has attracted the interest of researchers in many different fields.

To demonstrate the computing capabilities of the GPU, we chose two common physical models, the Euler equations and the somewhat simpler shallow-water equations. We have attacked these equations with an assembly of finite-volume methods ranging from the basic Lax–Friedrichs scheme to the sophisticated third-order CWENO scheme, which represents state-of-the-art in high-resolution methods. Using standard numerical test cases in square domains, we observe that moving the computations from the CPU to a GPU gives a speedup of at least one order of magnitude. This amazing speedup is possible because the methods we have considered are explicit. Therefore, each cell can be updated independently of its neighbours, using only local points in a texture representing the previous time-step. This makes the methods "brilliantly parallel" and ripe for the data-based stream processing of graphics hardware. On the other hand, we have seen that in order to exploit this potential speedup, the computational algorithms must be recast to graphics terms. In our experience this requires familiarity with both computer graphics and the underlying hardware. However, once the algorithm has been recast to computer graphics, any programmer with a background from scientific computing should be able to write the actual shaders.

In the following subsections we give a short discussion of how GPU computations can be applied to more complex problems. Moreover, we also present a current outlook for GPUs for solving PDEs in scientific and industrial problems.

More Complex Physics

Having seen the success of using GPUs for the shallow water and the Euler equations, one can easily envisage that GPU computations can be used to speed up the computation of other (and more complex) hyperbolic models. As an example, magneto-hydrodynamics in two spatial dimensions is described by a set of seven equations and therefore do not map directly into a single four-component texture. A simple solution would be to split the vector of unknowns between two textures, and similarly for the fluxes, reconstructed slopes, etc. The flux computations would then typically involve a single shader reading from both textures, whereas a componentwise reconstruction could be performed by running the same shader consecutively on the two textures. Similarly, the Euler equations in three spatial dimensions is a 5×5 system and thus could be represented as e.g., a one-component texture for density and a four-component texture for momentum and energy or into two three-component textures as done in [18]. Three-dimensional grids can either be represented directly using the recent feature of 3D textures or by mapping each it to a regular 2D texture [39, 18].

Real-life problems often involve complex internal and external boundaries. A efficient method for representing complex boundaries in fluid flow simulations was presented by Wu et al. [39, 55]. Their idea is to preprocess the computational domain and define texture coordinate offsets to identify the nodes that determine the values of nodes close to the boundaries. Although this approach is introduced for the incompressible Navier–Stokes equations, we believe that it can be adopted for conservation laws as well.

To Whom Will GPU-Computing Be Useful?

From one point of view, the GPU can be considered as a parallel computer of the SIMD (single-instruction, multiple data) type, except that for the GPU there are no expressed communication between the nodes/pipelines anywhere in the code. From a user's point of view, the main attraction with the GPUs is the price-performance ratio (or the ratio of price versus power consumption). Assuming that the memory available on a graphics card is not a limitation, a simple back-of-an-envelope comparison of speedup versus dollar favour the GPU over a cluster of PCs, viz.

Price of graphics card:	\$ 500
Price of 24 computers (\$ 1000 each):	\$ 24 000
Speedup of graphics card:	20 times
Speedup of cluster:	24 times
Dollar/speedup for graphics card:	\$ 25
Dollar/speedup for cluster:	\$ 1000

With these figures, purchasing a 24-pipeline programmable graphics card gives 40 times more speedup per dollar than a 24-node cluster.

Having seen the performance of one single GPU it is tempting to ask what can be gained by exploiting a *cluster* of GPUs. Today, there are several research groups (including us) that try to use clusters equipped with GPUs or even clusters consisting of PlayStations to perform large-scale computing, see e.g., [12]. Although early performance reports are very good, the fact that GPUs are based upon a different programming model may in the end prevent wide-spread use of GPUs for high-performance computing.

On the other hand, the use of graphics cards may be a key factor in bringing PDE simulations from batch mode to interactive mode for desktop-sized applications, thereby opening up for their use in real-time systems (for monitoring and control), computer games and entertainment industry.

Current Technology Trends (Afterword March 2007)

The study reported herein was performed in the period 2004 to early 2006. During that period, the most recent generations of GPUs from NVIDIA (6800 Ultra from 2004/2005 and 7800 GTX from 2005/2006) demonstrated amazing floating-point performance of about 54 Gflops and 165 Gflops, respectively, compared with the typical commodity CPUs; contemporary Intel Pentium 4 CPUs, for instance, had a theoretical performance of at most 15 Gflops.

As the book goes into press (March 2007), the performance of GPUs has increased even further to about 0.5 Tflops for the NVIDIA GeForce 8800 GTX, which has 128 pipelines and 768 MB memory. A bit simplified, the reason behind this tremendous increase in computing power is as follows: Due to the parallel nature of a GPU, it is possible to increase the performance simply by adding more computational units, thereby increasing the number of computations that are made in parallel. Moreover, the architecture of commodity CPUs has changed significantly in the sense that dual-core CPUs have become common. Currently quad-core CPUs are beginning to hit the market. As for the future, Intel has recently showed "Polaris", a prototype processor with 80 cores, each with its own programmer-managed memory. Last year, AMD bought ATI, the other main manufacturer of GPUs. Recently, AMD announced the development of a hybrid CPU-GPU processor (Fusion) and the Torrenza initiative aimed at easy integration of various hardware accelerator units (like graphics cards). This development points in the direction of heterogeneous computers consisting of traditional CPU type units in combination with specialised accelerator units.

Another important point is the appearance in the mass-marked of other data-parallel processors, in particular the Cell BE processor designed for PlayStation 3 and used in blade servers. The Cell BE processor consists of one general processor connected to eight specialised computational cores, which may remind of stream processors. With this tile architecture, the Cell processor should be quite well-suited for general-purpose computing.

The development in hardware is expected to be followed by a similar development in software tools. NVIDIA recently released CUDA (Compute Unified Device Architecture – `http://developer.nvidia.com/object/cuda.html`), which allows the programmer to treat graphics cards in the GeForce 8000 series as general data-parallel processors without using the graphical API. Similarly, companies like RapidMind and PeakStream are developing software tools for taking advantage of multi-core processors and data-parallel processors like GPUs and Cell.

Altogether, it therefore seems that data-parallel processors from the mass market may offer computing capabilities in the next few years that are hard to ignore, and that utilising these capabilities for scientific computing will be a very exciting field to work in for a scientist.

Acknowledgement. This research was funded in part by the Research Council of Norway under grant no. 158911/I30.

References

1. J. E. Aarnes, T. Gimse, and K.-A. Lie. An introduction to the numerics of flow in porous media using Matlab. In *this book.*
2. J. E. Aarnes, V. Kippe, K.-A. Lie, and A. B. Rustad. Modelling of multiscale structures in flow simulations for petroleum reservoirs. In *this book.*
3. P. Arminjon, D. Stanescu, and M.-C. Viallon. A two-dimensional finite volume extension of the Lax–Friedrichs and Nessyahu–Tadmor schemes for compressible flows. In M. Hafez and K. Oshima, editors, *Proceedings of the 6th International Symposium on CFD, Lake Tahoe*, volume IV, pages 7–14, 1995.
4. E. Audusse, F. Bouchut, M.-O. Bristeau, R. Klein, and B. Perthame. A fast and stable well-balanced scheme with hydrostatic reconstruction for shallow water flows. *SIAM J. Sci. Comp.*, 25:2050–2065, 2004.
5. Derek S. Bale, Randall J. Leveque, Sorin Mitran, and James A. Rossmanith. A wave propagation method for conservation laws and balance laws with spatially varying flux functions. *SIAM J. Sci. Comput.*, 24(3):955–978 (electronic), 2002.
6. F. Benkhaldoun and R. Vilsmeier, editors. *Finite volumes for complex applications.* Hermes Science Publications, Paris, 1996. Problems and perspectives.
7. F. Bianco, G. Puppo, and G. Russo. High-order central schemes for hyperbolic systems of conservation laws. *SIAM J. Sci. Comput.*, 21(1):294–322 (electronic), 1999.
8. A. J. Chorin and J. E. Marsden. *A mathematical introduction to fluid mechanics*, volume 4 of *Texts in Applied Mathematics.* Springer-Verlag, New York, third edition, 1993.

9. R. Courant and K. O. Friedrichs. *Supersonic Flow and Shock Waves.* Interscience Publishers, Inc., New York, N. Y., 1948.
10. T. Dokken, T. R. Hagen, and J. M. Hjelmervik. An introduction to general-purpose computing on programmable graphics cards. In *this book.*
11. M. Van Dyke. *An Album of Fluid Motion.* Parabolic Press, 1982.
12. Z. Fan, F. Qiu, A. Kaufman, and S. Yoakum-Stover. GPU cluster for high performance computing. In *SC '04: Proceedings of the 2004 ACM/IEEE conference on Supercomputing*, page 47, Washington, DC, USA, 2004. IEEE Computer Society.
13. R. Fernando, editor. *GPU Gems: Programming Techniques, Tips and Tricks for Real-Time Graphics.* Addison Wesley, 2004.
14. R. Fernando and M.J. Kilgard. *The Cg Tutorial: The Definitive Guide to Programmable Real-Time Graphics.* Addison-Wesley Longman Publishing Co., Inc., 2003.
15. E. Godlewski and P.-A. Raviart. *Numerical approximation of hyperbolic systems of conservation laws*, volume 118 of *Applied Mathematical Sciences.* Springer-Verlag, New York, 1996.
16. S. K. Godunov. A difference method for numerical calculation of discontinuous solutions of the equations of hydrodynamics. *Mat. Sb. (N.S.)*, 47 (89):271–306, 1959.
17. T. R. Hagen, J. M. Hjelmervik, K.-A. Lie, J. R. Natvig, and M. Ofstad Henriksen. Visual simulation of shallow-water waves. *Simul. Model. Pract. Theory*, 13(8):716–726, 2005.
18. T. R. Hagen, K.-A. Lie, and J. R. Natvig. Solving the Euler equations on graphical processing units. In V.N. Alexandrov, G.D. van Albada, P.M.A. Sloot, and J. Dongarra, editors, *Computational Science – ICCS 2006: 6th International Conference, Reading, UK, May 28-31, 2006, Proceedings, Part IV*, volume 3994 of *Lecture Notes in Computer Science (LNCS)*, pages 220–227. Springer Verlag, 2006.
19. A. Harten. High resolution schemes for hyperbolic conservation laws. *J. Comput. Phys.*, 49(3):357–393, 1983.
20. A. Harten, P. D. Lax, and B. van Leer. On upstream differencing and Godunov-type schemes for hyperbolic conservation laws. *SIAM Rev.*, 25(1):35–61, 1983.
21. R. Herbin and D. Kröner, editors. *Finite volumes for complex applications III.* Laboratoire d'Analyse, Topologie et Probabilités CNRS, Marseille, 2002. Problems and perspectives, Papers from the 3rd Symposium held in Porquerolles, June 24–28, 2002.
22. H. Holden and N. H. Risebro. *Front tracking for hyperbolic conservation laws*, volume 152 of *Applied Mathematical Sciences.* Springer-Verlag, New York, 2002.
23. G.-S. Jiang and E. Tadmor. Nonoscillatory central schemes for multidimensional hyperbolic conservation laws. *SIAM J. Sci. Comput.*, 19(6):1892–1917, 1998.
24. S. N. Kružkov. First order quasilinear equations with several independent variables. *Mat. Sb. (N.S.)*, 81 (123):228–255, 1970.
25. A. Kurganov, S. Noelle, and G. Petrova. Semidiscrete central-upwind schemes for hyperbolic conservation laws and Hamilton–Jacobi equations. *SIAM J. Sci. Comput.*, 23(3):707–740 (electronic), 2001.
26. A. Kurganov and E. Tadmor. New high-resolution semi-discrete central schemes for Hamilton–Jacobi equations. *J. Comp. Phys.*, 160:720–742, 2000.
27. Alexander Kurganov and Doron Levy. Central-upwind schemes for the Saint-Venant system. *M2AN Math. Model. Numer. Anal.*, 36(3):397–425, 2002.

28. L. D. Landau and E. M. Lifshitz. *Fluid mechanics.* Translated from the Russian by J. B. Sykes and W. H. Reid. Course of Theoretical Physics, Vol. 6. Pergamon Press, London, 1959.
29. P. D. Lax. Weak solutions of nonlinear hyperbolic equations and their numerical computation. *Comm. Pure Appl. Math.*, 7:159–193, 1954.
30. P.D. Lax and B. Wendroff. Systems of conservation laws. *Comm. Pure Appl. Math.*, 13:217–237, 1960.
31. R. J. LeVeque. *Numerical Methods for Conservation Laws.* Lectures in Mathematics ETH Zürich. Birkhäuser Verlag, Basel, second edition, 1994.
32. R. J. LeVeque. Balancing source terms and flux gradients in high-resolution Godunov methods: The quasi-steady wave-propagation algorithm. *J. Comput. Phys*, 146:346–365, 1998.
33. R. J. LeVeque. *Finite volume methods for hyperbolic problems.* Cambridge Texts in Applied Mathematics. Cambridge University Press, Cambridge, 2002.
34. D. Levy, G. Puppo, and G. Russo. Compact central WENO schemes for multidimensional conservation laws. *SIAM J. Sci. Comput.*, 22(2):656–672, 2000.
35. K.-A. Lie and S. Noelle. An improved quadrature rule for the flux-computation in staggered central difference schemes in multidimensions. *J. Sci. Comput.*, 18(1):69–81, 2003.
36. K.-A. Lie and S. Noelle. On the artificial compression method for second-order nonoscillatory central difference schemes for systems of conservation laws. *SIAM J. Sci. Comput.*, 24(4):1157–1174, 2003.
37. R. Liska and B. Wendroff. Composite schemes for conservation laws. *SIAM J. Numer. Anal.*, 35(6):2250–2271, 1998.
38. X.-D. Liu and E. Tadmor. Third order nonoscillatory central scheme for hyperbolic conservation laws. *Numer. Math.*, 79(3):397–425, 1998.
39. Y. Liu, X. Liu, and E. Wu. Real-time 3d fluid simulation on GPU with complex obstacles. In *Proceedings of Pacific Graphics 2004*, pages 247–256. IEEE Computer Society, 2004.
40. H. Nessyahu and E. Tadmor. Nonoscillatory central differencing for hyperbolic conservation laws. *J. Comput. Phys.*, 87(2):408–463, 1990.
41. S. Noelle, N. Pankratz, G. Puppo, and J. R. Natvig. Well-balanced finite volume schemes of arbitrary order of accuracy for shallow water flows. *J. Comput. Phys.*, 213(2):474–499, 2006.
42. J. D. Owens, D. Luebke, N. Govindaraju, M. Harris, J. Krüger, A. E. Lefohn, and T. J. Purcell. A survey of general-purpose computation on graphics hardware. In *Eurographics 2005, State of the Art Reports*, pages 21–51, August 2005.
43. M. Pharr, editor. *GPU Gems 2: Programming Techniques for High-Performance Graphics and General-Purpose Computation.* Addison-Wesley Professional, 2005.
44. R. D. Richtmyer and K. W. Morton. *Difference methods for initial-value problems.* Second edition. Interscience Tracts in Pure and Applied Mathematics, No. 4. Interscience Publishers John Wiley & Sons, Inc., New York-London-Sydney, 1967.
45. R. J. Rost. *OpenGL^R Shading Language.* Addison Wesley Longman Publishing Co., Inc., 2004.
46. M. Rumpf and R. Strzodka. Graphics processor units: new prospects for parallel computing. In A.M. Bruaset and A. Tveito, editors, *Numerical Solution of Partial Differential Equations on Parallel Computers*, volume 51 of *Lecture Notes in Computational Science and Engineering*, pages 89–134. Springer Verlag, 2006.

47. C.-W. Shu. Total-variation-diminishing time discretisations. *SIAM J. Sci. Stat. Comput.*, 9:1073–1084, 1988.
48. C.-W. Shu. Essentially non-oscillatory and weighted essentially non-oscillatory schemes for hyperbolic conservation laws. In *Advanced numerical approximation of nonlinear hyperbolic equations (Cetraro, 1997)*, volume 1697 of *Lecture Notes in Math.*, pages 325–432. Springer, Berlin, 1998.
49. V. A. Titarev and E. F. Toro. Finite-volume WENO schemes for three-dimensional conservation laws. *J. Comput. Phys.*, 201(1):238–260, 2004.
50. E. F. Toro. *Riemann solvers and numerical methods for fluid dynamics.* Springer-Verlag, Berlin, second edition, 1999.
51. E. F. Toro. *Shock-capturing methods for free-sufrace shallow flows.* Wiley and Sons Ltd., 2001.
52. E. F. Toro. Multi-stage predictor-corrector fluxes for hyperbolic equations. Technical Report NI03037-NPA, Isaac Newton Institute for Mathematical Sciences, 2003.
53. R. Vilsmeier, F. Benkhaldoun, and D. Hänel, editors. *Finite volumes for complex applications II.* Hermes Science Publications, Paris, 1999. Problems and perspectives, Papers from the 2nd International Conference held in Duisburg, July 19–22, 1999.
54. G. B. Whitham. *Linear and nonlinear waves.* Wiley-Interscience [John Wiley & Sons], New York, 1974. Pure and Applied Mathematics.
55. E. Wu, Y. Liu, and X. Liu. An improved study of real-time fluid simulation on GPU. *J. of Computer Animation and Virtual World*, 15(3–4):139–146, 2004.

A Higher-Order WENO Reconstructions

A popular approach for making higher-order reconstructions is the *essentially nonoscillatory* (ENO) approach. In Section 4.4 we saw how we could avoid the creation of spurious oscillations for a piecewise linear reconstruction by comparing one-sided slopes and picking the slope that gave the least oscillation. (This was done by the nonlinear limiter function).

The ENO approach is basically the same idea extended to higher-order polynomials constructed using divided differences. Assume for the moment that we have one-dimensional data $\{Q_i\}$. The original ENO idea starts by making a piecewise linear polynomial $P_i^1(x)$ based upon Q_i and Q_{i-1}. Next, we consider the two candidate stencils made out of either $\{Q_{i+1}, Q_i, Q_{i-1}\}$ or $\{Q_i, Q_{i-1}, Q_{i-2}\}$. The piecewise quadratic ENO polynomial $P_i^2(x)$ can be constructed using the three points that will create the least oscillatory polynomial. The correct stencil is picked by comparing the magnitude of the corresponding divided differences. This way, one can continue to recursively construct higher and higher order polynomials by adding points to the left or right, depending on the magnitude of the divided differences.

In the more recent *weighted ENO* (WENO) approach, the polynomials are constructed by weighting all possible stencils rather than choosing the least oscillatory. The idea of WENO is that stencils from smooth parts of the solution are given high weight and stencils from nonsmooth parts are given low weight. More details on (W)ENO reconstructions can be found in the survey paper [48].

There are essentially two different ways to reconstruct multidimensional data: genuinely multidimensional and dimension-by-dimension. Below we will give an example of each type. First we will introduce a multidimensional third-order centred WENO reconstruction. Then we present a dimension-by-dimension, fifth-order WENO reconstruction that is specifically tailored to give maximum resolution at the Gaussian integration points, as proposed by Titarev and Toro [49].

A.1 Third Order CWENO Reconstruction

Third order accuracy is obtained by using a piecewise biquadratic reconstruction

$$\begin{aligned}\widehat{Q}_{ij}(x,y,t^n) = Q_{ij}(x,y) &= V_{ij} + s_{ij}^{x}(x-x_i) + s_{ij}^{y}(y-y_j)\\ &+ \frac{1}{2}s_{ij}^{xx}(x-x_i)^2 + \frac{1}{2}s_{ij}^{yy}(y-y_j)^2 + s_{ij}^{xy}(x-x_i)(y-y_j).\end{aligned}$$

For smooth data, one can use an optimal polynomial $Q^{\mathrm{OPT}}(x,y)$ based upon a centred nine-point stencil with

$$\begin{aligned}V_{ij} &= Q_{ij}^n - \frac{1}{24}\left(\Delta x^2 s_{ij}^{xx} + \Delta y^2 s_{ij}^{yy}\right),\\ s_{ij}^{x} &= \frac{Q_{i+1,j}^n - Q_{i-1,j}^n}{2\Delta x}, \qquad s_{ij}^{xx} = \frac{Q_{i+1,j}^n - 2Q_{ij}^n + Q_{i-1,j}^n}{\Delta x^2},\\ s_{ij}^{y} &= \frac{Q_{i,j+1}^n - Q_{i,j-1}^n}{2\Delta y}, \qquad s_{ij}^{yy} = \frac{Q_{i,j+1}^n - 2Q_{ij}^n + Q_{i,j-1}^n}{\Delta y^2},\\ s_{ij}^{xy} &= \frac{Q_{i+1,j+1}^n + Q_{i-1,j-1}^n - Q_{i+1,j-1} - Q_{i-1,j+1}^n}{4\Delta x \Delta y}.\end{aligned} \tag{37}$$

For nonsmooth data, a direct application of this reconstruction will introduce spurious oscillations. We therefore present another compact reconstruction called CWENO (centred WENO) that was introduced by Levy, Puppo and Russo [34]. The reconstruction employs a weighted combination of four one-sided piecewise bilinear and a centred piecewise quadratic stencil

$$P_{ij}(x,y) = \sum_k w_{ij}^k P_{ij}^k(x,y), \qquad k \in \{\mathrm{NE,NW,SW,SE,C}\}. \tag{38}$$

By this weighting we seek to emphasize contributions from cell averages in smooth regions and diminish contributions from cell averages in nonsmooth regions.

The four bilinear stencils are

$$\begin{aligned}
P_{ij}^{\text{NE}}(x,y) &= Q_{ij}^n + \frac{Q_{i+1,j}^n - Q_{ij}^n}{\Delta x}(x-x_i) + \frac{Q_{i,j+1}^n - Q_{ij}^n}{\Delta y}(y-y_j),\\
P_{ij}^{\text{NW}}(x,y) &= Q_{ij}^n + \frac{Q_{ij}^n - Q_{i-1,j}^n}{\Delta x}(x-x_i) + \frac{Q_{i,j+1}^n - Q_{ij}^n}{\Delta y}(y-y_j),\\
P_{ij}^{\text{SW}}(x,y) &= Q_{ij}^n + \frac{Q_{ij}^n - Q_{i-1,j}^n}{\Delta x}(x-x_i) + \frac{Q_{ij}^n - Q_{i,j-1}^n}{\Delta y}(y-y_j),\\
P_{ij}^{\text{SE}}(x,y) &= Q_{ij}^n + \frac{Q_{i+1,j}^n - Q_{ij}^n}{\Delta x}(x-x_i) + \frac{Q_{ij}^n - Q_{i,j-1}^n}{\Delta y}(y-y_j),
\end{aligned} \tag{39}$$

and the centred stencil is taken to satisfy

$$P^{\text{OPT}}(x,y) = \sum_k C^k P^k(x,y), \qquad \sum_k C^k = 1, \qquad k \in \{\text{NE,NW,SW,SE,C}\}.$$

On smooth data, a third-order reconstruction is obtained if we choose the weights to be $C^k = 1/8$ for $k \in \{\text{NE,NW,SW,SE}\}$ and $C^{\text{C}} = 1/2$. In other words, we choose the centred stencil to be

$$P^c(x,y) = 2P^{\text{OPT}}(x,y) + \frac{1}{4}\sum_k P^k(x,y), \qquad k \in \{\text{NE,NW,SW,SE}\}.$$

To tackle discontinuous data, we will introduce a nonlinear weighting procedure, analogously to the limiter defined for the bilinear reconstruction in Section 4.4. The nonlinear weights in (38) are designed so that they are as close as possible to the optimal weights C^k for smooth data. For nonsmooth data, the largest contribution in the reconstruction comes from the stencil(s) that generates the least oscillatory reconstruction. The weights are given as

$$w_{ij}^k = \frac{\alpha_{ij}^k}{\sum_\ell \alpha_{ij}^\ell}, \qquad \alpha_{ij}^k = \frac{C^k}{(\beta_{ij}^k + \epsilon)^2}, \tag{40}$$

for $k, \ell \in \{\text{NE,NW,SW,SE,C}\}$. Here ϵ is a small parameter (typically 10^{-6}) that prevents the denominator from vanishing for constant data, and the β^k's are smoothness indicators that are responsible for detecting discontinuities or large gradients

$$\beta_{ij}^k = \sum_{|\alpha|=1,2} \int_{x_{i-1/2}}^{x_{i+1/2}} \int_{y_{j-1/2}}^{y_{j+1/2}} \Delta x^{2(|\alpha|-1)} \Big(\frac{d^{|\alpha|}}{dx^{\alpha_1} dy^{\alpha_2}} P^k(x,y) \Big)\, dxdy.$$

For the bilinear stencils the smoothness indicators read

$$\beta_{ij}^k = \Delta x^2 \Big((s_{ij}^{x,k})^2 + (s_{ij}^{y,k})^2 \Big), \qquad k \in \{\text{NE,NW,SW,SE}\}$$

with slopes given by (39), and for the quadratic stencil the indicator reads

$$\beta_{ij}^{\text{C}} = \Delta x^2 \Big((s_{ij}^x)^2 + (s_{ij}^y)^2 \Big) + \frac{\Delta x^4}{3} \Big(13(s_{ij}^{xx})^2 + 14(s_{ij}^{xy})^2 + 13(s_{ij}^{yy})^2 \Big)$$

with slopes given by (37).

Compared with the bilinear reconstruction from Section 4.4, the construction of the CWENO polynomials involves a large number or arithmetic operations. Moreover, a large number of variables are needed to represent the reconstruction in each grid cell: 14 polynomial coefficients from (37) and (39) and five weights (40). In the CPU implementation, we therefore chose to recompute the coefficients of the reconstructions rather than storing them. The evaluation of each edge flux involves four one-sided point values taken from the reconstructions in the two adjacent cells. Therefore, in order to compute all edge fluxes, we make a single pass through all grid cells (including one layer of ghost cells) and compute the edge flux between the current cell (i, j) and its neighbour to the east $(i+1, j)$ and to the north $(i, j+1)$. Altogether, this means that in order to save computer memory we recompute the coefficients and weights three times in each grid cell.

In the GPU implementation, we were not able to use the same approach due to the lack of temporary registers. We therefore had to split the computation of edge fluxes into two passes: one for the F-fluxes and one for the G-fluxes, meaning that we recompute the coefficients and weights four times in each grid cell. Moreover, by splitting the flux computation into two render passes, we introduce one additional context switch and extra texture fetches. Unfortunately, we therefore reduce the theoretical speedup by a factor between 1/2 and 3/4. We expect the number of temporary registers to increase in future generations of GPUs and thus become less limiting.

A.2 A Fifth Order WENO Reconstruction

We seek the point values $Q(x_{i\pm 1/2}, y_{j\pm\alpha})$ and $Q(x_{i\pm\alpha}, y_{j\pm 1/2})$. The reconstruction is performed in two steps: first by reconstructing averages over the cell edges from the cell-averages, then by reconstructing the point-values from the edge-averages. In each step we use a one-dimensional, piecewise WENO reconstruction consisting of three quadratic stencils

$$V(\xi) = \sum_{k=0}^{2} w_k Q^k(\xi).$$

In the first step, we start from the cell averages and reconstruct one-sided averages over the cell edges in the x-direction

$$Q^L_{ij} = \frac{1}{\Delta} y \int_{y_{j-1/2}}^{y_{j+1/2}} Q(x^-_{i+1/2}, y)\, dy,$$

$$Q^R_{ij} = \frac{1}{\Delta} y \int_{y_{j-1/2}}^{y_{j+1/2}} Q(x^+_{i-1/2}, y)\, dy.$$

To this end, we use linear combinations of three one-dimensional quadratic stencils centred at $(i+1, j)$, (i, j), and $(i-1, j)$. The corresponding smoothness indicators read

$$
\begin{aligned}
\beta^0_{ij} &= \frac{13}{12}\left(Q^n_{ij} - 2Q^n_{i+1,j} + Q^n_{i+2,j}\right)^2 + \frac{1}{4}\left(3Q^n_{ij} - 4Q^n_{i+1,j} + Q^n_{i+2,j}\right)^2,\\
\beta^1_{ij} &= \frac{13}{12}\left(Q^n_{i-1,j} - 2Q^n_{ij} + Q^n_{i+1,j}\right)^2 + \frac{1}{4}\left(Q^n_{i-1,j} - Q^n_{i+1,j}\right)^2,\\
\beta^2_{ij} &= \frac{13}{12}\left(Q^n_{i-2,j} - 2Q^n_{i-1,j} + Q^n_{ij}\right)^2 + \frac{1}{4}\left(Q^n_{i-2,j} - 4Q^n_{i-1,j} + 3Q^n_{ij}\right)^2.
\end{aligned}
$$

The optimal linear weights for the left extrapolated value Q^L_{ij} are

$$
C^0 = \frac{3}{10}, \quad C^1 = \frac{6}{10}, \quad C^2 = \frac{1}{10},
$$

and Q^L_{ij} becomes

$$
\begin{aligned}
Q^L_{ij} =& \frac{w^0_{ij}}{6}\left(2Q^n_{ij} + 5Q^n_{i+1,j} - Q^n_{i+2,j}\right) + \frac{w^1_{ij}}{6}\left(-Q^n_{i-1,j} + 5Q^n_{ij} + 2Q^n_{i+1,j}\right)\\
&+ \frac{w^2_{ij}}{6}\left(2Q^n_{i-2,j} - 7Q^n_{i-1,j} + 11Q^n_{ij}\right).
\end{aligned}
$$

By symmetry, the linear weights for Q^R_{ij} are

$$
C^0 = \frac{1}{10}, \quad C^1 = \frac{6}{10}, \quad C^2 = \frac{3}{10},
$$

and the edge average itself becomes

$$
\begin{aligned}
Q^R_{ij} =& \frac{w^0_{ij}}{6}\left(11Q^n_{ij} - 7Q^n_{i+1,j} + 2Q^n_{i+2,j}\right) + \frac{w^1_{ij}}{6}\left(2Q^n_{i-1,j} + 5Q^n_{ij} - Q^n_{i+1,j}\right)\\
&+ \frac{w^2_{ij}}{6}\left(-Q^n_{i-2,j} + 5Q^n_{i-1,j} + 2Q^n_{ij}\right).
\end{aligned}
$$

In the second step, we start from the edge averages defined above and reconstruct point values $Q(x^-_{i+1/2}, y_{j+\pm\alpha})$ at the Gaussian integration points. For the first integration point $(x_{i+1/2}, y_j - \Delta y/2\sqrt{3})$, the optimal weights are

$$
C^0 = \frac{210 - \sqrt{3}}{1080}, \quad C^1 = \frac{11}{18}, \quad C^2 = \frac{210 + \sqrt{3}}{1080},
$$

and the reconstructed point value reads

$$
\begin{aligned}
Q(x_{i+1/2}, y_{j-\alpha}) =& w^0_{ij}\left(Q^L_{ij} + \left(3Q^L_{ij} - 4Q^L_{i+1,j} + Q^L_{i+2,j}\right)\frac{\sqrt{3}}{12}\right)\\
&+ w^1_{ij}\left(Q^L_{ij} - \left(Q^L_{i+1,j} - Q^L_{i-1,j}\right)\frac{\sqrt{3}}{12}\right)\\
&+ w^2_{ij}\left(Q^L_{ij} - \left(3Q^L_{ij} - 4Q^L_{i-1,j} + Q^L_{i-2,j}\right)\frac{\sqrt{3}}{12}\right).
\end{aligned}
$$

Similarly, for the second Gaussian point $(x_{i+1/2}, y_j + \Delta y/2\sqrt{3})$

$$C^0 = \frac{210 + \sqrt{3}}{1080}, \quad C^1 = \frac{11}{18}, \quad C^2 = \frac{210 - \sqrt{3}}{1080},$$

and

$$\begin{aligned} Q(x_{i+1/2}, y_{j+\alpha}) =& w_{ij}^0 \Big(Q_{ij}^L - (3Q_{ij}^L - 4Q_{i+1,j}^L + Q_{i+2,j}^L) \frac{\sqrt{3}}{12} \Big) \\ &+ w_{ij}^1 \Big(Q_{ij}^L + (Q_{i+1,j}^L - Q_{i-1,j}^L) \frac{\sqrt{3}}{12} \Big) \\ &+ w_{ij}^2 \Big(Q_{ij}^L + (3Q_{ij}^L - 4Q_{i-1,j}^L + Q_{i-2,j}^L) \frac{\sqrt{3}}{12} \Big). \end{aligned}$$

To reconstruct the point values $Q(x^+_{i-1/2}, y_{j\pm\alpha})$, we start from Q^R_{ij} and repeat the second step. A completely analogous procedure is then used in the y-direction to construct the point values $Q(x_{i\pm\alpha}, y^\pm_{j+1/2})$.

An Introduction to the Numerics of Flow in Porous Media using Matlab

Jørg E. Aarnes, Tore Gimse, and Knut–Andreas Lie

Summary. Even though the art of reservoir simulation has evolved through more than four decades, there is still a substantial research activity that aims toward faster, more robust, and more accurate reservoir simulators. Here we attempt to give the reader an introduction to the mathematics and the numerics behind reservoir simulation. We assume that the reader has a basic mathematical background at the undergraduate level and is acquainted with numerical methods, but no prior knowledge of the mathematics or physics that govern the reservoir flow process is needed. To give the reader an intuitive understanding of the equations that model filtration through porous media, we start with incompressible single-phase flow and move step-by-step to the black-oil model and compressible two-phase flow. For each case, we present a basic numerical scheme in detail, before we discuss a few alternative schemes that reflect trends in state-of-the-art reservoir simulation. Two and three-dimensional test cases are presented and discussed. Finally, for the most basic methods we include simple Matlab codes so that the reader can easily implement and become familiar with the basics of reservoir simulation.

1 Introduction

For nearly half a century, reservoir simulation has been an integrated part of oil-reservoir management. Today, simulations are used to estimate production characteristics, calibrate reservoir parameters, visualise reservoir flow patterns, etc. The main purpose is to provide an information database that can help the oil companies to position and manage wells and well trajectories in order to maximise the oil and gas recovery. Unfortunately, obtaining an accurate prediction of reservoir flow scenarios is a difficult task. One of the reasons is that we can never get a complete and accurate characterization of the rock parameters that influence the flow pattern. And even if we did, we would not be able to run simulations that exploit all available information, since this would require a tremendous amount of computer resources that exceed by far the capabilities of modern multi-processor computers. On the other hand, we do not need, nor do we seek a simultaneous description of the flow scenario on all scales down to the pore scale. For reservoir management it is usually sufficient to describe the general trends in the reservoir flow pattern.

In the early days of the computer, reservoir simulation models were built from two-dimensional slices with 10^2–10^3 grid cells representing the whole reservoir. In contrast, reservoir characterizations today model the porous rock formations by the means of grid-blocks down to the meter scale. This gives three-dimensional models consisting of multi-million cells. Despite an astonishing increase in computer power, and intensive research on computation techniques, commercial reservoir simulators can seldom run simulations directly on geological grid models. Instead, coarse grid models with grid-blocks that are typically ten to hundred times larger are built using some kind of upscaling of the geophysical parameters. How one should perform this upscaling is not trivial. In fact, upscaling has been, and probably still is, one of the most active research areas in the oil industry (see e.g., [7, 14, 15, 32]). This effort reflects that it is a general opinion that, with the ever increasing size and complexity of the geological reservoir models, we cannot run simulations directly on these models in the foreseeable future.

Along with the development of better computers, new and more robust upscaling techniques, and more detailed reservoir characterizations, there has also been an equally significant development in the area of numerical methods. State-of-the-art simulators employ numerical methods that can take advantage of multiple processors, distributed memory workstations, adaptive grid refinement strategies, and iterative techniques with linear complexity. For the simulation, there exists a catalogue of different numerical schemes that all have their pros and cons. With all these techniques available we see a trend where methods are being tuned to a special set of applications, as opposed to traditional methods that were developed for a large class of differential equations. As an example, we mention the recent multiscale numerical methods [4, 5, 12, 19, 20, 25] that are specially suited for differential equations whose solutions may display a multiple scale structure. Although these methods resemble numerical schemes obtained from an upscaling procedure, they are somewhat more rigorous in the sense that they exploit fine-scale information in a mathematically more consistent manner.

It is possible that multiscale methods can help bridge the gap between the size of the geological grid and the size of the simulation grid in reservoir simulation. This type of contribution would certainly take reservoir simulation a big leap forward, and could simplify reservoir management workflow considerably. However, although upscaling is undoubtedly an important part of reservoir simulation technology, neither upscaling nor multiscale techniques will be discussed here. This part of the reservoir simulation framework is discussed separately elsewhere in this book [1]. Here our purpose is to present a self-contained tutorial on reservoir simulation. The main idea is to let the reader become familiar with the mathematics behind porous media flow simulation, and present some basic numerical schemes that can be used to solve the governing partial differential equations. Moreover, to ease the transition from theory to implementation, we supply compact Matlab codes for some of the presented methods applied to Cartesian grids. We hope that this material can give students or researchers about to embark on, for instance, a Master project or a PhD project, a head start.

We assume that the reader has knowledge of mathematics and numerics at the undergraduate level in applied mathematics, but we do not assume any prior knowledge of reservoir simulation. We therefore start by giving a crash course on the physics and mechanics behind reservoir simulation in Section 2. This section presents and explains the role of the various geophysical parameters and indicates how they are obtained. In Section 3 we present the basic mathematical model in its simplest form: the single-phase flow model giving an equation for the fluid pressure. Here we also present some numerical schemes for solving the pressure equation. Then, in Section 4 we move on to multiphase flow simulation. Multiphase flows give rise to a coupled system consisting of a (nearly) elliptic *pressure equation* and a set of (nearly) hyperbolic mass transport equations, so-called *saturation equations*. To balance all terms in these equations properly is a difficult task, and requires quite a bit of bookkeeping. Therefore, to enhance readability, we make some simplifying assumptions before we start to discretise the equations. Section 5 is therefore devoted to immiscible two-phase flow where gas is allowed to be dissolved in the oleic phase. These assumptions give rise to what is known as the *black-oil model*. We then show how to discretise both the pressure equation and the saturation equations separately, and explain how to deal with the coupling between the pressure equation and the saturation equations. Finally, in Section 6, we make some additional assumptions and present a Matlab code for a full two-phase flow simulator. For illustration purposes, the simulator is applied to some two-dimensional test-cases extracted from a strongly heterogeneous reservoir model that was used as a benchmark for upscaling techniques [14].

2 A Crash Course on the Physics and Mechanics behind Reservoir Simulation

The purpose of this section is to briefly summarise some aspects of the art of modelling porous media flow and motivate a more detailed study on some of the related topics. More details can be found in one of the general textbooks describing modelling of flow in porous media, e.g., [6, 11, 16, 21, 28, 30, 34]. For a newcomer in the field, we can in particular recommend the recent book by Chen, Huan, and Ma [13].

Over a period of millions of years, layers of sediments containing organic material built up in the area that is now below the North Sea. A few centimetres every hundred years piled up to hundreds and thousands of meters and made the pressure and temperature increase. Simultaneously, severe geological activity took place. Cracking of continental plates and volcanic activity changed the area from being a relatively smooth, layered plate into a complex structure where previously continuous layers were cut, shifted, or twisted in various directions. As the organic material was compressed, it eventually turned into a number of different hydrocarbons. Gravity separated trapped water and the hydrocarbon components. The lightest hydrocarbons (methane, ethane, etc.) usually escaped quickly, whilst the heavier oils moved slowly towards the surface. At certain cites, however, the geological activity had created and bent layers of low-permeable (or non-permeable) rock, so that the migrating hydrocarbons were trapped. These are today's oil and gas reservoirs in the North Sea, and they are typically found at about 1 000–3 000 meters below the sea bed.

Although we speak about a porous medium, we should remember it is solid rock. However, almost any naturally formed rock contains pores, and the distribution and volume fraction of such pores determine the rock properties, which in turn are the parameters governing the hydrocarbon flow in the reservoir.

2.1 Porosity

The rock *porosity*, usually denoted by ϕ, is the void volume fraction of the medium, that is, $0 \leq \phi < 1$. The porosity usually depends on the pressure; the rock is *compressible*, and the rock compressibility is defined by:

$$c_r = \frac{1}{\phi}\frac{d\phi}{dp}, \tag{1}$$

where p is the overall reservoir pressure. For simplified models, it is customary to neglect the rock compressibility and assume that ϕ only depends on the spatial coordinate. If compressibility cannot be neglected, it is common to use a linearization so that:

$$\phi = \phi_0\big(1 + c_r(p - p_0)\big). \tag{2}$$

For a North Sea reservoir, ϕ is typically in the range 0.1–0.3, and compressibility can be significant, as e.g., witnessed by the subsidence in the Ekofisk area. Since the dimension of the pores is very small compared to any interesting scale for reservoir simulation, one normally assumes that porosity is a piecewise continuous spatial function. However, ongoing research aims to understand better the relation between flow models on pore scale and on reservoir scale.

2.2 Permeability

The (absolute) *permeability*, denoted by $\mathbf{K}$, is a measure of the rock's ability to transmit a single fluid at certain conditions. Since the orientation and interconnection of the pores are essential for flow, the permeability is not necessarily proportional to the porosity, but $\mathbf{K}$ is normally strongly correlated to ϕ.

Rock formations like sandstones tend to have many large or well-connected pores and therefore transmit fluids readily. They are therefore described as permeable. Other formations, like shales, may have smaller, fewer or less interconnected pores and are hence described as impermeable. Although the SI-unit for permeability is m^2, it is commonly represented in Darcy (D), or milli-Darcy (mD). The precise definition of 1D ($\approx 0.987 \cdot 10^{-12}\,m^2$) involves transmission of a 1cp fluid (see below) through a homogeneous rock at a speed of 1cm/s due to a pressure gradient of 1atm/cm. Translated to reservoir conditions, 1D is a relatively high permeability.

In general, $\mathbf{K}$ is a tensor, which means that the permeability in the different directions depends on the permeability in the other directions. However, by a change of basis, $\mathbf{K}$ may sometimes be diagonalised, and due to the reservoir structure, horizontal and vertical permeability suffices for several models. We say that the medium is isotropic (as opposed to anisotropic) if $\mathbf{K}$ can be represented as a scalar function, e.g., if the horizontal permeability is equal to the vertical permeability. Moreover, due to transitions between different rock formations, the permeability may vary rapidly over several orders of magnitude, local variations in the range 1 mD to 10 D are not unusual in a typical field.

Production (or measurements) may also change the permeability. When temperature and pressure is changed, microfractures may open and significantly change the permeability. Furthermore, since the definition of permeability involves a certain fluid, different fluids will experience different permeability in the same rock sample. Such rock-fluid interactions are discussed below.

2.3 Fluid Properties

The void in the porous medium is assumed to be filled with the different *phases*. The volume fraction occupied by each phase is the *saturation* (s) of that phase. Thus,

$$\sum_{\text{all phases}} s_i = 1. \tag{3}$$

For practical reservoir purposes, usually only three phases are considered; aqueous (w), oleic (o), and gaseous (g) phase. Each phase contains one or more *components*. A hydrocarbon component is a unique chemical species (methan, ethane, propane, etc). Since the number of hydrocarbon components can be quite large, it is common to group components into psuedo-components. Henceforth we will make no distinction between components and pseudo-components.

Due to the varying and extreme conditions in a reservoir, the hydrocarbon composition of the different phases may change throughout a simulation (and may sometimes be difficult to determine uniquely). The mass fraction of component i in phase j is denoted by c_{ij}. In each of the phases, the mass fractions should add up to unity, so that for N different components, we have:

$$\sum_{i=1}^{N} c_{ig} = \sum_{i=1}^{N} c_{io} = \sum_{i=1}^{N} c_{iw} = 1. \tag{4}$$

Next we assign a *density* ρ and a *viscosity* μ to each phase. In general, these are functions of *phase pressure* p_i ($i = g, o, w$) and the composition of each phase. That is, for gas

$$\rho_g = \rho_g(p_g, c_{1g}, ..., c_{Ng}), \qquad \mu_g = \mu_g(p_g, c_{1g}, ..., c_{Ng}), \tag{5}$$

and similarly for the other phases. These dependencies are most important for the gas phase, and are usually ignored for the water phase.

The compressibility of the phase is defined as for rock compressibility:

$$c_i = \frac{1}{\rho_i}\frac{d\rho_i}{dp_i}, \quad i = g, o, w. \tag{6}$$

Compressibility effects are more important for gas than for oil and water. In simplified models, water compressibility is therefore usually neglected.

Due to interfacial tensions, the phase pressures are different, defining the *capillary pressure*,

$$p_{cij} = p_i - p_j, \tag{7}$$

for $i, j = g, o, w$. Although other dependencies are reported, it is usually assumed that the capillary pressure is a function of the saturations only.

Other parameters of importance are the bubble-point pressures for the various components. At given temperature, the bubble-point pressures signify the pressures where the respective phases start to boil. Below the bubble-point pressures, gas is released and we get transition of the components between the phases. For most realistic models, even if we do not distinguish between all the components, one allows gas to be dissolved in oil. For such models, an important pressure-dependent parameter is the solution gas-oil ratio R_s for the gas dissolved in oil at reservoir conditions. It is also common to introduce so-called formation volume factors (B_w, B_o, B_g) that model the pressure dependent ratio of bulk volumes at reservoir and surface conditions. Thermodynamical behaviour is, however, a very complex topic that is usually significantly simplified or even ignored in reservoir simulation. Thermodynamics will therefore not be discussed any further here.

2.4 Relative Permeabilities

Even though phases do not really mix, for macroscale modelling, we assume that all phases may be present at the same location. Thus, it turns out that the ability of one phase to move depends on the environment at the actual location. That is, the permeability experienced by one phase depends on the saturation of the other phases at that specific location, as well as the phases' interaction with the pore walls. Thus, we introduce a property called *relative permeability*, denoted by $k_{ri}, i = g, o, w$, which describes how one phase flows in the presence of the two others. Thus, in general, and by the closure relation (3), we may assume that

$$k_{ri} = k_{ri}(s_g, s_o), \tag{8}$$

where subscript r stands for *relative* and i denotes one of the phases g, o, or w. Thus, the (effective) permeability experienced by phase i is $\mathbf{K}_i = \mathbf{K}k_{ri}$. It is important to note that the relative permeabilities are nonlinear functions of the saturations, so that the sum of the relative permeabilities at a specific location (with a specific composition) is not necessarily equal to one. In general, relative permeabilities may depend on the pore-size distribution, the fluid viscosity, and the interfacial forces between the fluids. These features, which are carefully reviewed by Demond and Roberts [18], are usually ignored. Of greater importance to oil recovery is probably the temperature dependency [29], which may be significant, but very case-related.

Another effect is that due to the adsorption at pore walls and creation of isolated, captured droplets, the relative permeability curves do not extend all over the interval $[0, 1]$. The smallest saturation where a phase is mobile is called the *residual* saturation. The adsorption effects may vary, and this may have important effects. Particularly for simulation of polymer injection, adsorption will occur and have significant impact on the results. The adsorption of a component is usually a nonlinear function, assumed to depend on the rock matrix, and the concentration of the actual component. Microscopic rock-fluid interactions also imply that the rock absolute permeability is not a well-defined property. Liquids obey no-slip boundary conditions, whilst gases may not experience the same effects (Klinkenberg effect).

Obtaining measurements of the quantities discussed above, is very difficult. Particularly the relative permeability measurements are costly and troublesome [24]. Recently, the laboratory techniques have made great progress by using computer tomography and nuclear magnetic resonance (NMR) to scan the test cores where the actual phases are being displaced. Although standard experimental procedures exist for measuring two-phase relative permeabilities (systems where only two phases are present), there is still usually a significant uncertainty concerning the relevance of the experimental values found and it is difficult to come up with reliable data to be used in a simulator. This is mainly due to boundary effects. Particularly for three-phase systems, no reliable experimental technique exists. Thus, three-phase relative permeabilities are usually modelled using two-phase measurements, for which several theoretical models have been proposed. Most of them are based on Stone's model [33], where sets of two-phase relative permeabilities are combined to give three-phase data [17]. Recently, a mathematically more plausible model has been proposed [26, 27] for the case with no gravity (or with gravity and no counter-current flow). Contrary to the Stone models, the latter exhibits no elliptic regions for the resulting conservation laws.

The uncertainty regarding relative permeability is, however, modest compared to the uncertainty imposed by the sparsity of absolute permeability data. To measure **K**, core samples from the rock are used. These may be about 10 cm in diameter, taken from vertical test wells at every 25 cm of depth. Clearly these samples have negligible volume compared to the entire oil reservoir extending over several kilometres. From outcrops and mines it is known that rock permeability may vary widely, indicating that the modelling of permeability for a reservoir based on core samples is a tremendously difficult problem. Some additional information may be gained from seismic measurements, but with the technology available today, only large scale structures may be found by this technique. Therefore, we have very limited information about how the subsurface reservoirs look like. To meet this problem, stochastic methods have been applied extensively in the field of reservoir description; see Haldorsen and MacDonald [23] and the references therein. Moreover, when comparing and adjusting parameters as real-life production starts, permeability models can be tuned to fit the actual production data, and thereby hopefully improve the original models and predict future production better. Such approaches are called *history matching*.

2.5 Production Processes

Initially, a hydrocarbon reservoir is at equilibrium, and contains gas, oil, and water, separated by gravity. This equilibrium has been established over millions of years with gravitational separation and geological and geothermal processes. When a well is drilled through the upper non-permeable layer and penetrates the upper hydrocarbon cap, this equilibrium is immediately disturbed. The reservoir is usually connected to the the well and surface production facilities by a set of valves. If there were no production valves to stop the flow, we would have a "blow out" since the reservoir is usually under a high pressure. As the well is ready to produce, the valves are opened slightly, and hydrocarbons flow out of the reservoir due to over-pressure. This in turn, sets up a flow inside the reservoir and hydrocarbons flow towards the well, which in turn may induce gravitational instabilities. Also the capillary pressures will act as a (minor) driving mechanism, resulting in local perturbations of the situation.

During the above process, perhaps 20 percent of the hydrocarbons present are produced until a new equilibrium is achieved. We call this *primary production* by natural drives. One should note that a sudden drop in pressure also may have numerous other intrinsic effects. Particularly in complex, composite systems this may be the case, as pressure-dependent parameters experience such drops. This may give non-convective transport and phase transfers, as vapour and gaseous hydrocarbons may suddenly condensate.

As pressure drops, less oil and gas is flowing, and eventually the production is no longer economically sustainable. Then the operating company may start *secondary production*, by engineered drives. These are processes based on injecting water or gas into the reservoir. The reason for doing this is twofold; some of the pressure is rebuilt or even increased, and secondly one tries to push out more profitable hydrocarbons with the injected substance. One may perhaps produce another 20 percent of the oil by such processes, and engineered drives are standard procedure at most locations in the North Sea today.

In order to produce even more oil, *Enhanced Oil Recovery* (EOR, or tertiary recovery) techniques may be employed. Among these are heating the reservoir or injection of sophisticated substances like foam, polymers or solvents. Polymers are supposed to change the flow properties of water, and thereby to more efficiently push out oil. Similarly, solvents change the flow properties of the hydrocarbons, for instance by developing miscibility with an injected gas. In some sense, one tries to wash the pore walls for most of the remaining hydrocarbons. The other technique is based on injecting steam, which will heat the rock matrix, and thereby, hopefully, change the flow properties of the hydrocarbons. At present, such EOR techniques are considered too expensive for large scale commercial use, but several studies have been conducted and the mathematical foundations are being carefully investigated, and at smaller scales EOR is being performed.

One should note that the terms primary, secondary, and tertiary are ambiguous. EOR techniques may be applied during primary production, and secondary production may be performed from the first day of production.

3 Incompressible Single-Phase Flow

The simplest possible way to describe the displacements of fluids in a reservoir is by a single-phase model. This model gives an equation for the pressure distribution in the reservoir and is used for many early-stage and simplified flow studies. Single-phase models are used to identify flow directions; identify connections between producers and injectors; in flow-based upscaling; in history matching; and in preliminary model studies.

3.1 Basic Model

Assume that we want to model the filtration of a fluid through a porous medium of some kind. The basic equation describing this process is the continuity equation which states that mass is conserved

$$\frac{\partial(\phi\rho)}{\partial t} + \nabla\cdot(\rho v) = q. \tag{9}$$

Here the source term q models sources and sinks, that is, outflow and inflow per volume at designated well locations.

For low flow velocities v, filtration through porous media is modelled with an empirical relation called Darcy's law after the French engineer Henri Darcy.

Darcy discovered in 1856, through a series of experiments, that the filtration velocity is proportional to a combination of the gradient of the fluid pressure and pull-down effects due to gravity. More precisely, the volumetric flow density v (which we henceforth will refer to as flow velocity) is related to pressure p and gravity forces through the following gradient law:

$$v = -\frac{\mathbf{K}}{\mu}(\nabla p + \rho g \nabla z). \qquad (10)$$

Here $\mathbf{K}$ is the permeability, μ is the viscosity, g is the gravitational constant and z is the spatial coordinate in the upward vertical direction. For brevity we shall in the following write $G = -g\nabla z$ for the gravitational pull-down force.

In most real field geological reservoir models, $\mathbf{K}$ is an anisotropic diagonal tensor. Unfortunately, as mentioned in the introduction, commercial reservoir simulators can seldom run simulations directly on these geological models. Effects of small-scale heterogeneous structures are therefore incorporated into simulation models by designing effective permeability tensors that represent impact of small-scale heterogeneous structures on a coarser grid, as discussed in more detail elsewhere in this book [1]. These effective, or upscaled, permeability tensors try to model principal flow directions, which may not be aligned with the spatial coordinate directions, and can therefore be full tensors. Hence, here we assume only that $\mathbf{K}$ is a symmetric and positive definite tensor, whose eigenvalues are bounded uniformly below and above by positive constants.

We note that Darcy's law is analogous to Fourier's law of heat conduction (in which $\mathbf{K}$ is replaced with the heat conductivity tensor) and Ohm's law of electrical conduction (in which $\mathbf{K}$ is the inverse of the electrical resistance). However, whereas there is only one driving force in thermal and electrical conduction, there are two driving forces in porous media flow: gravity and the pressure gradient. Since gravity forces are approximately constant inside a reservoir domain Ω, we need only use the reservoir field pressure as our primary unknown. To solve for the pressure, we combine Darcy's law (10) with the continuity equation (9).

To illustrate, we derive an equation that models flow of a fluid, say, water (w), through a porous medium characterised by a permeability field $\mathbf{K}$ and a corresponding porosity distribution. For simplicity, we assume that the porosity ϕ is constant in time and that the flow can be adequately modelled by assuming incompressibility, i.e., constant density. Then the temporal derivative term in (9) vanishes and we obtain the following elliptic equation for the water pressure:

$$\nabla \cdot v_w = \nabla \cdot \left[-\frac{\mathbf{K}}{\mu_w}(\nabla p_w - \rho_w G) \right] = \frac{q_w}{\rho_w}. \qquad (11)$$

To close the model, we must specify boundary conditions. Unless stated otherwise we shall follow common practice and use no-flow boundary conditions.

Hence, on the reservoir boundary $\partial\Omega$ we impose $v_w \cdot n = 0$, where n is the normal vector pointing out of the boundary $\partial\Omega$. This gives us an isolated flow system where no water can enter or exit the reservoir.

In the next subsections we restrict our attention to incompressible flows and present several numerical methods for solving (11). To help the interested reader with the transition from theory to implementation, we also discuss some simple implementations in Matlab for uniform Cartesian grids. We first present a cell-centred finite-volume method, which is sometimes referred to as the *two-point flux-approximation (TPFA) scheme*. Although this scheme is undoubtedly one of the simplest discretisation techniques for elliptic equations, it is still widely used in the oil-industry.

3.2 A Simple Finite-Volume Method

In classical finite-difference methods, partial differential equations (PDEs) are approximated by replacing the partial derivatives with appropriate divided differences between point-values on a discrete set of points in the domain. Finite-volume methods, on the other hand, have a more physical motivation and are derived from conservation of (physical) quantities over cell volumes. Thus, in a finite-volume method the unknown functions are represented in terms of average values over a set of finite-volumes, over which the integrated PDE model is required to hold in an averaged sense.

Although finite-difference and finite-volume methods have fundamentally different interpretation and derivation, the two labels are used interchangeably in the scientific literature. We therefore choose to not make a clear distinction between the two discretisation techniques here. Instead we ask the reader to think of a finite-volume method as a conservative finite-difference scheme that treats the grid cells as control volumes. In fact, there exist several finite-volume and finite-difference schemes of low order, for which the cell-centred values obtained with the finite-difference schemes coincide with cell averages obtained with the corresponding finite-volume schemes. The two representations of the unknown functions will therefore also be used interchangeably henceforth.

To derive a set of finite-volume mass-balance equations for (11), denote by Ω_i a grid cell in Ω and consider the following integral over Ω_i:

$$\int_{\Omega_i} \left(\frac{q_w}{\rho_w} - \nabla \cdot v_w \right) dx = 0. \tag{12}$$

Invoking the divergence theorem, assuming that v_w is sufficiently smooth, equation (12) transforms into the following mass-balance equation:

$$\int_{\partial\Omega_i} v_w \cdot n \, d\nu = \int_{\Omega_i} \frac{q_w}{\rho_w} \, dx. \tag{13}$$

Here n denotes the outward-pointing unit normal on $\partial\Omega_i$. Corresponding finite-volume methods are now obtained by approximating the pressure p_w

with a cell-wise constant function $\mathbf{p}_w = \{p_{w,i}\}$ and estimating $v_w \cdot n$ across cell interfaces $\gamma_{ij} = \partial\Omega_i \cap \partial\Omega_j$ from a set of neighbouring cell pressures.

To formulate the standard two-point flux-approximation (TPFA) finite-volume scheme that is frequently used in reservoir simulation, it is convenient to reformulate equation (11) slightly, so that we get an equation on the form

$$-\nabla \cdot \lambda \nabla u = f, \tag{14}$$

where $\lambda = \mathbf{K}/\mu_w$. To this end, we have two options: we can either introduce a flow potential $u_w = p_w + \rho_w g z$ and express our model as an equation for u_w

$$-\nabla \cdot \lambda \nabla u_w = \frac{q_w}{\rho_w},$$

or we can move the gravity term $\nabla \cdot (\lambda \rho_w G)$ to the right-hand side. Hence, we might as well assume that we want to solve equation (14) for u.

As the name suggests, the TPFA scheme uses two points, the cell-averages u_i and u_j, to approximate the flux $v_{ij} = -\int_{\gamma_{ij}} (\lambda \nabla u) \cdot n \, d\nu$. To be more specific, let us consider a regular hexahedral grid with gridlines aligned with the principal coordinate axes. Moreover, assume that γ_{ij} is an interface between adjacent cells in the x–coordinate direction so that $n_{ij} = (1,0,0)^T$. The gradient ∇u on γ_{ij} in the TPFA method is now replaced with

$$\delta u_{ij} = \frac{2(u_j - u_i)}{\Delta x_i + \Delta x_j}, \tag{15}$$

where Δx_i and Δx_j denote the respective cell dimensions in the x-coordinate direction. Thus, we obtain the following expression for v_{ij}:

$$v_{ij} = \delta u_{ij} \int_{\gamma_{ij}} \lambda d\nu = \frac{2(u_j - u_i)}{\Delta x_i + \Delta x_j} \int_{\gamma_{ij}} \lambda d\nu.$$

However, in most reservoir simulation models, the permeability $\mathbf{K}$ is cell-wise constant, and hence not well-defined at the interfaces. This means that we also have to approximate λ on γ_{ij}. In the TPFA method this is done by taking a distance-weighted harmonic average of the respective directional cell permeabilities, $\lambda_{i,ij} = n_{ij} \cdot \lambda_i n_{ij}$ and $\lambda_{j,ij} = n_{ij} \cdot \lambda_j n_{ij}$. To be precise, the n_{ij}–directional permeability λ_{ij} on γ_{ij} is computed as follows:

$$\lambda_{ij} = (\Delta x_i + \Delta x_j) \left(\frac{\Delta x_i}{\lambda_{i,ij}} + \frac{\Delta x_j}{\lambda_{j,ij}} \right)^{-1},$$

Hence, for orthogonal grids with gridlines aligned with the coordinate axes, one approximates the flux v_{ij} in the TPFA method in the following way:

$$v_{ij} = -|\gamma_{ij}| \lambda_{ij} \delta u_{ij} = 2|\gamma_{ij}| \left(\frac{\Delta x_i}{\lambda_{i,ij}} + \frac{\Delta x_j}{\lambda_{j,ij}} \right)^{-1} (u_i - u_j). \tag{16}$$

Finally, summing over all interfaces to adjacent cells, we get an approximation to $\int_{\partial\Omega_i} v_w \cdot n \, d\nu$, and the associated TPFA method is obtained by requiring the mass balance equation (13) to be fulfilled for each grid cell $\Omega_i \in \Omega$.

In the literature on finite-volume methods it is common to express the flux v_{ij} in a more compact form than we have done in (16). Terms that do not involve the cell potentials u_i are usually gathered into an interface transmissibility t_{ij}. For the current TPFA method the transmissibilities are defined by:

$$t_{ij} = 2|\gamma_{ij}| \left(\frac{\Delta x_i}{\lambda_{i,ij}} + \frac{\Delta x_j}{\lambda_{j,ij}} \right)^{-1}.$$

Thus by inserting the expression for t_{ij} into (16), we see that the TPFA scheme for equation (14), in compact form, seeks a cell-wise constant function $\mathbf{u} = \{u_i\}$ that satisfies the following system of equations:

$$\sum_j t_{ij}(u_i - u_j) = \int_{\Omega_i} f \, dx, \qquad \forall \Omega_i \subset \Omega, \tag{17}$$

This system is clearly symmetric, and a solution is, as for the continuous problem, defined up to an arbitrary constant. The system is made positive definite, and symmetry is preserved, by forcing $u_1 = 0$, for instance. That is, by adding a positive constant to the first diagonal of the matrix $\mathbf{A} = [a_{ik}]$, where:

$$a_{ik} = \begin{cases} \sum_j t_{ij} & \text{if } k = i, \\ -t_{ik} & \text{if } k \neq i, \end{cases}$$

The matrix $\mathbf{A}$ is sparse, consisting of a tridiagonal part corresponding to the x-derivative, and two off-diagonal bands corresponding to the y-derivatives.

A short Matlab code for the implementation of (17) on a uniform Cartesian grid is given in Listing 1. In a general code, one would typically assemble the coefficient matrix by looping over all the interfaces of each grid block and for each interface, compute the associated transmissibilities and add them into the correct position of the system matrix. Similarly, the interface fluxes would be computed by looping over all grid block interfaces. Such an approach would easily be extensible to other discretisation (e.g., the MPFA method discussed in the next section) and to more complex grids (e.g., unstructured and faulted grids). Our code has instead been designed to be compact and efficient, and therefore relies entirely on the Cartesian grid structure to set up the transmissibilities in the coefficient matrix using Matlab's inherent vectorization.

In the code, `K` is a 3×`Nx`×`Ny`×`Nz` matrix holding the three diagonals of the tensor λ for each grid cell. The coefficient matrix `A` is sparse (use e.g., `spy(A)` to visualise the sparsity pattern) and is therefore generated with Matlab's built-in sparse matrix functions. Understanding the Matlab code should be rather straightforward. The only point worth noting is perhaps that the constant that we use to force $u_1 = 0$ in element `A(1,1)` is taken as the sum of the diagonals in λ. This is done in order to control that this extra equation does not have an adverse effect on the condition number of `A`.

Listing 1. TPFA finite-volume discretisation of $-\nabla(K(x)\nabla u) = q$.

```
function [P,V]=TPFA(Grid,K,q)

% Compute transmissibilities by harmonic averaging.
Nx=Grid.Nx; Ny=Grid.Ny; Nz=Grid.Nz; N=Nx*Ny*Nz;
hx=Grid.hx; hy=Grid.hy; hz=Grid.hz;
L = K.^(-1);
tx = 2*hy*hz/hx; TX = zeros(Nx+1,Ny,Nz);
ty = 2*hx*hz/hy; TY = zeros(Nx,Ny+1,Nz);
tz = 2*hx*hy/hz; TZ = zeros(Nx,Ny,Nz+1);
TX(2:Nx,:,:) = tx./(L(1,1:Nx-1,:,:)+L(1,2:Nx ,:,:));
TY(:,2:Ny,:) = ty./(L (2,:,1: Ny-1,:)+L(2,:,2:Ny,:));
TZ (:,:,2: Nz) = tz./(L (3,:,:,1: Nz-1)+L(3,:,:,2:Nz));

% Assemble TPFA discretization matrix.
x1 = reshape(TX(1:Nx,:,:),N,1); x2 = reshape(TX(2:Nx+1,:,:),N,1);
y1 = reshape(TY(:,1:Ny,:),N,1); y2 = reshape(TY(:,2:Ny+1,:),N,1);
z1 = reshape(TZ(:,:,1:Nz),N,1); z2 = reshape(TZ(:,:,2:Nz+1),N,1);
DiagVecs = [-z2,-y2,-x2,x1+x2+y1+y2+z1+z2,-x1,-y1,-z1];
DiagIndx = [-Nx*Ny,-Nx,-1,0,1,Nx,Nx*Ny];
A = spdiags(DiagVecs,DiagIndx,N,N);
A(1,1) = A(1,1)+sum(Grid.K(:,1,1,1));

% Solve linear system and extract interface fluxes.
u = A\q;
P = reshape(u,Nx,Ny,Nz);
V.x = zeros(Nx+1,Ny,Nz);
V.y = zeros(Nx,Ny+1,Nz);
V.z = zeros(Nx,Ny,Nz+1);
V.x(2:Nx ,:,:) = (P(1:Nx-1,:,:)-P(2:Nx,:,:)).*TX(2:Nx,:,:);
V.y (:,2: Ny,:) = (P(:,1:Ny-1,:)-P(:,2:Ny,:)).*TY(:,2:Ny,:);
V.z (:,:,2: Nz) = (P (:,:,1: Nz-1)-P(:,:,2:Nz)).*TZ(:,:,2:Nz);
```

Example 1. In the first example we consider a homogeneous and isotropic permeability $K \equiv 1$ for all $x \in \mathbb{R}^2$. We place an injection well at the origin and production wells at the points $(\pm 1, \pm 1)$ and specify no-flow conditions at the boundaries. These boundary conditions give the same flow as if we repeated the five-spot well pattern to infinity in every direction. The flow in the five-spot is symmetric about both the coordinate axes. We can therefore reduce the computational domain to a quarter, and use e.g., the unit box $\Omega = [0, 1]^2$. The corresponding problem is called a *quarter-five spot* problem, and is a standard test-case for numerical methods in reservoir simulation.

Figure 1 shows pressure contours. The pressure P is computed by the following lines for a 8×8 grid:

```
>> Grid.Nx=8; Grid.hx=1/Grid.Nx;
>> Grid.Ny=8; Grid.hy=1/Grid.Ny;
>> Grid.Nz=1; Grid.hz=1/Grid.Nz;
>> Grid.K=ones(3,Grid.Nx,Grid.Ny);
>> N=Grid.Nx*Grid.Ny*Grid.Nz; q=zeros(N,1); q([1 N])=[1 -1];
>> P=TPFA(Grid,Grid.K,q);
```

Here the command `P=TPFA(···)` assigns to `P` the first (the local variable `U`) of the four output variables `U, FX, FY`, and `FZ`.

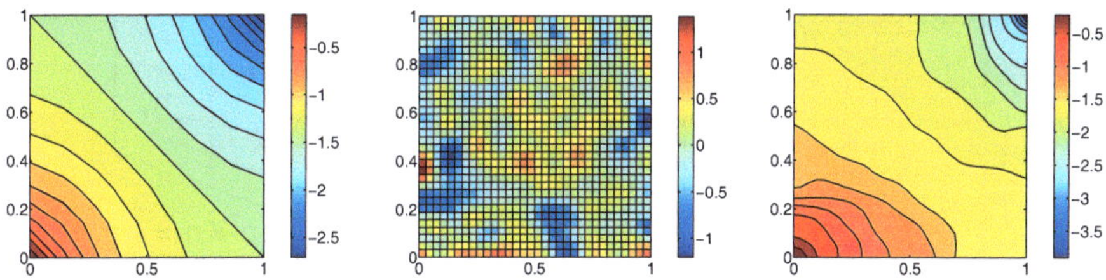

Fig. 1. The left plot shows pressure contours for a homogeneous quarter-five spot. The middle plot shows logarithm of the permeability for the heterogeneous quarter-five spot and the right plot the corresponding pressure distribution. As particles flow in directions of decreasing pressure gradient, the pressure decays from the injector in the lower-left to the producer in the upper-right corner.

Next we increase the number of grid-cells in each direction from eight to 32 and consider a slightly more realistic permeability field obtained from a log-normal distribution.

```
>> Grid.Nx=32; Grid.hx=1/Grid.Nx;
>> Grid.Ny=32; Grid.hy=1/Grid.Ny;
>> Grid.Nz=1; Grid.hz=1/Grid.Nz;
>> Grid.K=exp(5*smooth3(smooth3(randn(3,Grid.Nx,Grid.Ny))));
>> N=Grid.Nx*Grid.Ny*Grid.Nz; q=zeros(N,1); q([1 N])=[1 -1];
>> P=TPFA(Grid,Grid.K,q);
```

The permeability and pressure distribution is plotted in Figure 1.

3.3 Multipoint Flux-Approximation Schemes

The TPFA finite-volume scheme presented above is convergent only if each grid cell is a parallelepiped and

$$n_{ij} \cdot \mathbf{K} n_{ik} = 0, \qquad \forall \Omega_i \subset \Omega, \qquad n_{ij} \neq n_{ik}, \tag{18}$$

where n_{ij} and n_{ik} denote normal vectors into two neighbouring grid cells. A grid consisting of parallelepipeds satisfying (18) is said to be **K**-orthogonal. Orthogonal grids are, for example, **K**-orthogonal with respect to diagonal permeability tensors, but not with respect to full tensor permeabilities. Figure 2 shows a schematic of an orthogonal grid and a **K**-orthogonal grid.

If the TPFA method is used to discretise equation (14) on grids that are not **K**-orthogonal, the scheme will produce different results depending on the orientation of the grid (so-called grid-orientation effects) and will converge to a wrong solution. Despite this shortcoming of the TPFA method, it is still the dominant (and default) method for practical reservoir simulation, owing to its simplicity and computational speed. We present now a class of so-called *multi-point flux-approximation (MPFA) schemes* that aim to amend the shortcomings of the TPFA scheme.

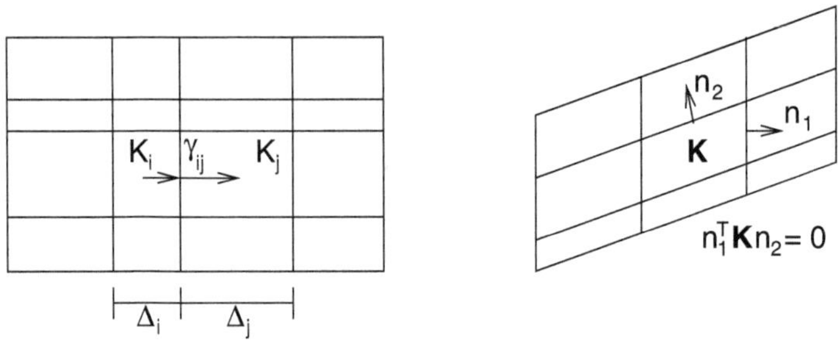

Fig. 2. The grid in the left plot is orthogonal with gridlines aligned with the principal coordinate axes. The grid in the right plot is a **K**-orthogonal grid.

Consider an orthogonal grid and assume that **K** is a constant tensor with nonzero off-diagonal terms. Moreover, for presentational simplicity, let γ_{ij} be an interface between two adjacent grid cells in the x–coordinate direction. Then for a given function u, the corresponding flux across γ_{ij} is given by:

$$\int_{\gamma_{ij}} v_w \cdot n_{ij}\, d\nu = -\int_{\gamma_{ij}} \big(k_{xx}\partial_x u + k_{xy}\partial_y u + k_{xz}\partial_z u\big)\, d\nu.$$

This expression involves derivatives in three orthogonal coordinate directions. Evidently, two point values can only be used to estimate a derivative in one direction. In particular, the two cell averages u_i and u_j can not be used to estimate the derivative of u in the y and z-directions. Hence, the TPFA scheme neglects the flux contribution from $k_{xy}\partial_y u$ and $k_{xz}\partial_z u$.

To obtain consistent interfacial fluxes for grids that are not **K**-orthogonal, one must also estimate partial derivatives in coordinate directions parallel to the interfaces. For this purpose, more than two point values, or cell averages, are needed. This leads to schemes that approximate v_{ij} using multiple cell averages, that is, with a linear expression on the form:

$$v_{ij} = \sum_k t_{ij}^k g_{ij}^k(\mathbf{u}).$$

Here $\{t_{ij}^k\}_k$ are the transmissibilities associated with γ_{ij} and $\{g_{ij}^k(\mathbf{u})\}_k$ are the corresponding multi-point pressure or flow potential dependencies. Thus, we see that MPFA schemes for (14) can be written on the form:

$$\sum_{j,k} t_{ij}^k g_{ij}^k(\mathbf{u}) = \int_{\Omega_i} f\, dx, \qquad \forall \Omega_i \subset \Omega. \tag{19}$$

MPFA schemes can, for instance, be designed by simply estimating each of the partial derivatives $\partial_\xi u$ from neighbouring cell averages. However, most MPFA schemes have a more physical motivation and are derived by imposing certain continuity requirements. We will now outline very briefly one such method,

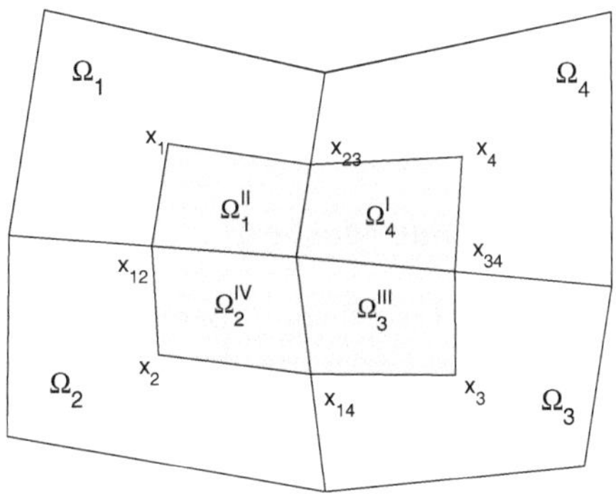

Fig. 3. The shaded region represents the interaction region for the O-method on a two-dimensional quadrilateral grid associated with cells Ω_1, Ω_2, Ω_3, and Ω_4.

called the O-method [2, 3], for irregular, quadrilateral, matching grids in two spatial dimensions.

The O-method is constructed by defining an interaction region (IR) around each corner point in the grid. For a two-dimensional quadrilateral grid, this IR is the area bounded by the lines that connect the cell-centres with the midpoints on the cell interfaces, see Figure 3. Thus, the IR consists of four sub-quadrilaterals ($\Omega_1^{\text{II}}, \Omega_2^{\text{IV}}, \Omega_3^{\text{III}}$, and Ω_4^{I}) from four neighbouring cells (Ω_1, Ω_2, Ω_3, and Ω_4) that share a common corner point. For each IR, define now

$$U_{\text{IR}} = \text{span}\{U_i^{\text{J}} : i = 1, \ldots, 4, \quad \text{J=I}, \ldots, \text{IV}\},$$

where $\{U_i^{\text{J}}\}$ are linear functions on the respective four sub-quadrilaterals. With this definition, U_{IR} has twelve degrees of freedom. Indeed, note that each U_i^{J} can be expressed in the following non-dimensional form

$$U_i^{\text{J}}(x) = u_i + \nabla U_i^{\text{J}} \cdot (x - x_i),$$

where x_i is the cell centre in Ω_i. The cell-centre values u_i thus account for four degrees of freedom and the (constant) gradients ∇U_i^{J} for additional eight.

Next we require that functions in U_{IR} are: (i) continuous at the midpoints of the cell interfaces, and (ii) flux-continuous across the interface segments that lie inside the IR. To obtain a globally coupled system, we first use (i) and (ii) to express the gradients ∇U_i^{J}, and hence also the corresponding fluxes across the IR interface segments, in terms of the unknown cell-centre potentials u_i. This requires solution of a local system of equations. Finally, the cell-centre potentials are determined (up to an arbitrary constant for no-flow boundary conditions) by summing the fluxes across all IR interface segments and requiring that the mass balance equations (13) hold. In this process, transmissibilities are assembled to obtain a globally coupled system for the unknown pressures over the whole domain.

We note that this construction leads to an MPFA scheme where the flux across an interface γ_{ij} depends on the potentials u_j in a total of six neighbouring cells (eighteen in three dimensions). Notice also that the transmissibilities $\{t_{ij}^k\}$ that we obtain when eliminating the IR gradients now account for grid-cell geometries in addition to full-tensor permeabilities.

3.4 A Mixed Finite-Element Method

As an alternative to the MPFA schemes, one can use mixed finite-element methods (FEMs) [10]. In mixed FEMs, the fluxes over cell edges are considered as unknowns in addition to the pressures, and are not computed using a (numerical) differentiation as in finite-volume methods. For the mixed FEMs there is little to gain by reformulating (11) into an equation on the form (14). We therefore return to the original formulation and describe how to discretise the following system of differential equations with mixed FEMs:

$$v = -\lambda(\nabla p - \rho G), \qquad \nabla \cdot v = q. \tag{20}$$

As before we impose no-flow boundary conditions on $\partial\Omega$. To derive the mixed formulation, we first define the following Sobolev space

$$H_0^{1,\mathrm{div}}(\Omega) = \{v \in (L^2(\Omega))^d : \nabla \cdot v \in L^2(\Omega) \text{ and } v \cdot n = 0 \text{ on } \partial\Omega\}.$$

The mixed formulation of (20) (with no-flow boundary conditions) now reads: find $(p, v) \in L^2(\Omega) \times H_0^{1,\mathrm{div}}(\Omega)$ such that

$$\int_\Omega v \cdot \lambda^{-1} u \, dx - \int_\Omega p \, \nabla \cdot u \, dx = \int_\Omega \rho G \cdot u \, dx, \tag{21}$$

$$\int_\Omega l \, \nabla \cdot v \, dx = \int_\Omega ql \, dx, \tag{22}$$

for all $u \in H_0^{1,\mathrm{div}}(\Omega)$ and $l \in L^2(\Omega)$. We observe again that, since no-flow boundary conditions are imposed, an extra constraint must be added to make (21)–(22) well-posed. A common choice is to use $\int_\Omega p \, dx = 0$.

In mixed FEMs, (21)–(22) are discretised by replacing $L^2(\Omega)$ and $H_0^{1,\mathrm{div}}(\Omega)$ with finite-dimensional subspaces U and V, respectively. For instance, in the Raviart–Thomas mixed FEM [31] of lowest order (for triangular, tetrahedral, or regular parallelepiped grids), $L^2(\Omega)$ is replaced by

$$U = \{p \in L^2(\Omega) : p|_{\Omega_i} \text{ is constant } \forall \Omega_i \in \Omega\}$$

and $H_0^{1,\mathrm{div}}(\Omega)$ is replaced by

$$\begin{aligned} V = \{&v \in H_0^{1,\mathrm{div}}(\Omega) : v|_{\Omega_i} \text{ have linear components } \forall \Omega_i \in \Omega, \\ &(v \cdot n_{ij})|_{\gamma_{ij}} \text{ is constant } \forall \gamma_{ij} \in \Omega, \text{ and } v \cdot n_{ij} \text{ is continuous across } \gamma_{ij}\}. \end{aligned}$$

Here n_{ij} is the unit normal to γ_{ij} pointing from Ω_i to Ω_j. The corresponding Raviart–Thomas mixed FEM thus seeks

$$(p, v) \in U \times V \text{ such that (21)–(22) hold for all } u \in V \text{ and } q \in U. \tag{23}$$

To express (23) as a linear system, observe first that functions in V are, for admissible grids, spanned by base functions $\{\psi_{ij}\}$ that are defined by

$$\psi_{ij} \in \mathcal{P}(\Omega_i)^d \cup \mathcal{P}(\Omega_j)^d \quad \text{and} \quad (\psi_{ij} \cdot n_{kl})|_{\gamma_{kl}} = \begin{cases} 1 & \text{if} \quad \gamma_{kl} = \gamma_{ij}, \\ 0 & \text{else,} \end{cases}$$

where $\mathcal{P}(K)$ is the set of linear functions on K. Similarly

$$U = \text{span}\{\chi_m\} \quad \text{where} \quad \chi_m = \begin{cases} 1 & \text{if} \quad x \in \Omega_m, \\ 0 & \text{else.} \end{cases}$$

Thus, writing $p = \sum_{\Omega_m} p_m \chi_m$ and $v = \sum_{\gamma_{ij}} v_{ij}\psi_{ij}$, allows us to write (23) as a linear system in $\mathbf{p} = \{p_m\}$ and $\mathbf{v} = \{v_{ij}\}$. This system takes the form

$$\begin{bmatrix} \mathbf{B} & -\mathbf{C}^T \\ \mathbf{C} & \mathbf{0} \end{bmatrix} \begin{bmatrix} \mathbf{v} \\ \mathbf{p} \end{bmatrix} = \begin{bmatrix} \mathbf{g} \\ \mathbf{f} \end{bmatrix}. \tag{24}$$

Here $\mathbf{f} = [f_m]$, $\mathbf{g} = [g_{kl}]$, $\mathbf{B} = [b_{ij,kl}]$ and $\mathbf{C} = [c_{m,kl}]$, where:

$$g_{kl} = \Big[\int_\Omega \rho G \cdot \psi_{kl}\, dx\Big], \qquad f_m = \Big[\int_{\Omega_m} f\, dx\Big],$$
$$b_{ij,kl} = \Big[\int_\Omega \psi_{ij} \cdot \lambda^{-1}\psi_{kl}\, dx\Big], \qquad c_{m,kl} = \Big[\int_{\Omega_m} \nabla \cdot \psi_{kl}\, dx\Big].$$

Note that for the current Raviart-Thomas finite elements, we have

$$c_{m,kl} = \begin{cases} 1 & \text{if } m = k, \\ -1 & \text{if } m = l, \\ 0 & \text{otherwise.} \end{cases}$$

The matrix entries $b_{ij,kl}$, on the other hand, depend on the geometry of the grid cells and the form of λ. In other words, the entries depend on whether λ is isotropic or anisotropic, whether λ is cell-wise constant or models subgrid variations in the permeability field.

We provide now a Matlab code implementing the Raviart-Thomas mixed FEM for the first-order system (20) on a regular hexahedral grid in three spatial dimensions. The code is divided into three parts: assembly of the **B** block; assembly of the **C** block; and a main routine that loads data (permeability and grid), assembles the whole matrix, and solves the system. Again we emphasise compactness and efficiency, by heavily relying on the Cartesian grid and analytical integration of the Raviart-Thomas basis functions to perform a direct assembly of the matrix blocks. For more general grids, one would typically perform an elementwise assembly by looping over all grid blocks and replace the integration over basis functions by some numerical quadrature rule.

Listing 2. Assembly of **B** for the lowest-order Raviart–Thomas elements.

```
function B=GenB(Grid,K)

Nx=Grid.Nx; Ny=Grid.Ny; Nz=Grid.Nz; N=Nx*Ny*Nz;
hx=Grid.hx; hy=Grid.hy; hz=Grid.hz;
L = K.^(-1);
ex=N-Ny*Nz; ey=N-Nx*Nz; ez=N-Nx*Ny;                    % Number of edges
tx=hx/(6*hy*hz); ty=hy/(6*hx*hz); tz=hz/(6*hx*hy);     % Transmissibilities

X1=zeros(Nx-1,Ny,Nz); X2=zeros(Nx-1,Ny,Nz);            % Preallocate memory
X0=L(1,1:Nx-1,:,:)+L(1,2:Nx,:,:);   x0=2*tx*X0(:);     % Main diagonal
X1(2:Nx-1,:,:)=L(1,2:Nx-1,:,:);     x1=tx*X1(:);       % Upper diagonal
X2(1:Nx-2,:,:)=L(1,2:Nx-1,:,:);     x2=tx*X2(:);       % Lower diagonal

Y1=zeros(Nx,Ny-1,Nz); Y2=zeros(Nx,Ny-1,Nz);            % Preallocate memory
Y0=L(2,:,1:Ny-1,:)+L(2,:,2:Ny,:); y0=2*ty*Y0(:);       % Main diagonal
Y1(:,2:Ny-1,:)=L(2,:,2:Ny-1,:);  y1=ty*Y1(:);          % Upper diagonal
Y2(:,1:Ny-2,:)=L(2,:,2:Ny-1,:);  y2=ty*Y2(:);          % Lower diagonal

Lz1=L(3,:,:,1:Nz-1); z1=tz*Lz1(:);                     % Upper diagonal
Lz2=L(3,:,:,2:Nz);   z2=tz*Lz2(:);                     % Lower diagonal
z0=2*(z1+z2);                                          % Main diagonal

B=[spdiags([x2,x0,x1],[-1,0,1],ex,ex),sparse(ex,ey+ez);...
   sparse(ey,ex),spdiags([y2,y0,y1],[-Nx,0,Nx],ey,ey),sparse(ey,ez );...
   sparse(ez,ex+ey),spdiags([z2,z0,z1],[-Nx*Ny,0,Nx*Ny],ez,ez)];
```

In Listing 2 we show a Matlab function that assembles the **B** matrix. Here λ is assumed to be a diagonal tensor and is represented as a 3×`Nx`×`Ny`×`Nz` matrix `K` with `Nx`, `Ny`, and `Nz` denoting the number of grid cells in the x, y, and z direction, respectively. The degrees of freedom at the interfaces (the fluxes) have been numbered in the same way as the grid-blocks; i.e., first in the x-direction, then in the y-direction, and finally in the z-direction. This gives **B** a hepta-diagonal structure as shown in Figure 4, where the three nonzero blocks correspond to the three components of v. Notice, however, that since the grid is K-orthogonal, we can make **B** tridiagonal by starting the numbering of each component of $v = (v_x, v_y, v_z)$ in the corresponding direction; i.e., number v_x as xyz, v_y as yxz, and v_z as zxy. Similar observations can be made for the assembly of **C**, which is given in Listing 3.

Let us now make a few brief remarks with respect to the implementation of the assembly of the matrix blocks **B** and **C**. First, since only a few components of **B** and **C** are nonzero, we store them using a sparse matrix format. The matrices are created in a block-wise manner using Matlab's built in sparse matrix functions. The function `sparse(m,n)` creates an $m \times n$ zero matrix and `spdiags(M,d,m,n)` creates an $m \times n$ sparse matrix with the columns of M on the diagonals specified by d. Finally, statements of the form `M(:)` turn the matrix M into a column vector.

A drawback with the mixed FEM is that it produces an indefinite linear system. These systems are in general harder to solve than the positive definite systems that arise, e.g., from the TPFA and MPFA schemes described in

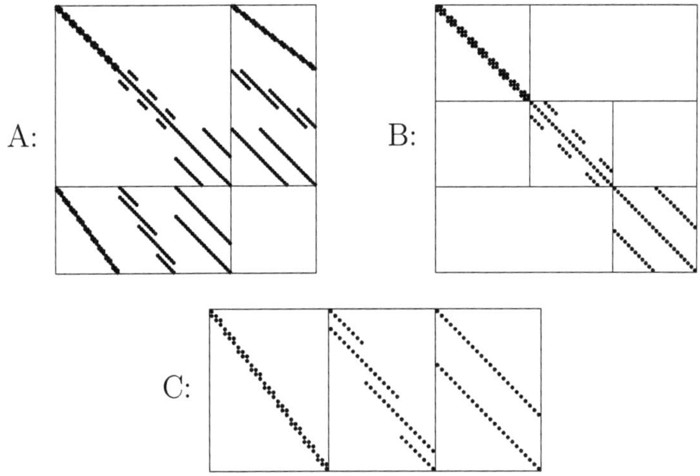

Fig. 4. Sparsity patterns for **A**, **B**, and **C** for a case with $4 \times 3 \times 3$ grid-blocks.

Listing 3. Assembly of **C** for the lowest-order Raviart–Thomas elements.

```
function C=GenC(Grid)

Nx=Grid.Nx; Ny=Grid.Ny; Nz=Grid.Nz;
C=sparse(0,0);                                          % Empty sparse matrix
Nxy=Nx*Ny; N=Nxy*Nz;                                    % Number of grid-points
vx=ones(Nx,1); vy=ones(Nxy,1); vz=ones(N,1);            % Diagonals

for i=1:Ny*Nz                                           % vx-block of C
  Cx=spdiags([vx,-vx],[-1,0]-(i-1)*Nx,N,Nx-1);          % create bidiagonal block
  C=[C,Cx];                                             % append to C
end

for i=1:Nz                                              % vy-block of C
  Cy=spdiags([vy,-vy],[-Nx,0]-(i-1)*Nxy,N,Nxy-Nx);      % create bidiagonal block
  C=[C,Cy];                                             % append to C
end

C = [C, spdiags([vz,-vz],[-Nxy,0],N,N-Nxy)];            % vz-block of C
```

Sections 3.2 and 3.3. In fact, for second-order elliptic equations of the form (11) it is common to use a so-called hybrid formulation. This method leads to a positive definite system where the unknowns correspond to pressures at grid-cell interfaces. The solution to the linear system arising from the mixed FEM can now be obtained from the solution to the hybrid system by performing only local algebraic calculations. This property, which is sometimes referred to as explicit flux representation, allows the inter-cell fluxes that appear in the saturation equation to be expressed as a linear combination of neighbouring values for the pressure.

Explicit flux representation is a feature that also the finite-volume methods enjoy. One of the main advantages is that it allows us to compute fully implicit solutions without computing the fluxes explicitly. Indeed, in the mixed FEM, which does not have an explicit flux representation, one has to solve the full indefinite linear system for the pressure equation alongside the linear system for the saturation equation in order to produce a fully implicit solution. We would like to comment, however, that whether or not a discretisation method for the pressure equation allows an explicit flux representation may mostly be regarded as a minor issue. Indeed, one is always free to use a sequential implicit solution strategy which is often faster and in most cases results in the same, or at least an equally accurate solution.

It is not within our scope here to discuss issues related to solving linear systems that arise from mixed FEM discretisations further. Indeed, for moderately sized problems we can rely on Matlab, and the possibly complex linear algebra involved in solving the sparse system (24) can be hidden in a simple statement of the form `x = A\b`. Readers interested in learning more about mixed FEMs, and how to solve the corresponding linear systems are advised to consult some of the excellent books on the subject, for instance [8, 9, 10].

Example 2. It is now time to consider our first reservoir having more than only a touch of realism. To this end we will simulate two horizontal slices of Model 2 from the 10th SPE Comparative Solution Project [14], which is publicly available on the net. The model dimensions are $1200 \times 2200 \times 170$ (ft) and the reservoir is described by a heterogeneous distribution over a regular Cartesian grid with $60 \times 220 \times 85$ grid-blocks. We pick the top layer, in which the permeability is smooth, and the bottom layer, which is fluvial and characterised by a spaghetti of narrow high-flow channels (see Figure 5). For both layers, the permeabilities range over at least six orders of magnitude. To drive a flow in the two layers, we impose an injection and a production well in the lower-left and upper-right corners, respectively.

Listing 4 contains a simulation routine that loads data, assembles the whole matrix, and solves the system (24) to compute pressure and fluxes (neglecting, for brevity, gravity forces so that $G = 0$) . For the flux, only the interior edges are computed, since the flux normal to the exterior edges is assumed to be zero through the no-flow boundary condition.

In Figure 5 we visualise the flow field using 20 streamlines imposed upon a colour plot of the logarithm of the permeability. For completeness, we have included the Matlab code used to generate each of the plots. Some of the streamlines in the plots seem to appear and disappear at the boundaries. This is a pure plotting artifact obtained when we collocate the staggered edge fluxes at the cell centres in order to use Matlab's `streamline` visualisation routine.

In the next sections we will demonstrate how the models and numerical methods introduced above for single-phase flow can be extended to more general flows involving more than one fluid phase, where each phase possibly contains more than one component.

Listing 4. Mixed finite-element discretisation of $v = -K\nabla p$, $\nabla \cdot v = q$ with lowest-order Raviart–Thomas elements applied to the top layer of the SPE-10 test case [14].

```
Grid.Nx=60; Grid.hx=20*.3048;            % Dimension in x-direction
Grid.Ny=220; Grid.hy=10*.3048;           % Dimension in y-direction
Grid.Nz=1;  Grid.hz= 2*.3048;            % Dimension in z-direction
Nx=Grid.Nx; Ny=Grid.Ny; Nz=Grid.Nz;      % Local variables
N=Nx*Ny*Nz;                              % Total number of grid points

Ex=(Nx-1)*Ny*Nz;                         % Number of edges in x-direction
Ey=Nx*(Ny-1)*Nz;                         % Number of edges in y-direction
Ez=Nx*Ny*(Nz-1);                         % Number of edges in z-direction
E=Ex+Ey+Ez;                              % Total number of edges in grid

q=zeros(E+N,1);                          % Right-hand side
q(E+1)=1;                                % Injection in block (1,1,1)
q(E+N)=-1;                               % Production in block (Nx,Ny,Nz)
load Udata; Grid.K=KU(:,1:Nx,1:Ny,1:Nz); % Load and extract permeability

B=GenB(Grid,Grid.K);                     % Compute B-block of matrix
C=GenC(Grid);                            % Compute C-block of matrix
A=[B,C';-C,sparse(N,N)];                 % Assemble matrix
A(E+1,E+1)=A(E+1,E+1)+1;
x=A\q;                                   % Solve linear system

v=x(1:E);                                % Extract velocities
vx=reshape(v(1:Ex),Nx-1,Ny,Nz);          %   x-component
vy=reshape(v(Ex+1:E-Ez),Nx,Ny-1,Nz);     %   y-component
vz=reshape(v(E-Ez+1:E),Nx,Ny,Nz-1);      %   z-component
p=reshape(x(E+1:E+N),Nx,Ny,Nz);          % Extract pressure
```

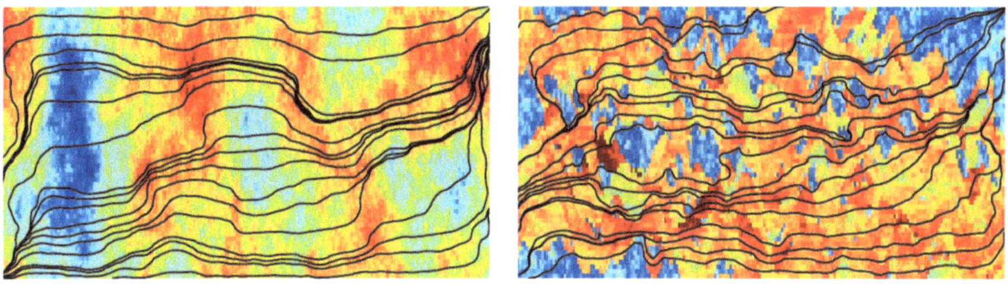

```
% Collocate velocities at cell centre
U = [zeros(1,Ny); vx; zeros(1,Ny)]; U=0.5*(U(1:end-1,:)+U(2:end,:));
V = [zeros(Nx,1), vy, zeros(Nx,1)]; V=0.5*(V(:,1:end-1)+V(:,2:end));

% Make grid and sampling points along diagonal
[Y,X]=meshgrid([1:Ny]*hy-0.5*hy,[1:Nx]*hx-0.5*hx);
sy = linspace(0.5*hy, (Ny-0.5)*hy, 20);
sx = (Nx-0.5)*hx/((Ny-0.5)*hy)*( (Ny-0.5)*hy - sy );
pcolor(Y,X,log10(squeeze(K(1,:,:)))); shading flat;

% Trace forward to producer and backward to injector
hp=streamline(Y,X,V,U,sy,sx); hn=streamline(Y,X,-V,-U,sy,sx);
set([hp, hn],'Color','k','LineWidth',1.5);
axis equal; axis tight; axis off;
```

Fig. 5. Streamlines and logarithm of permeability for two-dimensional simulation of the top and bottom layer of the SPE-10 test case. The source code for making the plots is included.

4 Multiphase and Multicomponent Flows

As for single-phase flow, the fundamental model describing the flow of a multiphase, multicomponent fluid is the conservation (or continuity) equations for each component ℓ:

$$\frac{\partial}{\partial t}\Big(\phi \sum_{\alpha} c_{\ell\alpha}\rho_\alpha s_\alpha\Big) + \nabla\cdot\Big(\sum_{\alpha} c_{\ell\alpha}\rho_\alpha v_\alpha\Big) = \sum_{\alpha} c_{\ell\alpha} q_\alpha, \qquad \alpha = w, o, g. \quad (25)$$

Here we recall that $c_{\ell\alpha}$ is mass fraction of component ℓ in phase α, ρ_α is the density of phase α, v_α is phase velocity, and q_α is phase source. As for single-phase flow, the phase velocities must be modelled. This is usually done by extending Darcy's law to relate the phase velocities to the phase pressures p_α

$$v_\alpha = -\mathbf{K}\frac{k_{r\alpha}}{\mu_\alpha}(\nabla p_\alpha - \rho_\alpha G). \quad (26)$$

In the multiphase version of Darcy's law (as opposed to the single-phase version (10)) we have used the relative permeabilities $k_{r\alpha}$ (see Section 2) to account for the reduced permeability of each phase due to the presence of the other phases.

4.1 Black-Oil Models

A large class of models that are widely used in porous media flow simulations are the black-oil models. The name refers to the assumption that the hydrocarbons may be described as two components: a heavy hydrocarbon component called "oil" and a light hydrocarbon component called "gas". The two components can be partially or completely dissolved in each other depending on the pressure and the temperature, forming either one or two phases (liquid and gaseous). In general black-oil models, the hydrocarbon components are also allowed to be dissolved in the water phase and the water component may be dissolved in the two hydrocarbon phases. The hydrocarbon fluid composition, however, remains constant for all times. The alternative, where the hydrocarbons are modelled using more than two components and hydrocarbon are allowed to change composition, is called a *compositional* model.

We will henceforth assume three phases and three components (gas, oil, water). By looking at our model (25)–(26), we see that we therefore so far have introduced 27 unknown physical quantities: nine mass fractions $c_{\ell\alpha}$ and three of each of the following quantities ρ_α, s_α, v_α, p_α, μ_α, and $k_{r\alpha}$. The corresponding numbers of equations are: three continuity equations (25), an algebraic relation for the saturations (3), three algebraic relations for the mass fractions (4), and Darcy's law (26) for each phase. This gives us only ten equations. Thus, to make a complete model of multiphase, multicomponent flow one must add extra closure relations.

In Section 2, we saw that the relative permeabilities are usually assumed to be known functions of the phase saturations. These functions are normally obtained from physical experiments, small-scale numerical simulations, or simply from known rock-type properties. The capillary pressure curves are also assumed to depend solely on the phase saturations, and can be obtained in a similar fashion. One usually employs oil-water and gas-oil capillary pressures:

$$p_{cow} = p_o - p_w, \qquad p_{cgo} = p_g - p_o.$$

The densities and viscosities are obtained from lab experiments and are related to the phase pressures. Summing up, this gives us a total of eleven closure relations. Finally, one can introduce six algebraic relations for $c_{\ell g}/c_{\ell o}$ and $c_{\ell g}/c_{\ell w}$ in the form of PVT models.

In the next section we consider the special case of immiscible flow. For more advanced models, we refer the reader to one of the general textbooks for reservoir simulation [6, 11, 16, 21, 28, 30, 13, 34].

5 Immiscible Two-Phase Flow

In this section we will consider the flow of two phases, one water phase w and one hydrocarbon phase. The water phase will consist of pure water, whereas the hydrocarbon phase generally is a two-component fluid consisting of dissolved gas and a residual (or black) oil. These assumptions translate to the following definitions for the mass fractions

$$\begin{aligned} c_{ww} &= 1, & c_{ow} &= 0, & c_{gw} &= 0, \\ c_{wo} &= 0, & c_{oo} &= \frac{m_o}{m_o + m_g}, & c_{go} &= \frac{m_g}{m_o + m_g}, \\ c_{wg} &= 0, & c_{og} &= 0, & c_{gg} &= 0. \end{aligned}$$

Here m_o and m_g are the masses of oil and gas, respectively. By adding the continuity equations for the oil and gas components, we obtain a continuity equation for the water-phase (w) and one for the hydrocarbon phase (o). Since $c_{oo} + c_{go} = 1$, both of these continuity equations have the following form:

$$\frac{\partial(\phi \rho_\alpha s_\alpha)}{\partial t} + \nabla \cdot (\rho_\alpha v_\alpha) = q_\alpha, \tag{27}$$

Expanding space and time derivatives, and dividing by the phase densities, we obtain an alternative formulation of the continuity equations (27):

$$\frac{\partial \phi}{\partial t} s_\alpha + \phi \frac{\partial s_\alpha}{\partial t} + \phi \frac{s_\alpha}{\rho_\alpha} \frac{\partial \rho_\alpha}{\partial t} + \nabla \cdot v_\alpha + \frac{v_\alpha \cdot \nabla \rho_\alpha}{\rho_\alpha} = \frac{q_\alpha}{\rho_\alpha}, \tag{28}$$

In the following we will rewrite the two continuity equations into a more tractable system of equations consisting of a pressure equation (as introduced for the single-phase model) and a saturation or fluid-transport equation.

5.1 The Pressure Equation

To derive the pressure equation, we introduce first, for brevity, the mobility of phase α: $\lambda_\alpha = k_{r\alpha}/\mu_\alpha$. Summing continuity equations (28) for the oil and water phases, letting $q = q_w/\rho_w + q_o/\rho_o$, and using the fact that $s_w + s_o = 1$, we deduce

$$\nabla \cdot (v_w + v_o) + \frac{\partial \phi}{\partial t} + \phi \frac{s_w}{\rho_w}\frac{\partial \rho_w}{\partial t} + \phi \frac{s_o}{\rho_o}\frac{\partial \rho_o}{\partial t} + \frac{v_w \cdot \nabla \rho_w}{\rho_w} + \frac{v_o \cdot \nabla \rho_o}{\rho_o} = q. \quad (29)$$

Introducing the *rock compressibility* defined by (1) and the *phase compressibilities* defined by (6), and inserting the expression for the Darcy velocities, we obtain

$$\begin{aligned} - \nabla \cdot \big[\mathbf{K}\lambda_w(\nabla p_w - \rho_w G) + \mathbf{K}\lambda_o(\nabla p_o - \rho_o G)\big] + c_r \phi \frac{\partial p}{\partial t} \\ - c_w \big[\nabla p_w \cdot \mathbf{K}\lambda_w(\nabla p_w - \rho_w G) - \phi s_w \frac{\partial p_w}{\partial t}\big] \\ - c_o \big[\nabla p_o \cdot \mathbf{K}\lambda_o(\nabla p_o - \rho_o G) - \phi s_o \frac{\partial p_o}{\partial t}\big] = q. \quad (30) \end{aligned}$$

In this equation we have three pressures: oil pressure p_o, water pressure p_w, and total pressure p. If we now treat, say, p_o as the primary variable, replace p_w with $p_o - p_{cow}$ (assuming the compressibilities are known) and express the total pressure as a function of p_o, equation (30) becomes a parabolic equation that can be solved for the oil-phase pressure p_o.

5.2 A Pressure Equation for Incompressible Immiscible Flow

Designing numerical methods for the pressure equation (30) that correctly account for flow dynamics and properly balance the spatial and temporal derivatives can be a very difficult task due to the many prominent scales that occur in porous media permeability. Since the temporal derivative terms are relatively small, equation (30) is nearly elliptic and can usually be discretised with numerical methods that are well suited for elliptic differential equations such as the ones described in Sections 3.2–3.4. The purpose of the current presentation is to help the reader become familiar with the basic flow equations, and to explain in detail some possible numerical methods for discretising the corresponding differential equations. We therefore make some simplifying assumptions before we consider numerical discretisation techniques.

To enhance readability, we henceforth assume that the rock and the two fluid phases are incompressible, i.e., $c_r = c_w = c_o = 0$. Equation (30) then reduces to

$$v = -\big[\mathbf{K}\lambda_w(\nabla p_w - \rho_w G) + \mathbf{K}\lambda_o(\nabla p_o - \rho_o G)\big], \qquad \nabla \cdot v = q.$$

In this equation, there are two unknown phase pressures, p_o and p_w. To eliminate one of them, it is common to introduce the capillary pressure

$p_{cow} = p_o - p_w$, which is assumed to be a function of water saturation s_w. Unfortunately, this leads to a rather strong coupling between the pressure equation and the saturation equation. We will therefore follow another approach; see [11] for more details. Instead of using the phase pressures p_w and p_o, we introduce a new *global pressure* p. The global pressure is defined to contain saturation-dependent pressure terms, thereby giving a better decoupling of the pressure and saturation equations . To this end, we first assume that the capillary pressure p_{cow} is a monotone function of the water saturation s_w and then define the global pressure as $p = p_o - p_c$, where the saturation-dependent complementary pressure p_c is defined by

$$p_c(s_w) = \int_1^{s_w} f_w(\xi) \frac{\partial p_{cow}}{\partial s_w}(\xi) d\xi. \tag{31}$$

Here the *fractional-flow* function $f_w = \lambda_w/(\lambda_w + \lambda_o)$ measures the water fraction of the total flow. Since $\nabla p_c = f_w \nabla p_{cow}$, we are able to express the total velocity $v = v_w + v_o$ as a function of the global pressure p:

$$v = -\mathbf{K}(\lambda_w + \lambda_o)\nabla p + \mathbf{K}(\lambda_w \rho_w + \lambda_o \rho_o)G. \tag{32}$$

Finally, introducing the total mobility $\lambda = \lambda_w + \lambda_o$ we obtain the following elliptic equation for the global pressure p:

$$-\nabla \cdot \big[\mathbf{K}\lambda\nabla p - \mathbf{K}(\lambda_w \rho_w + \lambda_o \rho_o)G\big] = q \tag{33}$$

To make the pressure equation complete we need to prescribe some boundary conditions. The default is to impose no-flow boundary conditions, but if the reservoir is connected to e.g., an aquifer, it might be possible to determine an approximate pressure distribution on the reservoir boundary.

5.3 The Saturation Equation

The pressure equation gives an equation for our first primary unknown p. To derive a complete model, we must also derive equations for the phase saturations s_w and s_o using the continuity equations of each phase. However, since $s_w + s_o = 1$, we need only one saturation equation and it is common practice to pick s_w as the second primary unknown. To connect the continuity equation for water to the pressure equation (33), we need to derive an expression for the phase velocity v_w in terms of the global and complementary pressures p and p_c. To this end, we will use what is called the total velocity formulation and express v_w in terms of the total velocity v and some additional terms that only depend on the saturation s_w. From Darcy's law (26) it follows that

$$\mathbf{K}\lambda_o \lambda_w \nabla p_{cow} = \lambda_o v_w - \lambda_w v_o + \mathbf{K}\lambda_o \lambda_w(\rho_o - \rho_w)G. \tag{34}$$

Inserting $v_o = v - v_w$ into this equation and dividing by λ, we obtain

$$v_w = f_w\big[v + \mathbf{K}\lambda_o \nabla p_{cow} + \mathbf{K}\lambda_o(\rho_w - \rho_o)G\big],$$

where we have used the fractional flow function introduced above. Finally, expanding $\nabla p_{cow} = \frac{\partial p_{cow}}{\partial s_w}\nabla s_w$ and invoking the incompressibility assumption, we arrive at an equation for the fluid transport having the following form:

$$\phi\frac{\partial s_w}{\partial t} + \nabla\cdot\Big(f_w(s_w)\big[v + d(s_w, \nabla s_w) + g(s_w)\big]\Big) = \frac{q_w}{\rho_w}. \tag{35}$$

To make a complete description of the flow, the continuity equation must be equipped with boundary conditions, e.g., no-flow conditions, and initial conditions $s_w(x,0) = s_w^0(x)$. Henceforth we will drop the subscript w.

Equation (35) is called the *saturation equation* and is generally a parabolic equation. However, on a reservoir scale, the terms $f(s)v$ and $f(s)g(s)$ representing viscous and gravity forces, respectively, usually dominate the term $f(s)d(s,\nabla s)$ representing capillary forces. The saturation equation will therefore usually have a strong hyperbolic nature and will require other discretisation techniques than those introduced for the almost-elliptic pressure equation (33). Examples of appropriate discretisation techniques will be presented below, but first we discuss how to solve the coupled system consisting of (35) and (33) or (30).

5.4 Solution Strategies for the Coupled System

The equations (33) and (35) derived in the previous section are called the fractional-flow model for immiscible two-phase flow. The model consists of an elliptic pressure equation (33) (or the more general parabolic equation (30), which has a similar elliptic nature) and a saturation or fluid transport equation (35) that has a certain hyperbolic nature. The equations are nonlinearly coupled. The coupling is primarily through the saturation-dependent mobilities λ_α in the pressure equation and through the pressure-dependent velocities v_α in the saturation equation. However, the equations are also coupled through other terms that depend on pressure or saturation, e.g., viscosities and capillary and complementary pressures.

A natural strategy for solving the coupled system is to make an implicit discretisation for each equation and simultaneously solve for the two primary unknowns p and s_w. This strategy is usually referred to as *fully implicit* solution and is a common solution method in industry due to its robustness. However, fully implicit solution is computationally expensive, since we need to solve a large nonlinear system of equations through some iterative procedure like e.g., the Newton–Raphson method. On the other hand, more efficient methods can be developed by using operator splitting to decouple the two equations. In this sequential approach, each equation is solved separately and one can therefore use very different methods to discretise the two fundamentally different equations. As an example, we mention the IMPES (implicit pressure, explicit saturation) method, which used to be quite popular in the industry. Currently, a more popular choice is to use a method called the adaptive implicit method (AIM), in which some grid blocks are solved fully implicitly while the others are treated with a sequential splitting method (IMPES). The method therefore gives robustness in problematic areas with large changes in pressure and saturations (like near a well bore), while at the same time giving high computational efficiency away from problem regions.

In the current paper, we will only present sequential splitting methods. For the global pressure and total velocity formulation (33) and (35) of incompressible and immiscible two-phase flow, a sequential splitting method can be designed as follows: First, the saturation distribution from the previous time step (or initial data) is used to compute the saturation-dependent coefficients in (33), before the equation is solved for global pressure and total velocity. Then, the total velocity v is kept constant as a parameter in (35), while the saturation is advanced in time. Next, the new saturation values are used to update the saturation-dependent coefficients in (33), and the pressure equation is solved again, and so on. Consequently, we can develop numerical schemes for (33) and (35) without being concerned about the nonlinear coupling between the two.

Whereas the advantage of a sequential method is its efficiency (and potential spatial accuracy), the disadvantage is the splitting errors introduced by decoupling the equations. In certain (gas-dominated) cases, splitting errors may lead to unphysical flow predictions unless inordinately small splitting steps are used. To remedy this potential pitfall, one can introduce an extra loop for each time step and iterate a few times (until convergence) between solving the pressure and saturation, before moving on the next time step. The corresponding method is sometimes called sequential implicit, indicating its intermediary nature. We will now use the sequential splitting method to develop and present a full simulator for our simple two-phase model. For presentational simplicity, the sequential implicit technique will not be employed, but can easily be included whenever necessary.

5.5 Discretising the Pressure Equation

To discretise the pressure equation (33) with a finite-volume method, terms corresponding to gravity must be moved over to the right-hand side. The single-phase formulation (13) is hence replaced with

$$- \int_{\partial \Omega_i} \mathbf{K}\lambda(s_w^k)\nabla p^{k+1} \, d\nu = \int_{\Omega_i} q \, dx - \int_{\partial \Omega_i} \mathbf{K}\Big(\lambda_w(s_w^k)\rho_w + \lambda_o(s_w^k)\rho_o\Big) G \cdot n \, d\nu, \quad (36)$$

where the superscript k denotes the time step. The integrals over the cell boundaries are computed with a TPFA scheme or an MPFA scheme, as presented for the single-phase flow in Sections 3.2 and 3.3.

For the mixed FEM, we need only to replace equations (21)–(22) with

$$\int_\Omega \left(\mathbf{K}\lambda(s_w^k)\right)^{-1} v^{k+1} \cdot u \, dx - \int_\Omega p^{k+1} \, \nabla \cdot u \, dx = \int_\Omega \left(\rho_w f_w(s_w^k) + \rho_o f_o(s_w^k)\right) G \cdot u \, dx, \tag{37}$$

$$\int_\Omega q \, \nabla \cdot v^{k+1} \, dx = \int_\Omega f q \, dx. \tag{38}$$

Thus, the pressure equation modelling two-phase flow may be discretised with the same methods that were used to discretise the single-phase pressure equation in Section 3. The main difference is that for two-phase flow the pressure is a dynamic function of saturation, and must therefore be solved repeatedly throughout a simulation.

5.6 Discretising the Saturation Equation

Traditionally, algorithms for solving the pressure equation (33) has accounted for the majority of the computational time in reservoir simulations. However, pressure solvers have improved a lot during the last years as a result of improved numerical linear algebra (e.g., multigrid and other iterative methods). It is therefore likely that the design of robust numerical schemes that balance viscous, gravity, and capillary forces in the saturation equations correctly, may be an equally challenging part of reservoir simulation in the future. The saturation equations are, for example, typically advection dominated on a reservoir scale, in particular in high-flow regions. This implies that propagating interfaces between oil and water, commonly referred to as saturation fronts, can be very sharp. Numerical diffusion may therefore dominate the capillary forces if one does not use a scheme with high resolution that allows accurate tracing of the saturation fronts.

Most commercial reservoir simulators use an implicit or semi-implicit time discretisation to evolve saturation profiles in time, while a conservative finite-volume method is used to resolve the spatial derivatives. Consider a cell Ω_i with edges γ_{ij} and associated normal vectors n_{ij} pointing out of Ω_i. Using the θ-rule for temporal discretisation, a finite-volume scheme takes the following general form

$$\frac{\phi_i}{\Delta t}\left(s_i^{k+1} - s_i^k\right) + \frac{1}{|\Omega_i|}\sum_{j \neq i}\left[\theta F_{ij}(s^{k+1}) + (1-\theta)F_{ij}(s^k)\right] = \frac{q_i(s_i^k)}{\rho}. \tag{39}$$

Here ϕ_i is the porosity in Ω_i, q_i denotes the source term, Δt is the time step, and s_i^k is the cell-average of the water saturation at time $t = t_k$,

$$s_i^k = \frac{1}{|\Omega_i|}\int_{\Omega_i} s(x, t_k) \, dx.$$

Finally, F_{ij} is a numerical approximation of the flux over edge γ_{ij},

$$F_{ij}(s) \approx \int_{\gamma_{ij}} f_w(s)_{ij} \big[v_{ij} + d_{ij}(s) + g_{ij}(s) \big] \cdot n_{ij} \, d\nu. \tag{40}$$

Here $f_w(s)_{ij}$ denotes the fractional-flow function associated with γ_{ij}, v_{ij} is the Darcy flux, d_{ij} the diffusive flux, and g_{ij} the gravitational flux across the edge. Different finite-volume schemes are now defined by the quadrature rule used for the edge integrals in (40) and by the way the integrand is evaluated. For a first-order scheme, it is common to use upstream weighting for the fractional flow, e.g.,

$$f_w(s)_{ij} = \begin{cases} f_w(s_i) & \text{if } v \cdot n_{ij} \geq 0, \\ f_w(s_j) & \text{if } v \cdot n_{ij} < 0, \end{cases} \tag{41}$$

and so on. If we now choose $\theta = 0$ in the temporal discretisation, we end up with a first-order, explicit scheme. Such a scheme is quite accurate but imposes stability restrictions on the time step in the form of a CFL condition. The time-step restrictions may be quite severe due to the near-singular nature of the forcing velocity field near wells and the possible presence of cells with arbitrarily small porosity. We will come back to this discussion in Section 6.1. To get rid of the time-step restriction, it is customary to use implicit schemes for which $\theta > 0$; for instance, $\theta = 1.0$ gives a fully implicit scheme. Unfortunately, this introduces other difficulties like excessive numerical diffusion for large time steps and the need for a fast solver for the corresponding nonlinear system of equations; see Section 6.2.

If, for instance, the impact of capillary forces is important, methods with high-order spatial resolution may be more appropriate to prevent the numerical diffusion from dominating the capillary diffusion. However, since most reservoir simulation studies are conducted on quite coarse grids, the use of high-resolution schemes, e.g., as discussed elsewhere in this book [22], has yet to become widespread in the reservoir engineering community. The primary benefits with the low-order finite-volume schemes are that they are quite robust and easy to implement.

6 A Simulator for Incompressible Two-Phase Flow

In this section we present a full simulator for a sequential IMPES formulation of an incompressible and immiscible two-phase flow system. We provide a Matlab code designed for simulating water injection into a reservoir filled with oil. However, to keep the presentation simple, and the implementation code short, we disregard gravity and neglect capillary forces (i.e., $\nabla p_{cow} = 0$). Under these assumptions, equations (33) and (35) reduce to the following system of equations:

Listing 5. TPFA finite-volume discretisation of $-\nabla(K\lambda(s)\nabla u) = q$.

```
function [P,V]=Pres(Grid,S,Fluid,q)

% Compute K*lambda(S)
[Mw,Mo]=RelPerm(S,Fluid);
Mt=Mw+Mo;
KM = reshape([Mt,Mt,Mt]',3,Grid.Nx,Grid.Ny,Grid.Nz).*Grid.K;

% Compute pressure and extract fluxes
[P,V]=TPFA(Grid,KM,q);
```

$$-\nabla \cdot \mathbf{K}\lambda(s)\nabla p = q, \tag{42}$$

$$\phi\frac{\partial s}{\partial t} + \nabla \cdot \big(f(s)v\big) = \frac{q_w}{\rho_w}. \tag{43}$$

Since we only inject water and produce whatever reaches our producers, the source term for the saturation equation becomes

$$\frac{q_w}{\rho_w} = \max(q,0) + f(s)\min(q,0).$$

To close the model, we must also supply expressions for the saturation-dependent quantities. Here we use simple analytical expressions:

$$\lambda_w(s) = \frac{(s^*)^2}{\mu_w}, \qquad \lambda_o(s) = \frac{(1-s^*)^2}{\mu_o}, \qquad s^* = \frac{s - s_{wc}}{1 - s_{or} - s_{wc}}.$$

Here s_{or} is the irreducible oil saturation, i.e., the lowest oil saturation that can be achieved by displacing oil by water, and s_{wc} is the connate water saturation, i.e., the saturation of water trapped in the pores of the rock during formation of the rock; see Section 2.

Listing 5 shows a Matlab code for solving (42) with the TPFA scheme. Observe that the only difference between this code and the single-phase code in Listing 1 is that the input `K`, which corresponds to $\mathbf{K}\lambda$, now depends on saturation. This dependence involves the viscosities μ_w and μ_o, the irreducible oil saturation s_{or}, and the connate water saturation s_{wc}. In a similar manner, the mixed finite-element method introduced in Section 3.4 can easily be extended to two-phase flows. In fact, the only point we need to change is the assembly of $\mathbf{B}$ in (24). In other words, the code in Listing 2 can be extended to two-phase flows similarly by letting `K` represent $\mathbf{K}\lambda$ as in Listing 5. The corresponding relative permeability module is given in Listing 6.

For the saturation equation, we use a finite-volume scheme on the form

$$s_i^{n+1} = s_i^n + (\delta_x^t)_i\Big(\max(q_i,0) - \sum_j f(s^m)_{ij}v_{ij} + f(s_i^m)\min(q_i,0)\Big),$$

where $(\delta_x^t)_i = \Delta t/(\phi_i|\Omega_i|)$. By specifying for time level m in the evaluation of the fractional flow functions, we obtain either an explicit ($m = n$) or a fully

Listing 6. Relative permeabilities of oil and water.

```
function [Mw,Mo,dMw,dMo]=RelPerm(s,Fluid)

S = (s-Fluid.swc)/(1-Fluid.swc-Fluid.sor);  % Rescale saturations
Mw = S.^2/Fluid.vw;                          % Water mobility
Mo =(1-S).^2/Fluid.vo;                       % Oil mobility

if (nargout==4)
  dMw = 2*S/Fluid.vw/(1-Fluid.swc-Fluid.sor);
  dMo = -2*(1-S)/Fluid.vo/(1-Fluid.swc-Fluid.sor);
end
```

Listing 7. Matrix assembly in upwind finite-volume discretisation of (43).

```
function A=GenA(Grid,V,q)

Nx=Grid.Nx; Ny=Grid.Ny; Nz=Grid.Nz; N=Nx*Ny*Nz;
N=Nx*Ny*Nz;                                      % number of unknowns
fp=min(q,0);                                     % production

XN=min(V.x,0); x1=reshape(XN(1:Nx,:,:),N,1);     % separate flux into
YN=min(V.y,0); y1=reshape(YN(:,1:Ny,:),N,1);     % - flow in positive coordinate
ZN=min(V.z,0); z1=reshape(ZN(:,:,1:Nz),N,1);     %   direction (XP,YP,ZP)
XP=max(V.x,0); x2=reshape(XP(2:Nx+1,:,:),N,1);   % - flow in negative coordinate
YP=max(V.y,0); y2=reshape(YP(:,2:Ny+1,:),N,1);   %   direction (XN,YN,ZN)
ZP=max(V.z,0); z2=reshape(ZP(:,:,2:Nz+1),N,1);   %

DiagVecs=[z2,y2,x2,fp+x1-x2+y1-y2+z1-z2,-x1,-y1,-z1]; % diagonal vectors
DiagIndx=[-Nx*Ny,-Nx,-1,0,1,Nx,Nx*Ny];           % diagonal index
A=spdiags(DiagVecs,DiagIndx,N,N);              % matrix with upwind FV stencil
```

implicit ($m = n+1$) scheme. Here v_{ij} is the total flux (for oil and water) over the edge γ_{ij} between two adjacent cells Ω_i and Ω_j, and f_{ij} is the fractional flow function at γ_{ij}, which we evaluate using the upwind discretisation in (41). Written in compact form, the two schemes read

$$S^{n+1} = S^n + (\delta_x^t)^T\big(\mathbf{A}f(S^m) + Q^+\big), \qquad m = n, n+1, \tag{44}$$

where S^n is the vector of cell-saturations s_i^n and $f(S^m)$ is the vector of fractional flow values $f(s_i^m)$. Furthermore, Q^+ and Q^- denote the positive an negative parts, respectively, of the vector representing q and $\mathbf{A}$ is a matrix implementing $[f(s)Q^- - \nabla\cdot(f(s)v)]$ on a cell-by-cell basis. A Matlab code for the assembly of $\mathbf{A}$, given a vector of saturations S, is presented in Listing 7.

6.1 An Explicit Solver

The explicit solver is obtained by using $m = n$ in (44). Explicit schemes are only stable provided that the time step Δt satisfies a stability condition (a so-called CFL condition). For the homogeneous transport equation (with $q \equiv 0$), the standard CFL condition for the first-order upwind scheme reads

$$\max_{s\in(0,1)} |f'(s)| \max_i (\delta_x^t)_i \sum_j |v_{ij}| \leq 2(1 - s_{wc} - s_{or}), \tag{45}$$

where $(\delta_x^t)_i = \Delta t/(\phi_i|\Omega_i|)$. For the inhomogeneous equation, we can derive a stability condition on Δt using a more heuristic argument. Physically, we require that $s_{wc} \leq s_i^{n+1} \leq 1 - s_{or}$. This implies that the discretisation parameters $|\Omega_i|$ and Δt must satisfy the following dynamic conditions:

$$s_{wc} \leq s_i^n + (\delta_x^t)_i \Big(\max(q_i, 0) - \sum_j f(s^n)_{ij} v_{ij} + f(s_i^n)\min(q_i, 0)\Big) \leq 1 - s_{or}.$$

This condition is saturation dependent and must therefore be enforced on every saturation timestep, and will generally give a time-varying time-step Δt^n. However, it is possible to derive an alternative stability condition that gives a new bound on the timestep only when the velocity field has been recomputed. To this end, note that since we have assumed incompressible flow, and employed a mass conservative scheme to solve the pressure equation, the interface fluxes v_{ij} satisfy the following mass balance property:

$$v_i^{\text{in}} = \max(q_i, 0) - \sum_j \min(v_{ij}, 0) = -\min(q_i, 0) + \sum_j \max(v_{ij}, 0) = v_i^{\text{out}}.$$

Thus, since $0 \leq f(s) \leq 1$ it follows that

$$-f(s_i^n)v_i^{\text{in}} \leq \max(q_i, 0) - \sum_j f(s^n)_{ij} v_{ij} + f(s_i^n)\min(q_i, 0) \leq (1 - f(s_i^n))v_i^{\text{in}}.$$

This implies that the general saturation dependent stability condition holds in Ω_i if the following inequality is satisfied:

$$\max\left(\frac{f(s_i^n) - 0}{s_i^n - s_{wc}}, \frac{1 - f(s_i^n)}{1 - s_{or} - s_i^n}\right) (\delta_x^t)_i v_i^{\text{in}} \leq 1. \tag{46}$$

Finally, to remove the saturation dependence from (46), we invoke the mean value theorem and deduce that (46) holds whenever

$$\Delta t \leq \frac{\phi|\Omega_i|}{v_i^{\text{in}} \max\{f'(s)\}_{0\leq s\leq 1}}. \tag{47}$$

This condition is saturation dependent only through the velocities v_{ij} and the timestep Δt need therefore only be modified each time we compute a new solution to the pressure equation.

The fully explicit upwind method incorporating the stability condition (47) is given in Listing 8. A few points in the implementation are worth noting. First of all, instead of coding the finite-volume stencil explicitly, we represent it as a matrix-vector product to make full use of Matlab's inherent efficiency for matrix-vector operations and vectorisation of loops. Second, in the derivation

Listing 8. Explicit upwind finite-volume discretisation of (43).

```
function S=Upstream(Grid,S,Fluid,V,q,T)

Nx=Grid.Nx; Ny=Grid.Ny; Nz=Grid.Nz;            % number of grid points
N=Nx*Ny*Nz;                                     % number of unknowns
pv = Grid.V(:).*Grid.por(:);                    % pore volume=cell volume*porosity

fi =max(q,0);                                    % inflow from wells
XP=max(V.x,0); XN=min(V.x,0);                    % influx and outflux, x-faces
YP=max(V.y,0); YN=min(V.y,0);                    % influx and outflux, y-faces
ZP=max(V.z,0); ZN=min(V.z,0);                    % influx and outflux, z-faces

Vi = XP(1:Nx,:,:)+YP(:,1:Ny,:)+ZP(:,:,1:Nz)-...  % total flux into
     XN(2:Nx+1,:,:)-YN(:,2:Ny+1,:)-ZN(:,:,2:Nz+1); %   each gridblock
pm = min(pv./(Vi(:)+fi));                        % estimate of influx
cfl = ((1-Fluid.swc-Fluid.sor)/3)*pm;            % CFL restriction
Nts = ceil(T/cfl);                               % number of local time steps
dtx = (T/Nts)./pv;                               % local time steps

A=GenA(Grid,V,q);                                % system matrix
A=spdiags(dtx,0,N,N)*A;                          % A * dt/|Omega_i|
fi =max(q,0).*dtx;                               % injection

for t=1:Nts
 [mw,mo]=RelPerm(S,Fluid);                       % compute mobilities
 fw = mw./(mw+mo);                               % compute fractional flow
 S = S+(A*fw+fi);                                % update saturation
end
```

of the saturation solver we have tacitly assumed that the porosities are all nonzero. This means that porosity fields generally must be preprocessed and values below a certain threshold replaced by a small nonzero value. Finally, to get an exact value for $\max\{f'(s)\}_{0\leq s\leq 1}$ is a bit cumbersome, but numerical approximations are readily available. Here we have expanded $f'(s)$ as follows:

$$\frac{\partial f}{\partial s} = \frac{\partial f}{\partial s^*}\frac{\partial s^*}{\partial s} = \frac{1}{1 - s_{wc} - s_{or}}\frac{\partial f}{\partial s^*},$$

and computed a rough approximation to the upper bound for $\partial f/\partial s^*$ that corresponds to the current choice of viscosities.

Example 3. Let us revisit the quarter-five spot from Example 1, but now assume that the reservoir is initially filled with pure oil. To produce the oil in the upper-right corner, we inject water in the lower left. We assume unit porosity, unit viscosities for both phases, and set $s_{or} = s_{wc} = 0$. In this nondimensional model, it takes one time unit to inject one pore-volume water, i.e., the time unit corresponds to the number of injected pore volumes of water.

Listing 9 contains a sequential splitting code for simulating the water injection problem up to water-breakthrough in the production well with the explicit upwind scheme. (For the homogeneous quarter-five spot it is known that breakthrough occurs when approximately 0.7 pore-volumes of water have been injected). The algorithm is quite simple. First, we set up the grid, the

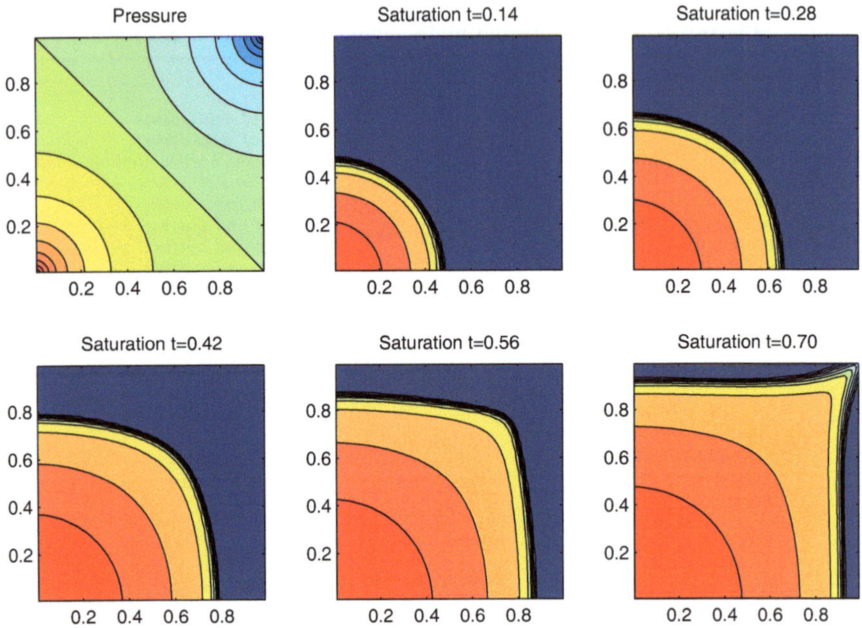

Fig. 6. Pressure and saturation profiles for the homogeneous quarter-five spot.

Listing 9. Two-phase, immiscible, incompressible simulator for a homogeneous quarter-five spot.

```
Grid.Nx=64; Dx=1; Grid.hx = Dx/Grid.Nx;          % Dimension in x-direction
Grid.Ny=64; Dy=1; Grid.hy = Dy/Grid.Ny;          % Dimension in y-direction
Grid.Nz=1; Dz=1; Grid.hz = Dz/Grid.Nz;           % Dimension in z-direction
N=Grid.Nx*Grid.Ny;                               % Total number of grid blocks
Grid.V=Grid.hx*Grid.hy*Grid.hz;                  % Cell volumes
Grid.K=ones(3,Grid.Nx,Grid.Ny,Grid.Nz);          % Unit permeability
Grid.por =ones(Grid.Nx,Grid.Ny,Grid.Nz);         % Unit porosity
Q=zeros(N,1); Q([1 N])=[1 -1];                   % Production/injection

Fluid.vw=1.0; Fluid.vo=1.0;                      % Viscosities
Fluid.swc=0.0; Fluid.sor=0.0;                    % Irreducible saturations

S=zeros(N,1);                                    % Initial saturation
nt = 25; dt = 0.7/nt;                            % Time steps
for t=1:nt
  [P,V]=Pres(Grid,S,Fluid,Q);                    % pressure solver
  S=Upstream(Grid,S,Fluid,V,Q,dt);               % saturation solver

  % plot filled contours at the midpoints of the grid cells
  contourf(linspace(Grid.hx/2,Dx-Grid.hx/2,Grid.Nx),...
           linspace(Grid.hy/2,Dy-Grid.hy/2,Grid.Ny),...
           reshape(S,Grid.Nx,Grid.Ny),11,'k');
  axis square; caxis([0 1]);                     % equal axes and color
  drawnow;                                       % force update of plot
end
```

fluid properties and generate the initial saturation distribution. Then, the solution is advanced in time by repeating the following two steps: (i) solve the pressure equation and compute edge velocities; (ii) using the fixed edge velocities, solve the fluid transport equation a time step Δt.

In Figure 6 we have plotted the initial pressure profile and the saturation profiles at five equally-spaced time levels as computed on a uniform 64×64 grid. The saturation profile consists of a leading shock-wave, in which water immediately displaces a fraction of the in situ oil. For a unit viscosity ratio, the post-shock saturation value equals $\sqrt{2}/2$. Behind the front, the water saturation is increasing monotonically towards the injector, meaning that more oil is gradually displaced as more water is injected. Close to the injector, the level curves are almost circular, corresponding to the circular symmetry in pressure at the injector. As more water is injected, the leading water front develops a finger extending toward and finally breaking through to the producer.

6.2 An Implicit Solver

Implicit schemes are unconditionally stable in the sense that there is no CFL condition on the size of the time step. On the other hand, implicit discretisation of (43) gives rise to *nonlinear* systems of equations. Such systems are normally solved with a Newton or a Newton–Raphson iterative method. These methods typically give second-order convergence, but it is well known that they can be very sensitive to the initial guess. It is therefore common to use an alternative technique to compute an initial approximation for each time step, and then continue with a Newton or Newton–Raphson method.

We shall employ a Newton–Raphson method to solve the implicit system. However, since we do not want to mix too many different methods, we assume that the initial approximation for each Newton–Raphson iteration (which will be the saturation field from the previous time step) is sufficiently close to the solution S^{n+1}. In practice, this means that one often has to start with small time steps after each update of the velocity field. In the code that we present in Listing 11 the timesteps are selected in a dynamic fashion in order to ensure convergence. That is, if the Newton–Raphson iteration has not converged in ten iterations, then the time step is simply decreased by a factor two so that the original timestep is split into two sub-timesteps. If the Newton–Raphson still does not converge in less than ten iterations, the original time step is split into four sub-timesteps, and so on. In commercial software, the maximum timestep is controlled by using an estimate of the time discretisation error. Here we have chosen the maximum timestep by convenience to be five days.

To derive the Newton–Raphson method for the implicit upwind discretisation, consider the following homogeneous equation (see (44)):

$$0 \equiv G(S^{n+1}) = S^{n+1} - S^n - (\delta_x^t)^T \left[\mathbf{A} f(S^{n+1}) + Q^+\right]. \qquad (48)$$

Listing 10. Implicit upwind finite-volume discretisation of (43).

```
function S=NewtRaph(Grid,S,Fluid,V,q,T);

N = Grid.Nx*Grid.Ny*Grid.Nz;                       % number of unknowns
A = GenA(Grid,V,q);                                % system matrix

conv=0; IT=0; S00=S;
while conv==0;
 dt = T/2^IT;                                      % timestep
 dtx = dt./(Grid.V(:)*Grid.por(:));                % timestep / pore volume
 fi = max(q,0).*dtx;                               % injection
 B=spdiags(dtx,0,N,N)*A;

 I=0;
 while I<2^IT;                                     % loop over sub-timesteps
  S0=S; dsn=1; it=0; I=I+1;

  while dsn>1e-3 & it<10;                          % I T E R A T I O N
   [Mw,Mo,dMw,dMo]=RelPerm(S,Fluid);               % mobilities and derivatives
   df=dMw./(Mw + Mo)-Mw./(Mw+Mo).^2.*(dMw+dMo);  % df_w/ds
   dG=speye(N)-B*spdiags(df,0,N,N);                % G'(S)

   fw = Mw./(Mw+Mo);                               % fractional flow
   G = S-S0-(B*fw+fi);                             % G(s)
   ds = -dG\G;                                     % increment ds
   S = S+ds;                                       % update S
   dsn = norm(ds);                                 % norm of increment
   it = it+1;                                      % number of N-R iterations
  end

  if dsn>1e-3; I=2^IT; S=S00; end                  % check for convergence
 end

 if dsn<1e-3; conv=1;                              % check for convergence
 else IT=IT+1; end                                 % if not converged, decrease
end                                                % timestep by factor 2
```

Assume now that we have obtained an approximation $\tilde{S}$ to the true fix-point S^{n+1}. Then, by a Taylor expansion, we have

$$0 = G(S^{n+1}) \approx G(\tilde{S}) + G'(\tilde{S})\big(S^{n+1} - \tilde{S}\big),$$

and a new, and hopefully better, approximation to S^{n+1} is obtained by $\tilde{S}+d\tilde{S}$, where $d\tilde{S}$ satisfies $-G'(\tilde{S})d\tilde{S} = G(\tilde{S})$. For the implicit scheme (48) we have

$$G'(S) = I - (\delta_x^t)^T \mathbf{A} f'(S),$$

where $f'(S) = [f'(s_i)]_i$. The code for the fully implicit upwind discretisation is given in Listing 10.

Example 4. In Example 3 we have seen the typical behaviour of a two-phase oil-water system. Let us now consider a reservoir with a much higher degree of realism by revisiting the two-dimensional SPE-10 models from Example 2. We consider an incompressible oil-water system, for which $s_{wc} = s_{or} = 0.2$, $\mu_w = 0.3$ cp, and $\mu_o = 3.0$ cp. The reservoir is initially filled with oil, meaning

Listing 11. Two-phase, immiscible, incompressible simulator for the top layer of the SPE-10 model.

```
Grid.Nx=60; Grid.hx=20*.3048;                       % Dimension in x-direction
Grid.Ny=220; Grid.hy=10*.3048;                      % Dimension in y-direction
Grid.Nz=1;  Grid.hz=2*.3048;                        % Dimension in z-direction
N=Grid.Nx*Grid.Ny*Grid.Nz;                          % Number of grid celles
Grid.V=Grid.hx*Grid.hy*Grid.hz;                     % Volume of each cells
Fluid.vw=3e-4; Fluid.vo=3e-3;                       % Viscosities
Fluid.swc=0.2; Fluid.sor=0.2;                       % Irreducible saturations
St = 5;                                             % Maximum saturation time step
Pt = 100;                                           % Pressure time step
ND = 2000;                                          % Number of days in simulation

Q=zeros(Grid.Nx,Grid.Ny,1);                         % Source term for injection
IR=795*(Grid.Nx*Grid.Ny/ (60*220*85));              %  and production. Total
Q(1,1,:)=IR; Q(Grid.Nx,Grid.Ny,:)=-IR; Q=Q(:);      %  rate scaled to one layer

load Udata; Grid.K=KU(:,1:Grid.Nx,1:Grid.Ny,1);     % Permeability in layer 1
Por=pU(1:Grid.Nx,1:Grid.Ny,1);                      % Preprocessed porosity in layer 1
Grid.por=max(Por(:),1e-3);

S=Fluid.swc*ones(N,1);                              % Initial saturation
Pc=[0; 1]; Tt=0;                                    % For production curves
for tp=1:ND/Pt;
  [P,V]=Pres(Grid,S,Fluid,Q);                       % Pressure solver
  for ts=1:Pt/St;
    S=NewtRaph(Grid,S,Fluid,V,Q,St);                % Implicit saturation solver
    subplot('position',[0.05 .1 .4 .8]);            % Make left subplot
    pcolor(reshape(S,Grid.Nx,Grid.Ny,Grid.Nz)');    % Plot saturation
    shading flat; caxis([Fluid.swc 1-Fluid.sor]);   %

    [Mw,Mo]=RelPerm(S(N),Fluid); Mt=Mw+Mo;          % Mobilities in well-block
    Tt=[Tt,(tp-1)*Pt+ts*St];                        % Compute simulation time
    Pc=[Pc,[Mw/Mt; Mo/Mt]];                         % Append production data
    subplot('position',[0.55 .1 .4 .8]);            % Make right subplot
    plot(Tt,Pc(1,:),Tt,Pc(2,:));                    % Plot production data
    axis([0,ND,-0.05,1.05]);                        % Set correct axis
    legend('Water_cut','Oil_cut');                  % Set legend
    drawnow;                                        % Force update of plot
  end
end
```

that the initial water saturation is equal the connate water saturation $s(x,0) \equiv s_{wc}$. The porosity is strongly correlated with the horizontal permeability and contains about 2.5% zero values. To avoid division by zero in these cells, we simply replace the zero values by a certain minimal nonzero value.

Listing 11 contains a simulation routine that loads data and computes the flow using sequential splitting. Since both the porosity and permeability has a strongly heterogeneous structure spanning several orders of magnitude it is not practical to use the explicit scheme due to the inordinately large number of time steps enforced by the stability condition. Instead we use an implicit scheme—the Newton–Raphson method presented in Listing 10.

Figures 7 and 8 show saturation and production profiles during the simulation for the top and bottom layers in the SPE-10 model. In the figures, oil and water cuts refer to the fractional flows of oil and water, respectively, into

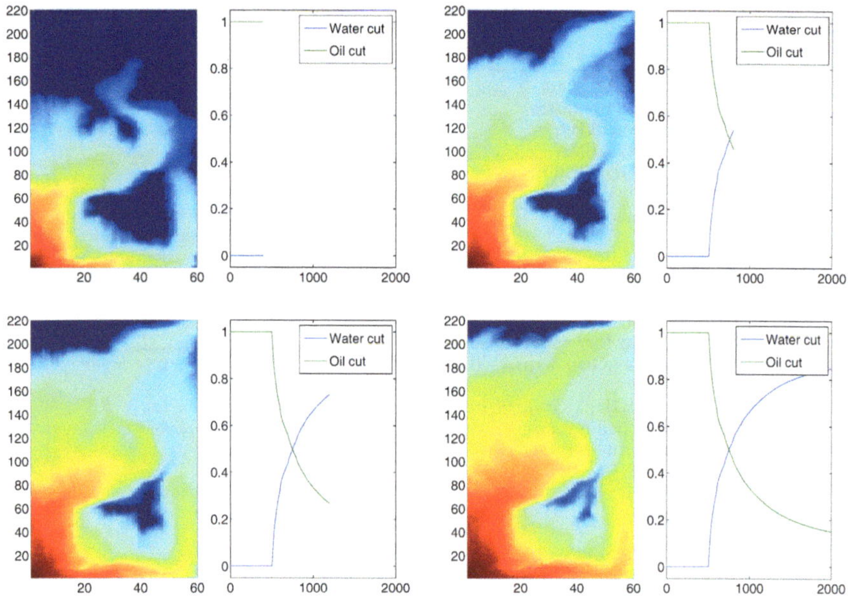

Fig. 7. Saturation and production profiles after 400, 800, 1200, and 2000 days of production for the Tarbert formation in the top layer of the SPE-10 model.

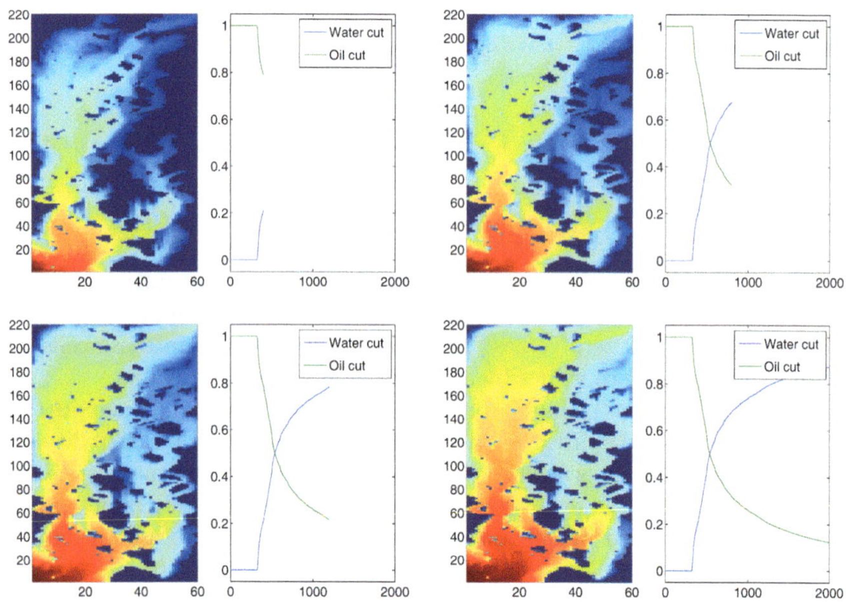

Fig. 8. Saturation and production profiles after 400, 800, 1200, and 2000 days of production for the Upper Ness formation in the bottom layer of the SPE-10 model.

the production well. Here we clearly see that the spaghetti of channelling systems that we have in the Upper Ness formation leads to a very different flow scenario from what we observed in the top layer, which has a quite smooth permeability distribution compared to the bottom layer.

Acknowledgement. This research was funded in part by the Research Council of Norway under grant no. 158908/I30.

References

1. J.E. Aarnes, V. Kippe, K.-A. Lie, and A.B. Rustad. Modelling of multiscale structures in flow simulations for petroleum reservoirs. In *this book*.
2. I. Aavatsmark, T. Barkve, Ø. Bøe, and T. Mannseth. Discretization on unstructured grids for inhomogeneous, anisotropic media. part i: Derivation of the methods. *Siam J. Sci. Comp.*, 19(5):1700–1716, 1998.
3. I. Aavatsmark, T. Barkve, Ø. Bøe, and T. Mannseth. Discretization on unstructured grids for inhomogeneous, anisotropic media. part ii: Discussion and numerical results. *Siam J. Sci. Comp.*, 19(5):1717–1736, 1998.
4. T. Arbogast. An overview of subgrid upscaling for elliptic problems in mixed form. In Z. Chen, R. Glowinski, and K. Li, editors, *Current trends in scientific computing*, Contemporary Mathematics, pages 21–32. AMS, 2003.
5. T. Arbogast. Analysis of a two-scale, locally conservative subgrid upscaling for elliptic problems. *SIAM J. Numer. Anal.*, 42(2):576–598, 2004.
6. K. Aziz and A. Settari. *Petroleum reservoir simulation.* Elsevier, London and New York, 1979.
7. J.W. Barker and S. Thibeau. A critical review of the use of pseudorelative permeabilities for upscaling. *SPE Reservoir Eng.*, 12(2):138–143, 1997.
8. D. Braess. *Finite elements: Theory fast solvers and applications in solid mechanics.* Cambridge University Press, Cambridge, 1997.
9. S.C. Brenner and L.R. Scott. *The mathematical theory of finite element methods.* Springer–Verlag, New York, 1994.
10. F. Brezzi and M. Fortin. *Mixed and hybrid finite element methods.* Computational Mathematics. Springer–Verlag, New York, 1991.
11. G. Chavent and J. Jaffre. *Mathematical models and finite elements for reservoir simulation.* North Holland, 1982.
12. Z. Chen and T.Y. Hou. A mixed multiscale finite element method for elliptic problems with oscillating coefficients. *Math. Comp.*, 72:541–576, 2003.
13. Zhangxin Chen, Guanren Huan, and Yuanle Ma. *Computational methods for multiphase flows in porous media.* Computational Science & Engineering. Society for Industrial and Applied Mathematics (SIAM), Philadelphia, PA, 2006.
14. M. A. Christie and M. J. Blunt. Tenth SPE comparative solution project: A comparison of upscaling techniques. *SPE Reservoir Eval. Eng.*, 4(4):308–317, 2001. url: www.spe.org/csp.
15. M.A. Christie. Upscaling for reservoir simulation. *JPT J. Pet. Tech.*, 48:1004–1010, 1996.
16. L.P. Dake. *Fundamentals of reservoir engineering.* Elsevier, Amsterdam, 1978.

17. M. Delshad and G.A Pope. Comparison of the three-phase relative permeability models. *Transp. Porous Media*, 4:59–83, 1979.
18. A.H. Demond and P.V. Roberts. An examination of relative permeability relations for two-phase flow in porous media. *Water Res. Bull.*, 23:617–628, 1987.
19. B. Engquist and W. E. The heterogeneous multi-scale methods. *Comm. Math. Sci.*, 1:87–133, 2003.
20. B. Engquist and W. E. Multiscale modeling and computation. *Notices Amer. Math. Soc.*, 50(9):1062–1070, 2003.
21. R.E. Ewing. *The mathematics of reservoir simulation.* SIAM, 1983.
22. T.R. Hagen, M.O. Henriksen, J.M. Hjelmervik, and K.-A. Lie. How to solve systems of conservation laws numerically using the graphics processor as a high-performance computational engine. In *this book.*
23. H.H. Haldorsen and C.J. MacDonald. Stochastic modeling of underground reservoir facies (SMURF). In *SPE Annual Technical Conference and Exhibition, 27-30 September, Dallas, Texas*, SPE 16751, 1987.
24. M. Honarpour and S.M. Mahmood. Relative-permeability measurements: An overview. *J. Petr. Tech*, 40:963–966, 1988.
25. T.Y. Hou and X-H. Wu. A multiscale finite element method for elliptic problems in composite materials and porous media. *J. Comput. Phys.*, 134:169–189, 1997.
26. R. Juanes and T.W. Patzek. Relative permeabilities for strictly hyperbolic models of three-phase flow in porous media. *Tranp. Porous Media*, 57(2):125–152, 2004.
27. R. Juanes and T.W. Patzek. Three-phase displacement theory: an improved description of relative permeabilities. *SPE J.*, 9(3):302–313, 2004.
28. L. Lake. *Enhanced oil recovery.* Prentice Hall, Inglewood Cliffs, NJ, 1989.
29. B.B. Maini and T. Okazawa. Effects of temperature on heavy oil-water relative permeability. *J. Can. Petr. Tech*, 26:33–41, 1987.
30. D.W. Peaceman. *Fundamentals of numerical reservoir simulation.* Elsevier, Amsterdam, 1977.
31. P.A. Raviart and J.M. Thomas. A mixed finite element method for second order elliptic equations. In I. Galligani and E. Magenes, editors, *Mathematical Aspects of Finite Element Methods*, pages 292–315. Springer–Verlag, Berlin – Heidelberg – New York, 1977.
32. P. Renard and G. de Marsily. Calculating equivalent permeability. *Adv. Water Resour.*, 20:253–278, 1997.
33. H.L. Stone. Estimation of three-phase relative permeability and residual oil. *J. Can. Petr. Tech*, 12:53–61, 1973.
34. S.M. Skjæveland and Jon Kleppe, editors. *SPOR Monograph, Recent advances in improved oil recovery methods for North Sea sandstone reservoirs.* Norwegian Petroleum Directorate, Norway, 1992.

Modelling of Multiscale Structures in Flow Simulations for Petroleum Reservoirs

Jørg E. Aarnes, Vegard Kippe, Knut–Andreas Lie, and Alf Birger Rustad

Summary. Flow in petroleum reservoirs occurs on a wide variety of physical scales. This poses a continuing challenge to modelling and simulation of reservoirs since fine-scale effects often have a profound impact on flow patterns on larger scales. Resolving all pertinent scales and their interaction is therefore imperative to give reliable qualitative and quantitative simulation results. To overcome the problem of multiple scales it is customary to use some kind of upscaling or homogenisation procedure, in which the reservoir properties are represented by some kind of averaged properties and the flow is solved on a coarse grid. Unfortunately, most upscaling techniques give reliable results only for a limited range of flow scenarios. Increased demands for reservoir simulation studies have therefore led researchers to develop more rigorous multiscale methods that incorporate subscale effects more directly.

In the first part of the paper, we give an overview of some of the many scales that are considered important for flow simulations. Next, we present and discuss several upscaling approaches that have played a role in the history of reservoir simulation. In the final part, we present some more recent approaches for modelling scales in the flow simulations based upon the multiscale paradigm. We conclude with a discussion of benefits and disadvantages of using multiscale methods, rather than using traditional upscaling techniques, in reservoir simulation.

1 Introduction

Simulation of petroleum reservoirs started in the mid 1950's and has become an important tool for qualitative and quantitative prediction of the flow of fluid phases. Today, computer models have a widespread use as a complement or even a competitor to field observations, pilot field and laboratory tests, well testing and analytical models. However, even though reservoir simulation can be an invaluable tool to enhance oil-recovery, the demand for simulation studies depends on many factors. For instance, petroleum fields vary in size from small pockets of hydrocarbon that may be buried just a few meters beneath the surface of the earth, to huge reservoirs stretching out several square kilometres beneath remote and stormy seas. This illustrates that the value of reservoir simulation studies depends on what kind of extra profit one can expect by increasing the oil-recovery from a given reservoir.

A typical North Sea reservoir is located some two kilometres under the sea floor and several hundred kilometres offshore with a sea depth between one hundred and three hundred meters. These reservoirs are often fairly large horizontally, but the vertical dimensions of the zones carrying hydrocarbon may be just a few meters. The rock structures are usually layered with different mixtures of rock types (called facies) in different layers forming an irregular pattern. In addition, geological activity has created faults and fractures that highly influence flow properties. On the other end of the length scale, the fluids in the reservoir are embedded in porous formations like water in a sponge, and flow through small channels formed by the void space between rock particles. The volume fraction of the porous media formations consisting of pores open to fluid flow (henceforth called porosity) is typically in the range 0.1–0.3. To produce petroleum from the reservoir, one drills wells into the rock formations carrying the petroleum resources. In simple terms, a well can be considered as a hole with a diameter of about ten inches extending several kilometres into the earth and through the reservoir. The hole is lined with a casing that is perforated at places where the well is to produce or inject fluids.

We shall let this type of reservoir provide a conceptual model of a petroleum reservoir on which we want to perform simulation studies. Flow processes in this type of reservoir involve a large gap in scales. On one end we have the kilometre scale of the reservoir. On the other end we have the micrometer scale of the pore channels. In between we find, for instance, the centimetre scale of the wells. To obtain a model of the reservoir, one builds geological models that attempt to reproduce the true geological heterogeneity in the reservoir rock. To create a geological model that properly reflects all pertinent scales that impact fluid flow is, however, impossible. Indeed, the size of a grid block in a typical geological grid-model is in the range 10–50 m in the horizontal direction and 0.1–1 m in the vertical direction. Thus, a geological model is clearly too coarse to resolve small scale features such as radiant flow around wells, and flows through narrow high permeable channels. Neither can we expect to resolve properly important geological features such as fractures and faults. Nevertheless, from a simulation point of view, geological models are too complex, i.e., they contain more information than we can exploit in simulation studies. Indeed, in simulation models we usually use a coarsened grid model, with the possible exception of regions in the proximity of wells, and variations in the geological model occurring at length scales below the simulation grid block scale are normally replaced with averaged or upscaled quantities.

Many fields of science face the problem of multiple scales and share the need for upscaling or homogenisation[1]. The main purpose of this chapter is to

---

[1]Homogenisation and upscaling are used interchangeably in the scientific literature to describe procedures for generating coarsened reservoir models. However,

Fig. 1. Layered geological structures as seen in these pictures typically occur on both large and small scales in petroleum reservoirs. Pictures are courtesy of Silje Søren Berg, University of Bergen.

discuss alternative strategies for modelling reservoir heterogeneity at the scale of the geological model in a coarsened model suitable for fluid flow simulation. We review and discuss some historically important upscaling techniques in Section 3. The basic motivation behind these upscaling regimes is to create simulation models that produce flow scenarios that are in close correspondence with the flow scenarios that one would obtain by running simulations directly on the geological models. Unfortunately, experience has shown that it is very difficult to design a robust upscaling scheme that gives somewhat reliable results for all kinds of flow scenarios. Therefore, due to the limitations of upscaling, a new type of methodology based on so-called *multiscale methods* has started to gain popularity. The multiscale paradigm generally refers to a class of methods that incorporate fine-scale information into a set of coarse-scale equations in a way that is consistent with the local properties of the differential operators. In Section 4 we present a recent class of methods that seek to incorporate subscale information more directly in the flow simulation as a means to avoid the use of upscaling. But before we discuss different upscaling and multiscale approaches for simulating flow in reservoirs, we will identify some scales of importance.

2 Models and Scales in Porous Media Flow

In petroleum reservoirs sedimentary structures are formed when sediments are deposited to form thin layers called laminae. Due to alternating layers of coarse and fine-grained material, laminae may exhibit large permeability contrasts on the mm-cm scale, but these effects are usually left unresolved

homogenisation is also the name of a specific mathematical theory (for asymptotic analysis of periodic structures) that has been applied, by coincidence, to upscaling geological models for reservoir simulation. Thus, to avoid confusion we will only use the term upscaling for this coarsening procedure.

in geological models. Laminae are stacked to form beds, which are the smallest stratigraphic units. The thickness of beds varies from millimetres to tens of meters, and different beds are separated by thin layers with significantly lower permeability. Beds are, in turn, grouped and stacked into parasequences or sequences (parallel layers that have undergone similar geologic history). Parasequences represent the deposition of marine sediments, during periods of high sea level, and they tend to be somewhere in the range from meters to tens of meters thick and have a horizontal extent of several kilometres.

The trends and heterogeneity of parasequences depend on the depositional environment. For instance, whereas shallow marine deposits may lead to rather smoothly varying permeability distributions with correlation lengths in the order of 10–100 meters, fluvial reservoirs may contain intertwined patterns of narrow high-flow channels on a background of significantly lower permeability. Fractures and faults, on the other hand, are created by stresses in the rock. A fault is a planar surface in the rock, across which the rocks have been displaced. A fracture is a crack or a surface of breakage, across which the rocks have not been displaced. These structures may have higher or lower permeability than the surrounding rocks. The reservoir geology can also consist of other structures like for instance shale layers, which consist of impermeable clays and are the most abundant sedimentary rock.

We have seen that geological structures in petroleum reservoirs occur at a range of length scales. Moreover, it is known that geological structures at all scales can have a profound impact on fluid flow. Hence, ideally geological features that span across all types of length scales should be reflected in the geological reservoir model. However, since a geological model can only cope with a certain range of scales, we need to discretise with respect to scale, try to identify the most important scales, and develop different models when studying different phenomena. Choosing scales is often done by intuition and experience, and it is hard to give very general guidelines, but an important concept in choosing model scales is the notion of representative elementary volumes (REVs). This concept is based on the idea that petrophysical flow properties (typically porosity and permeability) are constant on some intervals of scale, see Figure 2. REVs, if they exist, mark transitions between scales of heterogeneity, and present natural length scales for modelling.

In order to identify a range of length scales where REVs exist for porosity (or permeability), we move along the length-scale axis from the micrometer scale of the pores toward the kilometre scale of the reservoir. At the pore scale level the porosity is a rapidly oscillating function equal to zero (in solid rock) and one (in the pores). Hence, obviously no REVs can exist at this scale. At the next characteristic length scale, the core scale level, we find laminae deposits. Since the laminae consist of alternating layers of coarse and fine grained material, we cannot expect to find a common porosity value for the different rock structures. Moving further along the length scale axis we may find long thin layers, perhaps extending throughout the entire horizontal length of the reservoirs. Each of these individual layers may be nearly homogeneous since

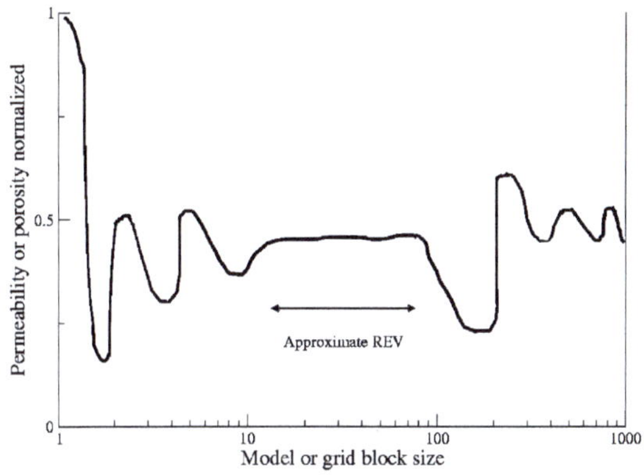

Fig. 2. Flow properties may be approximately constant on scale intervals.

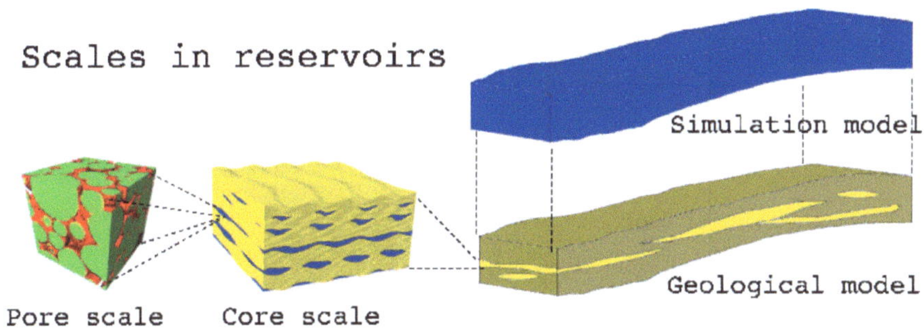

Fig. 3. Various scales in a reservoir.

they are created by the same geological process, and probably contain approximately the same rock types. Hence, at this level it sounds reasonable to speak of REVs. If we move to the high end of the length scale axis we start to group more and more layers into families of layers with different sedimentary structures, and porosity REVs will probably not exist.

The previous discussion gives some grounds to claim that reservoir rock structures contain scales where REVs may exist. However, from a general point of view the existence of REV's in porous media is highly disputable. For instance a faulted reservoir with faults distributed continuously both in length and size throughout the reservoir, will typically have no REV. Moreover, no two reservoirs are identical so it is difficult to capitalise from previous experience. Indeed, porous formations in reservoirs may vary greatly, also in terms of scales. Nevertheless, the concept of REVs can serve as a guideline when deciding what scales to model. In the following we will discuss four different model classes, as shown in Figure 3: pore scale models, core scale models, geological models and simulation models.

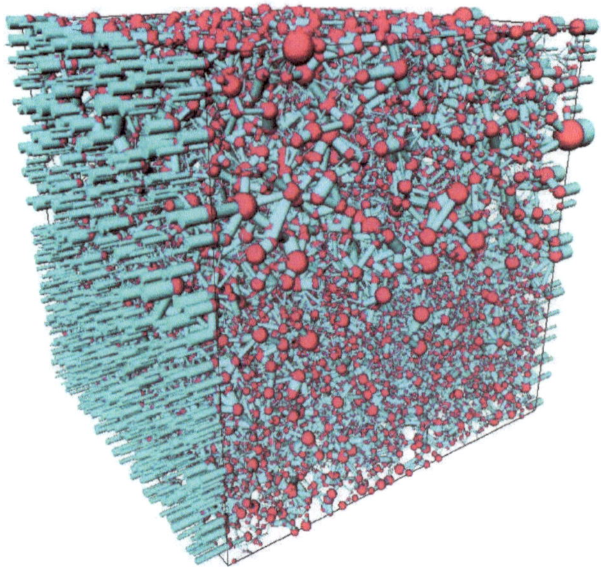

Fig. 4. Graphical illustration of a pore network.

2.1 Pore-Scale Model

Modelling flow at the pore scale is quite different from modelling flow at the reservoir scale. On the reservoir scale, the fundamental equations are continuity of fluid phases and Darcy's law, as we will see below. Darcy's law basically assumes that the primary forces driving the flow are the pressure gradient and gravity. Flow at the pore scale, on the other hand, is mainly dominated by capillary forces, although gravitational forces can still be important. We therefore need a different set of equations to model flow at the pore scale. Since it would be off-track to give a thorough presentation of pore-scale flow modelling, we give only a very simplified overview.

A pore-scale model (as illustrated in Figure 3) may be about the size of a sugar cube. At the pore scale the porous medium is usually represented by a graph (see e.g., [54]). A graph is a pair (V, E), where V is a set whose elements are called vertices (or nodes), and E is a subset of $V \times V$ whose elements are called edges. The vertices are taken to represent pores, and the edges represent pore-throats (i.e., connections between pores), see Figure 4.

The flow process where one fluid invades the void space filled by another fluid is generally described as an invasion–percolation process. Invasion is the process where a phase can invade a pore only if a neighbouring pore is already invaded. For each pore there is an entry pressure (i.e., the threshold pressure needed for the invading phase to enter the pore) depending on the size and shape of pores, the size of pore throats, as well as other rock properties. Among those pores neighbouring invaded pores, the invading phase will first invade the pore with lowest threshold pressure. This gives a way of updating the set of pores neighbouring invaded ones. Repeating the process establishes a recursive algorithm, determining the flow pattern of the invading phase. In the invasion process we are interested in whether a phase has a path through the model (i.e., percolates) or not, and the time variable is often not modelled at all. For pore networks this is misleading, because we are also interested in modelling the flow after the first path through the model has been established. After a pore has been invaded, the saturations in the pore will vary with pressures and saturations in the neighbouring pores (as well as in the pore itself). New pores may also be invaded after the first path is formed, so that we may get several paths through the model where the invading phase can flow. Once the invading phase percolates (i.e., has a path through the model) we can start estimating flow properties. As the simulation progresses we will have an increasing saturation for the invading phase, which can be used to estimate flow properties at different saturation compositions in the model.

As mentioned initially this overview was simplified. To elaborate on the underlying complexity we need to mention wettability. When two immiscible fluids (such as oil and water) contact a solid surface (such as the rock), one of them tends to spread on the surface more than the other. The fluid in a porous medium that preferentially contacts the rock is called the wetting fluid. Note that wettability conditions are usually changing throughout a reservoir. The flow process where the invading fluid is non-wetting is called drainage and is typically modelled with invasion–percolation. The flow process where the wetting fluid displaces the non-wetting fluid is called imbibition, and is more complex, involving effects termed film flow and snap-off. A further presentation is beyond the scope here, but the interested reader is encouraged to see [54] and references therein.

From an analytical point of view, pore-scale modelling is very important as it represents flow at the fundamental scale (or more loosely, where the flow really takes place), and hence provides the proper framework for understanding the fundamentals of porous media flow. From a practical point of view, pore-scale modelling has a huge potential. Modelling flow at all other scales may be seen as averaging of flow at the pore scale, and properties describing the flow at larger scales are usually a mixture of pore-scale properties. At larger scales, the complexity of flow modelling is often overwhelming, with large uncertainties in determining flow parameters. Hence being able to single out and estimate the various factors determining flow parameters is invaluable, and pore-scale models can be instrumental in this respect. However, to extrapolate properties from the pore scale to an entire reservoir is very challenging, even if the entire pore space of the reservoir was known (of course, in real life you will not be anywhere close to knowing the entire pore space of a reservoir).

Fig. 5. Three core plugs with diameter of one and a half inches, and a height of five centimetres.

2.2 Core-Scale Model

When drilling a well it is common to bring up parts of the rock from the well trajectory. Three such rock samples are shown in Figure 5. These rock samples are called cores or core plugs, and are necessarily confined (in dimension) by the radius of the well, although they lengthwise are only confined by the length of the well. Cores taken from a well can give detailed information at scales restricted only by human apparatus. For instance, using thin slices of the rock one may study the pore structure with an electron microscope with micrometer resolution. However, it is common to do also flow experiments on cores (e.g., flood them with water in a laboratory), thereby obtaining flow properties for the core, e.g., relative permeability curves and capillary pressure curves.

Properties from core-scale flow experiments are often used as input for the geological model, or directly for the simulation model. Cores should therefore ideally be representative for the heterogeneous structures that one may find in a typical grid block in the geological model. However, flow experiments are usually performed on relatively homogeneous cores that rarely exceed one meter in length. Cores can therefore seldom be classified as representative elementary volumes. For instance, cores may contain a shale barrier that blocks flow inside the core, but which does not extend much outside the well bore region. Thus, if the core was slightly wider, then there would be a passage past the shale barrier. Flow at the core scale level is also more influenced by capillary forces than flow on a reservoir field scale.

This discussion shows that the problem of extrapolating information from cores to build a geological model is largely under-determined. Supplementary pieces of information are also needed, and the process of gathering geological data from other sources is described next.

2.3 Geological Model

The two key characteristics that affect the fluid flow in a reservoir rock are its porosity and permeability. Porosity is the fraction of space not occupied by the rock, i.e., the volume fraction occupied by the phases (oil, water and gas). Permeability is a measure of how easy fluid flows through the rock. The main purpose of the geological model is therefore to provide the distribution of these petrophysical parameters, besides giving location and geometry of the reservoir. Thus, a geological model is a conceptual, three-dimensional representation of a reservoir and consists of a set of grid cells that each has a constant porosity and a corresponding constant permeability tensor. The tensor is represented by a matrix where the diagonal terms represent direct flow (i.e., flow in one direction caused by a pressure drop in the same direction), and the off-diagonal terms represent cross-flow (flow caused by pressure drop in directions perpendicular to the flow). Thus, a full tensor is needed to model local flow in directions at an angle to the coordinate axes. For example, in a layered system the dominant direction of flow will generally be along the layers. Thus, if the layers form an angle to the coordinate axes, then a pressure drop in, say, the x–coordinate direction will typically produce flow at an angle to x–direction. This type of flow can be modelled correctly only with a permeability tensor with nonzero off-diagonal terms.

Geological models are built using a combination of stratigraphy (the study of rock layers and layering), sedimentology (study of sedimentary rocks), and interpretation of measured data. Unfortunately, building a geological model for a reservoir is like finishing a puzzle where most of the pieces are missing. Ideally, all available information about the reservoir is utilised, but the amount of data available is limited due to costs of acquiring them. Seismic surveys give a sort of X–ray image of the reservoir, but they are both expensive and time consuming, and can only give limited resolution (you cannot expect to see structures thinner than ten meters from seismic data). Wells give invaluable information, but the results are restricted to the vicinity of the well. While seismic has (at best) a resolution of ten meters, information on a finer scale are available from well-logs. Well-logs are basically data from various measuring tools lowered into the well to gather information, e.g., radiating the reservoir and measuring the response. Even well-logs give quite limited resolution, rarely down to centimetre scale. Detailed information is available from cores taken from wells, where resolution is only limited by the apparatus at hand. The industry uses X-ray, CT-scan as well as electron microscopes to gather high resolution information from the cores. However, information from cores and well-logs are from the well or near the well, and extrapolating this information to the rest of the reservoir is subject to great uncertainty. Moreover, due to costs, the amount of data acquisitions made is limited. You cannot expect well-logs and cores to be taken from every well. All these techniques give separately small contributions that can help build a geological model. However, in the end we still have very limited information available considering that a petroleum reservoir can have complex geological features that span across all types of length scales from a few millimetres to several kilometres.

In summary, the process of making a geological model is generally strongly under-determined. It is therefore customary to use *geostatistics* to estimate the subsurface characteristics between the wells. Using geostatistical techniques one builds petrophysical realisations in the form of grid models that both honour measured data and satisfy petrophysical trends and heterogeneity. Since trends and heterogeneity in petrophysical properties depend strongly on the structure of sedimentary deposits, high-resolution petrophysical realisations are not built directly. Instead, one starts by building a facies model. A facies is the characteristics of a rock unit that reflects its origin and separates it from surrounding rock units. By supplying knowledge of the depositional environment (fluvial, shallow marine, deep marine, etc) and conditioning to observed data, one can determine the geometry of the facies and how they are mixed. In the second step, the facies are populated with petrophysical data and stochastic simulation techniques are used to simulate multiple realisations of the geological model in terms of high-resolution grid models for petrophysical properties. Each grid model has a plausible heterogeneity and can contain from a hundred thousand to a hundred million cells. The collection of all realisations gives a measure of the uncertainty involved in the modelling. Hence, if the sample of realizations (and the upscaling procedure that converts the geological models into simulation models) is unbiased, then it is possible to supply predicted production characteristics, such as the cumulative oil production, obtained from simulation studies with a measure of uncertainty.

2.4 From Geomodel to Simulation Model

For full sized reservoirs the traditional approach has been to model geological structures with a geological model, and fluid flow with a coarser simulation model. Indeed, core models and pore models are designed only to give input to the geological model and perhaps to derive flow parameters for the simulation model. As we have seen, geological models are designed to reproduce the true geological heterogeneity in the reservoir rock and possibly incorporate a measure of inherent uncertainty. The grid cells in the geomodels are considered as a representative volume having certain geometry and constant petrophysical properties like porosity and permeability. The petrophysical properties are parameters in the equations governing fluid flow, and the size of the grid cells are typically of order 10–50 m in the horizontal directions and 10 cm to 10 m in the vertical direction, allowing the geomodel to capture the most important variations and characteristics in the petrophysical parameters. However, rocks are heterogeneous at all levels and the current choice of scales means that geomodels are too coarse to capture small-scale geological structures like tidal deposital layers that occur on a centimetre or even smaller scale.

Geological structures on a finer scale than the resolution of the geomodel may have a strong impact on fluid flow. On the other hand, geomodels are typically too detailed for routine flow simulations. Indeed, flow simulations are usually performed on coarser models where the petrophysical properties in the geomodel are replaced with an average of some kind over regions that correspond to a grid block in the simulation model. There are several reasons for this. First of all, the global flow in the reservoir may be dominated by large-scale phenomena (occurring on a scale of 10-1000 m) and a coarser model may therefore be sufficient to predict the reservoir production with the required accuracy. Secondly, since coarse models contain fewer parameters, they are easier to history-match[2] than fine-grid models. Similarly, the results of very detailed flow simulations may be masked by the large uncertainty involved in reservoir modelling. Thus, by running flow simulations directly on geomodels we get more detailed flow scenarios, but they might not produce more accurate production characteristics.

The primary reason for not running simulations on geomodels, however, is that geomodels typically contain far too many grid-blocks from a numerical point of view and cannot be used directly because the memory and the computational time required are outside the limits of current computers. It is perhaps tempting to believe that with future increase in computing power, one will soon be able to simulate the flow in a full field reservoir model on a grid fine enough to capture rock properties at the smallest significant scale. The development so far, indicates that such an idea is wrong. The trend is rather to increase the number of scales and the sizes of models whenever more computing power or new numerical techniques become available. Indeed, the sizes and complexity of geomodels have increased continuously with the increase in computer memory and processing power. Hence, the need for handling different scales will most likely persist in the foreseeable future and there will be a continuing need for rescaling techniques in order to upscale and downscale models.

Upscaling techniques have primarily been used to go from geomodels to simulation models, but similar techniques are also used to go from e.g., a pore-scale model to a geomodel. By upscaling detailed geomodels one generates reduced simulation models that have fewer grid cells and possibly less irregular grid structures. The effective petrophysical properties are calculated in each cell of the simulation grids based on properties of the underlying geomodels. In this process, the aim is to preserve the small-scale effects in the large-scale computations (as well as possible). Systematic small-scale variations in permeability and porosity can have a significant effect on a larger scale, and this should be captured in the upscaled model. Upscaling techniques range from simple averaging techniques to relatively complex flow analyses. Although the latter type involves solving numerous local flow problems that

[2]History matching is the process of calibrating parameters (here porosity and permeability) to observations.

require a substantial computational effort, they do not always produce reliable results. Indeed, if the heterogeneity in the reservoir rock does not contain well-separated length-scales, this type of upscaling may lead to highly erroneous results, even for single-phase flow.

Downscaling techniques are typically used when going from a simulation model to a geomodel. For instance, one can be interested in refining coarse-scale modifications in petrophysical properties obtained during a history match on a simulation model in order to update the underlying geomodel. In this process, the aim is to preserve both the coarse-scale trends and the fine-scale heterogeneity structures.

2.5 Reservoir Simulation Model

The fundamental equations that model flow in porous media are the continuity equations. These equations ensure conservation of mass for the different fluids phases, and take the following form

$$\frac{\partial}{\partial t}\left(\phi\rho_i s_i\right) + \nabla\cdot\left(\rho_i v_i\right) = q_i. \tag{1}$$

Here each fluid phase is modelled by its density ρ_i, phase velocity v_i, and saturation s_i. The saturations are the volume fractions occupied by each phase and satisfy the closure relation $\sum_{i=1}^{n} s_i = 1$. Production and injection wells are modelled by the source term q_i. The phase velocities v_i are related to the phase pressure p_i through an empirical relation called Darcy's law

$$v_i = -\frac{\mathbf{K}k_{ri}}{\mu_i}\left(\nabla p_i - \rho_i G\right). \tag{2}$$

Here $\mathbf{K}$ is the rock permeability given from the geological model; μ_i is the viscosity of phase i; $k_{ri} = k_{ri}(s_1,\ldots,s_n)$ is the relative permeability of phase i, i.e., reduced permeability due to the presence of other phases; p_i is phase pressure; and $G = gn_z$ where g is the gravitational constant and n_z is the unit vector pointing in the downward vertical direction. It is also customary to introduce the phase mobility: $\lambda_i = k_{ri}(s_i)/\mu_i$.

The primary unknowns in simulation models are the phase saturations and the phase pressures. Darcy's law together with the continuity equations (1) gives us one equation for each phase. The closure relation $\sum_{i=1}^{n} s_i = 1$ for the phase saturations gives us one additional equation. This gives us a complete model for single-phase flow. The simplest model for single-phase flow is obtained by assuming incompressibility (constant density). In this case, the equations (1)–(2) simplify to an elliptic equation for the fluid pressure

$$-\nabla\cdot\left[\frac{\mathbf{K}}{\mu}(\nabla p - \rho G)\right] = \frac{q}{\rho}. \tag{3}$$

For multi-phase flow, however, we can eliminate only one of the saturations and extra closure relations in terms of capillary pressure functions are added.

That is, one assumes that the phase pressures are related through a function describing the difference between the phase pressures across fluid interfaces. These functions are assumed to depend solely on the phase saturations, and are obtained from experiments and simulations on core scale models, or from tables describing capillary pressure functions for different rock types with known properties. We should also mention that pore scale-modelling can provide capillary pressure curves, and provide an appropriate framework for understanding the capillary forces.

For two-phase flow we can eliminate one of the transport equations. We then get a coupled system consisting of pressure equation and an equation for the fluid transport. Moreover, if we assume that the two fluids, say, water (w) and oil (o) are incompressible, and introduce a so-called global pressure p (see e.g., [2] in this book) and a total velocity $v = v_o + v_w$, then the corresponding system reads

$$\nabla \cdot v = q, \quad \text{where} \quad v = -\mathbf{K}\big[\lambda \nabla p - (\lambda_w \rho_w + \lambda_o \rho_o) G\big], \tag{4}$$

$$\phi \frac{\partial s_w}{\partial t} + \nabla \cdot \big[f_w \big(v + \mathbf{K} \lambda_o \nabla p_{cow} + \mathbf{K} \lambda_o (\rho_w - \rho_o) G \big) \big] = \frac{q_w}{\rho_w}. \tag{5}$$

Here we have introduced the total mobility $\lambda = \lambda_w + \lambda_o$, the fractional flow of water $f_w = \lambda_w / \lambda$, the capillary pressure $p_{cow} = p_o - p_w$, and the accumulated contribution from the wells $q = q_w/\rho_w + q_o/\rho_o$.

The transport equation (5) contains three different terms that stem from three different types of forces: viscous forces give rise to the term $q_v = f_w v$, capillary forces give rise to $q_c = f_w \mathbf{K} \lambda_o \nabla p_{cow}$, and gravity forces give rise to $q_g = f_w \mathbf{K} \lambda_o (\rho_w - \rho_o) g$. The range of these different forces is influenced by the flow process and model. In particular, model size, flow-rate, heterogeneities and fluid system are all important ingredients in determining the balance of forces. There are no absolute rules as to what problems are dominated by the different forces, although some rules of thumb apply. Gas-oil systems are generally more influenced by gravity than oil-water systems due to larger density differences, and horizontally layered reservoirs with poor vertical communication are less influenced by gravity than reservoirs with layers that are tilted with respect to the vertical axis or have good vertical communication. Capillary effects are mainly small-scale phenomena, and are therefore more apparent in small-scale models, but capillary forces can play a dominant role also in large reservoir regions with strong heterogeneous structures. However, as a general rule we can say that viscous forces tend to be dominant on a reservoir (or field) scale, whereas capillary forces typically are dominant in core scale models. In fact, capillary forces are rarely included in simulation models, implying that small scale effects of capillary forces need to be adjusted for in the relative permeability curves.

In the introduction we discussed why reservoir simulation models are usually posed on a different length scale than the geomodel and how this introduces an upscaling of the grid models. Another reason for operating with different grid-models for the fluid simulation comes from the fact that geological grid models tend to follow the spatial structure of the geology. For (classical) numerical methods to behave properly, one would generally prefer that grid-blocks are as regular as possible. However, reservoir dimensions impose very different length scales in the vertical and horizontal direction. Moreover, layers, fractures and faults all give rise to complex physical geometries and grid-blocks that represent the underlying geology are rarely orthogonal, but tend to be skewed, irregular, and even have non-neighbour connections. These geometrical problems are also related to the multiscale structure of the reservoirs and lack of local grid resolution, but will not be discussed at all in the current paper. Instead, we will focus on the multiscale structures in permeability and discuss approaches for meeting the challenges they impose on the simulation. To ease the presentation, we will therefore henceforth assume that our reservoirs have a shoe-box geometry and consist of Cartesian grid-blocks.

In the oil industry, it is common to discretise the pressure equation with a *finite-volume method.* These methods consider the grid cells V_i as control volumes, and invoke the divergence theorem to yield

$$\int_{V_i} q\,dx = -\int_{\partial V_i} \mathbf{K}\left[\lambda\nabla p - (\lambda_w\rho_w + \lambda_o\rho_o)G\right]\cdot n\,d\nu.$$

Here n is the outward unit normal on ∂V_i. A fully discrete scheme is obtained by expressing the flux across the cell interfaces in terms of the pressure at the cell centres. For instance, the most basic finite volume method, the two-point flux approximation (TPFA) scheme, takes the following form:

$$\int_{V_i} q\,dx = \sum_j T_{ij}(p_i - p_j). \tag{6}$$

Here the sum is taken over all non-degenerate interfaces, i.e., over all j such that $\partial V_j \cap \partial V_i$ has a positive measure. The coupling factors T_{ij} are called transmissibilities and we will return to these later.

Similarly, the transport equation (5) is usually discretised with a finite volume method where upstream weighting is used to discretise the term representing viscous forces. Invoking the divergence theorem once again we have

$$\int_{V_i} \phi\frac{\partial S_w}{\partial t} + \int_{\partial V_i} f_w\big(v + \mathbf{K}\lambda_o\nabla p_{cow} + \mathbf{K}\lambda_o(\rho_w - \rho_o)G\big)\,d\nu = \int_{V_i} \frac{q_w}{\rho_w}\,dx.$$

By upstream weighting, it is meant that $f_w v$ is computed using the saturation on the upstream side of the cell interfaces. For a discussion of how to discretise the terms representing gravity and capillary forces, we refer the reader to the previous chapter in this book [2], where a more thorough introduction to the governing equations and corresponding numerical methods is given.

3 Upscaling for Reservoir Simulation

In the introduction we saw how reservoir modelling often involves an upscaling phase, where high-resolution geological models containing several million cells are turned into coarser grid models that are more suitable for fluid flow simulation. This process leads to many fundamental questions. For instance, do the partial differential equations modelling porous media flow at the coarse scale take the same form as the equations modelling flow at the subgrid level? And, if so, how do we honour the fine-scale heterogeneities at the coarse scale?

Even though upscaling has been a standard procedure in reservoir simulation for nearly four decades, nobody has answered these questions rigorously, except for cases with special heterogeneous formations such as periodic or stratified media. Most upscaling techniques are based on some kind of local averaging procedure in which effective properties in each grid-block are calculated solely from properties within the grid block. As such, the averaging procedure does not account for coupling with neighbouring coarse grid-blocks. This implies in particular that the upscaled quantities do not reflect the impact of global boundary conditions and large scale flow patterns in the reservoir. Because different flow patterns may call for different upscaling procedures, it is generally acknowledged that global effects must also be taken into consideration in order to obtain robust coarse-scale simulation models.

A related problem to upscaling is that of gridding. That is, what kind of grid do we want to use to represent our porous medium, what resolution do we need, and how do we orient the grid. To design robust simulation models, the grid should be designed so that the grid blocks capture heterogeneities on the scale of the block. This often implies that we need significantly more cells in a regular grid than if we use corner point grids in which the grid block corners are on straight, but possibly tilted lines, that extend from top-to-bottom of the reservoir. It is therefore much more common to use corner-point grids in reservoir simulation. In the vicinity of wells one might use a different type of grid, for instance some kind of radial grid, to account for a radial flow pattern in the near well region. The importance of obtaining a good grid representation of the porous medium should not be underestimated, but gridding issues will not be discussed any further here. We assume henceforth that we have a fixed coarse grid suitable for numerical simulation and focus on upscaling techniques for the pressure equation (4) and the saturation equation (5).

The literature on upscaling techniques is extensive, ranging from simple averaging techniques, e.g., [45], via local simulation techniques [11, 21], to multiscale methods [1, 18, 40, 42] and rigorous homogenisation techniques for periodic structures [12, 39, 44]. Some attempts have been made to analyse the upscaling process, e.g., [8, 64], but so far there is generally no theory or framework for assessing the quality of an upscaling technique. In fact, upscaling techniques are seldom rigorously quantified with mathematical error estimates. Instead, the quality of upscaling techniques is usually assessed by comparing upscaled production characteristics with those obtained from a reference solution computed on an underlying fine grid.

It is not within our scope to give a complete overview over the many upscaling techniques that have been applied in reservoir simulation. Instead, we refer the reader to the many review papers that have been devoted to this topic, e.g., [10, 20, 57, 62]. However, to understand why some upscaling methods work well for some flow scenarios, and not for others, whereas other upscaling techniques may work well for a completely different set of flow scenarios, we do need to discuss some basic concepts. To this end, we start by giving a brief introduction to upscaling rock permeability for the pressure equation (3) or (4) in Subsection 3.1. In Section 3.2 we discuss techniques for generating pseudo functions for upscaling the saturation equation (5). In Subsection 3.3 we present a different type of upscaling technique for the saturation equation that splits the variables into volume averaged and fluctuating components. In this approach various moments of the fluctuating components are used to derive a coarse scale equation for grid block saturations. Finally, in Subsection 3.4 we describe a class of upscaling techniques that assume steady state flow, i.e., that the time derivative term in the continuity equations can be neglected in the upscaling process.

Before we proceed with details, we should mention that upscaling is also applied to other quantities, but then mostly using very simple techniques. For instance, quantities such as saturation and porosity are upscaled using simple volume averaging. That is, for a grid block V we obtain porosity and saturations, denoted ϕ^* and s_i^*, respectively, by

$$\phi^* = \frac{1}{|V|}\int_V \phi(x)\,dx, \qquad s_i^* = \frac{1}{\phi^*|V|}\int_V s_i(x)\phi(x)\,dx.$$

Fluxes are upscaled similarly. Moreover, if a grid block is known to contain a certain fraction of different rock types (also often termed flow-units), then a so-called majority vote can be used to approximate the effective properties. In this method one identifies which rock type occupies the largest volume fraction of the grid block and simply assigns the properties associated with this rock type to the entire grid block. Note however, that such a simple approach is not robust, see, e.g., [65]. Actually, the problem of upscaling rock-types is closely related to that of upscaling permeability, but has not received much attention in the literature.

3.1 Single-Phase Upscaling

The process of upscaling permeability for the pressure equation (3) or (4) is often termed single-phase upscaling. Most single-phase upscaling techniques seek homogeneous block permeabilities that reproduce the same total flow through each coarse grid-block as one would get if the pressure equation was solved on the underlying fine grid with the correct fine-scale heterogeneous structures. However, to design upscaling techniques that preserve averaged fine-scale flow-rates is in general nontrivial because the heterogeneities at all scales have a significant effect on the large-scale flow pattern. A proper coarse-scale reservoir model must therefore capture the impact of heterogeneous structures at all scales that are not resolved by the coarse grid.

To illustrate the concept behind single-phase upscaling, let p be the solution that we obtain by solving

$$-\nabla \cdot \mathbf{K}\nabla p = q, \qquad \text{in } \Omega \tag{7}$$

on a fine grid with a suitable numerical method, e.g., a TPFA scheme of the form (6). To reproduce the same total flow through a grid-block V we have to find a homogenised tensor $\mathbf{K}_V^*$ such that

$$\int_V \mathbf{K}\nabla p \, dx = \mathbf{K}_V^* \int_V \nabla p \, dx. \tag{8}$$

This equation states that the net flow-rate v_V through V is related to the average pressure gradient $\nabla_V p$ in V through the upscaled Darcy law $v_V = -\mathbf{K}^* \nabla_V p$.

In the discrete case, the choice of an appropriate upscaling technique also depends on the underlying numerical method. For instance, if the pressure equation is discretised by a TPFA finite-volume scheme of the form (6), then grid-block permeabilities are used only to compute interface transmissibilities at the coarse scale. As a result, one needs correct coarse-scale transmissibilities to reproduce a fine-scale flow field in an averaged sense. Hence, instead of upscaled block-homogenised tensors $\mathbf{K}^*$, one should seek block transmissibilities T_{ij}^* satisfying

$$Q_{ij} = T_{ij}^* \left(\frac{1}{|V_i|} \int_{V_i} p \, dx - \frac{1}{|V_j|} \int_{V_j} p \, dx \right), \tag{9}$$

where $Q_{ij} = -\int_{\partial V_i \cap \partial V_j} (\mathbf{K}\nabla p) \cdot n \, d\nu$ is the total Darcy flux across $\partial V_i \cap \partial V_j$.

If all transmissibilities T_{ij}^* are positive, then the TPFA scheme defined by

$$\sum_{j:\partial V_i \cap \partial V_j \neq \emptyset} T_{ij}^* (p_i - p_j) = \int_{V_i} q \, dx,$$

will reproduce the net grid block pressures $p_l = \frac{1}{|V_l|} \int_{V_l} p \, dx$, and hence also the coarse grid fluxes Q_{ij}. Unfortunately, there is no guarantee that the transmissibilities defined by (9) are positive, or even exist. Neither can we guarantee that the upscaled permeability tensors defined by (8) are positive definite. Positive transmissibilities and positive definite permeability tensors ensure stability of TPFA finite-volume and finite-element methods, respectively. The possible absence of these properties illustrates that the fundamental problem of single-phase upscaling is ill-posed. However, this ill-posedness is usually not a subject of debate since upscaling methods are usually devised such that the occurrence of unphysical upscaled quantities is avoided. For a more thorough discussion of existence and uniqueness, we refer the reader to [64]. In the next subsections we review some standard single-phase upscaling methods.

Averaging Methods

The simplest form of upscaling permeability is to take some kind of average of the permeabilities at the subgrid level. A general class of averaging techniques is defined by the power average,

$$\mathbf{K}_V^{*,p} = \left(\frac{1}{|V|}\int_V \mathbf{K}(x)^p\,dx\right)^{1/p}, \quad -1 \le p \le 1.$$

Note that $p = 1$ and $p = -1$ correspond to the arithmetic and harmonic means respectively, while the geometric mean is obtained in the limit $p \to 0$.

The use of power averaging can be motivated by the so-called Wiener-bounds [63], which state that for a statistically homogeneous medium, the correct upscaled permeability will be bounded above and below by the arithmetic and harmonic mean, respectively. This result has also a more intuitive explanation. To see this, consider the one dimensional pressure equation:

$$-\partial_x(\mathbf{K}(x)\partial_x p) = 0 \quad \text{in } (0,1), \quad p(0) = p_0,\ p(1) = p_1.$$

It is easy to see that the solution of this equation induces constant Darcy velocity v. This implies that $\partial_x p$ must scale proportional to the inverse of $\mathbf{K}(x)$. Hence, we derive

$$\partial_x p = \frac{p_1 - p_0}{\mathbf{K}(x)}\left[\int_0^1 \frac{dx}{\mathbf{K}(x)}\right]^{-1} = \frac{p_1 - p_0}{\mathbf{K}(x)}\mathbf{K}_V^{*,-1}.$$

If we insert this expression into equation (8) we find that the correct upscaled permeability $\mathbf{K}_V^*$ is identical to the harmonic mean $\mathbf{K}_V^{*,-1}$.

This result, which shows that the harmonic mean gives the correct upscaled permeability, is generally valid only in one dimension, but it applies to a special case in higher dimensions as well. Indeed, consider the following problem:

$$\begin{aligned} -\nabla\cdot\mathbf{K}\nabla p &= 0 && \text{in } V = (0,1)^3, \\ p(0,y,z) &= p_0, && p(1,y,z) = p_1, \\ (-\mathbf{K}\nabla p)\cdot n &= 0 && \text{for } y,z \in \{0,1\}, \end{aligned} \tag{10}$$

where n is the outward unit normal on $\partial\Omega$.

Now, if $\mathbf{K}$ models a perfectly stratified isotropic medium with layers perpendicular to the x–axis (so that $\mathbf{K}(x,\cdot,\cdot)$ is constant for all x), then the solution models uniform flow in the x–coordinate direction. This means that for each pair $(y,z) \in (0,1)^2$ the solution $p_{y,z} = p(\cdot,y,z)$ solves

$$-\nabla\cdot\mathbf{K}\nabla p_{y,z}(x) = 0 \text{ in } (0,1), \quad p_{y,z}(0) = p_0,\ p_{y,z}(1) = p_1.$$

It follows that

$$-K(x)\nabla p = -(K(x)\partial_x p_{y,z}, 0, 0)^T = -\mathbf{K}_V^{*,-1}(p_1 - p_0, 0, 0)^T.$$

Hence, the correct upscaled permeability is equal to the harmonic mean.

Similarly, if **K** models instead a stratified isotropic medium with layers perpendicular to the y– or z–axis, then the correct upscaled permeability would be equal to the arithmetic mean. These examples show that averaging techniques can give correct upscaling in special cases, also in three dimensional space. However, if we consider the model problem (10) with a less idealised heterogeneous structures, or with the same heterogeneous structures but with other boundary conditions, then both the arithmetic and harmonic average will generally give wrong net flow-rates. Indeed, these averages give correct upscaled permeability only for cases with essentially one-dimensional flow.

To try to model flow in more than one direction, one could generate a diagonal permeability tensor with the following diagonal components:

$$k_{xx} = \mu_a^z(\mu_a^y(\mu_h^x)), \quad k_{yy} = \mu_a^z(\mu_a^x(\mu_h^y)), \quad k_{zz} = \mu_a^x(\mu_a^y(\mu_h^z)).$$

Here μ_a^ξ and μ_h^ξ represent respectively the arithmetic and harmonic means in the ξ-coordinate direction. Thus, in this method one starts by taking a harmonic average along grid cells that are aligned in one coordinate-direction. One then computes the corresponding diagonal by taking the arithmetic mean of all "one dimensional" harmonic means. This average is sometimes called the harmonic-arithmetic average and may give good results if, for instance, the reservoir is layered and the primary direction of flow is along the layers.

Despite the fact that averaging techniques can give correct upscaling in special cases, they tend to perform poorly in practise since the averages do not reflect the structure or orientation of the heterogeneous structures. It is also difficult to decide which averaging technique to use since the best average depends both on the heterogeneities in the media and the flow process we want to model (flow direction, boundary conditions, etc). To illustrate the dependence on the flow process we consider an example.

Example 1. Consider a reservoir in the unit cube $[0,1]^3$ with two different geomodels that each consist of a $8 \times 8 \times 8$ uniform grid-blocks and permeability distribution as depicted in Figure 6. We consider three different upscaling methods: harmonic average; arithmetic average; and harmonic-arithmetic average. The geomodels are upscaled to a single coarse grid-block, which is then subjected to three different boundary conditions:

BC1: $p = 1$ at $(x, y, 0)$, $p = 0$ at $(x, y, 1)$, no-flow elsewhere.
BC2: $p = 1$ at $(0, 0, z)$, $p = 0$ at $(1, 1, z)$, no-flow elsewhere.
BC3: $p = 1$ at $(0, 0, 0)$, $p = 0$ at $(1, 1, 1)$, no-flow elsewhere.

Table 1 compares the observed coarse-block rates with the flow-rate obtained by direct simulation on the $8 \times 8 \times 8$ grid. For the layered model, harmonic and harmonic-arithmetic averaging correctly reproduces the vertical flow normal to the layers for BC1. Arithmetic and harmonic-arithmetic averaging correctly reproduces the flow along the layers for BC2. The harmonic arithmetic averaging method also performs reasonably well for corner-to-corner flow (BC3).

Fig. 6. Plot of the logarithm of two permeability distributions: (left) a layered medium, and (right) a cube extracted from the lower part of the fluvial Upper Ness formation from the 10th SPE test case [19].

Table 1. Flow-rates relative to the reference rate Q_R on the fine grid.

	Model 1			Model 2		
	BC1	BC2	BC3	BC1	BC2	BC3
Q_H/Q_R	1	2.31e−04	5.52e−02	1.10e−02	3.82e−06	9.94e−04
Q_A/Q_R	4.33e+03	1	2.39e+02	2.33e+04	8.22	2.13e+03
Q_{HA}/Q_R	1	1	1.14	8.14e−02	1.00	1.55e−01

For model two, however, most of the methods produce some significant errors, and none of the methods are able to produce an accurate flow-rate for the test cases with boundary conditions specified by BC1 and BC3.

Averaging techniques can also be used to upscale transmissibility. One such method is the half-block method [46]. In this method each grid block is divided into two pieces (using the average surface between the two opposing faces), obtaining six upscaled permeabilities for each block. The transmissibility T_{ij} for the coarse grid interface, which essentially models the conductivity across the corresponding interface in the direction of the associated coordinate unit normal n_{ij}, is then obtained by taking the harmonic average of the half-block arithmetic averages on both sides of the interface. This method has the advantage of improving the model resolution at a low computational cost.

Numerical Pressure Computation Techniques

To approximate the combined effect of heterogeneous structures and imposed boundary conditions, the idea of inverting local pressure computations was suggested by Begg et al. [11]. In this approach one solves, for each grid block V, a set of homogeneous pressure equations on the form

$$-\nabla \cdot \mathbf{K}\nabla p = 0 \quad \text{in } V,$$

with prescribed boundary conditions. Thus, this method raises the immediate question: what kind of boundary conditions should be imposed?

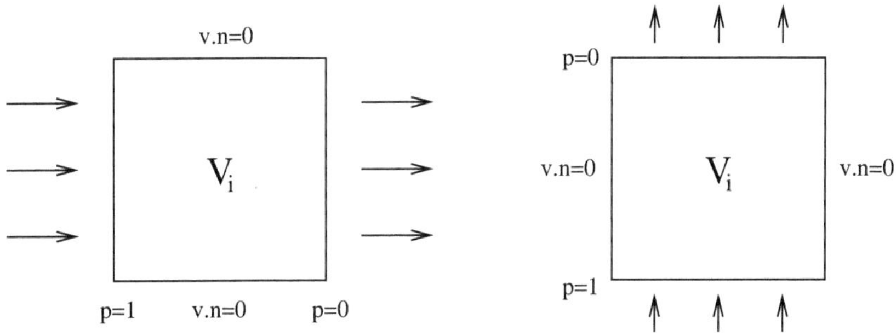

Fig. 7. A schematic of the pressure solver method with p=1 and p=0 along the inflow and outflow boundaries respectively, and no-flow boundary conditions elsewhere.

One alternative is to impose three different sets of boundary conditions for each block to create a pressure drop across the coarse cell in each of the three coordinate directions (see Figure 7). This gives us a set of flow-rates for each grid block that can be used to compute an effective diagonal permeability tensor with components

$$k_{xx} = -Q_x L_x / \Delta P_x, \qquad k_{yy} = -Q_y L_y / \Delta P_y, \qquad k_{zz} = -Q_z L_z / \Delta P_z.$$

Here Q_ξ, L_ξ and ΔP_ξ are respectively the net flow, the length between opposite sides, and the pressure drop in the ξ-direction inside V.

Another popular option is to choose periodic boundary conditions. That is, one assumes that each grid block is a periodic cell in a periodic medium and imposes full correspondence between the pressures and velocities at opposite sides of the block, e.g., to compute k_{xx} we impose the following boundary conditions:

$$p(1,y) = p(0,y) - \Delta p, \qquad p(x,1) = p(x,0),$$
$$v(1,y) = v(0,y), \qquad v(x,1) = v(x,0)$$

This approach yields a symmetric and positive definite tensor [21], and is usually more robust than the directional flow boundary conditions.

Improved accuracy of the upscaled permeability can be obtained if one uses an oversampling technique, in which the domain of the local flow problem is enlarged with a border region surrounding each grid block. In this approach we let the coarse grid-block V be embedded in a macro-block V' and solve the elliptic pressure equation (7) in V' with some prescribed boundary conditions on $\partial V'$. The flow is then solved in the whole enlarged region, but the averaging of the permeability tensor is performed only over the (original) coarse block. The motivation for using such a technique is to better account for permeability trends that are not aligned with the grid directions and possible large-scale connectivity in the permeability fields.

Example 2. We revisit the test-cases considered in Example 1, but now we compare harmonic-arithmetic averaging (HA) with the pressure computation

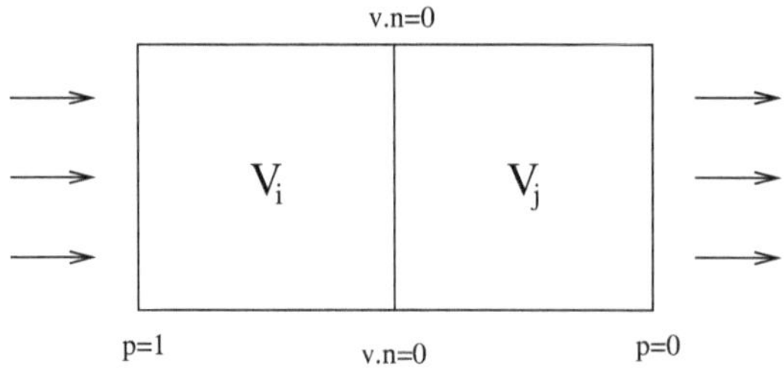

Fig. 8. A schematic of the pressure solver method for upscaling transmissibility.

techniques using; (D) the boundary conditions depicted in Figure 7; and (P) the periodic boundary conditions with a directional pressure drop. The latter method gives rise to full permeability tensors, but for the cases considered here the off-diagonal terms in the upscaled permeability tensors are small, and are therefore neglected for simplicity.

Table 2. Flow-rates relative to the reference rate Q_R on the fine grid.

	Model 1			Model 2		
	BC1	BC2	BC3	BC1	BC2	BC3
Q_{HA}/Q_R	1	1	1.143	0.081	1.003	0.155
Q_D/Q_R	1	1	1.143	1	1.375	1.893
Q_P/Q_R	1	1	1.143	0.986	1.321	1.867

Table 2 compares the observed coarse-block rates with the flow-rate obtained by direct simulation on the $8 \times 8 \times 8$ grid. For the layered model, all methods give rise to the same diagonal permeability tensor, and hence give exactly the same results. For model 2 we see that the numerical pressure computation methods give significantly better results than the HA averaging technique. Indeed, the worst results for the pressure computation methods, which were obtained for corner-to-corner flow, is within a factor 2, whereas the HA averaging technique underestimates the flow rates for BC1 and BC3 by almost an order of magnitude.

A similar approach can be used to derive upscaled transmissibilities. A simple way of doing this is illustrated in Figure 8. Here we create a flow across the interface between V_i and V_j by imposing a pressure drop between the grid block faces opposite to $\partial V_i \cap \partial V_j$. Thus, by solving the corresponding pressure solution in the two-block domain $V_i \cup V_j$ numerically, we can compute the average pressures p_i and p_j in V_i and V_j respectively, and derive an upscaled transmissibility by requiring that equation (9) holds.

Global and Local-Global Upscaling Techniques

The methods described so far have all been local in nature; the averaging techniques derive the upscaled permeability solely from local heterogeneous structures, whereas the pressure computation techniques try to account for flow responses by solving local problems with prescribed boundary conditions. The numerical examples show, however, that all the local methods may fail to capture the correct flow behaviour. For instance, none of the local methods were able to capture the correct flow behaviour for model 2 with boundary conditions prescribed by BC3. The main problem for the pressure computation techniques is, of course, that we do not know a priori the precise flow that will occur in a given region of the reservoir. Thus, it is generally not possible to specify the appropriate boundary conditions for the local flow problems in a unique manner (unless we already have solved the flow problem).

In order to account for global flow patterns, Holden and Nielsen [38] proposed to solve the pressure equation once on a fine grid, and extract information about the fine-grid flow pattern to compute correct upscaled transmissibilities. This approach may seem to contradict the purpose of upscaling, but it can be justified for the numerical treatment of two-phase flows. Indeed, for two-phase flow the pressure equation has to be solved several times throughout the simulation, and the cost of solving the pressure once (or even a few times) on a fine grid will typically be negligible compared with the total computational cost of the full multi-phase flow simulation. Another question is of course the robustness of such an upscaling with respect to changes in the well configuration and global boundary conditions. For some cases the upscaled transmissibilities may prove useful, while for other cases they may not. A similar technique avoiding the solution of the fine-grid problem was presented by Nielsen and Tveito [53].

Another way to avoid the full fine-grid computation was proposed by Chen et al. [16, 17] in the form of a local-global technique that uses global coarse-scale calculations to determine boundary conditions for local calculations that determine upscaled transmissibilities. Since the initial coarse-grid calculation may still give quite poor boundary conditions for the subgrid problems, they use an iteration procedure to ensure consistency between the local and global calculations.

A problem encountered in the global and local-global upscaling procedures is the occurrence of negative or anomalously large transmissibilities. Holden and Nielsen [38] avoid negative and very large transmissibilities by perturbing the transmissibilities through an iterative process, involving an optimisation problem, until all the transmissibilities are contained in some interval $[a,b] \subset \mathcal{R}^+$. Chen et al. [16, 17] observed that unphysical transmissibilities occur mainly in low-flow regions. Therefore, instead of using an optimisation procedure that alters all transmissibilities in the reservoir, they use a thresholding procedure, where negative and very large transmissibilities are replaced with transmissibilities obtained by a local pressure computation technique. Since the transmissibilities are altered only in low-flow regions, the perturbation will have limited impact on the total flow through the reservoir.

Another class of methods that account for global boundary conditions are multiscale methods. One can think of these methods as a way of upscaling the flow field instead of computing coarse scale reservoir properties (e.g., porosity and permeability) while trying to preserve important trends in the fine-scale flow pattern. Two examples of such methods will be discussed in Section 4.

Other Techniques

There are many upscaling techniques that we have not discussed. For instance, we have not said anything about a class of methods that are based on the homogenisation theory (see e.g., [12, 39, 44]). Homogenisation is a rigorous mathematical theory for asymptotic analysis in periodic structures. A relevant result states that for a periodic medium with permeability $\mathbf{K}(\frac{x}{\varepsilon})$, there exists a constant symmetric and positive definite tensor $\mathbf{K}_0$ such that the solutions p_ε and $v_\varepsilon = -\mathbf{K}(\frac{x}{\varepsilon})\nabla p_\varepsilon$ to the following elliptic problem

$$-\nabla \cdot \mathbf{K}(\frac{x}{\varepsilon})\nabla p_\varepsilon = q$$

converges uniformly as $\varepsilon \to 0$ to the corresponding solutions of a homogenised equation of the form

$$-\nabla \cdot \mathbf{K}_0 \nabla p_0 = q.$$

Thus, if we consider each coarse grid-block as a cell in an infinite periodic medium, then homogenisation theory can be used to derive corresponding homogenised tensors for each grid-block. The use of homogenisation theory to upscale grid-block permeability for porous media flow has, however, been a subject of debate. A common question is why one should study a periodic medium when no natural medium is periodic? A natural response to this question is that the homogenisation theory provides the mathematical tools capable of proving the existence and uniqueness of the solution, and gives some verification that the governing equations at the macroscopic level take the same form as (7), which governs porous media flow at the continuous level.

Among other techniques for single-phase upscaling that have not been discussed above are the fast, but less robust, real-space renormalisation techniques [47], and the use of geostatistical methods and Monte-Carlo analysis to generate reservoir descriptions at the reservoir simulation scale directly [34]. Neither have we mentioned how the definition of the coarse-grid geometry can be linked to upscaling through non-uniform coarsening approaches [24], or the use of elastic grids [32]. For a deeper discussion of these, and other related, issues, we ask the reader to consult [20, 57, 62] and the references therein.

3.2 Pseudo Functions for Upscaling the Saturation Equation

Single-phase upscaling alone is often not sufficient to capture large-scale heterogeneity effects in a multi-phase system; for instance, when the flow follows narrow high-flow channels that penetrate the coarse grid blocks. In such cases, only a small portion of a grid block may be subjected to flow, and the fluid displacement cannot be characterised by single-phase upscaling, even in an averaged sense. Similar problems arise in the presence of long and thin shale barriers or high-permeable layers. Thus, in addition to capturing the macroscopic effect of the rock permeability, we need to incorporate the large-scale effects of the relative permeabilities.

Whereas the problem of upscaling permeability for solving the pressure equation on a coarse scale is fairly well understood, derivation of upscaling techniques for the saturation equation has only been moderately successful. In fact, a robust methodology that can be used to upscale the saturation equation reliably for general heterogeneous formations still does not exist. This is obviously a very difficult task, since we need to trace sharp saturation fronts and at the same time model high-flow channels. Indeed, consider the saturation equation in its simplest form (neglecting gravity and capillary forces)

$$\phi \frac{\partial s}{\partial t} + \nabla \cdot f_w(s) v = 0, \tag{11}$$

and suppose that it is discretised with the following upstream-weighted finite-volume method:

$$\bar{\phi}\, \frac{s_i^{n+1} - s_i^n}{\Delta t} + \sum_j F^{us}(s_i, s_j) Q_{ij} = 0 \qquad \forall V_i \subset \Omega. \tag{12}$$

Here s_i^n is the average saturation in grid-block V_i at time t_n, $\bar{\phi}$ denotes the arithmetic mean of the porosity and Q_{ij} is the total flux across the interface between grid-blocks V_i and V_j obtained from the coarse grid solution of the pressure equation. The numerical fractional flux function F^{us} is given as

$$F^{us}(s_i, s_j) = \begin{cases} f_w(s_i), & \text{if } Q_{ij} \geq 0, \\ f_w(s_j), & \text{if } Q_{ij} < 0. \end{cases}$$

In this scheme, the only way to properly account for the discrepancy between the scales of the coarse grid and the fine grid, is to define special fractional flow functions F_{ij} (so-called pseudo functions) that represent the averaged flux over each coarse grid interface. An intrinsic feature with this approach to two-phase flow upscaling is that pseudo functions depend, not only on the heterogeneous porous media structures, but also on the saturation distribution within the grid block and on the reservoir flow history, which in turn depends on well locations and global boundary conditions. All these effects must be accounted for in the macroscopic reservoir characterisation to achieve robust coarse-scale models.

Pseudo functions can generally be divided in two categories: (vertical) equilibrium pseudo functions and dynamic pseudo functions. Equilibrium pseudo functions have traditionally been used to reduce the dimension of the system under static flow conditions, e.g., from three to two spatial dimensions and are reminiscent of the steady-state methods discussed below. Dynamic pseudo functions, on the other hand, aim at reproducing subscale effects under dynamic flow conditions by upscaling relative permeabilities or fractional flow functions. Below we briefly review some of the most important techniques that have been proposed for generating dynamic coarse-grid pseudo functions.

Dynamic Pseudo Functions for Two-Phase Flow

The main idea behind dynamic pseudo functions is to do simplified fine-scale flow simulations that mimic the flow patterns of the case that we want to model. An intrinsic feature with this approach to two-phase flow upscaling is that pseudo functions depend, not only on the heterogeneous porous media structures, but also on the saturation distribution within the grid block and on the reservoir flow history, which in turn depends on well locations and global boundary conditions. All these effects must be accounted for in the macroscopic simulation model to achieve reliable results. It is therefore generally not sufficient to perform local, or even extended local, flow simulations [37]. This makes generating reliable pseudo functions without too much computational effort a very challenging task.

The use of dynamic pseudo functions for two-phase flow simulation goes back to Jacks et al. [41], and Kyte and Berry [51]. Their objective was to design relative permeability functions in coarse grid-blocks that account for subgrid flow patterns and scale discrepancies to reduce numerical dispersion. Since then, many different methods have been developed for generating dynamic pseudo functions. Here we will describe two basic types: pseudo functions from individual phase flow-rates and pseudo functions based upon averaged total mobility.

The individual phase flow-rate methods include the methods of Jack et al. [41], the Kyte and Berry method [51], the flux-weighted potential method [35], and the pore-volume weighted method. All methods are based upon the upscaled Darcy's law and vary only in the way the fine-grid simulations are factored into averaged quantities. Although, these methods have received some criticism for being unreliable (and for being limited to cases where capillary or gravity equilibrium can be assumed at the coarse scale) [10], the individual phase flow-rate methods have been widely used for upscaling of two-phase flow in reservoir simulation. To describe the essential ingredients in these methods, assume that we have a coarse grid overlaying a fine grid, as illustrated in Figure 9. Moreover, we assume that the flow has been computed on the fine grid such that the saturation and pressure history is known in each fine-grid cell along with derived quantities like relative permeability and flow-rates. Based on this, the aim is to define pseudo functions that produce a coarse-grid solution equal to the averaged fine-grid solution. The functions are dynamic in the sense that they are saturation (and pressure) dependent.

The method of Jacks et al. [41] (which is sometimes called the weighted relative permeability method) defines an upscaled relative permeability function for each coarse grid interface by taking a transmissibility-weighted average of relative permeabilities in cells on the upstream side. To be precise, assume that $Q_{ij} > 0$, so that there is a net flow from V_i to V_j and denote by $T_{i,kl}$ the transmissibility for a fine grid interface that lies on $\partial V_i \cap \partial V_j$, i.e., between two fine-grid cells $U_k \subset V_i$ and $U_l \subset V_j$. The upscaled relative permeability associated with $\partial V_i \cap \partial V_j$ is now defined as follows:

$$\bar{k}_{rw,ij} = \frac{\sum_{kl} T_{i,kl} k_{rw,k}(s_k)}{\sum_{kl} T_{i,kl}}.$$

Thus, the sum is taken over all interfaces $(\partial U_k \cap \partial U_l) \subset (\partial V_i \cap \partial V_j)$. The suffix k in $k_{rw,k}$ indicates that there may be different relative permeabilities in each layer (for different rock types). If $Q_{ij} < 0$ then saturations from the corresponding grid cells in V_j are used.

Similar methods to the method of Jacks et al. [41] first average the phase pressure p_α or the phase potential $\Phi_\alpha = p_\alpha - \rho_\alpha g D$ as will be described below and then calculate upscaled pseudo-relative permeabilities by inverting Darcy's law for each phase. For water,

$$\bar{k}_{rw} = -\frac{\bar{\mu}_w \bar{q}_w \Delta x}{\bar{K}(\Delta \bar{p}_w - g\bar{\rho}_w \Delta D)}.$$

Here D is the average depth of the coarse block, $\bar{q}_w$ is the spatial average of the fine-grid water fluxes, and $\bar{K}$ is an upscaled permeability obtained by using a single-phase upscaling technique. The coarse-grid viscosity and density are computed as pore-volume weighted or flow-rate weighted averages.

In the Kyte and Berry method the pseudo water phase-pressures $\bar{p}_w$ are obtained by computing the weighted average along the centre column of the coarse grid block,

$$\bar{p}_w = \frac{\sum_{j=J_1}^{J_2} \left[K k_{rw} \delta y \big(p_w - g\rho_w (d - D) \big) \right]_j}{\sum_{j=J_1}^{J_2} \left[k_{rw} K \delta y \right]_j}. \tag{13}$$

In the equation, values with subscript j refer to midpoint (or cell-average) values on the underlying fine-grid along centre column in the coarse block (i.e., $i = \frac{1}{2}(I_1 + I_2)$, see Figure 9). The weighting factor is the phase permeability multiplied by the thickness of each layer. The dynamic behaviour of the pseudo functions follows by tabulating each averaged quantity as a function of the corresponding average saturation. The model can easily be extended to three dimensions, see [61]. By using the procedure on each cell-interface on the coarse grid, one should, in principle, obtain ideal pseudo functions that reproduce the underlying fine-scale simulation in an averaged sense.

Let us now analyse the method in some detail. Notice first that since distinct weighting factors are used for different phases in (13), nonzero capillary

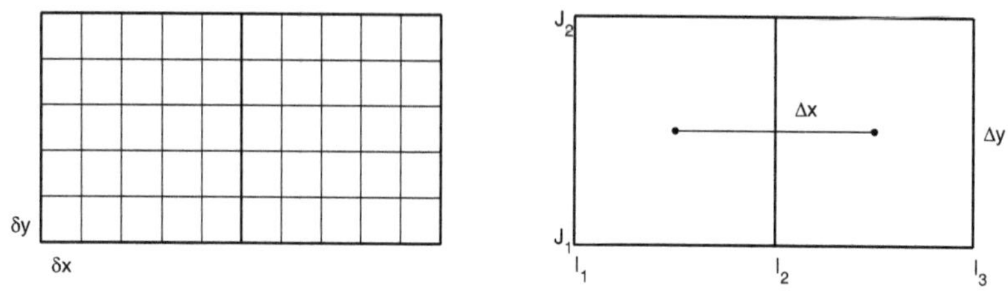

Fig. 9. Schematic of the coarse and the fine grid used in calculation of dynamic pseudo functions.

pressures may be introduced on the coarse grid even if the phase pressures are equal on the fine grid. Moreover, the weighing factors are directionally dependent, leading to pseudo functions that depend on the flow direction (and thus the history of the reservoir). More alarming, however, is the fact that pseudo-relative permeabilities may be negative or have arbitrary large magnitudes. Negative $\bar{k}_{rw}$ values occur if the average flux $\bar{q}_w$ has the same sign as the increment in the phase potential $\Delta\bar{p}_w - g\rho_w g\Delta D$, and if this increment tends to zero, then $|\bar{k}_{rw}| \to \infty$. The Kyte and Berry method can therefore be very sensitive to small perturbations in the flow conditions.

Alternative methods can be derived by using different weighting factors in to compute the phase pressures. The pore-volume weighted method reads

$$\bar{p}_w = \frac{\sum_{j=J_1}^{J_2}\sum_{i=I_1}^{I_2}\left[\phi\delta x\delta y(p_w - g\rho_w(d-D))\right]_{j,i}}{\sum_{j=J_1}^{J_2}\sum_{i=I_1}^{I_2}\left[\phi\delta x\delta y\right]_{j,i}},$$

and similarly, the flux-weighted potential method [35]

$$\bar{k}_{rw} = -\frac{\mu\bar{q}_w\Delta x}{\bar{K}_x\Delta\Phi_w}, \qquad \bar{\Phi}_w = \frac{\sum_{j=J_1}^{J_2}\left[q_w\Phi_w\right]_j}{\sum_{j=J_1}^{J_2}\left[q_w\right]_j},$$

where the sums are taken over the fine-grid column in the centre of the coarse block.

To avoid some of the problems associated with the individual phase flow-rate methods, Stone [60] suggested upscaling the fractional-flow function $f_o(S)$ based on averaging the total mobilities λ at the fine scale,

$$\bar{f}_o = \frac{\sum_{j=J_1}^{J_2}\left[q_t f_o\right]_j}{\sum_{j=J_1}^{J_2}\left[q_t\right]_j}, \qquad \bar{\lambda} = \frac{\sum_{j=J_1}^{J_2}\left[T_x\lambda\right]_j}{\sum_{j=J_1}^{J_2}\left[T_x\right]_j}.$$

Here the sum is taken over the fine-grid column on the upstream side of the coarse-grid interface. Stone's formula assumes constant pressure drop in the layers and neglects gravity and capillary pressure. His method for is therefore usually inadequate in cases with significant gravity or capillary pressure effects, or in cases with large local variations in total mobility.

Several authors [9, 36, 48] have proposed methods similar to e.g., Stone's method, with "better" averaging of the total mobility. One possibility is to do local potential calculations with e.g., periodic boundary conditions and invert the solution with respect to the phase mobilities. These total mobility methods are more robust than the Kyte and Berry method in the sense that infinite values can be avoided and negative values occur less frequently (but can still occur if the net flows of the two phases are in opposite directions). In summary, both the Kyte and Berry method and the method proposed by Stone (and variants of the respective methods) have their drawbacks. The Kyte and Berry method is computationally expensive, and may be highly sensitive to small perturbations in the flow conditions. Total mobility methods like Stone's method are also computationally expensive, but at least these methods tend to be more robust than the Kyte and Berry method.

3.3 Volume Averaged Equations

As an alternative to using pseudo functions, a more general framework based on using higher moments to develop volume-averaged equations (VAE) has been proposed by several authors, e.g., [23, 28, 52]. In this approach the basic idea is to express the unknown quantities in terms of average and fluctuating components. To illustrate, consider the simplified saturation equation (11) with unit porosity $\phi = 1$. Furthermore, for any given variable $\nu(x)$, write

$$\nu(x) = \overline{\nu} + \tilde{\nu}(x),$$

where $\bar{\nu}$ denotes a volume-averaged quantity (constant within the averaging region), and $\tilde{\nu}(x)$ denotes a spatially fluctuating quantity with zero mean in the averaging region. Substituting now the volume-average expansions for s, v, and f_w into (11), and then averaging the resulting equations gives

$$\overline{s}_t + \tilde{s}_t + \overline{v} \cdot \nabla \overline{f_w} + \overline{v} \cdot \nabla \tilde{f}_w + \tilde{v} \cdot \nabla \overline{f_w} + \tilde{v} \cdot \nabla \tilde{f}_w = 0. \tag{14}$$

Averaging each term in this equation, noting that the average of terms with only one fluctuating component vanishes, we obtain the following volume averaged equation for $\overline{s}$:

$$\overline{s}_t + \overline{v} \cdot \nabla \overline{f_w} + \overline{\tilde{v} \cdot \nabla \tilde{f}_w} = 0 \tag{15}$$

By subtracting (15) from (14) one obtains a corresponding equation for the fluctuating part $\tilde{s}$:

$$\tilde{s}_t + \overline{v} \cdot \nabla \tilde{f}_w + \tilde{v} \cdot \nabla \overline{f_w} + \tilde{v} \cdot \nabla \tilde{f}_w = \overline{\tilde{v} \cdot \nabla \tilde{f}_w}. \tag{16}$$

The equation for the fluctuating component can be used to generate equations for various moments of the fine-scale equations, see e.g., [22].

Assuming a unit mobility ratio, Efendiev et al. [26, 28] derive a single coarse-grid equation from (15) on the form

$$\frac{\partial \overline{s}}{\partial t} + \nabla \cdot G(x, \overline{s}) = \nabla \cdot D(x,t) \nabla \overline{s}, \tag{17}$$

where $G(x, \overline{s}) = \overline{\overline{v} f(\overline{s})}$ and $D(x,t)$ is a history dependent function that models the cross term $\tilde{v} \cdot \nabla \tilde{f_w}$. If we compare this equation with (11) we observe that (17) contains an extra diffusion term. The general form of the coarse-scale equation (17) has been obtained by many authors, e.g., [22, 23, 30, 52], and the history dependence in the diffusion term has been rigorously verified using homogenisation theory, see e.g., [30, 44]. This shows that the correct form of the coarse-scale saturation equation does not, unlike the coarse-scale pressure equation, take the same form as the equation that models the phase transport at the continuous level. This observation gives some justification for developing coarse-scale models that attempt to incorporate the history dependence.

Unfortunately, the implementation of VAEs into a simulator can be a bit complicated due to the nonlocal memory effect. As a remedy for this difficulty, Efendiev and Durlofsky [27] introduced a generalised convection-diffusion model for modelling subgrid transport in porous media. Here they observe that the nonlocal dispersivity in (17) imparts a somewhat convective character to the model. Therefore, to eliminate the history-dependence in (17), they propose a generalised convection-diffusion model where they introduce a convective correction term so that the convective term in (17) is on the form

$$G(x, \overline{s}) = \overline{v} f(\overline{s}) + m(x, \overline{s}).$$

The convective correction term $m(x, \overline{s})$ is determined from local subgrid computations, as is the diffusion term $D(x, \overline{s})$.

Durlofsky has investigated the relation between upscaling methods based on the VAE methodology and existing upscaling techniques for two-phase flow, with an emphasis on pseudo-relative permeability generation [10, 51, 55, 60], the non-uniform coarsening approach[3] [24], and the use of higher moments of the fine-scale variables to correlate upscaled relative permeabilities [22]. It is illustrated that some of the benefits and limitations with the various approaches can be understood and quantified with respect to volume averaged equations. Specifically, it is demonstrated that for flows without gravity and capillary forces, the VAE approach shows a higher degree of process independence than the traditional pseudo-relative permeability methods. However, to our knowledge, all published literature on VAEs and so-called generalised convection-diffusion models for two-phase flow neglect capillary and gravity forces. In addition, the methodology has for the most part been tested on cases with unit mobility ratio [23, 22, 25, 28]. In recent work, Efendiev and Durlofsky [26, 27] make an effort to generalise the VAE approach to two-phase

[3]The non-uniform coarsening method tries to generate grids that are coarsely gridded in low-flow regions and more finely gridded in high-flow regions. Thus, rather than modelling subgrid effects explicitly, the non-uniform coarsening methodology attempts to minimise errors caused by subgrid effects by introducing higher grid resolution in regions where subgrid terms would otherwise be important.

flow with general mobility ratios, and obtain promising results. However, further research is needed to validate the method and investigate if it can be extended to include capillary and gravity forces.

3.4 Steady-State Equilibrium Upscaling for Two-Phase Flow

Dynamic pseudo methods are numerical in nature and tend to ignore the underlying physics. The focus is on adjusting parameters to match a numerical solution, rather than basing the upscaling on effective physical properties. Steady-state upscaling (see e.g., [58]) is an alternative where the essential assumption is steady-state flow, i.e., that the time derivative in equation (1) is zero. Away from the well, this assumption leads us to a steady-state equation for each phase on the form on the following form:

$$\nabla \cdot \left[\frac{\mathbf{K} k_{ri}}{\mu_i} (\nabla p_i - \rho_i G) \right] = 0. \tag{18}$$

It may seem very restrictive to impose steady-state flow. However, away from saturation fronts and in regions with low flow we may have small saturation changes with respect to time, and small changes in fractional flow. Disregarding the temporal derivative in (1) can therefore be justified in these types of flow regions.

To derive effective relative permeabilities from equation (18) we first need to determine the saturation distribution inside each coarse grid block. This will be done from various physical principles, depending on the flow regime. An upscaled relative permeability that corresponds to the prescribed subgrid saturation distribution is then obtained by upscaling the phase permeability $\mathbf{K} k_{ri}$ using a pressure-computation single-phase upscaling technique, and dividing the result with the corresponding upscaled permeability tensor $\mathbf{K}$. Thus, steady-state relative permeability curves are defined by the following formula:

$$\bar{k}_{rw} = \bar{\mathbf{K}}^{-1} \overline{\mathbf{K} k}_{rw}. \tag{19}$$

This technique may thus be viewed as an extension to two-phase flows of the pressure computation technique for single-phase upscaling.

To model the subgrid saturation, additional assumptions are made. The precise assumptions will depend on the scale that we wish to model, the flow process (e.g., the magnitude of the flow velocity) as well as an assessment of which physical effects that are important. Assuming for the moment that we have a way of extracting an appropriate saturation field on the subgrid from the coarse scale solution, it is clear that also other saturation-dependent physical parameters, such as the capillary pressure, may be upscaled. Thus, as opposed to dynamic pseudo functions, we see that these methods do not seek numerical devices that reproduce subscale flow effects. Instead, steady-state upscaling techniques seek effective physical properties and therefore have a more physical motivation. On the other hand, steady-state methods do not correct for grid effects and numerical dispersion like dynamic pseudo functions tend to.

Steady-state upscaling methods may be put into three categories that are linked to the balance of three forces: viscous forces, capillary forces, and gravitational forces which give rise to flows q_v, q_c, and q_g (see (5) and the subsequent discussion). Understanding the balance of these forces may indicate which method to warrant. It is often not clear how to quantify these three forces for a particular modelling problem, and they will usually vary both with time and space. However, we can try to investigate the role of the different forces by comparing the respective contributions to the saturation equation

$$\phi \partial_t s_w = -\nabla \cdot \left[q_v + q_c + q_g \right]. \tag{20}$$

For certain types of models it may be possible to choose representative values for q_v, q_c and q_g. Thus, if we normalise them with respect to their sum and place them in a ternary diagram, then the position in the diagram may serve as a guideline deciding on which steady-state method to apply, see, e.g., [59]. Another possibility is to measure the balance of these three forces using laboratory experiments. This type of experiment is usually performed on core samples with constant saturation fraction injected through an inlet boundary, and pressure controlled production at an outlet boundary, where the outlet boundary is at the opposite face of the inlet boundary. By comparing the observed flow-rates at the outlet boundary with corresponding flow-rates obtained from numerical simulation studies, possibly through some tuning of relevant parameters, it is possible to say something about the importance of including the various forces in the upscaled model. Note that care need to be exercised when using laboratory experiments. The balance of forces are sensitive to scale, and also to whether reservoir conditions are used during a core flooding experiment. We now present how to determine the subgrid saturations for the different steady-state methods.

Capillary Equilibrium Methods

In these steady-state upscaling methods one assumes, in addition to the steady-state assumption, that capillary pressure is constant in space and time. In practise, approximately constant capillary pressure occurs mainly in regions with very low flow-rates. This implies that viscous and gravity forces are dominated by capillary forces. Viscous forces, and sometimes also gravity forces, are therefore neglected in capillary equilibrium steady-state methods.

In capillary equilibrium steady-state methods, the saturation in each grid-cell within the block to be upscaled is determined by inverting the capillary pressure curve. Indeed, for a two-phase system the capillary pressure is modelled as a strictly monotone function of one of the phases, so it is invertible on its range. The capillary pressure function is often scaled by the so-called Leverett J-function. This function takes the following form:

$$J(s_w) = \frac{p_c\sqrt{\frac{\mathbf{K}}{\phi}}}{\sigma cos(\theta)}, \tag{21}$$

where σ is the interfacial tension between the phases, and θ is the contact angle between the phases. The contact angle and interfacial tension are usually measured in a laboratory. The Leverett J-function is usually a known (tabulated) function of s_w. Now, since the capillary pressure is a known constant within each grid block and $\mathbf{K}$ and ϕ are known subgrid quantities, we can invert (21) to obtain the subgrid saturation distribution

$$s_w = J^{-1}\left(\frac{\bar{p}_c\sqrt{\frac{\mathbf{K}}{\phi}}}{\sigma cos(\theta)}\right). \tag{22}$$

With the J-function scaling, the capillary pressure curve for each grid block will be scaled according to the ratio between permeability and porosity, and tend to give large saturation contrasts in the model.

Once the subgrid saturation is known, we solve the steady-state equation (18) for each phase, and use (19) to derive effective relative permeabilities. Gravity forces can be included in the model by replacing p_c in (22) with $\varrho = p_c + (\rho_w - \rho_o)gz$, where z is vertical position in the model. This follows by observing that in absence of viscous forces we have $v_w = f_w\mathbf{K}\lambda_o\nabla\varrho$. Thus, conceptually the mapping defined by ϱ allows us to include gravity forces in the capillary pressure function. However this approach should be used only if capillary forces are dominant since the Leverett J-function does not model effects of gravity. Finally, we remark that the assumption that viscous forces are negligible can be relaxed, as is shown in [7].

Steady-State Upscaling Methods for Viscous Dominated Flows

The basic assumption in the capillary equilibrium methods was that the flow was dominated by capillary forces, or equivalently that the Darcy velocity v is so small that the viscous forces can be neglected. At the other end of the rate scale we have the *viscous limit*, which can be interpreted as the limit when capillary and gravitational forces tend to zero relative to the viscous forces. This can be the case in, for instance, nearly horizontal high permeable layers with restricted vertical extent. For a more in depth discussion on the applicability of viscous limit steady-state the reader may consult [31]. By convention, one often refers to this as the limit when the Darcy velocity goes to infinity.

In the viscous-limit approach to steady-state upscaling methods, one neglects gravity and capillary forces and assumes constant fractional flow. In addition, it is assumed that the pressure gradient across a grid-block is the same for both phases. If k_{ro} is nonzero, we now can divide the flow-rate of the phases with each other (given by Darcy's law) to derive

$$\frac{v_w}{v_o} = \frac{\frac{-Kk_{rw}}{\mu_w}\nabla p}{\frac{-Kk_{ro}}{\mu_o}\nabla p} = \frac{k_{rw}\mu_o}{k_{ro}\mu_w}.$$

The relative permeability curves are assumed to be known functions of the phase saturations and rock type (i.e., functions of x). With the additional equation that saturations add to one, solving for saturations is just a matter of solving two equations with two unknowns $s_w(x)$ and $s_w(x)$,

$$s_w + s_o = 1, \qquad k_{rw}(s_w, x) = \left(\frac{v_w\mu_w}{v_o\mu_o}\right)k_{ro}(s_o, x).$$

Note that saturations obtained are unique when the relative pressure curves are strictly monotonic. Moreover, if we have a single set of relative permeability curves, the subgrid saturations will be constant within the coarse block. As in the capillary equilibrium methods, effective relative permeabilities are obtained from the subgrid saturation distribution by using (19) and an effective permeability tensor $\bar{\mathbf{K}}$ is obtained by solving the steady-state single-phase equation.

It should be noted that for numerical implementation a strategy presented in [50] should be preferred. Rather that upscaling each phase separately, it is suggested to upscale the total mobility λ_t, and at the final step calculate the individual relative permeabilities using fractional flow. This approach has two advantages, the first being that you only need to solve the steady-state single phase equation once, the other being that you reduce contrast in the model, softening the numerical properties of the equation.

General Steady-State Methods

The two steady-state upscaling methods described above apply to cases with either very low or very high flow-rates. This may apply to certain regions of an oil-reservoir, but one will always have regions with somewhat moderate flow-rates. Moreover, many grid blocks will have some cells with low permeability and some blocks with high permeability. Hence, one will often have both high and low flow-rates inside the same grid block. In these situations, neither the capillary equilibrium method nor the viscous limit method tend to perform well. Still, the basic steady-state upscaling methodology that involves upscaling steady-state equations of the form (18) and using (19) to derive effective relative permeabilities may still be used. However, this requires full flow simulations for each grid block and is therefore computationally expensive. Moreover, since we can not exploit physical principles to derive upscaled parameters, we need to determine an initial saturation distribution for the simulations, as well as fluid injection rates on the boundary. On the upside we have that no physical assumptions are required other than steady-state, so that effects from gravity, phase transfer, and compressibility can be included.

Final Remarks on Two-Phase Upscaling

As with all mathematical tools for handling simulation of flow, the value of upscaling techniques will ultimately be judged by its success in applications to practical problems. Upscaling two-phase flow reliably has proved to be a very difficult task. In fact, although pseudo methods are widely used, Barker and Thibeau [10] argue that existing pseudo methods can only be used to upscale two-phase flow reliably for cases where gravity equilibrium (gravity forces dominate capillary and viscous forces) or capillary equilibrium can be assumed at the coarse-grid scale. One of the reasons why it is so hard to upscale reservoir flow scenarios reliably is that different types of reservoir structures require different upscaling methodologies. Hence, in principle one should divide grid blocks into different categories that each represents a particular type of representative elementary volumes. This implies that upscaling should ideally be linked to gridding. Of the methods discussed in this section, only pseudo methods tend to correct for grid effects, meaning that simulation results differ if you change the grid resolution. Relative permeability curves obtained using pseudo methods may therefore be effective for reproducing the production history of a field, but may perform poorly if the production strategy of the field is changed (e.g., if a new well is drilled). Conclusively, pseudo and steady-state methods both have their drawbacks. Pseudo-functions have been used by the industry for quite some time, so their limitations are well-known. Methods based on the VAE methodology are still immature for real field applications.

4 Multiscale Methods for Solving the Pressure Equation

Several types of multiscale methods have recently begun to emerge in various disciplines of science and engineering. By definition, multiscale methods is a label for techniques that model physical phenomena on coarse grids while honouring small-scale features that impact the coarse-grid solution in an appropriate way. Mathematical-oriented multiscale methods are often geared toward solving partial differential equations with rapidly varying coefficients. These methods try to incorporate subgrid information by utilising solutions of local flow problems to build a set of equations on a coarser scale. This is similar to the approach of flow-based upscaling methods, but the multiscale methods are different in the sense that they ensure consistency of the coarse-scale equations with the underlying differential operator. In practice this means that the solution of the coarse-scale system is consistent with the differential operator also on the finer scale on which the local flow problems are solved, and therefore gives an accurate solution on this scale as well.

To accomplish this, the multiscale methods may express the fine-scale solution as a linear combination of basis functions, where the basis functions are solutions to local flow problems. Here we describe and discuss two methods

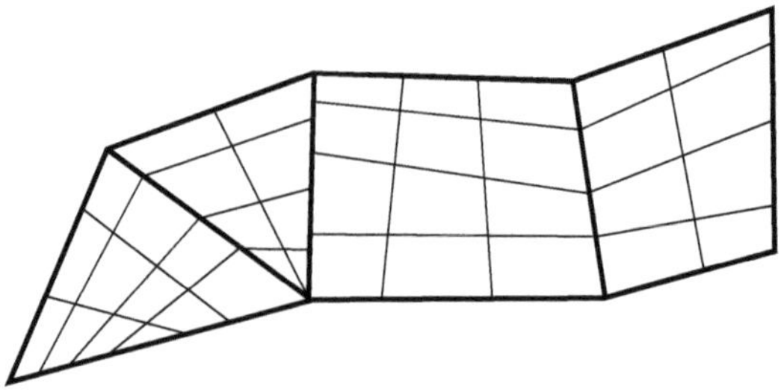

Fig. 10. Schematic reservoir model with a coarse grid superimposed on the fine grid.

of this type that are designed for solving pressure equations arising in reservoir simulation. The methods we shall consider are the the multiscale finite-volume method (MsFVM) [42, 43], and the multiscale mixed finite-element method (MsMFEM) [18, 1]. Both methods generate mass-conservative velocity fields on the subgrid scale, while solving only a coarse-scale system globally. For a more thorough discussion and comparison of various multiscale methods, we refer the reader to [49].

The starting point for both methods is a subdivision of the reservoir into polyhedral grid-blocks, with a coarse grid superimposed on the fine grid, e.g., as shown in Figure 10. Usually each coarse grid-block is a connected union of fine grid-blocks, but the coarse grid may be any connected non-overlapping partition of the reservoir, provided quantities may be properly mapped from the underlying fine grid. We shall assume that the reservoir data ($\mathbf{K}$ and λ) are given on the fine grid, and since this grid must be suitable for numerical discretisation of (23) using standard numerical techniques, it can not be completely arbitrary. However, in the multiscale methods we use the fine grid only to solve local problems. The properties that enter the coarse-scale equations are extracted from these local solutions alone, hence the multiscale formulations are in a sense independent of coarse grid geometry.

As a consequence of this, we can allow very general coarse grids. For instance, we can allow grids that are locally structured, but globally unstructured. This convenient feature allows us to use the subgrid to model physical features such as fractures and faults, or to tune the grid geometry to advanced well architectures. Moreover, few restrictions are placed on the geometry of each coarse grid-block. This is particularly true for the MsMFEM, which has demonstrated accurate results with almost arbitrarily shaped grid-blocks [4, 5]. Such a flexibility makes multiscale methods amenable for solving problems that demand unstructured grids, or grids with high grid-block aspect ratios.

Before describing each of the methods in more detail, we specify the model equations we will work with. For simplicity, we consider the elliptic pressure equation only, i.e., we assume incompressible flow. The methods can be extended to solve the parabolic equation arising from compressible flow using, for instance, an implicit Euler or the Crank-Nicholson method to discretise the time derivative. Moreover, we will disregard gravity effects, since these are straightforward to include, but complicate notation. After these simplifications, the pressure equation for two-phase flow (4) reads

$$v = -\mathbf{K}\lambda\nabla p, \qquad \nabla \cdot v = q, \tag{23}$$

and we assume this is to be solved in a domain Ω subject to no-flow boundary conditions $v \cdot n = 0$ on $\partial\Omega$.

4.1 The Multiscale Finite-Volume Method

The MsFVM, introduced by Jenny et al. [42, 43], is essentially a finite-volume scheme that extracts a mass-conservative velocity field on a fine scale from a coarse-grid solution. The method employs numerical subgrid calculations (analogous to those in [40]) to derive a multi-point finite-volume stencil for solving (23) on a coarse grid. The method then proceeds and reconstructs a mass-conservative velocity field on a fine grid as a superposition of local subgrid solutions, where the weights are obtained from the coarse-grid solution.

Computation of the Coarse-Scale Solution

The derivation of the coarse-scale equations in the MsFVM is essentially an upscaling procedure for generating coarse-scale transmissibilities. The first step is to solve a set of homogeneous boundary-value problems of the form

$$-\nabla \cdot \mathbf{K}\lambda\nabla\phi_i^k = 0, \quad \text{in } R, \qquad \phi_i^k = \nu_i^k, \quad \text{on } \partial R, \tag{24}$$

where R are so-called interaction regions as illustrated in Figure 11 and ν_i^k are boundary conditions to be specified below. The subscript i in ϕ_i^k denotes a corner point in the coarse grid (x_i in the figure) and the superscript k runs over all corner points of the interaction region (x^k in the figure). Thus, for each interaction region associated with e.g., a hexahedral grid in three dimensions, we have to solve a total of eight local boundary-value problems of the form (24). The idea behind the MsFVM is to express the global pressure as a superposition of these local pressure solutions ϕ_i^k. Thus, inside each interaction region R, one assumes that the pressure is a superposition of the local subgrid solutions $\{\phi_i^k\}$, where k ranges over all corner points in the interaction region (i.e., over the cell centres of the coarse grid-blocks).

We define now the boundary conditions ν_i^k in (24). At the corner points of the interaction region, the boundary condition ν_i^k satisfies $\nu_i^k(x^l) = \delta_{kl}$, where δ_{kl} is the Kronecker delta function. Assuming we are in three dimensions, the corner-point values $\nu_i^k(x^l)$ are extended to the edges of the interaction region by linear interpolation. To specify the boundary conditions on each face F of the interaction region, one solves a two-dimensional boundary problem of the form

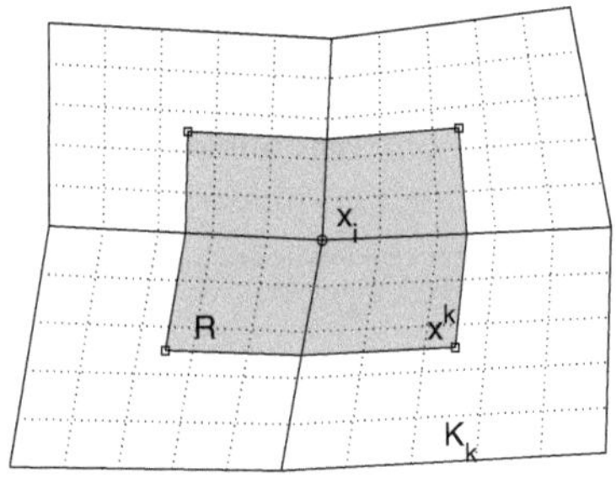

Fig. 11. The shaded region represents the interaction region R for the MsFVM, where x_i denotes corner points and x^k the midpoints of the coarse grid-blocks K_k. The midpoints x^k are the corner points of the interaction region R.

$$-\nabla \cdot \mathbf{K}\lambda\nabla\nu_i^k = 0 \quad \text{in } F, \tag{25}$$

with the prescribed piecewise-linear boundary conditions on the boundary of F (the edges of the interaction region that connect the corner points). In two dimensions it is not necessary to interpolate between corner points, and the boundary conditions $\nu_i^k(x)$ are obtained by solving (25) subject to the corner point values along the edges of the interaction region.

Having computed the local pressure solutions ϕ_i^k, we now show how to obtain a global solution as a superposition of the local subgrid solutions. To this end, we observe that the cell centres x^k constitute a corner point for eight interaction regions (in a regular hexahedral grid). Moreover, for all corner points x_i of the coarse grid, the corresponding boundary conditions ν_i^k for the different pressure equations coincide on the respective faces of the interaction regions that share the corner point x^k. This implies that the base function

$$\phi^k = \sum_i \phi_i^k \tag{26}$$

is continuous (in a discrete sense). In the following construction, the base functions defined in (26) will serve as building blocks that are used to construct a global "continuous" pressure solution.

Thus, define now the approximation space $V^{\text{ms}} = \text{span}\{\phi^k\}$ and observe that all base functions vanish at all but one of the grid-block centres x^k. This implies that, given a set of pressure values $\{p^k\}$, there exists a unique extension $\{p^k\} \to p \in V^{\text{ms}}$ with $p(x^k) = p^k$. This extension is defined by

$$p = \sum_k p^k \phi^k = \sum_{i,k} p^k \phi_i^k. \tag{27}$$

We derive now a multi-point finite-volume stencil by assembling the flux contribution across the grid-block boundaries from each base function. Thus, let

$$f_{k,l} = -\int_{\partial K_l} n \cdot \mathbf{K}\lambda\nabla\phi^k \, ds$$

be the local flux out of grid-block K_l induced by ϕ^k. The MsFVM for solving (23) then seeks constant grid-block pressures $\{p^k\}$ satisfying

$$\sum_k p^k f_{k,l} = \int_{K_l} q\,dx \qquad \forall l. \tag{28}$$

We now explain how to reconstruct a mass-conservative velocity field on the underlying subgrid from the coarse-grid solution.

Reconstruction of a Conservative Fine-Scale Velocity Field

To reconstruct a mass-conservative velocity field on a fine scale, notice first that the expansion (27) produces a mass-conservative velocity field on the coarse scale. Unfortunately, this velocity field will not preserve mass across the boundaries of the interaction regions. Thus, to obtain a velocity field that also preserves mass inside the coarse grid-blocks, one solves

$$v_l = -\mathbf{K}\lambda\nabla p_l, \quad \nabla\cdot v_l = \frac{1}{|K_l|}\int_{K_l} q\;dx \quad \text{in } K_l, \tag{29}$$

with boundary conditions obtained from (27), i.e.,

$$v_l = -\mathbf{K}\lambda\nabla p \quad \text{on } \partial K_l, \tag{30}$$

where p is the expanded pressure defined by (27).

Hence, in this approach the subgrid fluxes across the coarse-grid interfaces that we obtain from p are used as boundary conditions for a new set of local subgrid problems. If these subgrid problems are solved with a conservative scheme, e.g., a suitable finite-volume method, then the global velocity field $v = \sum_{K_l} v_l$ will be mass conservative. Moreover, since the fluxes correspond to interfaces that lie in the interior of corresponding interaction regions, it follows that the boundary condition (30) is well defined. Note, however, that since the subgrid problems (29)–(30) are solved independently, we loose continuity of the global pressure solution, which is now defined by $p = \sum_{K_l} p_l$.

It is possible to write the new global fine-scale solutions p and v as linear superpositions of (a different set of) local base functions, see [42] for details. This may be convenient when simulating two-phase flows with low mobility ratios. This is because low mobility ratios lead to a slowly varying mobility field. As a result, it is sometimes possible to achieve high accuracy with a unique set of base functions. Thus, by representing p and v as a linear superposition of local base functions, solving the subgrid problems (24)–(25) and (29)–(30) becomes part of an initial preprocessing step only.

It is worth noting that the present form of the MsFVM, which was developed by Jenny et al. [42], does not model wells at the subgrid scale. Indeed, the source term in (29) is equally distributed within the grid-block. Thus, in order to use the induced velocity field to simulate the phase transport, one has to treat the wells as a uniform source within the entire well block. However, to get a more detailed representation of flow around wells one needs only to replace (29) in grid blocks containing a well with the following equation:

$$v_l = -\mathbf{K}\lambda\nabla p_l, \qquad \nabla \cdot v_l = q \quad \text{in } K_l.$$

This completes the description of the MsFVM. The next section is devoted to a multiscale method with similar properties that is based on a so-called mixed formulation of the elliptic pressure equation (23); see [2].

4.2 A Multiscale Mixed Finite-Element Method

For finite-volume methods and finite-element methods the velocities are derived quantities of the unknown pressures, computed using some kind of numerical differentiation. In this section we consider a method that obtains the velocity directly. The underlying idea is to consider both the pressure and the velocity as unknowns and express them in terms of basis functions. The corresponding method is called a mixed finite-element method. In the lowest-order mixed method one uses piecewise-constant scalar basis functions for the pressure and piecewise-linear basis functions for the velocity (this is usually referred to as the Raviart–Thomas basis functions, see [56]).

In mixed finite-element discretisations of elliptic equations on the form (23), one seeks a solution (p, v) to the mixed equations

$$\int_\Omega u \cdot (\mathbf{K}\lambda)^{-1} v \, dx - \int_\Omega p \, \nabla \cdot u \, dx = 0, \tag{31}$$

$$\int_\Omega l \, \nabla \cdot v \, dx = \int_\Omega ql \, dx, \tag{32}$$

in a finite-dimensional product space $U \times V \subset L^2(\Omega) \times H_0^{1,\text{div}}(\Omega)$. If the subspaces $U \subset L^2(\Omega)$ and $V \subset H_0^{1,\text{div}}(\Omega)$ are properly balanced (see, e.g., [13, 14, 15]), then p and v are defined (up to an additive constant for p) by requiring that (31)–(32) holds for all $(l, u) \in U \times V$.

In MsMFEMs one constructs a special approximation space for the velocity v that reflects the important subgrid information. For instance, instead of seeking velocities in a simple approximation space spanned by base functions with linear components, as in the Raviart–Thomas method, one computes special multiscale base functions ψ in a manner analogous to the MsFVM, and defines a corresponding multiscale approximation space by $V^{\text{ms}} = \text{span}\{\psi\}$. An approximation space for the pressure p that reflects subgrid structures can be defined in a similar manner. However, there are reasons for not doing so.

First, the pressure equation (23) models incompressible flow. This means that the pressure is never used explicitly in the flow simulation, except possibly to determine well-rates through the use of an appropriate well-model. It is therefore often sufficient to model pressure on the coarse scale as long

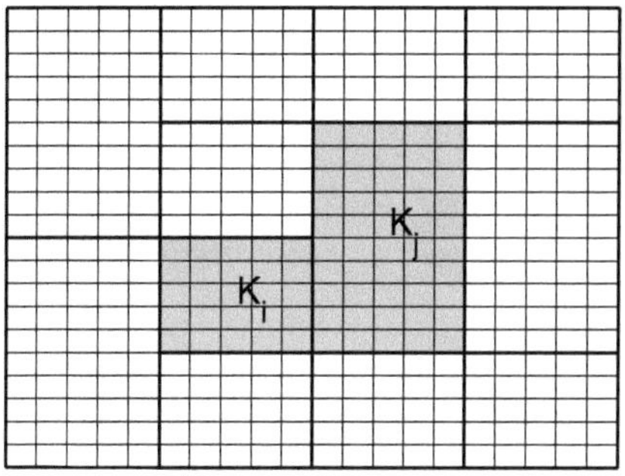

Fig. 12. Schematic of the coarse and fine grid for the MsMFEM. The shaded region denotes the support of the velocity base function associated with the edge between the two cells K_i and K_j.

as we can obtain a detailed velocity field without knowing how the pressure behaves at the subgrid scale. This is one of the nice features with mixed finite-element methods. Indeed, by treating the pressure and velocities as separate decoupled variables, we gain some freedom in terms of choosing the resolution of the two approximation spaces U and V more independently. In other words, the computational effort can be spent where it is most needed. On the other hand, the approximation spaces can not be chosen arbitrarily. Indeed, the convergence theory for mixed finite element methods, the so-called *Ladyshenskaja–Babuška–Brezzi* theory (see [13, 14, 15]) states that the approximation spaces must satisfy a relation called the *inf-sup* condition, or the LBB condition. By defining a multiscale approximation space also for the pressure variable, it can be quite difficult to show that this relation holds. Therefore, in the following we assume, unless stated otherwise, that U is the space of piecewise constant functions, i.e.,

$$U = \{p \in L^2(\Omega) : p|_K \text{ is constant for all } K \in \mathcal{K}\}.$$

Observe that this space is spanned by the characteristic functions with respect to the coarse grid-blocks: $U = \text{span}\{\chi_m : K_m \subset \mathcal{K}\}$ where

$$\chi_m(x) = \begin{cases} 1 & \text{if } x \in K_m, \\ 0 & \text{else.} \end{cases}$$

Next, we We define the approximation space V for the velocity.

Approximation Space for the Velocity

Consider a coarse grid that overlays a fine (sub)grid, for instance as illustrated in Figure 12. For the velocity we associate one vector of base functions with each non-degenerate interface γ_{ij} between two neighbouring grid-blocks K_i and K_j. To be precise, for each interface γ_{ij} we define a base function ψ_{ij} by

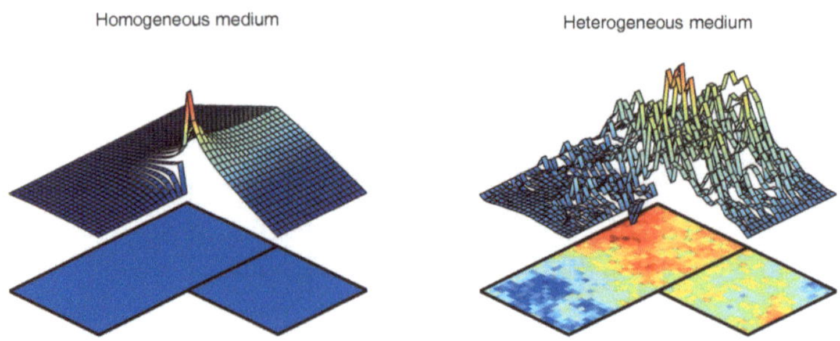

Fig. 13. The figure depicts the x-component of two MsMFEM base functions for an interface between two rectangular (two-dimensional) grid-blocks that are not penetrated by a well. The base function to the left corresponds to homogeneous permeability, whereas the base function to the right corresponds to random coefficients.

$$\psi_{ij} = -\mathbf{K}\lambda\nabla\phi_{ij}, \qquad \text{in } K_i \cup K_j, \tag{33}$$

where ϕ_{ij} is determined by

$$(\nabla \cdot \psi_{ij})|_{K_i} = \ell(x) / \int_{K_i} \ell(x)\, dx, \tag{34}$$

$$(\nabla \cdot \psi_{ij})|_{K_j} = -\ell(x) / \int_{K_j} \ell(x)\, dx. \tag{35}$$

with no-flow boundary conditions along the edges $\partial K_i \cup \partial K_j \backslash \gamma_{ij}$.

The function $\ell(x)$ can be defined in various ways. One option that was proposed by Chen and Hou [18] is to simply choose $\ell(x) = 1$. This option produces mass-conservative velocity fields on the coarse scale, but the subgrid velocity field is mass conservative only if we treat the wells as a uniform source term within grid-blocks containing wells. To model flow around wells at the subgrid scale, one can for instance choose $\ell(x) = q(x)$ in coarse blocks penetrated by an injection or production well. Another possibility is to choose $\ell(x) = 1$ and then to reconstruct a velocity field that models near-well flow patterns from the coarse grid solution. This final option is analogous to the way one models flow around wells with the MsFVM.

We are here primarily interested in getting a velocity field that can be used to model flows in heterogeneous porous media on a subgrid scale. To this end we must solve the subgrid problems (33)–(35) with a conservative scheme, for instance the Raviart–Thomas mixed finite-element method. Moreover, it is generally acknowledged that it is important to model the flow around wells correctly to obtain reliable production forecasts from reservoir simulations. In the current MsMFEM we have the possibility to do this in a rigorous manner by incorporating near-well flow patterns in the base functions. We will therefore choose $\ell(x) = 1$ away from the wells and $\ell(x) = q(x)$ in grid blocks penetrated by wells. As discussed in [1, 6], it is not necessary to change the approximation space for p even if we change $l(x)$ in blocks penetrated by wells.

Figure 13 shows the x-component of the velocity base functions for the MsMFEM obtained for two cases with homogeneous and random permeability, respectively. We see that the base function corresponding to random coefficients fluctuates rapidly and hence reflects the fine-scale heterogeneous structures. Note also that the base functions ψ_{ij} will generally be time dependent since they depend on λ, which is time dependent through $s_w(x,t)$. This indicates that one has to regenerate the base functions for each time step. However, for nearly incompressible fluids, the total mobility λ usually varies significantly only in the vicinity of propagating saturation fronts. It is therefore usually sufficient to regenerate a small portion of the base functions at each time step, see [1]. Similar observations have been made for the MsFVM, see [42].

Computing the Coarse Scale Solution

In the MsFVM the only information from the subgrid calculations that was explicitly used in the coarse grid equations was the flux contribution across the coarse grid interfaces from the respective base functions. This implies that a lot of potentially valuable information is disregarded in the coarse scale equations. For the MsMFEM, all information (apart from the subgrid pressure solutions) is exploited and enters directly into the coarse-scale system. Indeed, the MsMFEM seeks

$$p \in U,\ v \in V^{ms} \quad \text{such that (31)–(32) holds for} \quad \forall l \in U,\ \forall u \in V^{ms}.$$

This leads to a linear system of the form

$$\begin{bmatrix} \mathbf{B} & -\mathbf{C}^T \\ \mathbf{C} & \mathbf{0} \end{bmatrix} \begin{bmatrix} \mathbf{v} \\ \mathbf{p} \end{bmatrix} = \begin{bmatrix} \mathbf{0} \\ \mathbf{f} \end{bmatrix}, \tag{36}$$

where

$$\begin{aligned} \mathbf{B} &= [\int_{\Omega} \psi_{ij} \cdot (\lambda \mathbf{K})^{-1} \psi_{kl}\, dx], \\ \mathbf{C} &= [\int_{K_m} \operatorname{div}(\psi_{kl})\, dx], \\ \mathbf{f} &= [\int_{K_m} f\, dx]. \end{aligned}$$

The linear system (36) is indefinite, which generally makes it harder to solve than the symmetric, positive definite systems arising from, e.g., the two-point flux approximation (6). This is not particular to the multiscale version of the mixed finite-element method, but a well-known characteristic of mixed finite-element methods in general. Fortunately it is possible to reformulate the discrete form of the mixed equations (31)–(32) to eventually obtain a symmetric, positive definite linear system by applying a technique usually referred to as hybridisation [15].

The idea is to first remove the constraint that the normal velocity must be continuous across block interfaces and integrate (23) to get a weak form containing jump terms at block boundaries. Continuity of the normal component is then reintroduced by adding an extra set of equations, where the pressure at the interfaces plays the role of Lagrange multipliers. This procedure does not change the velocity solution, nor the grid cell pressures, but enables the recovery of pressure values at the cell interfaces, in addition to inducing the desired change in structure of the linear system.

In our multiscale setting, the mixed-hybrid problem is to find

$$(p, v, \pi) \in (U \times \tilde{V}^{ms} \times \Pi),$$

where

$$U \subset L^2(\Omega), \quad \tilde{V}^{ms} \subset (L^2(\Omega))^2, \quad \text{and } \Pi \subset L^2(\{\partial K\}),$$

such that

$$\int_\Omega u \cdot (\mathbf{K}\lambda)^{-1} v \, dx - \int_\Omega p \, \nabla \cdot u \, dx + \sum_{\kappa \in \{\partial K\}} \int_\kappa [u \cdot n] \, \pi \, ds = 0, \tag{37}$$

$$\int_\Omega l \, \nabla \cdot v \, dx = \int_\Omega ql \, dx, \tag{38}$$

$$\sum_{\kappa \in \{\partial K\}} \int_\kappa [v \cdot n] \, \mu \, ds = 0, \tag{39}$$

holds for all $(l, u, \mu) \in (U \times \tilde{V}^{ms} \times \Pi)$. The square brackets denote the jump in a discontinuous quantity. The new approximation space for the Darcy velocity is still defined by (34)–(35), but since $\tilde{V}^{ms}$ is not required to be a subspace of $H_0^{1,\text{div}}$, the matrices $\mathbf{B}$ and $\mathbf{C}$ in the resulting linear system

$$\begin{bmatrix} \mathbf{B} & -\mathbf{C}^T & \mathbf{\Pi}^T \\ \mathbf{C} & \mathbf{0} & \mathbf{0} \\ \mathbf{\Pi} & \mathbf{0} & \mathbf{0} \end{bmatrix} \begin{bmatrix} \mathbf{v} \\ \mathbf{p} \\ \pi \end{bmatrix} = \begin{bmatrix} \mathbf{0} \\ \mathbf{f} \\ \mathbf{0} \end{bmatrix}, \tag{40}$$

will have a block-diagonal structure. The new matrix $\mathbf{\Pi}$ is defined by

$$\mathbf{\Pi} = \Big[\int_{\partial K_m} \psi_{kl} \cdot n \, ds\Big],$$

and block reduction of the system (40) yields

$$\mathbf{M}\pi = \tilde{\mathbf{f}}, \tag{41}$$

where

$$\mathbf{M} = \mathbf{S}(\mathbf{C}^T\mathbf{D}^{-1}\mathbf{C} - \mathbf{B})\mathbf{S}^T,$$
$$\tilde{\mathbf{f}} = \mathbf{S}\mathbf{C}^T\mathbf{D}^{-1}\mathbf{f},$$
$$\mathbf{S} = \mathbf{\Pi}\mathbf{B}^{-1}, \qquad \text{and} \qquad \mathbf{D} = \mathbf{C}\mathbf{B}^{-1}\mathbf{C}^T.$$

Here $\mathbf{B}$ is block-diagonal, so $\mathbf{B}^{-1}$ can be computed block-by-block. Moreover, $\mathbf{D}$ is diagonal[4] and can therefore easily be inverted, and $\mathbf{M}$ is symmetric positive definite (see, e.g., [15]). Hence, $\mathbf{M}$ can be computed explicitly, and the linear system can be solved using one of the many efficient methods specialised for symmetric positive-definite linear systems.

4.3 Examples

Both the MsMFEM and the MsFVM solve a coarse-scale equation globally while trying to resolve fine-scale variations by using special multiscale basis functions. The basis functions are local solutions of the elliptic equation, and the therefore overall accuracy only depends weakly on the coarse grid size. We shall demonstrate this with an example.

Example 3. Consider a horizontal, two-dimensional reservoir with fluvial heterogeneity. An injection well is located in the centre and a producer is located in each of the four corners. The permeability on the 220×60 fine grid is taken from the bottom layer of Model 2 in the Tenth SPE Comparative Solution Project [19]. The dataset has many narrow and intertwined high-permeability channels that greatly impact the flow pattern. We solve the pressure equation using both the MsFVM and the MsMFEM with various coarse grid dimensions, and employ an upstream finite-volume method for the saturation equation on the underlying fine grid. A reference solution is computed using a two-point flux approximation scheme on a grid that is refined four times in each direction. Figure 14 shows the resulting saturation fields at dimensionless time $t = 0.3$PVI. The solutions are quite difficult to distinguish visually. We therefore measure errors in the saturation fields by

$$\delta(S) = \frac{\epsilon(S)}{\epsilon(S_{\text{ref}})}, \qquad \epsilon(S) = \frac{\|S - \mathcal{I}(S_{\text{ref}}^{4\times})\|_{L^1}}{\|\mathcal{I}(S_{\text{ref}}^{4\times})\|_{L^1}}.$$

Here the operator $\mathcal{I}(\cdot)$ represents mapping from the refined to the original grid. The results displayed in Table 3 verify that the accuracy is indeed relatively insensitive to coarse grid size, although there is a slight degradation of solution quality as the grid is coarsened.

---

[4]This may be seen by observing that $\mathbf{X}\mathbf{Y}^{\mathbf{T}}$ will be diagonal for any two matrices $\mathbf{X}$ and $\mathbf{Y}$ having the sparsity structure of $\mathbf{C}$, . The result then follows from the fact that multiplication by $\mathbf{B}^{-1}$ does not change the sparsity structure of $\mathbf{C}$.

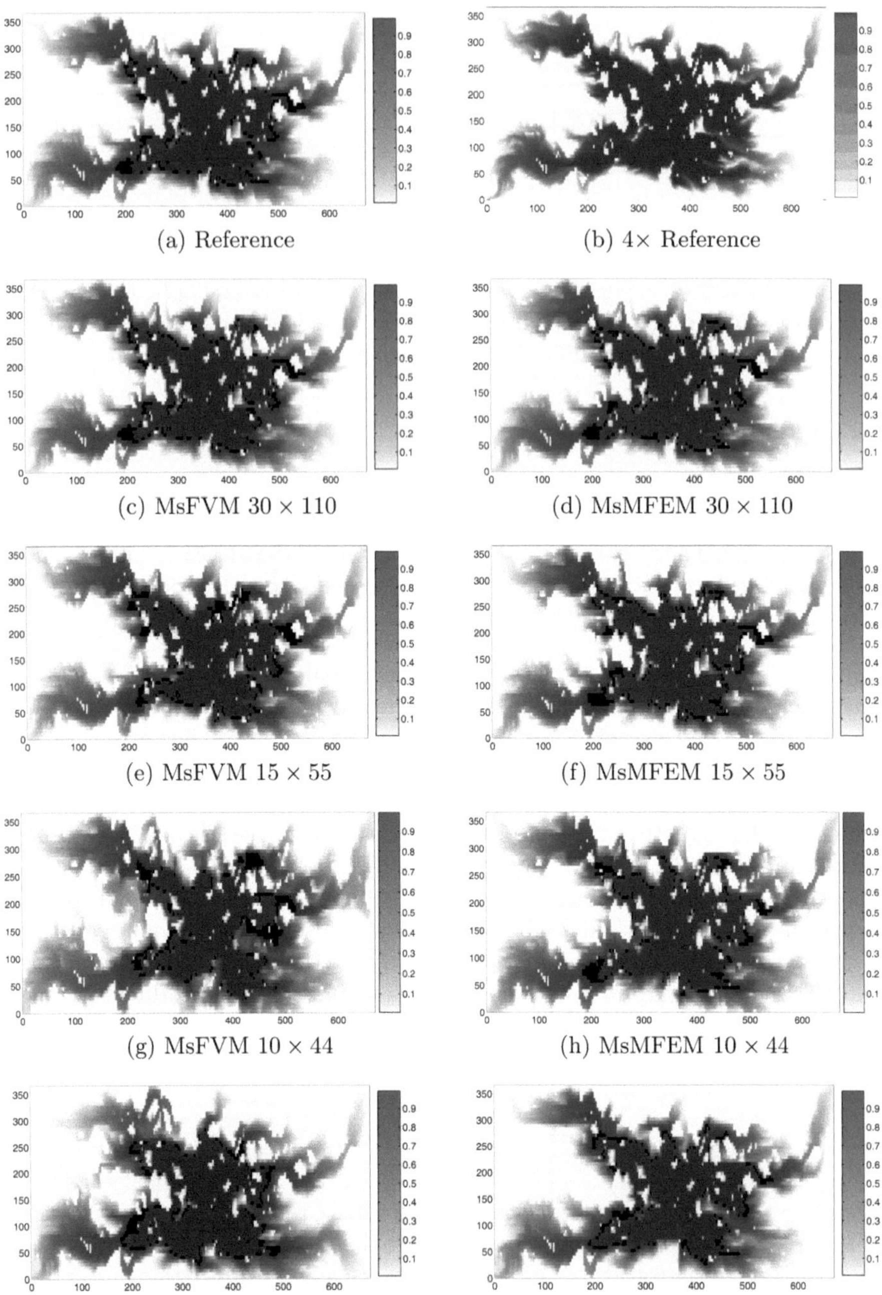

(a) Reference

(b) 4× Reference

(c) MsFVM 30 × 110

(d) MsMFEM 30 × 110

(e) MsFVM 15 × 55

(f) MsMFEM 15 × 55

(g) MsFVM 10 × 44

(h) MsMFEM 10 × 44

(i) MsFVM 5 × 11

(j) MsMFEM 5 × 11

Fig. 14. MsMFEM and MsFVM solutions for various coarse grids on the bottom layer of the SPE 10 test case. The two figures at the top show the reference solution on the original fine and the 4× refined grids.

Table 3. $\delta(S)$ for various coarse grids.

	30×110	15×55	10×44	5×11
MsMFEM	1.0916	1.2957	1.6415	1.9177
MsFVM	1.0287	1.6176	2.4224	3.0583

Boundary Conditions

The accuracy of the multiscale methods is largely determined by the boundary conditions for the local problems. In fact, it can be shown that by using properly scaled restrictions of a global fine-scale solution as boundary conditions for the basis functions, it is possible to replicate the fine-scale solution by solving the pressure equation on the coarse grid with the MsMFEM [1, 4] or the MsFVM [29]. For simulations with many time steps, computation of an initial fine-scale solution may be justified. Indeed, if the pressure field is reasonably stable throughout the simulation, it is possible to exploit the initial solution to obtain proper boundary conditions at subsequent time-steps [1, 43]. We usually refer to such boundary conditions as *global* boundary conditions, and they have been shown to give very accurate results, e.g., for the SPE10 model [3, 43].

Example 4. We now demonstrate in Table 4 that the multiscale methods with global boundary conditions are capable of replicating a fine-scale solution. Since we here use the TPFA to compute the fine-grid solution, we have $\delta(S) \equiv 1$ only for the MsFVM. If the lowest-order Raviart–Thomas mixed finite-element method had been used to compute the fine-grid solution and basis functions for the MsMFEM, we would get $\delta(S) \equiv 1$ for the MsMFEM instead.

Table 4. $\delta(S)$ for various coarse grids, boundary conditions determined using a fine-scale solution.

	30×110	15×55	10×44	5×11
MsMFEM	0.9936	0.9978	0.9938	0.9915
MsFVM	1.0000	1.0000	1.0000	1.0000

Multiscale Methods used as Upscaling Methods

Although the multiscale methods provide velocities on whatever we decide to use as the fine-grid, they can also be utilised as upscaling methods. Indeed, the velocities can easily be integrated along grid lines of an arbitrary coarse grid. The multiscale methods can therefore be viewed as seamless upscaling methods for doing fluid-transport simulations on user-defined coarse grids.

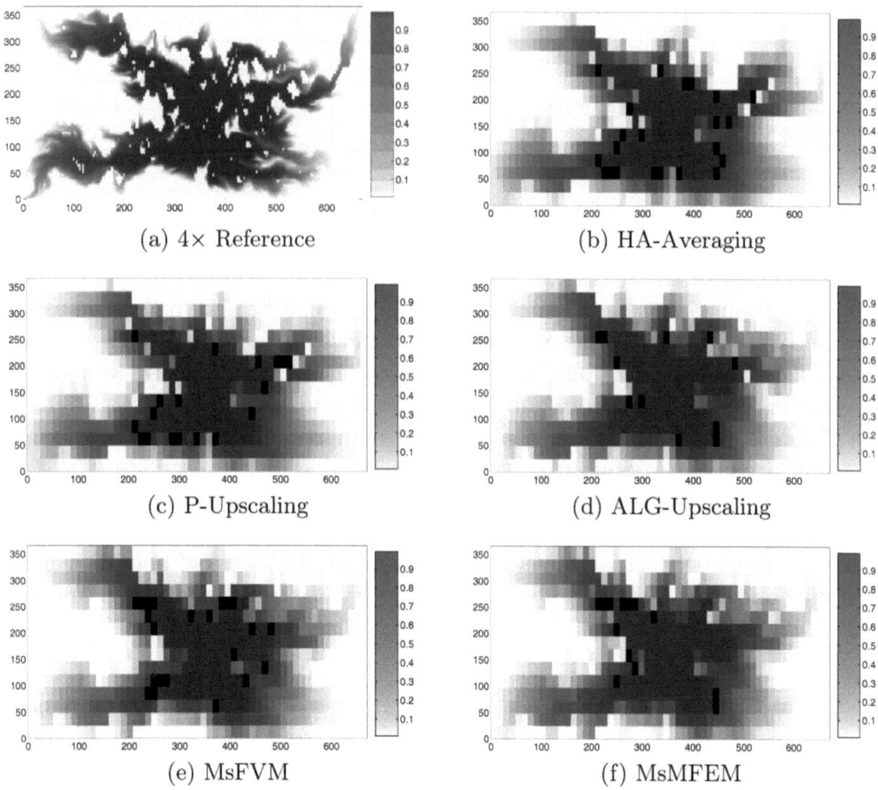

(a) 4× Reference
(b) HA-Averaging
(c) P-Upscaling
(d) ALG-Upscaling
(e) MsFVM
(f) MsMFEM

Fig. 15. Saturation profiles on a 15×55 grid obtained by the various upscaling and multiscale methods.

Next we show that the multiscale methods give comparable accuracy to standard upscaling techniques on fixed coarse grids if the saturation equation is solved on the coarse grid.

Example 5. To validate the multiscale methods as upscaling methods, we again use the data from the bottom layer of the SPE10 model and solve the saturation equation on the same coarse grid that is used in the multiscale methods. For comparison, we also compute solutions obtained by using some of the upscaling methods described in Section 3. Figure 15 shows saturations for the 15×55 coarse grid. Table 5 shows the corresponding error measure $\delta_c(S)$, where the subscript c indicates that errors are computed on the coarse (upscaled) grid. We see that the multiscale methods perform slightly better than the upscaling methods. A more substantial improvement of saturation accuracy cannot be expected, since the listed upscaling methods for the pressure equation are quite robust, and generally perform well.

Table 5. $\delta_c(S)$ for various coarse grids.

	30×110	15×55	10×44	5×11
MsMFEM	0.9891	0.9708	0.9556	0.9229
MsFVM	0.9932	0.9675	1.0351	1.1439
ALG-Upscaling	0.9813	0.9826	1.0223	1.0732
P-Upscaling	1.0313	1.0216	1.1328	1.2371
HA-Averaging	1.0371	1.0654	1.2300	1.8985

Table 6. $\delta_c(S)$ for various coarse grids, saturations computed on the original grid.

	30×110	15×55	10×44	5×11
MsMFEM	1.0775	1.2849	1.8564	2.9680
MsFVM	1.0190	1.7140	2.8463	5.9989
ALG-Upscaling	1.1403	1.5520	2.3834	4.4021
P-Upscaling	1.6311	2.6653	4.1277	6.5347
HA-Upscaling	1.6477	2.9656	4.8572	15.3181

We now reverse the situation from Example 5 and compute saturations on the fine grid also for the upscaling methods. This requires that we downscale the velocity field by reconstructing a conservative fine-scale velocity field from a coarse-scale solution. For this we employ the procedure described in [33]. After having solved the saturation equation on the fine grid, we map the saturations back to the upscaled grid and compute the errors there. From the results in Table 6 we see that the multiscale methods give much better results than simple averaging and purely local upscaling. The local-global method, on the other hand, gives good accuracy, but computationally it is also much more expensive than the multiscale methods.

4.4 Concluding Remarks

As we have seen, the multiscale methods are capable of producing accurate solutions on both the fine and coarse scales, and may therefore be utilised as either robust upscaling methods or efficient approximate fine-scale solution methods. Having fine-scale velocities available at a relatively low computational cost gives great flexibility in the choice of solution method for the saturation equation. In light of the limited success of upscaling methods for the saturation equation, it seems beneficial to perform the transport on the finest grid affordable, and a fine-scale velocity field may aid in choosing such a grid as well as ensure the accuracy of the resulting solution.

Although we have shown only very simple examples, we would like to emphasise the advantage multiscale methods have in their flexibility with respect to grids. In particular this is true for the MsMFEM, since it avoids the dual-grid concept. For the MsMFEM each coarse-grid block is simply a connected union of fine-grid blocks, which makes the method virtually grid-independent

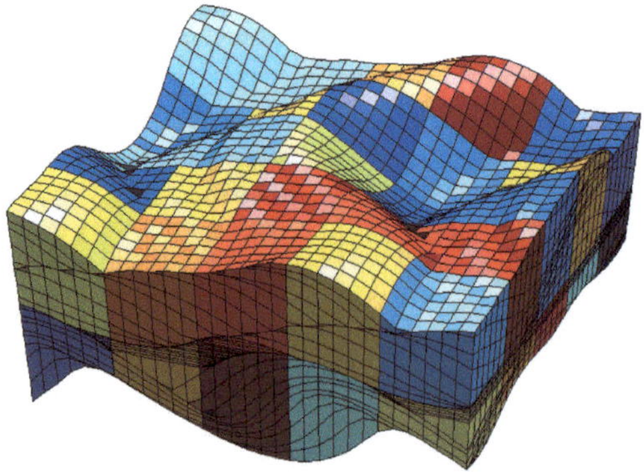

Fig. 16. The fine-grid is a corner-point grid, which is the industry standard for reservoir modeling and simulation. The coarse-grid blocks are given by different colours.

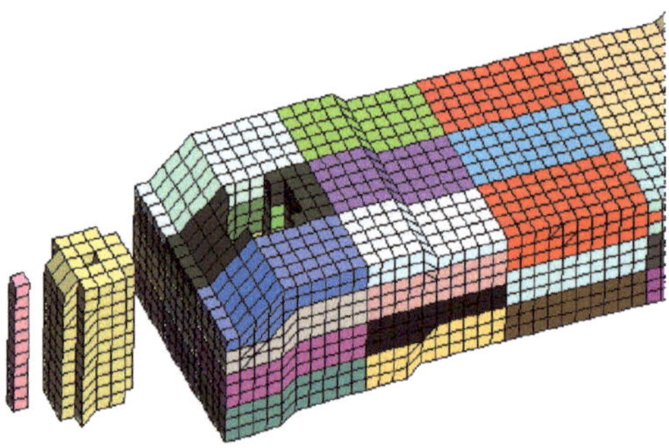

Fig. 17. The coarse grid is tuned to model a well and the near-well region. Away from the well regions the coarse-grid blocks are aligned with the physical layers.

given a fine-scale solver. Therefore it is straightforward to perform multiscale simulations on models with rather complicated fine-grids, such as the one displayed in Figure 16; see [5] for more details.

Furthermore, since the only real constraint on the geometry of coarse-grid blocks seems to be that it must allow computation of the local solutions, we may tune the coarse-grid to model particular reservoir features or wells without worrying about grid-block shapes. Figure 17 shows an example where the coarse-grid blocks are aligned with the physical layers. A vertical well is modelled by a narrow block, and the near-well region is modelled by another block with a hole. Even such exotic grid-blocks are unproblematic for the MsMFEM, and the multiscale simulations yield accurate results.

Finally, it is worth mentioning that the multiscale methods are flexible not only with regard to grids, but also with regard to fine-grid solution methods. The fine-scale numerical method may be chosen independently for each of the local problems, and thus a higher-order method or local grid-refinement, may be employed for higher accuracy in regions of interest. Such approaches will not significantly impact the efficiency of the methods, since multiscale efficiency mainly comes from adaptively updating only a small portion of the local solutions. The multiscale methods may therefore have the potential to combine accuracy and efficiency in an ideal way.

Acknowledgement. This research was funded in part by the Research Council of Norway under grants no. 158908/I30 and no. 152732/S30.

References

1. J.E. Aarnes. On the use of a mixed multiscale finite element method for greater flexibility and increased speed or improved accuracy in reservoir simulation. *Multiscale Model. Simul.*, 2(3):421–439, 2004.
2. J.E. Aarnes, T. Gimse, and K.-A. Lie. An introduction to the numerics of flow in porous media using Matlab. In *this book.*
3. J.E. Aarnes, V. Kippe, and K.-A. Lie. Mixed multiscale finite elements and streamline methods for reservoir simulation of large geomodels. *Adv. Water Resour.*, 28(3):257–271, 2005.
4. J. E. Aarnes, S. Krogstad, and K.-A. Lie. A hierarchical multiscale method for two-phase flow based upon mixed finite elements and nonuniform coarse grids. *Multiscale Model. Simul.*, 5(2):337–363, 2006.
5. J. E. Aarnes, S. Krogstad, and K.-A. Lie. Multiscale mixed/mimetic methods on corner-point grids. *Comput. Geosci.*, submitted.
6. J.E. Aarnes and K.-A. Lie. Toward reservoir simulation on geological grid models. In *Proceedings of the 9th European Conference on the Mathematics of Oil Recovery.* EAGE, Cannes, France, 2004.
7. O. J. Arntzen. Higher-order terms in the capillary limit approximation of viscous-capillary flow. *Transp. Porous Media*, 57:17–34, 2004.
8. J.W. Barker and P. Dupouy. An analysis of dynamic pseudo relative permeability methods. In *Proceedings of the 5th ECMOR conference, Leoben, Austria.* 1996.
9. J.W. Barker and F.J. Fayers. Transport coefficients for compositional simulations with coarse grids in heterogeneous media. *SPE Adv. Tech Series*, 2(2):103–112, 1994.
10. J.W. Barker and S. Thibeau. A critical review of the use of pseudorelative permeabilities for upscaling. *SPE Reservoir Eng.*, 12(2):138–143, 1997.
11. S.H. Begg, R.R. Carter, and P. Dranfield. Assigning effective values to simulator gridblock parameters for heterogeneous reservoirs. *SPE Reservoir Eng.*, 4(4):455–463, 1989.
12. A. Benesoussan, J.-L. Lions, and G. Papanicolaou. *Asymptotic Analysis for Periodic Structures.* Elsevier Science Publishers, Amsterdam, 1978.

13. D. Braess. *Finite elements: Theory fast solvers and applications in solid mechanics.* Cambridge University Press, Cambridge, 1997.
14. S.C. Brenner and L.R. Scott. *The mathematical theory of finite element methods.* Springer–Verlag, New York, 1994.
15. F. Brezzi and M. Fortin. *Mixed and hybrid finite element methods.* Computational Mathematics. Springer–Verlag, New York, 1991.
16. Y. Chen and L.J. Durlofsky. Adaptive local-global upscaling for general flow scenarios in heterogeneous formations. *Transp. Porous Media*, 62(2):157–185, 2006
17. Y. Chen, L.J. Durlofsky, M. Gerritsen, and X.H. Wen. A coupled local-global upscaling approach for simulationg flow in highly heterogeneous formations. *Adv. Water Resour.*, 26:1041–1060, 2003.
18. Z. Chen and T.Y. Hou. A mixed multiscale finite element method for elliptic problems with oscillating coefficients. *Math. Comp.*, 72:541–576, 2003.
19. M. A. Christie and M. J. Blunt. Tenth SPE comparative solution project: A comparison of upscaling techniques. *SPE Reservoir Eval. Eng.*, 4(4):308–317, 2001. url: www.spe.org/csp.
20. M.A. Christie. Upscaling for reservoir simulation. *JPT J. Pet. Tech.*, 48:1004–1010, 1996.
21. L.J. Durlofsky. Numerical calculations of equivalent gridblock permeability tensors for heterogeneous porous media. *Water Resour. Res.*, 27(5):699–708, 1991.
22. L.J. Durlofsky. Use of higher moments for the description of upscaled, process independent relative permeabilities. *SPE J.*, 2(4):474–484, 1997.
23. L.J. Durlofsky. Coarse scale models of two-phase flow in heterogeneous reservoirs: Volume averaged equations and their relation to existing upscaling techniques. *Comput. Geosci.*, 2:73–92, 1998.
24. L.J. Durlofsky, R.C. Jones, and W.J. Milliken. A nonuniform coarsening approach for the scale-up of displacement processes in heterogeneous porous media. *Adv. Water Resour.*, 20:335–347, 1997.
25. Y. Efendiev. *The Multiscale Finite Element Method and its Applications.* PhD thesis, California Institute of Technology, 1999.
26. Y. Efendiev and L.J. Durlofsky. Numerical modeling of subgrid heterogeneity in two phase flow simulations. *Water Resour. Res.*, 38:1128, 2002.
27. Y. Efendiev and L.J. Durlofsky. A generalized convection-diffusion model for subgrid transport in porous media. *Multiscale Model. Simul.*, 1:504–526, 2003.
28. Y. Efendiev, L.J. Durlofsky, and S.H. Lee. Modeling of subgrid effects in coarse scale simulations of transport in heterogeneous porous media. *Water Resour. Res.*, 36(8):2031–2041, 2000.
29. Y. Efendiev, V. Ginting, T. Hou, and R. Ewing. Accurate multiscale finite element methods for two-phase flow simulations. *J. Comp. Phys.*, 220(1):155–174, 2006.
30. Y. Efendiev and B. Popov. On homogenization of nonlinear hyperbolic equations. *Commun. Pure and Appl. Anal.*, 4:297–311, 2005.
31. S. Ekrann and J. O. Aasen. Steady-state upscaling. *Transp. Porous Media*, 41(3):245–262, 2000.
32. M.H. Garcia, A.G. Journel, and K. Aziz. Automatic grid generation and adjustment method for modeling reservoir heterogeneities. Technical report, Stanford Center for Reservoir Forecasting, 1990.

33. Y. Gautier, M. J. Blunt, and M. A. Christie. Nested gridding and streamline-based simulation for fast reservoir performance prediction. *Comput. Geosci.*, 3:295–320, 1999.
34. J.J. Gómez-Hernández. *A Stochastic Approach to the Simulation of Block Conductivity Conditioned Upon Data Measured at a Smaller Scale.* PhD thesis, Stanford University, CA, 1991.
35. R.E. Guzman, D. Giordano, F.J. Fayers, A. Godi, and K. Aziz. Evaluation of dynamic pseudo functions for reservoir simulation. In *SPE Annual Technical Conference and Exhibition, 6-9 October, Denver, Colorado*, SPE 35157, 1996.
36. T.A. Hewett and R.A. Behrens. Scaling laws in reservoir simulation and their use in a hybrid finite difference/streamtube approach to simulating the effects of permeability heterogeneity. In L.W. Lake, H.B. Caroll, and T.C. Wesson, editors, *Reservoir Characterization II.* Academic Press, San Diego, Ca, 1991.
37. T.A. Hewett, K. Suzuki, and M.A. Christie. Analytical calculation of coarse-grid corrections for use in pseudofunctions. *SPE J.*, 3(3):293–304, 1998.
38. L. Holden and B.F. Nielsen. Global upscaling of permeability in heterogeneous reservoirs; the output least squares (ols) method. *Trans. Porous Media*, 40(2):115–143, 2000.
39. U. Hornung. *Homogenization and Porous Media.* Springer–Verlag, New York, 1997.
40. T.Y. Hou and X-H. Wu. A multiscale finite element method for elliptic problems in composite materials and porous media. *J. Comput. Phys.*, 134:169–189, 1997.
41. H. Jacks, O.J.E. Smith, and C.C. Mattax. The modeling of a three-dimensional reservoir with a two-dimensional reservoir simulator - the use of dynamic pseudo functions. *SPE J.*, pages 175–185, 1973.
42. P. Jenny, S. H. Lee, and H. A. Tchelepi. Multi-scale finite-volume method for elliptic problems in subsurface flow simulation. *J. Comput. Phys.*, 187:47–67, 2003.
43. P. Jenny, S. H. Lee, and H. A. Tchelepi. Adaptive multiscale finite-volume method for multiphase flow and transport in porous media. *Multiscale Model. Simul.*, 3(1):50–64, 2004.
44. V.V. Jikov, S.M. Kozlov, and O.A. Oleinik. *Homogenization of differential operators and integral functionals.* Springer–Verlag, New York, 1994.
45. A.G. Journel, C.V. Deutsch, and A.J. Desbarats. Power averaging for block effective permeability. In *SPE California Regional Meeting, 2-4 April, Oakland, California*, SPE 15128, 1986.
46. M.J. King, D.G. MacDonald, S.P. Todd, and H. Leung. Application of novel upscaling approaches to the Magnus and Andrew reservoirs. In *European Petroleum Conference, 20-22 October, The Hague, Netherlands*, SPE 50643, 1998.
47. P.R. King. The use of renormalization for calculating effective permeability. *Transp. Porous Media*, 4:37–58, 1989.
48. P.R. King, A.H. Muggeridge, and W.G. Price. Renormalization calculations of immiscible flow. *Transp. Porous Media*, 12:237–260, 1993.
49. V. Kippe, J. E. Aarnes, and K.-A. Lie. A comparison of multiscale methods for elliptic problems in porous media flow. *Comput. Geosci.*, submitted.
50. A. T. Kumar and G. R. Jerauld. Impacts of scale-up on fluid flow from plug to gridblock scale in reservoir rock. In *SPE/DOE Improved Oil Recovery Symposium, 21-24 April, Tulsa, Oklahoma*, SPE 35452, 1996.

51. J.R. Kyte and D.W. Berry. New pseudofunctions to control numerical dispersion. *SPE J.*, pages 269–276, August 1975.
52. P. Langlo and Magne S. Espedal. Macrodispersion for two-phase, immiscible flow in porous media. *Adv. Water Resour.*, 17:291–316, 1994.
53. B. F. Nielsen and A. Tveito. An upscaling method for one-phase flow in heterogeneous reservoirs; a Weighted Output Least Squares (WOLS) approach. *Comput. Geosci.*, 2:92–123, 1998.
54. P.-E. Øren, S. Bakke, and O.J. Arntzen. Extending predictive capabilities to network models. *SPE J.*, 3(4):324–336, 1998.
55. G.E. Pickup and K.S. Sorbie. Development and application of a new two-phase scaleup method based on tensor permeabilities. *SPE J.*, pages 369–382, 1996.
56. P.A. Raviart and J.M. Thomas. A mixed finite element method for second order elliptic equations. In I. Galligani and E. Magenes, editors, *Mathematical Aspects of Finite Element Methods*, pages 292–315. Springer–Verlag, Berlin – Heidelberg – New York, 1977.
57. P. Renard and G. de Marsily. Calculating equivalent permeability. *Adv. Water Resour.*, 20:253–278, 1997.
58. E.H. Smith. The influence of small scale heterogeneity on average relative permeability. In L.W. Lake, H.B. Carroll, and T.C. Wesson, editors, *Reservoar Characterization II*. Academic Press, 1991.
59. K. D. Stephen, G. E. Pickup, and K. S. Sorbie. The local analysis of changing force balances in immiscible incompressible two-phase flow. *Transp. Porous Media*, 45:63–88, 2001.
60. H.L. Stone. Rigorous black-oil pseudo functions. In *SPE Symposium on Reservoir Simulation, 17-20 February, Anaheim, California*, SPE 21207, 1991.
61. G.W. Thomas. An extension of pseudofunction concepts. In *SPE Reservoir Simulation Symposium, 15-18 November, San Francisco, California*, SPE 12274, 1983.
62. X.-H. Wen and J.J. Gómez-Hernández. Upscaling hydraulic conductivities in heterogeneous media: An overview. *J. Hydrol.*, 183:ix–xxxii, 1996.
63. O. Wiener. *Abhandlungen der Matematisch.* PhD thesis, Physischen Klasse der Königlichen Sächsischen Gesellscaft der Wissenschaften, 1912.
64. X.H. Wu, Y. Efendiev, and T.Y. Hou. Analysis of upscaling absolute permeability. *Discrete Contin. Dyn. Syst. Ser. B*, 2(2):185–204, 2002.
65. P. Zhang, G. E. Pickup, and M. A. Christie. A new upscaling approach for highly heterogenous reservoirs. In *SPE Reservoir Simulation Symposium, 31 January-2 Feburary, The Woodlands, Texas*, SPE 93339, 2005.

Modelling of Stratified Geophysical Flows over Variable Topography

Torbjørn Utnes

Summary. In the present chapter a short review is given of the mathematical formulation relevant for geophysical flow modelling, and in addition computational examples are shown for some specific flow cases. These examples are described in some detail in order to illustrate useful methods to handle such problems in practice. The emphasis is on more local geophysical flows, including stratified flow over variable topography.

1 Introduction

The aim of the present chapter is twofold: first to give a short review of the mathematical formulation relevant for geophysical flow modelling – including common approximations and other modelling considerations, and then to focus on a few specific flow cases. These applications, which include stratified flows over mountainous terrain as a main example, are treated in some detail in order to illustrate useful methods to handle such geophysical problems in a practical way.

There is a vast number of publications covering the different fields of classical geophysical fluid dynamics. General reviews and background information may be found in several textbooks [19, 34, 39, 45]. In the present context, with emphasis on more local scale geophysical flows, we may divide the most relevant literature in atmospheric and waterly flows as follows: Analysis related to local scale atmospheric flows over hills and mountainous terrain may be found e.g., in [2, 4, 23, 25], while simulations for such cases are shown in [10, 13, 14, 41, 50, 53]. Similarly, for local waterly flows, analysis and laboratory experiments may be found e.g., in [6, 42], while modelling and applications ranging from local tidal flow to circulations in various restricted waters are given e.g., in [5, 33, 36, 49].

The present text is arranged in the following way: In Section 2 the mathematical formulation is reviewed, basically consisting of the classical conservation laws for mass, momentum and energy. Possible simplifications are then derived for certain kinds of flows (Section 3). This includes simplifications due to scale differences in the horizontal and vertical directions (the hydrostatic approximation), simplifications due to relatively weak density variations (the Boussinesq approximation), and the inclusion of rotational effects or not, depending on the temporal and spatial scales under consideration. In order to handle the resulting equations in a practical way, we introduce the so-called Reynolds-averaging in Section 4. This in turn leads to the question of turbulence modelling, which is briefly discussed in Section 5. Useful and simple turbulence models are also introduced in that section. Well-posed formulations require initial and boundary conditions, and this is dealt with in Section 6 for some relevant cases.

All of these items (Sections 2–6) are finally brought together and illustrated in Section 7. Here two different classes of geophysical flow problems are modelled in detail, namely a simple one-dimensional model for vertical variations and a more complicated three-dimensional model for local stratified flow. In the first case a conventional Galerkin finite element formulation is applied, and this is written out in detail for illustration. In the latter case a so-called algebraic projection formulation is first applied in order to solve the Reynolds equations, and then a mixed finite element method is used for the spatial discretisation.

Finally, in Section 8 we show some computational examples to illustrate the potential of the models. The emphasis is on the three-dimensional model derived in Section 7, which is here applied to predict flow around a hill in neutral and stratified flow. It is also indicated how this model can be used in a local weather prediction system, thus illustrating one of the many useful ways to apply such models.

2 Governing Equations

In this section we give a short description of the fundamental equations governing geophysical flows. The basic principles are the conservation of mass, momentum and energy. Geophysical flows are usually high-Reynolds number, turbulent flows. Other characteristics are that the effect of the earth's rotation is often important, and is accounted for by the so-called Coriolis acceleration. However, in some local scale analysis there may be exceptions where rotational effects are less important, depending on the Rossby number (defined in (5)). Stratification effects due to temperature and/or salinity variations are also characteristic properties for these kinds of flows.

The governing equations described in this section will later be simplified and modified (see Sections 3 and 4) in order to make numerical solutions manageable for realistic problems.

2.1 Equations of Motion

The equations of motion are the Navier–Stokes equations in a rotating coordinate frame, including the continuity equation. The latter is an equation for mass conservation, and may be written

$$\frac{\partial \rho}{\partial t} + \nabla \cdot (\rho \mathbf{u}) = 0 \,, \tag{1}$$

where $\mathbf{u} = (u_1, u_2, u_3)$ is the velocity and ρ is the density. The equation for conservation of momentum may be written in the form

$$\frac{\partial \rho \mathbf{u}}{\partial t} + \nabla \cdot (\rho \mathbf{u} \otimes \mathbf{u}) + f \mathbf{k} \times \rho \mathbf{u} = -\nabla p - \rho g \mathbf{k} + \nabla \cdot \boldsymbol{\tau} \,, \tag{2}$$

where p is the pressure, g is the gravity constant, $\mathbf{k}$ is the unit vector acting in the vertical upward direction, and the viscous stress tensor is given by

$$\tau_{ij} = \mu \left(\frac{\partial u_i}{\partial x_j} + \frac{\partial u_j}{\partial x_i} \right) , \tag{3}$$

where μ is the dynamic viscosity ($\mu = \rho \nu$; ν is the kinematic viscosity). The earth's rotation is represented by the Coriolis parameter $f = 2\Omega \sin\theta$, where θ is the latitude and $\Omega \approx 7.3 \times 10^{-5}$ $[s^{-1}]$ is the angular velocity of the earth. Equation (2) is the so-called 'conservative form' of the momentum equation. An alternative 'non-conservative form' is given by

$$\rho \frac{D \mathbf{u}}{D t} + f \mathbf{k} \times \rho \mathbf{u} = -\nabla p - \rho g \mathbf{k} + \nabla \cdot \boldsymbol{\tau} \,. \tag{4}$$

This form is obtained by employing the continuity equation (1) in the momentum equation (2) and perform some re-formulations, see [16].

The effects of the earth's curvature has been neglected in these equations, which is a good assumption when the actual domain is of the order of 1000 km or smaller; larger domains require the use of spherical coordinates [26].

The importance of the earth's rotation for the fluid motion depends on the Rossby number

$$Ro = \frac{U}{2\Omega L} \,, \tag{5}$$

where U and L represent typical velocity and length scales, respectively. Provided $Ro \leq 1$ the rotational effect may be regarded as important. If the period L/U is much less than the earth's rotational period, the fluid will not sense the earth's rotation over that time scale. In such cases the Coriolis terms may be neglected in the equations.

2.2 Stratification

The previous equations are sufficient provided that the density may be treated as constant. However, generally density is a function of several variables, depending on the actual kind of flow under consideration. In such cases the momentum and continuity equations are insufficient to close the dynamical system. A state equation must be applied to find the density, and conservation equations for temperature and/or salinity must be added.

An equation for temperature (T) can be derived from the first law of thermodynamics, which is basically a statement of energy conservation. For waterly, incompressible flow the temperature equation takes the form

$$\frac{DT}{Dt} = \kappa \nabla^2 T \,, \tag{6}$$

where κ is the thermal diffusivity. The density of near-coastal waters is strongly influenced by salinity (S) in addition to temperature, and a conservation equation for this must also be added in such cases:

$$\frac{DS}{Dt} = 0 \,. \tag{7}$$

The density can then be found from a state equation. For waterly flow an algebraic state equation may be written in the form (see [26])

$$\rho = \rho_o - \beta_T (T - T_o) + \beta_S (S - S_o) \,, \tag{8}$$

where $\beta_T = 0.07$, $\beta_S = 0.8$, $(T_o, S_o) = (273K, 35\,\text{psu})$, and $\rho_o = \rho(T_o, S_o)$. More accurate estimates are given e.g., in [7].

For the atmosphere the relation between pressure, specific volume and temperature, $pv = RT$, must be taken into consideration. It is convenient then to introduce a new variable, the potential temperature defined by

$$\theta = T \left(\frac{p_o}{p} \right)^{R/c_p} \,. \tag{9}$$

Here p_o refers to the pressure near sea level, $p_o = 1000$ mb, R is the gas constant, and c_p is the specific heat at constant pressure. The potential temperature plays a similar role in the atmosphere as temperature does in waterly flow. The energy equation for atmospheric flow may then be written in the form

$$\frac{D\theta}{Dt} = \frac{\theta}{T} \kappa \nabla^2 T \,, \tag{10}$$

where radiation effects have been neglected for the present purpose. It should be noted, however, that radiation effects are in general very important for atmospheric flows, and are then included as source terms in the energy equation (see e.g., [45]).

The atmospheric density is given from the state equation for dry air:

$$\rho = \frac{p}{R\theta} \left(\frac{p_o}{p} \right)^{R/c_p} \,. \tag{11}$$

3 Approximations

In principle the previous equations may be used to define any geophysical problem together with appropriate initial and boundary conditions. However, for practical and numerical reasons we introduce several simplifications. These simplifications are of two different kinds: First we introduce mathematical approximations to the equations based on physical rationalisations. This is described in the following section. Then the equations are 'filtered', i.e. averaged in some way in order to separate the means from the turbulent oscillations. The equations may then be solved on a mean scale only, hence reducing the numerical size of the problem by several orders of magnitude. However, this approach gives rise to the question of turbulence modelling, which is addressed in Section 4.

3.1 Approximations for the Continuity Equation

The continuity equation (1) can be simplified with good approximation in many types of geophysical flows. The strongest approximation leads to the incompressibility condition, which follows from assuming the flow to be divergence-free. This is a good assumption for waterly flow, but may be less accurate for certain types of air flow. In such cases the spatial density variations are retained, while sound waves may be eliminated by neglecting the time variation of the density in (1). This simplification is called the anelastic approximation (cf. [11]).

Incompressible Flow

The assumption of incompressible flow implies that the continuity equation is written as

$$\nabla \cdot \mathbf{u} = 0\,. \tag{12}$$

We suppose that the spatial distribution of $\mathbf{u}$ is characterised by a length scale L (meaning that $\mathbf{u}$ varies only slightly over a distance L), and that the variations of $|\mathbf{u}|$ have a magnitude U. Then the order of magnitude of spatial derivatives of $\mathbf{u}$ is U/L, and the velocity distribution is approximately divergence-free or solenoidal if

$$|\nabla \cdot \mathbf{u}| \ll U/L\,.$$

By rewriting the continuity equation (1) in the form

$$\frac{D\rho}{Dt} + \rho\,(\nabla \cdot \mathbf{u}) = 0\,,$$

it then follows immediately that

$$\left|\frac{1}{\rho}\frac{D\rho}{Dt}\right| \ll \frac{U}{L}\,,$$

which means that the relative time change of the density of a particle is much less than an inverse time scale U/L, and the fluid behaves as if it were incompressible. Hence the terms divergence-free and incompressible flow are often used synonymously. The conditions required for a velocity distribution to be divergence-free or incompressible are shown in [3] to be:

$$U^2/c^2 \ll 1, \quad N^2L^2/c^2 \ll 1, \quad gL^2/c^2 \ll 1\,, \tag{13}$$

where c is the speed of sound and N is a dominant oscillatory frequency of the flow. For water $c \approx 1500$ m/s, and these conditions are therefore satisfied with good approximation in oceanographic flow. However, for large scale air flow, e.g., dynamical meteorology, the last condition above may not be satisfied.

Anelastic Approximation

The gravity condition in (13) may be avoided if only the time-dependent part in the continuity equation (1) is neglected, giving

$$\nabla \cdot (\rho \mathbf{u}) = 0\,. \tag{14}$$

This is the anelastic approximation, which is more general than the incompressibility assumption. To be valid, this approximation requires only the first two of the conditions in (13). This approximation has been used in several studies (cf. [2]), and gives solutions that are similar in character to the incompressible system. It may be a good choice for many kinds of air flow modelling.

3.2 The Boussinesq Approximation

The Boussinesq approximation consists of taking the density to be a constant in computing rates of change of momentum from accelerations, but taking full account of density variations only when they give rise to buoyancy forces, i.e., when there is a multiplying factor g in the vertical component of the momentum equation. We may split density and pressure into two parts:

$$\rho = \rho_o + \rho_d; \; p = p_o + p_d\,, \tag{15}$$

where $(\rho_d,\ p_d)$ refer to the dynamic part, and $(\rho_o,\ p_o)$ to the hydrostatic part. The latter is defined via the hydrostatic equilibrium

$$\nabla p_o = -\rho_o g \mathbf{k}\,. \tag{16}$$

The Boussinesq approximation is valid provided $|\rho_d/\rho_o| \ll 1$. Normally this is a good approximation for all watery fluids and for many applications to the atmosphere, see [2, 19].

By introducing the Boussinesq approximation, the momentum equation is approximated as

$$\frac{D\mathbf{u}}{Dt} + f\mathbf{k} \times \mathbf{u} = -\frac{\nabla p}{\rho_o} - \frac{\rho}{\rho_o} g\mathbf{k} + \frac{1}{\rho_o} \nabla \cdot \boldsymbol{\tau} \,. \tag{17}$$

This approximation implies that g is replaced by $g(\rho/\rho_o)$, i.e., an error of relative magnitude $|\rho_d/\rho_o|$, which is very small in our actual cases.

We may subtract the hydrostatic part (16) from the momentum equation and obtain the following alternative form:

$$\frac{D\mathbf{u}}{Dt} + f\mathbf{k} \times \mathbf{u} = -\frac{\nabla p_d}{\rho_o} - \frac{\rho_d}{\rho_o} g\mathbf{k} + \frac{1}{\rho_o} \nabla \cdot \boldsymbol{\tau} \,. \tag{18}$$

Note that in this formulation the density variation appears only via ρ_d in the gravity term.

3.3 The Hydrostatic Approximation

In classical geophysical flows the horizontal length scale (L) is much larger than the vertical length scale (h), i.e.

$$h/L \ll 1 \,.$$

On a large scale this implies that the flow is predominantly horizontal, and the vertical acceleration is small compared to the gravity acceleration. The vertical component of the equation of motion can then be reduced to a simple hydrostatic balance:

$$\rho g = -\frac{\partial p}{\partial z} \,. \tag{19}$$

This is the hydrostatic approximation. If the Boussinesq approximation is applied in addition, the equations for horizontal motion may be written in the following form:

$$\frac{Du}{Dt} - fv = -g\frac{\partial}{\partial x}(\zeta - q) + \frac{1}{\rho_o}(\nabla \cdot \boldsymbol{\tau})_x \,, \tag{20}$$

$$\frac{Dv}{Dt} + fu = -g\frac{\partial}{\partial y}(\zeta - q) + \frac{1}{\rho_o}(\nabla \cdot \boldsymbol{\tau})_y \,, \tag{21}$$

where the stratification is represented by

$$q = \int_z^\zeta \frac{\rho_o - \rho}{\rho_o} dz \,, \tag{22}$$

and ζ represents the free surface elevation. In these equations the stress tensor is usually treated in a special way by distinguishing between horizontal and vertical parts, due to scale arguments. See e.g., [1, 29, 39] for more information; practical applications of this idea include much of the literature on three-dimensional ocean current modelling (cf. [5, 49]).

By restricting attention to oceanography for the moment, we may use the incompressible continuity equation

$$\frac{\partial u}{\partial x} + \frac{\partial v}{\partial y} + \frac{\partial w}{\partial z} = 0\,. \tag{23}$$

From this equation we may derive an expression for the free surface elevation ζ. This is performed by integrating (23) vertically from the bottom $z = -h$ to the free surface $z = \zeta$, and use the kinematic boundary conditions (Section 6.4) at the bottom and free surface to get

$$\frac{\partial \zeta}{\partial t} + \frac{\partial (H\overline{u})}{\partial x} + \frac{\partial (H\overline{v})}{\partial y} = 0\,, \tag{24}$$

where

$$\overline{u_i} = \frac{1}{H}\int_{-h}^{\zeta} u_i dz\,, \quad \text{for } (i = 1, 2)\,, \tag{25}$$

and $H = h + \zeta$ is the total depth.

The equations above are usually the starting point for numerical modelling of ocean dynamics.

4 Filtered Equations

Theoretically the most exact approach to turbulence simulation is to solve the Navier–Stokes equations. This is called direct numerical simulation (DNS). A DNS must include all scales of importance, from the energy-containing large scales and down to the dissipative small scales. By using scale arguments one may estimate the number of grid points required in a DNS as proportional to $Re^{3/4}$ for each of the three directions [47], where Re is the characteristic Reynolds number of the flow. The time step is related to the grid size, and the cost of a simulation may therefore scale as Re^3. This implies that DNS can be carried out only for relatively low Reynolds numbers.

Geophysical flows are usually characterised as (very) high-Reynolds number flows. As an example, for local atmospheric flow with a typical length scale of $L = 10^3$ m and a wind speed of the order of $U = 10$ m/s, the Reynolds number becomes $Re = UL/\nu \approx 10^9$. This implies that direct numerical simulation is presently of little practical use for such flows. In order to solve larger problems, the governing equations then need to be filtered in some way, and some kind of turbulence modelling is therefore necessary.

While large-eddy simulations (LES) are of great interest in many situations, it seems that Reynolds-averaged Navier–Stokes (RANS) models still remain the most viable means for solution of complex high-Reynolds flows, cf. [44]. From a modelling point of view there is a close relation between LES and RANS, and one may in fact use a RANS model for LES and vise versa.

Reynolds-averaging is the traditional way to derive filtered equations from the original governing equations. The variables are decomposed into mean and fluctuating parts, e.g., for the velocity

$$\mathbf{u} = \overline{\mathbf{u}} + \mathbf{u}', \tag{26}$$

and similarly for the other variables [40, 44]. Here $\overline{\mathbf{u}}$ is the Reynolds-averaged part and $\mathbf{u}'$ the remaining fluctuating part. When this decomposition is placed into the Navier–Stokes equations, the resulting Reynolds equations resemble the original equations, although the variables now represent averaged values instead of instantaneous values. For convenience we will use the same notation as before (i.e., $\mathbf{u}$ instead of $\overline{\mathbf{u}}$ etc.), although the quantities are now to be understood as Reynolds-averaged quantities. The Reynolds-averaged equations of motion then take the form:

$$\frac{D\mathbf{u}}{Dt} + f\mathbf{k} \times \mathbf{u} = -\frac{1}{\rho}\nabla p - g\mathbf{k} + \nabla \cdot \mathbf{R}, \tag{27}$$

where

$$\mathbf{R} = -\overline{\mathbf{u}' \otimes \mathbf{u}'}, \tag{28}$$

is the Reynolds stress tensor, which needs to be modelled. Note that mathematically this quantity takes a similar role as the stress tensor, see equation (4). Hence, the governing equations keep their original structure in the Reynolds-averaged form.

4.1 Eddy Viscosity and Diffusivity

A simple and useful way to represent the Reynolds stress tensor is the eddy viscosity concept which assumes that, in analogy to the viscous stress in laminar flows, the turbulent stress is proportional to the mean velocity gradient. This concept may be expressed as

$$R_{ij} = \nu_T \left(\frac{\partial u_i}{\partial x_j} + \frac{\partial u_j}{\partial x_i} \right) - \frac{2}{3} K \delta_{ij}, \tag{29}$$

where ν_T is the turbulent or eddy viscosity, and K the turbulent kinetic energy. The isotropic part of the Reynolds stress may be included into the mean pressure, i.e., by re-defining $p \leftarrow p + 2/3K$. The Reynolds-averaged momentum equations then get a form similar to the original momentum equation (4), namely

$$\frac{D\mathbf{u}}{Dt} + f\mathbf{k} \times \mathbf{u} = -\frac{1}{\rho}\nabla p - g\mathbf{k} + \nabla \cdot \left[\nu_T(\nabla \mathbf{u} + \nabla \mathbf{u}^T)\right]. \tag{30}$$

The Reynolds-averaged continuity equation takes the same form as before:

$$\frac{\partial \rho}{\partial t} + \nabla \cdot (\rho \mathbf{u}) = 0. \tag{31}$$

Similarly, the Reynolds-averaged temperature or potential temperature equations may be written in the form

$$\frac{D\theta}{Dt} = \nabla \cdot (\gamma_T \nabla \theta) , \tag{32}$$

where γ_T represents the eddy diffusivity for turbulent mixing of heat. The relation between eddy viscosity (ν_T) and diffusivity (γ_T) is given by the turbulent Prandtl number

$$\sigma_T = \nu_T / \gamma_T . \tag{33}$$

This quantity turns out to be a function of the local gradient Richardson number [25, 26]:

$$Ri = -\frac{g}{\rho_o} \frac{\partial \rho / \partial z}{(\partial u_\tau / \partial z)^2} , \tag{34}$$

where $u_\tau = (u_1, u_2)$ is the horizontal velocity, z is the vertical coordinate, and ρ_o is a mean or reference density. An approximation partly based on data is of the form [31, 32]

$$\sigma_T = \sigma_{T0} \exp(1.5 Ri) , \tag{35}$$

where σ_{T0} refers to the neutral case ($Ri = 0$).

Transport equations of other scalar quantities such as sediment, plankton, etc. are of the same kind as (32), with the possible addition of an advection term for sinking velocity.

In classical geophysical flows, with a generally large difference between horizontal and vertical scales, the eddy viscosity is usually anisotropic with respect to horizontal and vertical directions. It is therefore common practice to distinguish between vertical and horizontal turbulence (see [26, 39]), and estimate the turbulence by use of different models for these two regimes.

Possible ways to model the eddy viscosity are discussed in the following section.

5 Turbulence Modelling

The Reynolds-averaged equations derived in the previous section imply a closure problem, i.e., there are more unknowns than equations due to the Reynolds stress. This makes it necessary to introduce some kind of turbulence modelling. By assuming the eddy viscosity concept (29), it remains to model the eddy viscosity in a reasonable way. In the following two actual turbulence models are specified; both of these are subsequently shown in numerical implementations.

5.1 One-Equation Model

The eddy viscosity ν_T has dimension $[L^2/T]$, where L is a length scale and T a time scale. This indicates that the eddy viscosity may be expressed by a turbulent velocity u_t and a turbulent length ℓ_t, namely

$$\nu_T \propto u_t \ell_t \,.$$

The turbulent kinetic energy is defined by

$$K = \frac{1}{2}\left(\overline{u_1'^2} + \overline{u_2'^2} + \overline{u_3'^2}\right),$$

where u_i' (i=1,2,3) represent the turbulent velocity fluctuations. Hence $\sqrt{K}$ may be a meaningful scale for the turbulent velocity fluctuations. From the eddy viscosity relation above the following results:

$$\nu_T = C_\mu^{1/4} K^{1/2} \ell \,, \tag{36}$$

where the constant $C_\mu \approx 0.09$. This is known as the Kolmogorov–Prandtl expression because both of these pioneers introduced it independently [43].

Reasonable empirical description of the turbulent length scale ℓ is possible for relatively simple flows, and is often used especially for the vertical dimension in classical geophysical flow modelling. Together with a model equation for the turbulent kinetic energy this forms the so-called one-equation models.

An equation for the turbulent kinetic energy may be derived from the Navier–Stokes equation, and the resulting equation is closed by introducing empirical modelling of diffusion and dissipation terms. The modelled form of the turbulent kinetic energy equation is given in the form [43]

$$\frac{DK}{Dt} = \frac{\partial}{\partial z}\nu_T \frac{\partial K}{\partial z} + \nu_T \left(\frac{\partial \mathbf{u}}{\partial z}\right)^2 + \frac{\nu_T}{\sigma_T}\frac{g}{\rho_o}\frac{\partial \rho}{\partial z} - C_\mu^{3/4}\frac{K^{3/2}}{\ell} \,. \tag{37}$$

For simple flows the length scale (ℓ) typically goes like (κz) close to the boundary, where κ is the von Karman constant and z is the normal distance from the boundary surface. For air flow a typical form in neutral flow may be

$$\ell_o = \frac{\kappa z}{1 + \kappa z/\lambda} \,, \tag{38}$$

where λ is an asymptotic value. For ocean flow one form of the length scale is [27]

$$\ell_o = \kappa z (1 - z/H) \,. \tag{39}$$

The turbulent length scale is usually modified (reduced) as a function of the Richardson number. Two such empirical forms are (cf. [13, 22])

$$\ell = \frac{\ell_o}{1 + 5Ri} \,, \tag{40}$$

and

$$\ell = \ell_o \exp(-8\,Ri) \,. \tag{41}$$

It should be noted that the present kind of one-equation model can only estimate a vertical eddy viscosity. Any horizontal eddy viscosity must be modelled by some other method, e.g., a simple Smagorinsky model (see [40]).

5.2 Two-Equation Model

Several kinds of two-equation turbulence models have been developed; two of the most well-known which have also been applied in geophysical flows are the (K, ℓ) model [35] and the (K, ϵ) model [46, 43]. Both of thesc models have much in common, and behave rather similar.

Two-equation models are 'complete' in a certain sense, and specification of flow-dependent quantities like a length scale are not required. Hence such models are more general than a one-equation model, and can be used in somewhat more complicated flows. However, the limitations of these models should also be kept in mind: They cannot predict secondary flows, and may be inaccurate for flows far removed from simple shear flow. In the following we give a short description of the (K, ϵ) model.

In near-equilibrium turbulent flows there is a known relation between the dissipation (ϵ), the turbulent kinetic energy (K), and the turbulent length scale (ℓ). This allow us to use an equation for the dissipation as a means to obtain both ϵ and ℓ. Together with the K-equation this introduces the so-called (K, ϵ) closure, which is one of the most useful eddy-viscosity models. A closer analysis of this model may be found in [37]. Examples of use in geophysical flows may be found e.g., in [8, 46, 50].

The model is defined via the following two transport equations for turbulent kinetic energy and dissipation, and an algebraic relation between these two quantities to obtain the turbulent eddy viscosity:

$$\frac{DK}{Dt} = \nabla \cdot (\nu_T \nabla K) + P_k + G_\rho - \epsilon \,, \tag{42}$$

$$\frac{D\epsilon}{Dt} = \nabla \cdot \left(\frac{\nu_T}{\sigma_\epsilon} \nabla \epsilon \right) + \frac{\epsilon}{K} \left(C_1 P_k + C_3 G_\rho \right) - C_2 \frac{\epsilon^2}{K} \,, \tag{43}$$

$$\nu_T = C_\mu \frac{K^2}{\epsilon} \,, \tag{44}$$

where production and stratification terms are given by

$$P_k = \nu_T \left(\frac{\partial u_i}{\partial x_j} + \frac{\partial u_j}{\partial x_i} \right) \frac{\partial u_i}{\partial x_j} \,, \quad G_\rho = \frac{\nu_T}{\sigma_T} \frac{g}{\rho} \frac{\partial \rho}{\partial z} \,. \tag{45}$$

This is the standard high-Reynolds number formulation, with constants given in Table 1. Several generalisations are possible in order to improve some of the

Table 1. Conventional constants in the (K, ϵ) model

C_μ	C_1	C_2	C_3	σ_ϵ
0.09	1.44	1.92	0.8	1.3

shortcomings of the standard (K, ϵ) model. Interesting non-linear extensions are described e.g., in [18, 24, 44].

6 Boundary Conditions

In addition to the model equations, we need to specify appropriate initial and boundary conditions in order to get a well-defined formulation. In the following section we discuss some of the most common and relevant boundary conditions connected to atmospheric and ocean flows.

6.1 Wall Boundary

Bottom or land boundaries constitute wall boundaries for waterly flow, while for air flow similar wall boundaries are represented by land/ground and sea surface. No-slip conditions are generally the desired boundary conditions along walls. However, due to the high Reynolds number of the actual flows, this will require a dense grid close to such boundaries. Usually, at these high Reynolds numbers, with a flow field that is only slowly time-dependent, the classical logarithmic assumption on the tangential velocity will be a fairly good approximation close to the wall (cf. [43]). This assumption leads to the following boundary condition for the tangential velocity u_τ:

$$\nu_T \frac{\partial u_\tau}{\partial n} = \frac{\tau_w}{\rho} = C_w |u_\tau| u_\tau \,, \tag{46}$$

where n is the unit normal pointing into the domain, τ_w is the wall shear stress, and the friction coefficient C_w can be determined from the 'law of the wall' assumption, see [49]. In addition, the wall boundary condition for the normal component of the velocity is set to zero:

$$u_n = 0 \,. \tag{47}$$

Wall boundary conditions for other variables may typically be Dirichlet conditions. This includes specification of the potential temperature

$$\theta = \hat{\theta}_w \,. \tag{48}$$

and specification of turbulence parameters at a small distance z from the wall:

$$K = \frac{u_*^2}{{C_\mu}^{1/2}} \,, \qquad \epsilon = \frac{u_*^3}{\kappa z} \,. \tag{49}$$

In these expressions the friction velocity u_* is defined as

$$u_* = \sqrt{\tau_w / \rho} \,. \tag{50}$$

6.2 Inflow Boundary

The flow into the domain must be specified as a Dirichlet condition

$$\mathbf{u} = \hat{\mathbf{u}}_{in} \,. \tag{51}$$

Inflow boundary conditions for other variables are specified similarly:

$$\theta = \hat{\theta}_{in} \,, \quad K = \hat{K}_{in} \,, \quad \epsilon = \hat{\epsilon}_{in} \,. \tag{52}$$

6.3 Open Boundary

These conditions have to be specified along the outer boundary of the mesh, where the model domain is cut off e.g., at the mouth of a fjord. Usually a normal traction condition may be specified, and we can often assume that a hydrostatic assumption is fairly good:

$$-\frac{p}{\rho_o} + 2\nu_T \frac{\partial u_n}{\partial n} \approx -\frac{p_o}{\rho_o} \,. \tag{53}$$

Another way to specify the normal traction is simply to update the left hand side of (53) by using the last known values of the variables. This technique is well suited in the context of a finite element formulation.

Due to rotational effects we may also need to specify the flow direction. This can often be done by setting the tangential velocity equal to zero:

$$u_\tau = 0 \,. \tag{54}$$

In the oceanographic case one must in addition specify the time-dependent elevation along the open boundary, i.e.

$$\zeta = \zeta_o \,. \tag{55}$$

Open boundary conditions for other variables are typically free Neumann conditions:

$$\frac{\partial \theta}{\partial n} = 0 \,, \quad \frac{\partial K}{\partial n} = 0 \,, \quad \frac{\partial \epsilon}{\partial n} = 0 \,. \tag{56}$$

6.4 Free Surface Boundary

In case of waterly flows it is necessary to specify boundary conditions on the free surface.

The kinematic boundary condition requires that no fluid particles cross the inter-facial boundary, i.e., no spray or mixing occurs. Thus, if $\zeta(x, y; t)$ is the equation of the surface elevation above the equilibrium level z_o, then the vertical coordinate, z, of a particle on the free surface is

$$z = z_o + \zeta(x, y; t) \,. \tag{57}$$

The requirement that a fluid particle on the surface will remain there as it moves along is equivalent to

$$\frac{D}{Dt}\,(z - z_o - \zeta) = 0\,. \tag{58}$$

Since the coordinate z is independent of (x, y, t), we obtain the following kinematic boundary condition on the vertical velocity ($w = \partial z/\partial t$):

$$w = \frac{\partial \zeta}{\partial t} + u\frac{\partial \zeta}{\partial x} + v\frac{\partial \zeta}{\partial y}\,. \tag{59}$$

The dynamic boundary condition states that pressure differences across a free surface cannot be supported without disrupting the surface, unless a normal surface stress exists to balance the difference. Intermolecular forces exert such stresses via a surface tension. At longer wavelengths the capillary terms are negligible, and the dynamic boundary condition for surface gravity waves states continuity of pressure across the interface as

$$p = p_a\,, \tag{60}$$

where p_a is the atmospheric pressure. It should be noted that this dynamic condition may not be specified in its present form when using the full 3D formulation.

Free surface boundary conditions for the horizontal velocity (u_τ) can be specified as follows: The tangential surface stress provides boundary conditions for the horizontal velocity components via the Neumann condition

$$\nu_T \frac{\partial u_\tau}{\partial z} = \frac{\tau_s}{\rho_o}\,, \tag{61}$$

where τ_s is the wind induced surface stress and u_τ the tangential (or approximately horizontal) velocity.

7 Modelling Approach

In the previous sections we have described relevant model equations and boundary conditions for some geophysical problems. The main numerical demand is to solve these equations in an accurate, efficient and robust way. Among all these equations, the foremost demand is to solve the equations of motion, or the Reynolds-averaged equations, in a satisfactory way. Coupled methods may be robust, but their efficiency depends heavily on advanced general equation solvers. Projection formulations may be less robust for certain kinds of problems, but are attractive especially because the demand on equation solvers are reduced. It seems that projection methods have their best performance for non-stationary, high-Reynolds flows (cf. [48]), and this is characteristic for geophysical flows.

In the following sections we show examples of two different model formulations; the first is a one-dimensional simplified model, relevant for some upper-layer wind-induced problems. This may be regarded as an introduction to the methodology applied in the second example, which is a fully three-dimensional (non-hydrostatic) model, presently without a free surface. Selected results from this model will be shown in a subsequent case study.

In the following examples the numerical methods are based on finite element formulations. Other methods (finite difference or finite volume methods) could as well have been used, and all of these methods are well suited to solve the actual equation systems.

7.1 A Simple Model for Vertical Variations

One of the simplest and yet relevant models is given by assuming horizontal variations to be negligible compared to vertical ones. This may be relevant e.g., when studying wind-induced current in some open ocean situations (e.g. [28, 22]). In this case the hydrostatic equations reduce to the following simplified equations for the horizontal velocity components (20), (21):

$$\frac{\partial u}{\partial t} - fv = \frac{\partial}{\partial z}\left(\nu_T \frac{\partial u}{\partial z}\right), \tag{62}$$

$$\frac{\partial v}{\partial t} + fu = \frac{\partial}{\partial z}\left(\nu_T \frac{\partial v}{\partial z}\right). \tag{63}$$

In these equations ν_T represents the vertical turbulent eddy viscosity.

Stratification effects are modelled only via salinity variations, assuming a situation of wind-induced current in fjord- or near-coastal areas where salinity may be the main source of stratification effects:

$$\frac{\partial S}{\partial t} = \frac{\partial}{\partial z}\left(\frac{\nu_T}{\sigma_T}\frac{\partial S}{\partial z}\right). \tag{64}$$

In this case the state equation simplifies to

$$\rho = \rho_0 + \alpha S, \tag{65}$$

where ρ_0 refers to fresh water density, and $\alpha \approx 0.8$.

The one-equation turbulence model from (37) reduces to

$$\frac{\partial K}{\partial t} = \frac{\partial}{\partial z}\left(\nu_T \frac{\partial K}{\partial z}\right) + \nu_T\left[\left(\frac{\partial u}{\partial z}\right)^2 + \left(\frac{\partial v}{\partial z}\right)^2\right] + \frac{g}{\rho_o}\frac{\nu_T}{\sigma_T}\frac{\partial \rho}{\partial z} - C_\mu^{3/4}\frac{K^{3/2}}{\ell}, \tag{66}$$

and the same length scale as given in (39), (41).

The equations for (u, v, S, K) may be solved by use of a finite element method, and we illustrate the methodology in the following. By application of a weighted residual method the following formulation is obtained:

$$\int \left(\Phi \frac{\partial \mathbf{u}}{\partial t} + \frac{\partial \Phi}{\partial z} \nu_T \frac{\partial \mathbf{u}}{\partial z} \right) d\Omega + \int \Phi (\mathbf{f} \times \mathbf{u}) d\Omega = \oint \Phi \left(\nu_T \frac{\partial \mathbf{u}}{\partial z} \right) d\Gamma \,, \tag{67}$$

$$\int \left(\Phi \frac{\partial S}{\partial t} + \frac{\partial \Phi}{\partial z} \frac{\nu_T}{\sigma_T} \frac{\partial S}{\partial z} \right) d\Omega = \oint \Phi \left(\frac{\nu_T}{\sigma_T} \frac{\partial S}{\partial z} \right) d\Gamma \,, \tag{68}$$

$$\int \left(\Phi \frac{\partial K}{\partial t} + \frac{\partial \Phi}{\partial z} \nu_T \frac{\partial K}{\partial z} \right) d\Omega = \oint \Phi \left(\nu_T \frac{\partial K}{\partial z} \right) d\Gamma + \int \Phi \left(S_k - \frac{C_\mu^{3/4}}{\ell} K^{3/2} \right) d\Omega \,, \tag{69}$$

where the momentum equation has been written in compact form by introducing the horizontal velocity $\mathbf{u} = (u, v)$, and $\mathbf{f} = f\mathbf{k}$ where $\mathbf{k}$ is the vertical upward unit vector. In these expressions Φ represent global weighting functions. The integrals are taken over the domain Ω, while boundary integrals ($\oint$) result from integration by parts of the diffusion terms. These boundary integrals are used to implement Neumann type boundary conditions in a natural way.

The only non-zero boundary integral here is that for the momentum equation (67), which takes the following form when the actual boundary conditions (46) are included:

$$\oint \Phi \left(\nu_T \frac{\partial \mathbf{u}}{\partial z} \right) d\Gamma = \int_F \Phi(\tau_x^S, \tau_y^S)/\rho_0 \, d\Gamma - \int_B \Phi \, C_B |\mathbf{u}| (u, v) \, d\Gamma \,, \tag{70}$$

where F denotes the free surface and B the bottom.

In case of the salinity equation (68) the boundary integral is zero at the free surface, while a Dirichlet condition is applied at the bottom. For turbulent kinetic energy Dirichlet boundary conditions are used both at the free surface and bottom, hence the boundary integral in (69) is neglected in that equation.

The space discretisation of these equations is performed by use of a Galerkin finite element method, and the time integration is a two-level semi-implicit formulation. This implies that diffusion and Coriolis terms are treated semi-implicitly by use of a theta formulation [30], while other terms are treated explicitly. The resulting algebraic equations have symmetric coefficient matrices, and are solved efficiently by use of a preconditioned (SSOR) conjugate gradient method.

7.2 The Fully Three-Dimensional Case

Discrete Projection Formulation by a Finite Element Method

Here we describe a non-hydrostatic, fully three-dimensional model. This kind of model is well suited e.g., for applications to flow over variable topography. In the present case it is formulated without a free surface, and has been used in a nested model system for local weather prediction [13].

In the following approach we apply the conservative form of the momentum equation together with the anelastic continuity equation:

$$\frac{\partial \mathbf{q}}{\partial t} + \nabla \cdot (\mathbf{u}\mathbf{q}) = -\nabla p' - \rho' g \mathbf{k} + \nabla \cdot (\mu_T \nabla \mathbf{u}) - f\mathbf{k} \times \mathbf{q}\,, \tag{71}$$

$$\nabla \cdot \mathbf{q} = 0\,, \tag{72}$$

where $\mathbf{q} = \rho_o \mathbf{u}$, and ρ_o is the hydrostatic density. This amounts to applying the Boussinesq approximation in the momentum equations and using ρ_o instead of ρ in the anelastic continuity equation; which may be regarded as a good approximation for a large group of geophysical problems.

A mixed finite element method (FEM) of these equation may be written in the following form:

$$M\dot{q} + A(u)q + Du + Cp = b + r + \Gamma\,, \quad C^T q = 0\,, \tag{73}$$

where the variables (q, u, p) are to be interpreted as nodal vectors, and the global matrices are given by

$$M = \int \Phi \Phi^T, \;\; A = \int \Phi u \nabla \Phi^T, \;\; D = \int \nabla \Phi \mu_T \nabla \Phi^T, \tag{74}$$

$$C = -\int \nabla \Phi \Psi^T, \;\; C^T = -\int \Psi \nabla \Phi^T. \tag{75}$$

The buoyancy term acting in the vertical direction is given by

$$b = -\int \Phi \rho' g \mathbf{k}\,, \tag{76}$$

and the Coriolis contribution is

$$r = -\int \Phi f \rho_o (\mathbf{k} \times \mathbf{u})\,. \tag{77}$$

The boundary integral Γ represents a pseudo traction integral, and results from partial integration of the diffusion and pressure terms:

$$\Gamma = \oint \Phi \left(\mu_T \frac{\partial \mathbf{u}}{\partial n} - p\,\mathbf{n} \right). \tag{78}$$

In these expressions Φ and Ψ represent basis functions for velocity and pressure, respectively.

We introduce a 2-level semi-implicit time discretisation

$$M \Delta q^{n+1} + \frac{\Delta t}{2} \tilde{D} \Delta u^{n+1} + \Delta t C p^{n+1} = -\Delta t (A q^n + \tilde{D} u^n + b + r + \Gamma)\,, \tag{79}$$

$$C^T q^{n+1} = 0\,, \tag{80}$$

where $\Delta u^{n+1} = u^{n+1} - u^n$, $\Delta q^{n+1} = q^{n+1} - q^n$, and $\tilde{D}$ is a modified diffusion matrix containing additional contribution from upstream diffusion terms, defined as:

$$\tilde{D} = \int \frac{\partial \Phi}{\partial x_i} \tilde{\mu}_T \frac{\partial \Phi^T}{\partial x_j}, \text{ where } \tilde{\mu}_T = \rho \left(\nu_T \delta_{ij} + \frac{\Delta t}{2} u_i u_j \right). \tag{81}$$

The actual upstream effect may be derived systematically by use of a Taylor-Galerkin formulation, cf. [54] for a derivation.

In the computations we linearise the generalised diffusion term, i.e., the last known values of the velocity and eddy viscosity are used. This makes it possible to decouple the three components of the momentum equation, thus permitting a sequential solution procedure of these equations.

The discretised Reynolds equations (73) may be solved efficiently by use of a projection method. In the following we chose a $Q_1 Q_o$ discretisation for the velocity-pressure interpolation, i.e., multi-linear velocity and constant pressure on the element level. It may be noted that this element type does not satisfy the conventional LBB stability condition (after the mathematicians Ladyzhenskaya, Babuska, and Brezzi), but is 'slightly unstable and highly usable' in the words of Gresho and Sani [21]. This element may also be stabilised [21], and is simple to use because of its low order interpolation.

In order to satisfy the anelastic approximation it is also convenient to assume Q_o interpolation for the density. The following projection formulation is applied:

$$M \Delta \tilde{q}^{n+1} + \frac{\Delta t}{2} \tilde{D} \Delta \tilde{u}^{n+1} = -\Delta t \left(A^n q^n + \tilde{D} u^n \right) + \Delta t \left(b + r + \Gamma \right), \tag{82}$$

$$\Delta t \left(C^T M_L^{-1} C \right) p^{n+1} = C^T \tilde{q}^{n+1}, \tag{83}$$

$$M_L \Delta q^{n+1} = -\Delta t C p^{n+1}, \tag{84}$$

where $\tilde{q}^{n+1}, \tilde{u}^{n+1}$ are prediction values, and M_L represents the lumped (diagonalised) mass matrix. This formulation is similar to Gresho and Chan's "Projection 1 method" [20], here extended to the anelastic case. The boundary is divided into two parts, $\Gamma_D \bigcup \Gamma_N$, for Dirichlet and Neumann boundary conditions, respectively.

Along Γ_N traction conditions are specified, while the velocity is specified along Γ_D. Both prediction and final velocity are given the same specified values:

$$\tilde{u}^{n+1} = \hat{u}^{n+1}, \quad u^{n+1} = \hat{u}^{n+1}, \text{ along } \Gamma_D . \tag{85}$$

Along wall boundaries logarithmic elements may be applied for the velocity with no-slip condition. The basis function normal to the wall then satisfies the non-dimensional velocity condition

$$\frac{u_\tau}{u_*} = \frac{1}{\kappa} ln \left(\frac{z}{z_o} \right), \tag{86}$$

where u_τ is the velocity tangentially to the wall, u_* is the friction velocity, z_o is the roughness parameter, and z is the wall element hight. This formulation is explained in [51].

Note that the pressure equation is derived from the discretised system, and includes conditions from the original system. It is therefore unnecessary to specify explicit pressure boundary conditions, except possibly to fix the pressure at one point. When traction is specified for part of the boundary, no pressure specification is necessary.

In the FEM formulation above we have formally introduced a *group* interpolation, i.e., the same interpolation is used for the variables q and u, although $q = \rho_o u$. Such group formulations are extensively used by Fletcher [17], especially in compressible formulations. In the present context where Q_o interpolation is used for the density (and pressure), this formulation actually reduces to the conventional interpolation on the element basis.

In the practical implementation we need to do some re-writing. For example, we treat $\tilde{u}^{n+1}$ as unknown by writing $\tilde{q}^{n+1} = \rho_o \tilde{u}^{n+1}$, since the hydrostatic density ρ_o is time independent and known. If the dynamic part of the density had been included, we would need to linearise such group terms. The density may then be multiplied by matrix elements when the element matrices are constructed, so that the final form of e.g., equation (82) will be

$$\left(M_\rho + \frac{\Delta t}{2} \tilde{D} \right) \Delta \tilde{u}^{n+1} = -\Delta t \left(A^n q^n + \tilde{D} u^n \right) + \Delta t \left(b + r + \Gamma \right), \qquad (87)$$

where M_ρ includes the density contribution in the mass matrix.

FEM Discretisation of Other Scalar Equations

Scalar equations for temperature or potential temperature (32) may be written in FEM discretised form as

$$M\dot{\phi} + A(u)\phi + D_\phi \phi = 0 \,, \qquad (88)$$

where no source contribution is assumed. The corresponding time-discretised FEM formulation may be written

$$\left(M + \frac{\Delta t}{2} \tilde{D}_\phi \right) \Delta \phi^{n+1} = -\Delta t \left(A + \tilde{D}_\phi \right) \phi^n, \qquad (89)$$

where the time discretisation has been performed as for the momentum equation, namely by use of a Taylor-Galerkin formulation [54]. This again results in a second-order upstream diffusion effect indicated by the added tilde above the diffusion term (see (81) for specification).

Boundary conditions are given in Section 6 as specified temperature along ground and inflow boundaries, and zero normal derivative along outflow boundaries.

The two-equation (K, ϵ) model (42), (43) may be written in FEM discretised form as

$$M\dot{K} + A(u)K + D\,K = M\left[P_k + G_\rho - \epsilon\right], \tag{90}$$

$$M\dot{\epsilon} + A(u)\epsilon + D_\epsilon \epsilon = M\overline{\{\epsilon/K\}}\left[C_1 P_k + C_3 G_\rho - C_2\epsilon\right], \tag{91}$$

where the variables (K, ϵ) are interpreted as nodal vectors, and similarly (P_k, G_ρ) represent the nodal values of the source terms in (45)). The ratio $\overline{\{\epsilon/K\}}$ indicates element averaged values, assumed due to the fact that K and ϵ are interpolated by the same basis functions.

A semi-implicit time discretisation of these equations yields

$$\left[M + \frac{\Delta t}{2}\left(\tilde{D} + \overline{\{\epsilon/K\}}M\right)\right]\Delta K^{n+1} = -\Delta t\left(A + \tilde{D}\right)K^n$$
$$+\Delta t\left(MP_k^n + MG_\rho^n - M\epsilon^n\right), \tag{92}$$

$$\left[M + \frac{\Delta t}{2}\left(\tilde{D}_\epsilon + C_2\overline{\{\epsilon/K\}}M\right)\right]\Delta\epsilon^{n+1} = -\Delta t\left(A + \tilde{D}_\epsilon\right)\epsilon^n$$
$$+\Delta t M\overline{\{\epsilon/K\}}\left(C_1 P_k^n + C_3 G_\rho^n - C_2\epsilon^n\right). \tag{93}$$

In the numerical computations we linearise the coefficient matrices by calculating the generalised diffusion and the ratio $\overline{\epsilon/K}$ from the last known values at time t_n. This implies that the eddy viscosity is also time lagged, and the turbulence equations can be solved in a sequential way, similar to the momentum equations.

Boundary integrals from the diffusion terms have been neglected by assuming zero normal derivatives along open boundaries, which is reasonable for these variables. The wall boundary conditions are given in (49), and inflow conditions are specified as Dirichlet conditions.

Note that the negative part of the source term is treated semi-implicitly due to stability reasons. The source term P_k is numerically simplified by calculating the velocity gradients at the element midpoints. The algebraic relation for the eddy viscosity, $\nu_T = C_\mu \frac{K^2}{\epsilon}$, is satisfied pointwise at the nodes. A possible modification is to write $\nu_T = C_\mu K/\overline{\{\epsilon/K\}}$, which may improve the stability.

8 Some Applications

Results from different applications are shown; most of these are based on the FEM solution procedure described in Section 7.2.

8.1 Turbulent Flow around a Hill: Neutral and Stratified Flow

Hunt and Snyder [23] documented a series of laboratory experiments on flows over a three-dimensional model hill. The following examples are based on two of these experiments, using a hill shape

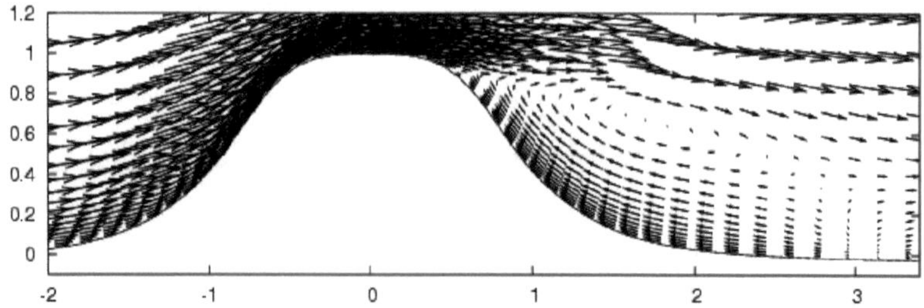

Fig. 1. Neutral flow around a hill; velocity along symmetry plane.

$$h(x,y) = H\left[\frac{1.04}{1+r^4} - \frac{0.083}{1+(r-r_1)^2/a_1^2} - 0.03\right].$$

where $r = \sqrt{(x^2+y^2)}/H$, $r_1 = 20.3/H$, $a_1 = 7.6/H$, and the hill height is given by $H = 22.9$ cm. A symmetry section of of this shape is illustrated in Figure 1.

The inflow profiles for velocity, turbulent kinetic energy, turbulent length scale and dissipation are specified as follows:

$$u = \frac{u_*}{\kappa}\ln(z/z_o)\,, \quad K = \max\left[C_\mu^{-0.5}u_*^2(1-z/\delta)^2,\, K_{min}\right],$$

$$\ell = \frac{\kappa z}{1+4z/\delta}\,, \quad \epsilon = C_\mu^{0.75}K^{1.5}/\ell\,.$$

For the inflow conditions the boundary layer height is $\delta = 0.3H$. The Reynolds number is $Re = 10^4$ based on the hill height, and a small roughness is assumed, $z_o/H = 0.0003$.

The computational domain is given by $-6 \le x/H \le 7$, $-5 \le y/H \le 5$, $0 \le z/H \le 7$, and the present computations are performed on a structured mesh consisting of (89 x 89 x 35) nodes, with elements clustered around the hill and towards the boundary layer.

The numerical model for these examples is a fully three-dimensional finite element model based on the formulation in Section 7.2, in the following called *Simra*, cf. [13].

Neutral Flow

The first example is without stratification, using initial conditions similar to the inflow boundary conditions above, distributed inside the domain.

This flow is characterised by two symmetric eddies downstream of the hill, as illustrated in Figure 2; and along the symmetry plane these eddies combine to a backflow circulation as shown in Figure 1. A simple way to characterise these eddies is to specify their centre at the downstream hill surface, and to identify the recirculation point downstream along the symmetry plane. A

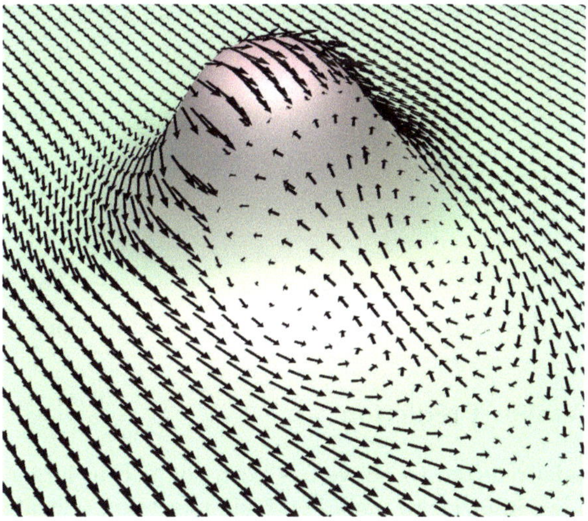

Fig. 2. Neutral flow around a hill; perspective. Velocity field close to ground.

Table 2. Comparison of experimental and numerical results

Case	Vortex centre $(x_c, y_c)/H$	Recirculation x_R/H
Experiment	1.3 ± 0.7	3.5
Computation	1.2 ± 0.7	3.2

comparison between experimental results and computations is shown in Table 2 for these quantities, and indicates a reasonable agreement. Provided that the grid refinement is fine enough, the main reasons for deviations in such flows may often be attributed to inaccuracies in the turbulence model.

Stratified Flow

The nature of the flow may change dramatically when stratification is introduced. This is illustrated in Figure 3 where an exponential potential temperature profile is specified at inflow, namely

$$\theta = T_o \exp\{N^2 z/g\}.$$

Here N is the buoyancy frequency, and we chose a Froude number $U/(NH) = 1$, where H is the hill height and U the free velocity.

Figure 3 shows that the eddies from the neutral flow case have disappeared, and the flow is attached to the hill along the downstream side while an internal wave is developing downstream. The length of this wave is approximately $L/H \approx 6$, and corresponds well with the experimental result given by $L/H = 6.0 \pm 1.5$ [23]. On the other hand, the experimental results indicate a small recirculating zone close to the hill side which is not predicted by the model.

Fig. 3. Stratified flow, N H/U = 1.0. Potential temperature along symmetry plane.

8.2 Local Weather Prediction

The Simra model referred to above has been coupled to larger scale weather prediction models in order to include terrain effects on a local scale. Here we show an example of this nesting applied to a local mountainous terrain around an airport. The operational weather prediction model for this case has a grid spacing of 10 km, and gives input to a regional scale model using 1 km grid spacing. The local scale Simra model is finally nested into the system using a mesh of about (0.1 – 0.2) km grid spacing horizontally, and down to 2 m vertically. The latter nesting is a down-scaling of the regional scale model, and we use stationary boundary conditions from the regional model.

This system has been described in more detail elsewhere [13], and here we only show some results in order to illustrate the kind of simulations performed on the smallest scale. Figure 4 illustrates the local terrain which is situated around an actual airport (Værnes, near Trondheim, Norway). The horizontal dimensions of the local domain are (12 x 12) km, and vertically the mesh reaches about 2 km above sea level. The mesh has about 500 000 nodes, and the simulation was run in a parallelised version.

A main intention of the actual model system is to predict the flight conditions close to the airport, and in a longer perspective to be of practical use for aviation traffic, especially related to hazardous wind conditions.

One of the most relevant parameters for turbulence prediction is the turbulence intensity ($\sqrt{K}/U_\infty$). This quantity is shown in Figure 5 along a horizontal section of about 150 m above the airport (sea level). Figure 6 shows the same quantity in a vertical section along a landing flight trajectory. These figures illustrate the effect of mountainous terrain, where areas of stronger turbulence follow in the wake behind such terrain. In addition, stratification induces some of the wave-like effects indicated by the figures.

There exists only a few point measurements to compare with, however for this particular case subjective registrations of turbulence were also obtained from a small aircraft. The general trend from these registrations seems to agree reasonable with the simulated result. Further details are given in [13].

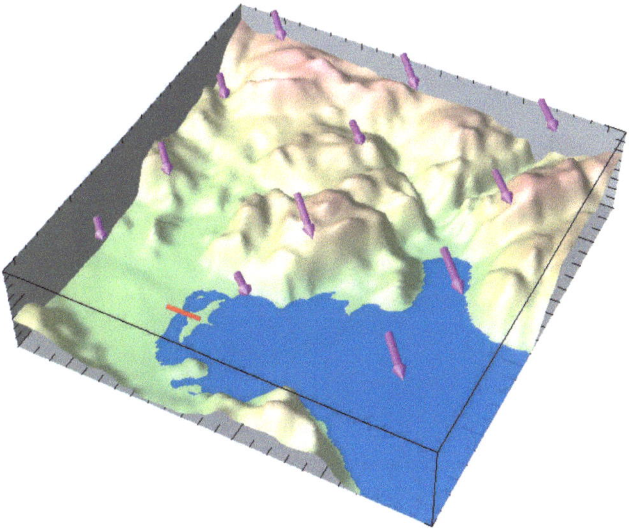

Fig. 4. Terrain over Værnes; the airport is indicated by a dense line. South-east wind at 500 m height is illustrated by arrows. The horizontal dimensions of the domain are (12 × 12) km. The maximum terrain height is about 500 m.

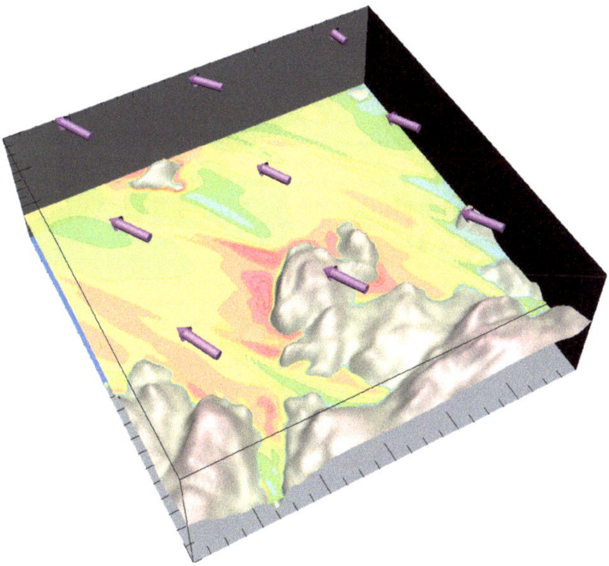

Fig. 5. Predicted turbulence intensity at 150 m height. Fairly strong turbulence ($\sqrt{K}/U_{max} \approx 0.14$) is indicated in the wake behind mountainous terrain.

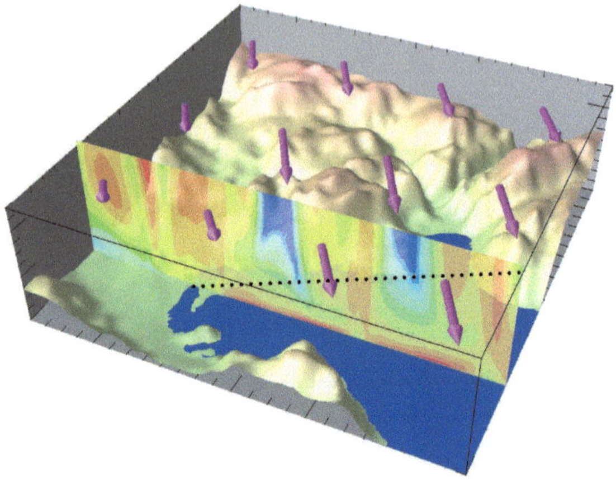

Fig. 6. Predicted turbulence intensity in a vertical section along the approach to landing from west. The dotted line indicates the incoming flight trajectory. The wave-like turbulence variations are effects from the upstream mountainous terrain.

In order to use this model system in a prediction mode, all models have been parallelised, and are presently being tested for that purpose. Such systems may be used operational to forecast possible hazardous wind events related to flight conditions, and thereby increase aviation safety.

The Rossby number for the local prediction is of the order $Ro \approx 10$, hence the rotational effect is negligible. This has also been verified by comparing results with and without Coriolis terms included in the momentum equation.

8.3 Hydrostatic vs Non-Hydrostatic Modelling Applied to Upwelling in a Narrow Basin

Our final example is an application to wind induced current in a closed water basin with upper layer stratification. The specifications are as follows:

The length (L) and depth (H) of the basin are $(L, H) = (1000, 50)$ m, and the initial upper layer of light water is $h = 10$ m, with a density difference of $\Delta\rho = 4\ kg/m^3$. A constant wind stress of $\tau_s = 5$ Pa is applied to the surface. This implies a *Wedderburn number* of the size

$$W = \frac{\Delta\rho g h^2}{\tau_s L} = 0.8\,.$$

From simplified theoretical considerations this implies that the density interface (also called pycnocline) is expected to reach the surface during the first half internal wave period (see [38]). This situation is referred to as upwelling, and is expected when $W < 1$. The internal wave period is estimated from linear theory as

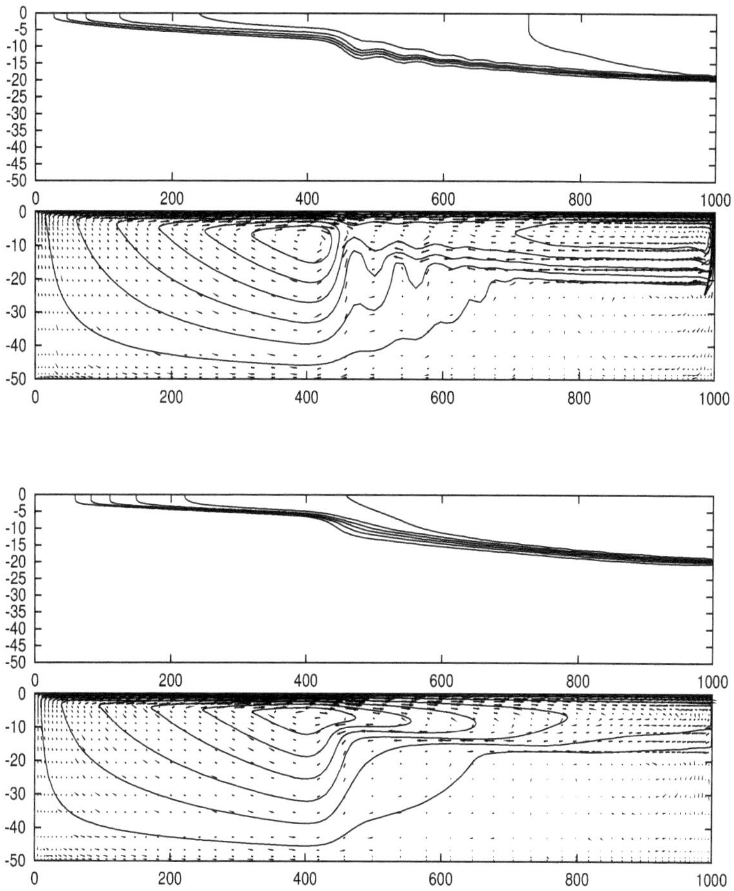

Fig. 7. Wind induced internal wave in a narrow basin. Comparison of hydrostatic and non-hydrostatic predictions. Upper two figures: Non-hydrostatic prediction, density interface and velocity field. Lower two figures: Hydrostatic prediction, density interface and velocity field.

$$T = \frac{2L}{\sqrt{gh\Delta\rho/\rho_o}} ,$$

where ρ_o is the reference (or mean) density.

Numerical simulations have been performed for this case by use of two different models: The non-hydrostatic Simra model and a hydrostatic model; the latter is based on the formulation in Section 3.3, and is similar to the one used in [49]. Figure 7 illustrates results after a time of $T/2$, i.e., half a period according to the estimate above. Two sets of figures are shown: The two upper figures are for the non-hydrostatic case, and the two lower figures are for the corresponding hydrostatic case. The upper figure of each set illustrates isolines for density contours, while the lower figures show current fields indicated by arrows and stream function contours. Typical tendencies for both cases are a fairly sharp vertical density interface, with a jet-like return velocity on the right side of the basin, just above the density interface. At this time instant the jet has reached the middle of the basin, where we notice some small scale disturbances and a relatively pronounced downstream velocity, while the density interface is tilted up-wards.

We note that the two different solutions are fairly similar. Some small scale velocity disturbances seem to be smoothed in the hydrostatic case, and this is reflected in the density field. However, the two models use different turbulence closures, and the small scale difference may as well be caused by this difference. In addition, a certain difference in time evolution may be observed between the two results. However, in conclusion it seems that the hydrostatic approximation is well suited to simulate upwelling, even though the vertical velocity component is an important part in such situations.

References

1. Apel, J.R.: Principles of Ocean Physics. Academic Press, London New York Sydney Tokio (1990).
2. Baines, P.G.: Topographic Effects in Stratified Flows. Cambridge University Press, Cambridge (1998).
3. Batchelor, G.K.: An Introduction to Fluid Dynamics. Cambridge University Press, Cambridge (1967).
4. Belcher, S.E., Hunt, J.C.R.: Turbulent flow over hills and waves. Annu. Rev. Fluid Mech. (1998), **30**, 507–538.
5. Blumberg, A.F., Herring, H.J.: Circulation modelling using orthogonal curvilinear coordinates. In: Nihoul, J.C.J., Jamart, B.M. (eds) Three-dimensional Models of Marine and Estuarine Dynamics. Elsevier Oceanography Series, Amsterdam (1987).
6. Boyer, D.L., Davies, P.A.: Laboratory studies of orographic effects in rotating and stratified flows. Annu. Rev. Fluid Mech. (2000), **32**, 165–202.
7. Bryan, K., Cox, M.D.: An approximate equation of state for numerical models of ocean circulation. J. Phys. Oceanogr. (1972), **2**, 510–514.
8. Burchard, H., Petersen, O., Rippeth, T.P.: Comparing the performance of the k-ϵ and the Mellor-Yamada two-equation turbulence models. J. Geophys. Res. (1998), **103**, 10543–10554.
9. Chorin, A.J., Marsden, J.E.: A Mathematical Introduction to Fluid Mechanics. Springer, New York Berlin Heidelberg Tokyo (1984).
10. Dornback, A., Shumann, U.: Numerical simulation of turbulent convective flow over wavy terrain. Boundary-Layer Meteorology (1993), **65**, 323–355.
11. Durran, D.R.: Numerical methods for wave equations in geophysical fluid dynamics. Springer, New York Heidelberg (1999).
12. Eidsvik, K.J.: Some contributions to the uncertainty of sediment transport predictions. Continental Shelf Res. (2004), **24**, 739–754.

13. Eidsvik, K.J., Holstad, A., Lie, I., Utnes, T.: A prediction system for local wind variations in mountainous terrain. Boundary-Layer Meteorology (2004), **112**, 557–586.
14. Eidsvik, K.J., Utnes, T.: Flow separation and hydraulic transitions over hills modelled by the Reynolds equations. J. Wind Engineering and Industrial Aerodynamics (1997), **67 & 68**, 403–413.
15. Ferziger, J.H.: Large eddy simulation. In: Gatski, T.B., Hussaini, Y., Lumly, J.L. (eds) Simulation and Modeling of Turbulent Flows. Oxford University Press, New York Oxford (1996).
16. Ferziger, J.H:, Peric, M.: Computational Methods for Fluid Dynamics. Springer, Berlin Heidelberg New York (1997).
17. Fletcher, C.A.J.: The group finite element method. Comput. Methods Appl. Mech. Engrg. (1983), **37**, 225–244.
18. Gatsky, T.B., Speziale, C.G.: On explicit algebraic stress models for complex turbulent flows. J. Fluid Mech. (1993), **254**, 59–78.
19. Gill, A.E.: Atmosphere-Ocean Dynamics. Academic Press, New York London Sydney Tokyo Toronto (1982).
20. Gresho, P.M., Chan, S.T.: On the theory of semi-implicit projection methods for viscous incompressible flow and its implementation via a finite element method that also introduces a nearly consistent mass matrix. Part 2: Implementation. Int. J. Numer. Methods Fluids (1990), **11**, 621–659.
21. Gresho, P.M., Sani, R.L.: Incompressible flow and the finite element method. Advection-diffusion and isothermal laminar flow. Wiley, Chichester New York Weinheim Brisbane Singapore Toronto (1998).
22. Holmedal, L.E., Utnes, T.: Physical-biological interactions and their effect on phytoplancton blooms in fjords and near-coastal waters. Submitted (2004).
23. Hunt, J.C.R., Snyder, W.H.: Experiments on stably and neutrally stratified flow over a model three-dimensional hill. J. Fluid Mech. (1980), **96**, 671–704.
24. Johansson, A.V.: Engineering turbulence models and their development, with emphasis on explicit algebraic Reynolds stress models. In: Oberlack, M., Busse, F.H. (eds) Theories of turbulence. Springer, Wien New York (2002).
25. Kaimal, J.C., Finnegan, J.J.: Atmospheric Boundary Layer Flows. Oxford University Press, Oxford (1994).
26. Kowalik, Z., Murty, T.S.: Numerical Modeling of Ocean Dynamics. World Scientific, Advanced Series on Ocean Engineering, Vol 5, Singapore New Jersey London Hong Kong (1993)
27. Kranenburg, C.: A K-Model for stably stratified nearly horizontal turbulent flows. Delft University of Technology (1985), Report No 4–85.
28. Kundu, P.K.: A Numerical Investigation of Mixed-Layer Dynamics. J. Phys. Oceangr. (1980), **10**, 220–236.
29. Kundu, P.K., Cohen, I.M.: Fluid Mechanics, Third Edition, Chapter 14. Elsevier Academic Press, Amsterdam Boston Heidelberg London (2004)
30. Lambert, J.D.: Numerical methods for ordinary differential systems. Wiley, Chichester New York Brisbane Toronto Singapore (1991)
31. Launder, B.E.: On the effects of a gravitational field on the turbulent transport of heat and momentum. J. Fluid Mech. (1975), **67**, 569–581.
32. Leendertse, J.J., Liu, S.K.: A three-dimensional model for the estuaries and coastal seas. Vol II, Aspects of Computation. RAND Corporation. R-1764-OWRT.

33. Lynch, D.R., Werner, F.E.: Three-dimensional hydrodynamics on finite elements. Part II: Non-linear time-stepping model. Int. J. Numer. Methods Fluids (1991), **12**, 507–533.
34. Mellor, G.L.: Introduction to Physical Oceanography. Springer, New York Berlin Heidelberg (1996).
35. Mellor, G.L., Yamada, T.: Development of a turbulence closure model for geophysical fluid problems. Pev. Geophys. Space Phys. (1982), **20**, 851–875.
36. Moe, H., Ommundsen, A., Gjevik, B.: A high resolution tidal model for the area around The Lofoten Islands, northern Norway. Continental Shelf Res. (2002), **22**, 485–504.
37. Mohammadi, B., Pironneau, O.: Analysis of the K-Epsilon Turbulence Model. Wiley, Chichester New York Brisbane Toronto Singapore (1994).
38. Monismith, S.: An experimental study of the upwelling responce of stratified reservoirs to surface shear stress. J. Flid Mech. (1986), **171**, 407–439.
39. Pedlosky, J.: Geophysical Fluid Dynamics. Springer, New York Heidelberg Berlin London Paris Tokyo (1987).
40. Pope, S.B.: Turbulent Flows. Cambridge University Press, Cambridge (2001).
41. Raitby, G.D., Stubley, G.D., Taylor, P.A.: The Askervein Hill project: A finite control volume prediction of three-dimensional flows over hill. Boundary-Layer Meteorology (1987), **39**, 247–267.
42. Riley, J.J., Lelong, M.P.: Fluid motions in the presence of strong stable stratification. Annu. Rev. Fluid Mech. (2000), **32**, 613–657.
43. Rodi, W.: Turbulence models and their applications in hydraulics. IAHR Monograph, Int. Assoc. for Hydraulic Res., Delft (1984).
44. Speziale, C.G.: Modelling of turbulent transport equations. In: Gatski, T.B., Hussaini, Y., Lumly, J.L. (eds) Simulation and modeling of turbulent flows. Oxford University Press, New York Oxford (1996).
45. Stull, R.B.: An Introduction to Boundary Layer Meteorology. Kluwer Academic Publishers, Dordrecht Boston London (1988).
46. Svensson, U.: A mathematical model of the seasonal thermocline. Rep. 1002, Dep. of Water Resour. Eng., Univ. of Lund, Sweden (1978).
47. Tennekes, H., Lumley, J.L.: A First Course in Turbulence. MIT Press (1976).
48. Turek, S.: Efficient Solvers for Incompressible Flow Problems: An Algorithmic and Computational Approach. Computer Science and Engineering, **6**, Springer, Berlin (1999).
49. Utnes, T., Brørs, B.: Numerical modelling of 3-D circulation in restricted waters. Applied Math. Modelling (1993), **17**, 1–14.
50. Utnes, T., Eidsvik, K.J.: Turbulent flows over mountainous terrain modelled by the Reynolds equations. Boundary-Layer Meteorology (1996), **79**, 393–416.
51. Utnes, T., Meling, T.S.: Treatment of turbulent wall boundary conditions using linear-logaritmic elements. Comput. Methods Appl. Mech. Engrg. (1998), **169**, 123–134.
52. White, F.M.: Viscous Fluid Flow. McGraw-Hill, Second Edition New York (1991).
53. Xu, D., Ayotte, K.W., Taylor, P.A.: Development of a non-linear finite difference model for turbulent boundary-layer flow over topography. Boundary-Layer Meteorology (1994), **70**, 341–367.
54. Zienkiewicz, O.C., Taylor, R.L.: The Finite Element Method. Volume 3: Fluid Dynamics. Butterworth-Heinemann, Oxford Auckland Boston Johannesburg Melbourne New Delhi (2000).

Part III

Optimization

Overview

In everyday language, optimization is about what is best. It deals with actions, processes, and methods that make something as perfect, functional, or effective as possible. Mathematical optimization denotes the branch of scientific computing that utilizes mathematical models, resolution methods, and algorithms for solving optimization problems. Applications are found almost everywhere, e.g., in mathematics, the physical, chemical, and biological sciences, engineering, architecture, operations research, economics, and management.

Central to mathematical optimization is the concept of a *mathematical program*. It consists of an objective function to be minimized or maximized, plus a set of constraints, all defined over a number of variables. Mathematical programs come in many guises, depending on the characteristics of the objective, the structure of the feasible region defined by the constraints, and the domain of variables. A particular case is *linear programming* (LP), where the objective and the constraints are linear. Highly efficient resolution methods have been developed for LP, and a tool industry of LP solvers has emerged. Such tools routinely solve LPs, maybe with millions of variables and constraints, to find the optimal solution to some real-life application. In many applications, a LP formulation is adequate for the optimization problem at hand. In these applications, modeling is often the most important challenge.

If LP is not an adequate formulation, non-linear mathematical programs may still be effectively solvable. For non-convex programs, where either the objective or the feasible region, or both are non-convex, there is normally a difficult search landscape with many local optima. A steepest-descent type method is inadequate, as it will get trapped in a local optimum. Strategies for exploring a larger part of the search landscape are needed.

In many real-world applications, discrete choice is an inherent part of the optimization problem and discrete variables must be used in a mathematical formulation. Typical examples are found in Operations Research, where discrete variables are needed to model discrete choice, for instance regarding machine allocations and sequencing of jobs in a production schedule. If all or some of the variables in a mathematical program are restricted to discrete domains, the computational complexity of the problem usually changes drastically for the worse. According to complexity theory, a computationally efficient algorithm for the so-called **NP**-hard problems does not exist. For large-size instances and realistic response times, one often needs to give up the quest for optimality and resort to approximation methods.

Optimization and Operations Research have a long history in several parts of SINTEF. At SINTEF Applied Mathematics, optimization activities originate from the mid 1980s. At the time, a new department on Knowledge-Based Systems originated from the Department of Computer-Aided Design. Soon, one of the focal areas became computationally hard planning problems in manufacturing and the process industries. Heuristic search techniques were developed, e.g., to make optimized production schedules in aluminum foundries, an important part of Norwegian process industries. In the mid 1990s, the optimization group merged with geometry and numerical simulation groups to form SINTEF Applied Mathematics. For the past ten years or so, the group has focused on applications with hard optimization problems where there is a need to develop tailored resolution algorithms. The strong algorithmic focus is common to SINTEF Applied Mathematics as a whole. Another common denominator is the focus on software reuse. Early on, the optimization group developed its own basic optimization software library called SCOOP.

The largest market area for the optimization group has been, and still is, transportation, a particularly important part of the Norwegian economy. A central problem to transportation of goods and people is the *Vehicle Routing Problem* (VRP), a generalization of the well-known Traveling Salesman Problem. Industrial, rich variants of the VRP, both in land-based and sea-based transportation, have been studied through numerous research projects over more than a decade. This research has been conducted in the context of industrial contracts for end users and tool vendors, as well as more basic research efforts funded by the European Commission, the Research Council of Norway, and internally by SINTEF.

A generic VRP solver named SPIDER is a concrete software product of this research. SPIDER is a component of several routing tools. In addition to making research based innovations for our clients, an important role for SINTEF is to commercialize and spawn new companies based on successful results. In 1999 a company called GreenTrip AS (now: SPIDER Solutions AS) was established by SINTEF to commercialize SPIDER. The company has the largest market share of routing tools in Norway, and now starts expanding into the international market. There is currently a strategic buildup of activities in maritime transportation that is based on competence and software from three SINTEF units including Applied Mathematics. Forestry, manufacturing, and the financial sector are other important market areas for the optimization group. The health and petroleum sectors are main targets for expansion.

The Optimization part of this book consists of five high-quality contributions. Together, they exemplify a broad spectrum of real-life applications of mathematical optimization. Four chapters describe operations research, whereas one is about novel applications of integer programming to image analysis. All chapters focus on modeling challenges. The important theme of stochastic optimization is a major part of one paper. Three chapters have development of efficient algorithms as a major theme. Computational testing is an integral part of four of the works reported.

The Industrial Vehicle Routing contribution by Hasle and Kloster gives an introduction to the Vehicle Routing Problem as it is defined in the scientific literature, and briefly describes exact methods and heuristics. Based on experience from applied VRP research at SINTEF Applied Mathematics alluded to above, important aspects of industrial VRPs are surveyed and contrasted with idealizations that are often made in academic research. The generic, uniform, heuristic algorithmic approach implemented in SPIDER for solving rich

VRPs is described, along with computational results on standard benchmarks from the OR literature. Some comments are also given on the development and commercialization of SPIDER through a spin-off company. The chapter concludes with important topics for further research.

Forestry is important to Norwegian economy. Lately, forestry in Norway, as in other countries, has been faced with new and demanding regulations related with bio-diversity, recreational value, and other environmental concerns. In forestry, environmental objectives must be balanced with economy in a sustainable way over a time horizon of decades or even centuries. The chapter by Stølevik et al. explains the increasing importance of the Long-term Forest Treatment Scheduling Problem and presents work at SINTEF Applied Mathematics for solving it. The authors present a rich optimization model, contrast it with existing literature, and describe a heuristic method based on the tabu search metaheuristic for solving it. A brief introduction to local search and metaheuristics is given for readers unfamiliar with these topics. As there are no relevant benchmarks, the method has been assessed through test data from a real-life case in Norway. The method is implemented in a system called Ecoplan that has been deployed at a Norwegian forest owner organization.

Image analysis has many important applications, e.g., remote sensing, medicine, and industrial inspection. Integer programming is a very recent approach. The chapter by Dahl and Flatberg describes results from collaboration between the Center of Mathematics for Applications, a center of excellence in applied mathematics organized at the University of Oslo, and SINTEF Applied Mathematics. The authors discuss image segmentation and image reconstruction, two important subproblems in image analysis. A general mathematical framework for analysis of two-dimensional discrete images is given. The two problems are formulated as integer linear programs. An exact method based on Lagrangian decomposition and a subgradient technique for solving the Lagrangian dual is presented and discussed. The method has been tested on a set of synthetic test problems and compared with a well-known commercial integer programming solver. Results are good and have already motivated more research in this direction.

The two last chapters both deal with the industrially important and scientifically challenging topic of supply chain coordination. The contribution by Christiansen and Grønhaug, both representing The Norwegian University of Science and Technology (NTNU) in Trondheim, concerns supply chain design in metallurgical industries. Globalization and increased competition has put pressure on profit margins. The companies need to optimize their supply chains. In collaboration with the Norwegian company Elkem, the largest producer of silicon metal in the world, the authors have developed a mixed integer model for supply chain design. The model supports strategic decisions regarding plant locations, product mixes, production capacities, and production of by-products. The latter issue is particularly important and needs to be taken into account in these industries, as by-products such as microsilica have considerable value. The chapter describes the application in some detail, and presents a Mixed Integer Linear Programming model. The model has been tested on real data from Elkem's silicon division, utilizing a commercial MILP solver. Testing shows that the value of by-products has significant influence on optimal supply chain design. The model has now been implemented at Elkem.

The final chapter by Tomasgard et al. gives an integrated view on how to model value chains in the petroleum industry. In particular, the paper describes the natural gas value chain, as seen from an upstream petroleum company. Such a company needs to design and operate its fields in an optimized way, taking long term contracts, short term markets, and transportation capacity into consideration. The authors gradually introduce complexities and level of detail in their model. Gas transportation is full of non-linearities, and the chapter suggests how to deal with them. Stochastic programming is introduced to deal with uncertainty. The authors arrive at a comprehensive stochastic portfolio optimization model. The work was performed at NTNU, in collaboration with its "twin department" at SINTEF in Trondheim. More detailed versions of the models introduced are the bases of decision-support systems currently in use by industry for their operations on the Norwegian continental shelf.

Lately, this editor has perceived an increased interest in optimization from industry and the public sector. It is my hope that this book will contribute to a strengthening of this trend, and a stronger demand for research on applied mathematics in general. Equally important, the editors hope that this book will motivate more students and researchers in academia to mathematical studies of real-life problems. In case, this book is a small, but important contribution to the SINTEF vision: "Technology for a better society".

Geir Hasle
Chief Scientist
Editor, Optimization

Industrial Vehicle Routing

Geir Hasle and Oddvar Kloster

Summary. Solving the Vehicle Routing Problem (VRP) is a key to efficiency in transportation and supply chain management. The VRP is an NP-hard problem that comes in many guises. The VRP literature contains thousands of papers, and VRP research is regarded as one of the great successes of OR. Vehicle routing decision support tools provide substantial savings in society every day, and an industry of routing tool vendors has emerged. Exact methods of today cannot consistently solve VRP instances with more than 50–100 customers in reasonable time, which is generally a small number in real-life applications. For industrial problem sizes, and if one aims at solving a variety of VRP variants, approximation methods is the only viable approach. There is still a need for VRP research, particularly for large-scale instances and complex, rich VRP variants. In this chapter, we give a brief general introduction to the VRP. We then describe how industrial requirements motivate extensions to the basic, rather idealized VRP models that have received most attention in the research community, and how such extensions can be made. At SINTEF Applied Mathematics, industrial variants of the VRP have been studied since 1995. Our efforts have led to the development of a generic VRP solver that has been commercialized through a spin-off company. We give a description of the underlying, rich VRP model and the selected uniform algorithmic approach, which is based on metaheuristics. Finally, results from computational experiments are presented. In conclusion, we point to important issues in further VRP research.

Key words: Transportation, Logistics, Optimization, VRP, Modeling, Approximation, Metaheuristics, Routing Tool.

1 Introduction

Efficient transportation logistics is increasingly important in society of today. In the EU, the transportation sector amounts to more than 10 % of the GDP and employs 10 million people [26]. In Norway, there are some 17,000 transportation companies with total revenue of 44 billion NOK. In 2002, the volume of lorry based road transport was 12.7 billion ton-kilometers. Total capacity utilization was 46.7 % [45]. Transportation demand is not geographically balanced in Norway. This is one important reason for the low utilization. Moving to the global picture, lack of coordination is another major reason for low efficiency. With few exceptions, transportation planning is performed manually today. However, sophisticated software tools for route design are now implemented at an increasing rate.

Solving *The Vehicle Routing Problem* (VRP) is a key to efficient transportation management and supply-chain coordination. In broad terms, it deals with the optimal assignment of a set of transportation orders to a fleet of vehicles and the sequencing of stops for each vehicle. The objective is to minimize total transportation costs. Often, it is a combination of fleet acquisition / depreciation costs and driving costs for the routing plan. The VRP has a large number of real-life applications and comes in many guises, depending on the type of operation, the time frame for decision making, the objective, and the types of constraint that must be adhered to. Outside of transportation logistics, the VRP has less intuitive but still important applications, e.g., in robotics and VLSI design.

The VRP is a computationally hard, discrete optimization problem. It is highly unlikely that an algorithm that solves any VRP instance to optimality in reasonable time will ever be found. According to most qualified researchers, there exists no VRP algorithm that guarantees to find an optimal solution in polynomial time of instance size for an arbitrary instance. From a mathematical perspective, this is somewhat dissatisfactory. Depending on VRP variant, exact methods today have a practical size limit of 50 – 100 orders. We cannot expect that this limit will be increased substantially. On the other hand, our ability to find good solutions to practical variants of the VRP, given a reasonable response time, has increased tremendously since the problem was introduced by Dantzig and Ramser in 1959 [18]. This is not only due to a general increase in computing power, but also to substantial methodological improvements in exact methods and heuristics for the VRP.

VRP research is regarded as one of the great successes of Operations Research. The results have lead to a tool industry in route design and fleet management, through which VRP research results yield huge economical and environmental savings. Vendors of routing tools with optimization typically claim a cost savings potential of 5 % – 30 %. The huge transportation volumes illustrated above will make even 2 % savings substantial.

The "engine" of an advanced routing tool is a VRP solver, i.e., a software component with functionality for modeling instances of targeted variants of the VRP and finding feasible, optimized solutions to a given instance. The effect of a given routing tool is highly dependent on the quality of its VRP solver. First, the applicability and flexibility of the tool is determined by the *richness* of the underlying VRP model, i.e., to which degree it expresses aspects of real-life routing applications. Second, the algorithmic performance of the VRP solver, for instance measured as quality of the objective as a function of CPU time for typical instances, determines the logistics improvement potential of the routing tool. Of course, there are other important aspects that influence tool utility, e.g., the quality of input data, the user interface, interfaces to relevant information systems, and organizational issues, to name a few.

A current trend in VRP research is to study *rich VRP models* and larger-size problems. This trend partly stems from an application pull, but also from forces within the scientific community. From a pragmatic perspective, the "simpler" variants of the VRP are regarded as being solved. Hence the VRP research community turns to more challenging variants. The current bulk of work on rich and large-size VRPs, as well as the rate of algorithmic improvements exhibited in the literature, lead us to believe that VRP will be studied by many researchers for a long time ahead. As a result, we expect further, significant advances in our ability to solve the VRP. The road from scientific improvements via tool industry to effects in end users is short.

As a contract research organization, one of SINTEF's main goals is to create research based innovations for our clients. It is also to commercialize and spawn new companies based on results from our research. SINTEF Applied Mathematics has performed applied research on the VRP since 1995. This research has been conducted in the context of industrial contracts for end users and tool vendors, as well as more basic research funded by the European Commission, the Research Council of Norway, and internally by SINTEF. A generic VRP solver named SPIDER is a concrete software product of this research. SPIDER is a component of several routing tools. In 1999 a company called GreenTrip AS was established by SINTEF to commercialize SPIDER.

In this chapter, we shall provide a general treatment of the VRP and relevant resolution methods, but also use SPIDER as a vehicle to illustrate applied VRP research, and to give insights into how a generic VRP solver is composed. Modeling and algorithmic issues related with rich variants of the VRP will be focused.

The chapter is organized as follows. In Section 2, we give an introduction to the VRP and present basic VRP variants. We proceed to briefly describe resolution approaches and give pointers to literature. In Section 3, we give some general insights into industrial applications of the VRP, and present how one accommodates some of the richer aspects of industrial routing problems in an extended formulation. In Section 4 we describe VRP research at SINTEF and the SPIDER solver for rich VRPs. The section contains a model description, and a survey of the unified algorithmic approach to rich VRPs that is utilized by SPIDER. The section is concluded by comments on the common empirical evaluation approach in VRP research, as well as some results of computational experiments with SPIDER. Section 5 concludes this chapter with a summary and pointers to important issues for future VRP research.

2 The Vehicle Routing Problem

The Vehicle Routing Problem (VRP) was first described in the Operations Research literature in [18]. Since then, thousands of papers have been written on variants of the VRP. The basic, classical Capacitated VRP (CVRP) is informally described as follows:

> *A number of identical vehicles with a given capacity are located at a central depot. They are available for servicing a set of customer orders, (all deliveries, or, alternatively, all pickups). Each customer order has a specific location and size. Travel costs between all locations are given. The goal is to design a least-cost set of routes for the vehicles in such a way that all customers are visited once and vehicle capacities are adhered to.*

The CVRP has two basic extensions: the Distance Constrained VRP (DVRP), and the VRP with Time Windows (VRPTW). For the DVRP, there is an additional constraint on each route: a maximum length or duration of each route must not be exceeded. In the VRPTW, customer service must start within a given time period specific to each customer, and there is a duration constraint on all routes.

2.1 Definition of Basic VRP Variants

More formally, the classical CVRP is defined on a simple digraph $\mathbf{G} = (\mathcal{N}, \mathcal{A})$, where $\mathcal{N} = \{0\} \cup \mathcal{C} \cup \{n+1\}$. $\mathcal{C} = \{1, \ldots, n\}$ represents the set of customers. 0 and $n+1$ both represent the depot[1]. The arcs in $\mathcal{A} \subseteq \mathcal{N} \times \mathcal{N}$ represent travel possibilities between nodes. $\mathbf{G}$ is usually complete. A non-negative travel cost c_{ij} is associated with each arc. A transportation order exists for each customer, each with a non-negative demand d_i in some capacity dimension. It says that the demand must be delivered from the depot to the customer[2]. $\mathcal{V} = \{1, \ldots, K\}$ indexes a set of identical vehicles, each with capacity q, that are available at the depot. In a consistent CVRP we have $q \geq d_i, i \in \{1, \ldots, n\}$. The goal in CVRP is to find a solution consisting of K *circuits*, i.e., one round trip *tour* for each vehicle, starting and stopping at the depot, with minimal total travel cost. All customers must be served exactly once (i.e., all transportation orders must be serviced, and no split deliveries are allowed), and the total demand for each tour must not exceed q.

If the matrix $[c_{ij}]$ is symmetric, the CVRP is called a Symmetric Capacitated VRP (SCVRP). In this case, the circuits may be traversed in either direction. Otherwise, it is an Asymmetric CVRP (ACVRP), where tour direction has significance. The objective to be minimized is the sum of arc costs c_{ij} over all tours.

---

[1] As we shall see, this is convenient for mathematical formulation.

[2] Equivalently, all orders specify pickups from the customers to the depot.

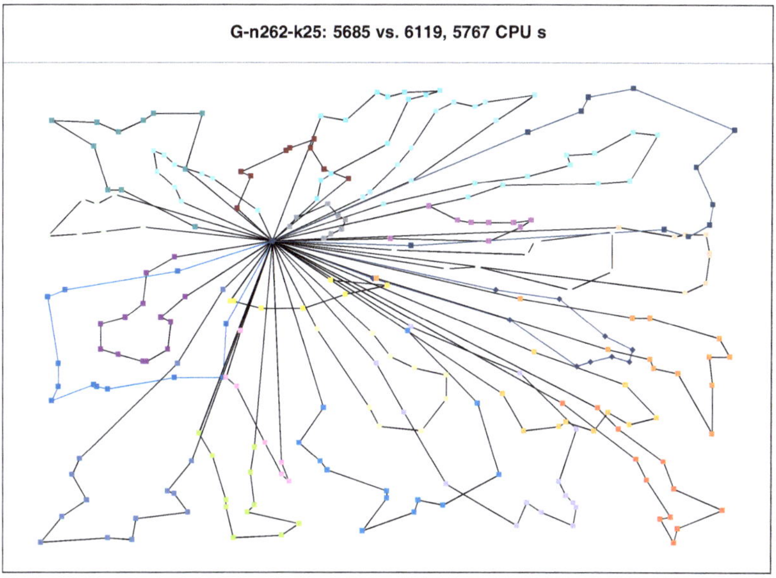

Fig. 1. The best known solution to the CVRP test instance G-N262-K25.

The optimal solution often requires fewer than K tours. Initial fixed costs for non-empty tours can be added, in which case there will be an explicit reward for solutions with fewer than K tours. For the CVRP, the minimum number of tours is obviously determined by solving a bin packing problem, as vehicle capacity is the only constraint that determines this number. This is computationally efficient even for instances with many customers. For most other VRP variants, this is not a valid approach.

In the literature, customer locations are often represented as points in the Euclidean plane, and the travel cost c_{ij} is modeled as Euclidean distance. Hence, the problem is symmetric and the triangle inequality holds. This is clearly an idealized model of real-life applications. For the DVRP, the distance constraint is often interpreted as a temporal constraint and time is used for travel cost. In case, the duration of service at each customer is easily included in arc costs, as explained for the VRPTW below.

Figure 1 shows an example of a solution to a Euclidean CVRP, namely the best known solution (June 2006) to the standard, synthetic CVRP test instance G-n262-k25 [29]. The instance has 261 customers and a limit of 25 vehicles.

The CVRP and DVRP are obvious generalizations of the Traveling Salesman Problem (TSP), where the goal is to find a minimum cost Hamiltonian cycle between given cities, i.e., a round trip that visits each city once. According to complexity theory (see for instance [27]), the TSP belongs to a class of problem types called **NP**-hard problems. Most serious researchers believe that **NP**-hard problems are not effectively solvable. Here, effectively solvable means that there exists a deterministic algorithm that guarantees to find the optimal solution within a computing time that is bounded by a polynomial of instance size, i.e., the number of cities for the TSP. Since the TSP is **NP**-hard in the strong sense[3], the CVRP, DVRP, and VRPTW are also **NP**-hard in the strong sense.

Recently, the VRPTW generalization of the CVRP has been much studied [16], [7], [8]. In addition to capacity and route length constraints there are individual time windows $[a_i, b_i]$ for each customer, and possibly also on the depot. The interpretation of time window is as follows. Service at customer i must start before b_i and cannot start before a_i in a feasible solution. A vehicle is allowed to wait at a customer upon early arrival. No extra cost is incurred for such waiting. In the classical formulation, the objective is simply the total travel cost. An alternative, richer model includes vehicle startup costs in the objective. Another model has a hierarchical objective, where the primary objective is to minimize the number of vehicles used, and the secondary is the total driving costs. The hierarchical objective is formulated as a single objective by adding penalties of sufficient size to the arc from the depot to the first customer for all non-empty tours.

For conciseness, and to illustrate how to model discrete optimization problems, we now present a well-known (see for instance [47]) mathematical formulation of the VRPTW.

$$\text{minimize} \sum_{k \in \mathcal{V}} \sum_{(i,j) \in \mathcal{A}} c_{ij} x_{ij}^k \tag{1}$$

$$\sum_{k \in \mathcal{V}} \sum_{j \in \mathcal{N}} x_{ij}^k = 1, \forall i \in \mathcal{C} \tag{2}$$

$$\sum_{(i,j) \in \mathcal{A}} d_i x_{ij}^k \leq q, \forall k \in \mathcal{V} \tag{3}$$

$$\sum_{j \in \mathcal{N} \setminus \{0\}} x_{0j}^k = 1, \forall k \in \mathcal{V} \tag{4}$$

$$\sum_{i \in \mathcal{N}} x_{ih}^k - \sum_{j \in \mathcal{N}} x_{hj}^k = 0, \forall h \in \mathcal{C}, \forall k \in \mathcal{V} \tag{5}$$

$$\sum_{i \in \mathcal{N}} x_{i,n+1}^k = 1, \forall k \in \mathcal{V} \tag{6}$$

$$x_{ij}^k (s_i^k + t_{i,j} - s_j^k) \leq 0, \forall (i,j) \in \mathcal{A}, \forall k \in \mathcal{V} \tag{7}$$

$$a_i \leq s_i^k \leq b_i, \forall i \in \mathcal{N}, \forall k \in \mathcal{V} \tag{8}$$

$$x_{ij}^k \in \{0,1\}, \forall (i,j) \in \mathcal{A}, \forall k \in \mathcal{V} \tag{9}$$

---

[3]This means that, even given a limit on the size of input numbers to TSP instances, there exists no optimization algorithm for the TSP where the computational effort is limited by a polynomial, unless **P** = **NP**.

In the multicommodity flow oriented VRPTW formulation above, where each vehicle is modeled as a separate commodity, we use the same notation as in the CVRP definition above. In addition, $[t_{ij}]$ is the travel time matrix, where we assume that t_{ij} includes the duration of service at customer i.

There are two types of variables. x_{ij}^k is a binary decision variable for each combination of vehicle and arc $(i,j) \in \mathcal{A}$. In a solution, $x_{ij}^k = 1$ if vehicle k travels directly from node i to node j, and 0 otherwise. There are $K\,|\mathcal{A}|$ such variables. s_i^k is a real variable that determines the exact start time of service at each customer and the depot. If vehicle k does not visit customer i, s_i^k is undefined. There are $n+2$ such variables.

The objective (1) to be minimized is simply the sum of the travel costs over all arcs that are used in the solution. Constraints (2) express that all customers must be served exactly once. (3) ensure that no tours violate the vehicle capacity constraints. (4) say that there is exactly one tour (possibly empty) per vehicle. (5) ensure that, if a given vehicle arrives at a customer, it also departs from that customer. (6) say that all tours end at the depot. (7) express that if a vehicle travels directly from customer i to customer j, the arrival time at customer j must allow for the travel between i and j. The time window constraints are formulated in (8). Eqs. (9) say that the x_{ij}^k variables are binary.

The VRPTW model is non-linear due to the integrality constraints (9), but also due to the quadratic terms in (7). (9) are critical, but (7) are easily linearized using the so-called "big M" trick. The resulting set of constraints is shown below, where M is a sufficiently large constant:

$$s_i^k + t_{i,j} - s_j^k \leq M(1 - x_{ij}^k), \forall (i,j) \in \mathcal{A}, \forall k \in \mathcal{V} \qquad (10)$$

The VRPTW formulation above with (7) replaced by (10) is an example of a Mixed Integer Linear Program (MILP), i.e., a linear program where some of the variables are constrained to be integral. It is easy to express this formulation in a MILP solver, commercial or otherwise. However, only instances with a small number of orders or a particularly simple structure will be solved to optimality in realistic time. For more complex instances, simple relaxations that are solved efficiently will provide lower bounds on the value of the optimal solution, but such bounds are generally weak[4].

If there is a feasible solution to the VRPTW instance, there will typically be a range of values for each s_i^k customer arrival time variable that render the whole solution feasible. In the VRPTW and other time-oriented routing problems, one normally does not care about the precise scheduling of each trip. In this interpretation, the VRPTW is a pure discrete optimization problem. In contrast, *Vehicle Scheduling Problems* also include decisions on precise arrival times, see for instance [25].

---

[4] An example is the LP-relaxation, which results from removing constraints (9) from the formulation above.

2.2 Exact Resolution Methods for the VRP

Most variants of the VRP have in principle been solved by exact methods such as Lagrangian relaxation, Branch & Cut, and Column Generation [4], [20]. As mentioned above, such methods have limited capability for solving VRPs: They typically cannot consistently solve instances with more than 100 orders in reasonable time. However, they are extremely useful for studying the nature of VRPs. Also, they may happen to solve even large size instances to optimality in a short time. For a thorough treatment of exact methods for the VRP, and a general textbook on the VRP, the interested reader is referred to [47]. For a general textbook on Integer Programming, see for instance [49] and [37]. We now turn to a discussion of heuristics for the VRP.

2.3 Heuristic Resolution Methods for the VRP

The VRP is a very hard optimization problem. For large-size instances, exact resolution methods generally take too much time for practical purposes. One therefore has to resort to approximation methods. Approximation methods can be based on exact methods. For instance, Branch & Bound can return an approximate solution at any time after the first feasible solution has been found. This approximate solution also comes with an upper bound on the difference to the optimal solution value. However, finding the first feasible solution may take too much time[5], and the optimality gap typically remains large for a long time. Particularly for problems that are **NP**-hard in the strong sense, non-systematic methods that basically come with no guarantees are good alternatives even for moderate size instances if the goal is to find high quality solutions in limited response time. Such methods are called *heuristics.*

A fundament for heuristic methods in discrete optimization is *Local Search* (LS). The formal basis of Local Search is not widespread knowledge. Metaheuristics are not well known in large parts of the applied mathematics community. We refer to Section 3.2 in [44] for a brief introduction to Local Search and metaheuristics. For a thorough treatment of local search, we refer to [1]. For an introduction to and survey of metaheuristics, we refer to [33] and [30].

2.4 Illustration of Local Search and Metaheuristics for the CVRP

We now illustrate the use of Local Search and metaheuristics for the CVRP defined in Section 2.1. Let a CVRP instance be defined by a set of customers $\{(i, d_i)\}_{i=0}^{n}$, where index $i = 0$ represents the depot and d_i is the customer demand. The matrix $[c_{ij}]_{i,j=0}^{n}$ gives the distances between nodes. There is a maximum of K vehicles (tours) available, each vehicle has capacity q. We assume there is no initial cost associated with using a vehicle.

[5] Generally, one does not have control over the response time for getting a feasible solution in Branch & Bound.

We use the *giant tour* representation of a solution, i.e., a representation where all tours in the solution are linked into one giant tour, see [13], [5], [41], and [34]. Duplicates of the depot are used as tour delimiters[6]. Hence, we represent a solution $\mathbf{s}$ by the sequence of customer indices

$$(0, o_{11}, \ldots, o_{1n_1}, 0, \ldots, 0, o_{K1}, \ldots 0, o_{Kn_K}, 0),$$

where subsequences corresponding to tours can be empty.

Local Search with Relocate for the CVRP

The RELOCATE neighborhood operator simply relocates a customer to a new position in the plan. We represent a move in the RELOCATE neighborhood by the pair (f, t), $f \neq t$, $0 \leq f, t \leq n + K - 1$. The semantics of the move (f, t) is that the order placed in position f in the sequence is relocated to position t. The size of the RELOCATE neighborhood is obviously $O(n^2)$. To each move we associate a delta objective value $\Delta(f, t)$ which is the cost savings associated with the move. The savings is simply the sum of costs for the arcs removed by the RELOCATE move minus the sum of costs for the arcs added by the RELOCATE move. Let $o(p)$ be the index of the order in position p in the current solution. We then have:

$$\begin{aligned}\Delta(f, t) = & c_{o(f-1),o(f)} + c_{o(f),o(f+1)} + c_{o(t-1),o(t)} \\ & - \left(c_{o(f-1),o(f+1)} + c_{o(t-1),o(f)} + c_{o(f),o(t)}\right)\end{aligned}$$

This calculation takes constant time, so the calculation of savings for all moves and selection of the best takes $O(n^2)$ time. We must also ensure feasibility. For each move, we check whether the vehicle capacity constraint is adhered to after the move. The feasibility check takes constant time for each move if one updates a record of the total capacity of orders allocated to each tour.

We are now equipped with a fast implementation of local search for the CVRP using the RELOCATE neighborhood. However, for large-size CVRP instances, $O(n^2)$ time can be forbidding. Effective and fast focusing techniques will be needed to select promising RELOCATE moves.

To start this Local Search procedure, we need to have an initial solution. The above Local Search is based on movement between feasible solutions, so we need a feasible initial solution[7].

Construction Heuristics for the CVRP

An easy way is to construct an initial solution from scratch, while maintaining feasibility. For this, we need a *List of unserviced orders*, and an INSERTION

[6] A unique giant tour representation is defined if one sorts the tours according to their first customer index.

[7] Local Search is sometimes extended to allow search into the infeasible region of the search space. Also, it may include search over incomplete solutions.

neighborhood. Without going into the details, we use an INSERTION neighborhood that focuses on a specific tour which is currently being built. For all unserviced orders, it finds the best (lowest additional cost) feasible insertion point in the tour, and then picks the unserviced order that has the lowest such cost to be inserted next. If no unserviced order has a feasible insertion point in the current tour, a new tour is opened for insertion. The new tour is seeded with the most critical unserviced customer. We define a simple, temporal criticality measure for customer i as $\zeta_i = 1 - (b_i - a_i)/b_i$, where we use the notation from Section 2.1 for customer time windows. The idea is to use customers that are tightly time constrained as seeds[8]. Thus, tours are built one by one and in a greedy fashion. In a tightly constrained problem instance, the construction procedure leaves some orders unserviced if there is a limit on the number of tours, simply because there is not enough vehicle capacity to accommodate all orders. It is straightforward to see that this simple, *sequential construction heuristic* needs $O(n^3)$ time.

Alternatively, one regards the List of unserviced orders as a predefined ordered list, defining the sequence in which the orders will be inserted in the solution. This variant requires $O(n^2)$ time. The quality of the resulting initial solution depends critically on the insertion sequence. Initial sorting based on distance and/or demand for capacity will be advantageous. A third, computationally even cheaper variant is construction by the NEAREST ADDITION heuristic. Here, one selects the remaining order that is closest to the last one that was inserted. NEAREST ADDITION also has $O(n^2)$ worst case time complexity, but it is on average faster than ordered cheapest insertion.

The above alternative procedures all sequential construction heuristics, i.e., tours are opened and built one by one. A new tour is opened only when no more customers can be inserted in a feasible way. In contrast, *parallel construction heuristics* generally start with a number of tours that are seeded with a single customer, and construct the tours of the initial solution in parallel. The choice of seeds is critical. Potvin et al. [40] report that a parallel version of Solomon's I1 sequential insertion heuristic for the VRPTW [43] outperforms the original.

See [11] for a treatment of construction heuristics for the VRP.

2.5 Industrial Methods for the VRP

In an industrial setting, unless one is faced with a specific application where problem size is small or a particular problem structure makes the VRP extraordinarily easy to solve, the only viable algorithmic approach is heuristics. If the goal is to make a generic solver for VRP, for instance as a part of an industrial transport optimization tool covering a variety of applications, this is particularly true. VRP heuristics can be based on Local Search based metaheuristics, population based methods, or hybrids.

[8]Geographical spread of tours and capacity demand are natural properties that could be taken into consideration when choosing seeds.

Above, we have discussed classical variants of the VRP, namely CVRP, DVRP and VRPTW. These VRP variants capture essential characteristics of many real-life routing problems. Most often, however, there are important aspects that require a richer model for representational adequacy. In Section 3, we discuss common Vehicle Routing Problem extensions found in industrial applications. Our basis is the VRPTW. Finally, we present a mathematical formulation that covers important parts of these extensions.

3 Applied Vehicle Routing

In this section, we shall address general issues related with modeling and resolution of real-life applications of the Vehicle Routing Problem. There are applications within transportation service providers and companies that manage a transportation fleet for servicing their internal needs. Such enterprises are found in many industries, including food and beverage, newspapers, mail service companies, waste management companies, and courier companies, to name a few. There are applications across transportation modes. Arguably, most applications exist in road based freight transportation. This is the area with the most optimization based decision-support systems. In ship routing, very few implementations are found today. However, maritime applications possibly hold the largest savings potential for routing technology. For a survey of ship routing, see [12].

In basic and applied research in transportation optimization, there is now a strong trend towards formulating and solving richer models. This trend partly originates from external requirements of tool vendors and end users, but also from within the scientific community itself. Many idealized models that have been studied extensively over the past 4-5 decades are, despite their theoretical computational complexity, regarded as being solved from a pragmatic point of view. This fact, together with recent methodological advances and the general increase in computing power, motivates a shift to novel, scientifically even more challenging and industrially more adequate problem formulations.

In real-life VRP applications, there are extensions to the basic, rather idealized models such as the VRPTW. Such extensions are related with the type of operation and various idiosyncrasies on the supply, demand, and infrastructure sides. Below, we discuss describe some fundamental and important non-standard aspects that are often seen in industrial applications, aspects that need to be included in the model for representational adequacy. Our discussion is structured according to central concepts in vehicle routing. Depending on the algorithmic starting point, some extensions are accommodated through trivial extensions to the model and the resolution algorithms. Others drastically change the nature of the problem.

For an example of a comprehensive, rich VRP model we refer to [19]. For a treatment of VRP generalizations motivated by industrial applications, i.e., Rich VRPs, we refer to [9], [10].

3.1 Fleet, Vehicle and Driver

For most basic VRP variants, vehicles are identical and vehicle acquisition costs are ignored. The objective is to minimize total distance. In slightly richer variants, e.g., the VRPTW, vehicles are still identical, but the objective is hierarchical: first minimize the number of vehicles, and then minimize the total distance. As we have seen, this hierarchical objective can be modeled with a single objective.

Very often in practice, the fleet of vehicles is not homogeneous. Rather, the company has a mixture of vehicle types at its disposition. A company either has its own fleet, or vehicles and drivers are hired. Extra capacity through transportation service providers is often available to meet demand peaks. Depending on the type of application, fleet acquisition and depreciation costs will be included in the objective.

If fleet costs are important, for instance if the problem is to design a transportation infrastructure or determine the optimal fleet size for a given demand, a heterogeneous fleet will make the VRP more difficult. In this case, an important criterion will be the optimal composition of the fleet, i.e., how many vehicles of each available type should be used. The so-called *Fleet Size and Mix VRP* (FSMVRP) [24] is defined to find an optimal balance between vehicle acquisition / depreciation costs and driving costs for a non-homogeneous fleet.

In many countries, the working time of commercial drivers is regulated by legislation[9]. The rules are often complex. In particular for long-haul transportation, such rules will seriously constrain operations. A routing decision support system that does not have the relevant driving time regulations as a part of its VRP model will create routing plans that are illegal. A sophisticated routing system decouples decisions on vehicles and drivers and provides plans where several drivers co-operate, share the driving of a vehicle, and switch between vehicles in an optimized and feasible way according to driving time regulations.

3.2 Depots, Tours, Start and Stop Locations

Basic VRP variants are defined on a single depot. The goal is to design a set of tours, one for each vehicle, that starts and stops at the depot. In practice, there are often be *multiple depots* and not necessarily a fixed connection between vehicles and depots. Vehicles can start and stop at arbitrary locations, e.g., the driver's home address. Some times, vehicles perform *multiple tours* over the planning horizon. The latter is accommodated in basic VRP models by simply defining a number of tours for every vehicle. However, temporal precedence constraints must be added between these tours unless one defines non-overlapping time frames for them a priori. Such temporal precedences will

[9]For instance, there are common driving time rules covering EU and the EEA.

seriously degrade the performance of heuristic resolution methods. Multiple depots and multiple tours further complicate the CVRP because the search space will be extended.

Customer orders in transportation are not necessarily precisely located. Rather, goods shall be picked up / delivered at one of *alternative locations*. This extension adds another degree of freedom to the already computationally very hard classical VRP.

3.3 Order Types, Type of Operation

For the basic VRP variants we have defined above, orders are either all pick-ups from customers to the depot, or deliveries from the depot to customers. Moreover, *split deliveries* are not allowed, there will be only one visit to each customer in a feasible solution. Transportation orders are defined on nodes in an abstract graph that defines the travel possibilities and travel costs between points. If splitting of orders is allowed, there is an extra degree of freedom in the VRP in that customers can be served by several visits from different vehicles. In general, this makes the problem harder. The *Split Delivery VRP* (SDVRP) is the resulting VRP formulation.

For its local distribution around a transportation terminal, a transportation service provider will receive orders to ship goods from a manufacturer to a consumer. Some orders must go through the terminal. Hence they can be modeled as a combination of a pickup and a delivery order. Others should be executed by going directly from the pickup point to the delivery point. Courier service providers and a "dial-a-ride" passenger transportation service providers typically have only "direct" transportation orders between pickup and delivery points.

The presence of direct orders will fundamentally change the nature of the VRP. Capacity utilization of the vehicle will not vary monotonically over the duration of a tour. The pickup task of a direct order must be executed by the same vehicle as its associated delivery task. There is a causal, temporal precedence constraint between the pickup and the delivery. The *Pickup and Delivery Problem* (PDP) is defined as a generalization of the VRP, to capture operations that have direct orders. In passenger transportation, there will typically be constraints on the duration between pickup and delivery for each passenger, or, more appropriately, constraints on the extra ride time relative to travel time for direct travel from the pickup point to the delivery point. The *Dial-A-Ride Problem* (DARP) is defined as such a generalization of the PDP.

In service operations, "repairmen" visit a number of customers. Normally, there is no loading or unloading of goods involved, and capacity is no issue. Such operations are easily modeled as a standard VRP. However, non-standard constraints or objective components are often important, e.g., regarding customer waiting time.

In applications such as snow removal, gritting, and even garbage collection and mail delivery in densely populated areas, transportation tasks are more naturally defined on arcs in a road network rather than on nodes representing customers. Such VRPs are called *Arc Routing Problems*. There are many variants, depending on whether all arcs or a subset must be visited, and whether vehicle capacities are important. Arc routing problems can be transformed to node routing problems, and vice versa. However, special algorithmic approaches to arc routing problems have been developed. These will typically be more efficient than node routing approaches to a transformed problem. For a survey of arc routing problems and their resolution methods, see [23].

In food and beverage industries, orders are often periodic over a longer time horizon, say a month. Replenishment of goods at food retailers will often take place according to repetitive patterns. Such patterns are often complex, and there are typically alternatives. In the *Periodic VRP* (PVRP), the goal is to find the optimal combination of routes over a longer time period, given alternative patterns.

A related generalization of the VRP is *Inventory Routing*. It captures the combination of transportation management and inventory management. Rather than repetitive patterns for replenishment (or pickup), transportation orders will be determined by inventory levels at producers and/or consumers.

3.4 Distances, Travel Times, and Service Times

In major parts of the VRP literature, distances are Euclidean. Vehicle velocities are assumed to be constant over time. These are not realistic assumptions for most real-life routing applications. In road based transport, the actual road network characteristics often have to be taken into consideration for reasonably accurate distance, time and travel cost calculations. Electronic road network and address databases are commercially available, e.g., through Geographical Information Systems. Their accuracy and level of detail are not sufficient for all types of application. A good example is newspaper delivery in urban areas. In particular, realistic speed information is generally missing from commercial road databases. The speed information provided is often the speed limit associated with each road link. In urban areas, rush hour traffic will seriously affect speed. Speeds typically vary between different types of vehicle. It is generally very costly to acquire missing address and road network information. GPS tracking of operations is one remedy. Currently, *time-varying* and *vehicle type specific* travel times are important issues in applied VRP research.

Constant service times specific to each customer are included even in the most basic VRP variants. Service times often varies between orders to the same customers, depending on order volume and the number of consecutive visits to the same customer. An initial "setup" / "rig down" time will typically accrue for the first order, whereas only the "real" service time will accrue for additional orders at the same address. In some applications, service time is the most constraining time factor. Hence, an adequate service time model will be critical to the overall adequacy. With a service time model as alluded to above, the sequencing of orders within each tour, and also the allocation of orders to tours, will be further compounded by sequence dependent service times.

3.5 Waiting Time, Time Windows and Capacity Constraints

In the classical VRPTW, waiting before the opening of the time window at a customer does not have an explicit cost. Many companies want to include a separate *waiting cost* objective component that will tend to reduce such waiting. Also in VRPTW, time windows are hard. A feasible plan must start service at all customers strictly within the defined time windows. Real life is normally not that crisp. Rather, there is a preferred visit time or time interval, and penalties for times outside of that, according to a given function. This type of temporal constraint is called *soft time windows*. We remember that, in a common interpretation of the VRPTW, one does not care about precise arrival times at customers, as long as a feasible set of arrival times exists. For the VRP with soft time windows, one has to define precise arrival times. Hence, for each tour in the plan, one must not only solve a sequencing problem, but also a scheduling problem to arrive at a minimum cost set of scheduled routes. Depending on the form of the penalty function, the scheduling subproblem for a given sequence of orders in a tour is solved by efficient methods such as linear programming, see e.g., the paper by Dumas et al. [25].

Another time window extension is *multiple time windows*. In many applications, there are alternative opening times for a delivery customer. Pickups take place at alternative days of the week. These are examples of motivations for multiple time windows. It is obviously a generalization of the (single) time window constraint. Relative to the VRPTW, the multiple time windows generalization introduces a new degree of freedom in the VRP: namely the selection of one of the alternative time windows for each order.

Akin to soft time windows, there are *soft capacity constraints*. Vehicles will have a nominal capacity in various dimensions: weight, volume, number of pallet places etc.. The generalization to multiple dimensions is straightforward. However, capacity limits are generally not crisp in practice. A separate but related issue is vehicle loading. If goods are discrete, e.g., packages, a full vehicle load is hardly attainable. With fragile goods there are constraints on stacking. Loading is a separate discrete optimization problem. In delivery, the loading sequence for a vehicle must be closely related to the sequence of deliveries to avoid excessive unloading times for most types of vehicle. For work on VRP with loading constraints we refer to [28].

3.6 Idiosyncratic Constraints and Objectives

A wide variety of idiosyncratic constraints are found in VRP applications. In transportation of livestock from farms to slaughterhouses [38], animals without horns must not be transported together with animals that have horns. A farm that recently has had animal sickness problems must be visited at the end of the route. Breeding farms must be visited first. In maritime bulk transport of chemicals, there are important constraints on the type of chemicals that can be loaded in neighboring compartments. Also, certain temporal sequences of chemicals in the same compartment are not allowed without expensive cleansing operations. See [12] for a survey of papers. In transport of liquids, there are typically constraints on the minimal filling level of tanks due to problems with sloshing.

Some customer orders require special equipment on the vehicle, or special certificates for the driver. Customers often want to be serviced by the same driver every day. Access to the loading/unloading ramp at a customer often excludes the use of certain types of vehicle.

Many idiosyncratic constraints are adequately modeled as *compatibility constraints* (or incompatibility constraints) between basic entities such as customers, type of load, vehicles, and drivers. Such constraints are generally easy to model and will in fact speed up the resolution process by reducing the search space.

More complex constraints, e.g., driving time restrictions, can be modeled through a penalty technique similar to Lagrangian relaxation. In a heuristic resolution method, it is computationally cheaper to calculate a restriction penalty than performing a full feasibility check according to complex driving time rules. Computational complexity issues should not be used as an excuse to reduce requirements on representational adequacy, however.

In our encounters with end users of routing tools, we have also met objective criteria that relate to the visual appearance of a routing plan in a map. Many users do not want to see routing plans with geographically overlapping tours, even if such plans often have lower "real" cost. Again, a penalty approach can be appropriate for such criteria. It is appropriate to include the user in the decision loop. In case, interactive user interfaces and functionality for manual changes and locking will be needed.

3.7 Stochastics and Dynamics

Industrial vehicle routing is about planning of future events that take place in an uncertain environment. By nature, the problem contains stochastic variables. Hence, the VRP is inherently a stochastic optimization problem. If stochastic variables are included in the formulation, the problem is denoted a *Stochastic VRP*. The objective in a stochastic VRP will typically be to minimize the expected cost value. The inclusion of stochastic variables in a VRP of course makes resolution more complex. One way of resolving stochastic VRPs is scenario generation, i.e., one uses the distribution of stochastic variables to generate a number of deterministic realizations. Each realization is resolved using deterministic methods.

The introduction of advanced Information and Communication Technology systems such as Enterprise Resource Planning, GPS tracking, and mobile communication with drivers has enabled a new level of dynamics in route planning and dispatching. Predictive planning will be based on more or less uncertain information, e.g., on travel times, service times, and cargo sizes. Additional orders will arrive, and existing orders may be canceled. As the routing plan is being executed, new information will be revealed. *Dynamic VRP* refers to a situation and a problem-solving strategy where input data change while the VRP instance is being solved.

Stochastic and dynamic routing are not the same, but they are related. A Dynamic VRP may or may not be based on a stochastic formulation. A stochastic formulation is beneficial to dynamic routing, particularly if one has access to high quality information on the distributions of stochastic variables. A simpler but more crude strategy is to revert to reactive planning based on a deterministic, predictive plan.

The literature on stochastic and dynamic VRP is fairly large, and rapidly growing. The early literature on stochastic routing included research on optimal decision policies under certain assumptions. An example is the determination of optimal waiting and positioning policies for emergency vehicles. More recent work on stochastic VRP is geared towards dynamic routing. We refer to [21] for a focused survey.

3.8 Response Time

In an industrial setting, response time requirements are very important. In tactical and strategic planning, response times of hours or days are often acceptable. In dynamic routing, commitments typically have to be made in a few seconds. Response time requirements strongly influence the design of resolution methods. *Predictability* of response time is equally important. In industrial vehicle routing, a method that guarantees to produce a solution within a certain time frame is preferable to a method that will predominantly produce higher quality solutions in the same time, but has an erratic response time characteristic.

3.9 The Common OR Approach versus a Holistic Approach

Above, we have illustrated some real-life aspects one has to consider to be industrially relevant in vehicle routing. As indicated, many of these aspects have been studied also in the academic part of vehicle routing research.

The conventional approach in OR is to focus on specific extensions to the VRP and study these in isolation. For instance, VRPTW is the extension of the DVRP with time window constraints. Hence, the VRPTW is obviously a generalization of the DVRP, as any DVRP instance is also a VRPTW instance[10]. As complexity theory is based on worst case analyses, the VRPTW must be at least as hard as the DVRP. The DVRP is again a generalization of the CVRP, as it includes one more type of constraint and is otherwise identical. Another example of a generalization of the CVRP is the PDP alluded to in Section 3.3 above. The PDPTW is a further generalization of both the PDP and the VRPTW.

VRP research has since its start in 1959 studied a fairly large number of generalizations of the classical CVRP. Hence, a taxonomy of VRPs has emerged. The common OR approach, which is basically reductionistic in its nature, has been highly successful in terms of understanding the effect of VRP model extensions. It has lead to a basic understanding of the nature of each variant, and to the development of corresponding effective and efficient algorithms. However, this approach will be hard to continue as the number of extensions increase.

An alternative approach is to study a more general and richer model that is supposed to include a large number of industrial aspects. The primary goal is to develop robust algorithms that on average give good performance. The research described below is based on the latter approach. Before we move on to our description of the rich VRP model and the uniform algorithmic approach in SPIDER, we turn to a mathematical formulation of an extended VRP variant.

3.10 A Mathematical Formulation for a Richer VRP

Above, we have briefly described a large number of real-life extensions to the CVRP. In Section 2.1, we presented a mathematical formulation of the relatively "poor" VRPTW, based on vehicle flow. Let us now illustrate how to extend this formulation to accommodate a small selection of the industrial aspects we have mentioned.

We include two basic generalizations: heterogeneous fleet, and alternative time windows. For the non-homogeneous fleet part, we introduce vehicle specific driving costs c_{ij}^k, travel times t_{ij}^k, vehicle acquisition / depreciation costs e^k, and capacities q^k. A number W_i of alternative time windows, each with lower limit a_{im} and upper limit b_{im}, $m = 1, \ldots, W_i$ is allowed for each customer. Moreover, we change the objective so as to minimize the sum of driving costs and acquisition / depreciation costs. The resulting problem is a Fleet Size and Mix VRP with Multiple Time Windows (FSMVRPMTW). The formulation is shown in (11)–(22) below.

[10] A DVRP instance is a VRPTW instance where all customer time windows span the full time horizon.

$$\text{minimize} \sum_{k\in\mathcal{V}} \sum_{(i,j)\in\mathcal{A}} c_{ij}x_{ij}^k + \sum_{k\in\mathcal{V}} \sum_{j\in\mathcal{C}} e_k x_{0j}^k \tag{11}$$

$$\sum_{k\in\mathcal{V}} \sum_{j\in\mathcal{N}} x_{ij}^k = 1, \forall i \in \mathcal{C} \tag{12}$$

$$\sum_{(i,j)\in\mathcal{A}} d_i x_{ij}^k \leq q^k, \forall k \in \mathcal{V} \tag{13}$$

$$\sum_{j\in\mathcal{N}\setminus\{0\}} x_{0j}^k = 1, \forall k \in \mathcal{V} \tag{14}$$

$$\sum_{i\in\mathcal{N}} x_{ih}^k - \sum_{j\in\mathcal{N}} x_{hj}^k = 0, \forall h \in \mathcal{C}, \forall k \in \mathcal{V} \tag{15}$$

$$\sum_{i\in\mathcal{N}} x_{i,n+1}^k = 1, \forall k \in \mathcal{V} \tag{16}$$

$$x_{ij}^k(s_i^k + t_{i,j} - s_j^k) \leq 0, \forall (i,j) \in \mathcal{A}, \forall k \in \mathcal{V} \tag{17}$$

$$x_{ij}^k(s_i^k - a_{im})y_{im} \geq 0, \forall m, \forall i \in \mathcal{N}, \forall k \in \mathcal{V} \tag{18}$$

$$x_{ij}^k(b_{im} - s_i^k)y_{im} \geq 0, \forall m, \forall i \in \mathcal{N}, \forall k \in \mathcal{V} \tag{19}$$

$$\sum_{m=1}^{W_i} y_{im} = 1, \forall i \in \mathcal{C} \tag{20}$$

$$x_{ij}^k \in \{0,1\}, \forall (i,j) \in \mathcal{A}, \forall k \in \mathcal{V} \tag{21}$$

$$y_{im} \in \{0,1\}, \forall m, \forall i \in \mathcal{N} \tag{22}$$

Compared with the VRPTW-formulation in Section 2.1, there are a few differences. In the objective (11) we have added a second term that corresponds to the vehicle acquisition / depreciation costs. We use the first edge of a non-empty tour as an indicator so these costs will accrue only for the vehicles that are used. In (13), the vehicle capacity constraints are now based on potentially different vehicle sizes[11]. The time window constraints (8) in the VRPTW formulation are split in 2 constraint sets: (18) and (19). (18) represents the lower time bound constraints, and (19) are the upper time bound constraints. An additional set of $\sum_{i=1}^{n} W_i$ binary (see (22)) decision variables y_{im} is used for selecting one of the W_i alternative time windows for customer i. Again, the time window constraints are non-linear, but they are easily linearized by use of the big M trick described in Section 2.1. (20) is a constraint set that makes sure that exactly one of the alternative time windows is selected for each customer.

Our extension of the VRPTW formulation to the FSMVRPMTW illustrates that some types of extension are trivial to formulate[12], whereas others

[11]The extension to multiple capacity dimensions is trivial and left to the reader.

[12]For instance, by just adding indices for constants such that constraints or objective components will be more specific.

are more fundamental and will require more substantial modeling efforts such as the addition of new integral decision variables. However, even trivial model extensions may require fundamental changes to the algorithmic approach.

4 The SPIDER Rich VRP Solver

4.1 Historical and Industrial Background

Since 1995, SINTEF Applied Mathematics (SAM) has performed applied research in vehicle routing. This research is driven by end user requirements emerging from contracts that span several industries and two transportation modalities: road based goods transportation and ship routing. A general philosophy of reuse of knowledge and software toolkits was developed at SAM during the early 90ies. Hence, an early goal of developing a generic VRP solver soon emerged, mainly aimed at road based transportation. The goal was reachable only through a combination of industrial contracts and more basic VRP research efforts funded by the Research Council of Norway, the European Commission, and SINTEF itself. Important results have also been reached by students from the University of Oslo, NTNU in Trondheim, Molde University College, as well as students and visiting researchers from several universities and research groups abroad.

The first version of the generic VRP Solver called SPIDER was finished in 1997 through a research contract with the Norwegian telecom company Telenor Mobil. Further development and first steps towards industrialization took place in the EU Framework Program IV project GreenTrip (Esprit 20603) in the period 1996-1999. At this point, SINTEF decided to commercialize the results by spawning a new company called GreenTrip AS (now: Spider Solutions AS) just before the start of the new millennium. Spider Solutions has built a route design and control system called Designer based on the SPIDER solver. They have roughly 1/3 of the Norwegian market, with 3-4 serious international competitors. Designer is used in freight forwarding, food and drink industries, newspapers, postal companies, distribution of oil, and distribution of animal fodder. Figure 2 shows parts of the user interface.

SPIDER consists of two main components: Planner and Guider. Planner is the VRP Solver, whereas Guider is a road topology model that provides travel time, distance, and travel cost data to Planner, in a highly efficient way. The Guider services are based road and address data from commercial GIS. Planner is again based on SAM's generic Discrete Optimization Problem solver called SCOOP (SINTEF Constrained Optimization Object Package).

In 2000, SINTEF won a proposal for strategic research in vehicle routing and related topics. The TOP Program lasted for 4 years (2001-2004) and included activities in the Norwegian TOP Network, consisting of University of Oslo, University of Bergen, Molde University College, NTNU in Trondheim, and several SINTEF research groups. The bulk of the work was done at SAM.

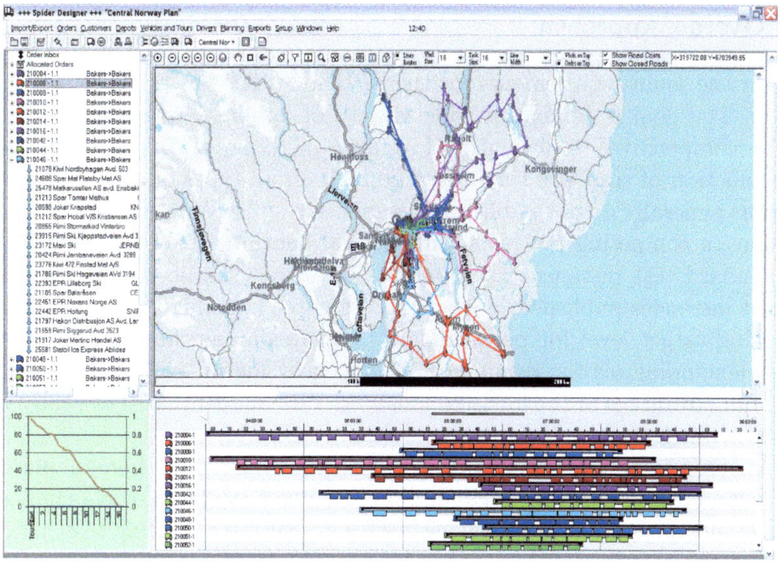

Fig. 2. Snapshot of the SPIDER Designer Routing Tool.

The main focus was industrial vehicle routing, including substantial activities in dynamic shortest path problems in the context of rich VRPs.

TOP was successful in developing new VRP models and resolution methods. Several new world records on standard test instances from the Operations Research literature were obtained. From TOP, a number of survey articles and technical papers have been published in well-reputed journals. The Norwegian network of research groups in transportation optimization, and operations research in general, was strengthened. Due to TOP, the interaction with the international research community is now at a substantially higher level than before. Most importantly, major parts of the results from TOP were implemented in SPIDER as they were produced. Hence, results were disseminated and industrially exploited, even while the TOP Program was running.

At the time of writing, SPIDER is further developed through two major research projects: DOiT and EDGE. In DOiT (2004-2007), one of the main goals is to develop an industrial approach to dynamic and stochastic routing. For details on the method for solving dynamic VRPs with use of stochastic information, we refer to [22].

EDGE (2005-2008) is a Norwegian-Finnish collaboration project, where the main goal is to provide more effective methods for huge scale VRPs and Arc Routing Problems[13]. The industrial motivation comes from applications where a large number of households in an area must be served, e.g., in waste management, mail delivery, or the final tier of newspaper distribution.

[13] "Huge-scale" meaning tens or hundreds of thousands of orders.

4.2 Overall Approach

One of the main requirements for the SPIDER VRP Solver is modeling flexibility. The goal is to span a wide variety of applications across industries. To be competitive, resolution performance in terms of routing plan quality as a function of response time must be high on all types of application. Tool vendors generally do not publish performance results. Without access to information on competing products, one natural benchmark source is the variety of standard test instances used in the literature. The best known solutions to test instances published by the VRP research community constitute a demanding target level for performance[14]. Academic benchmarks are particularly demanding, as the results are produced by highly skilled scientists that take full advantage of the focused, idealized VRP model that they study when implementing a solver.

Another important benchmark is the performance of human planners. However, comparison with manual planning poses methodological challenges. Human planners often violate constraints or optimize according to other criteria than the ones included in the model. In a before / after type of comparison, it is difficult to mask the effect of improvements that are external to the automated VRP optimization functionality. Often, confidentiality and business concerns seriously constrain publication of such comparative results.

In the development of SPIDER, three fundamental design decisions were taken initially. First, we decided to base SPIDER on a *single, rich, and generic* VRP model, rather than taking a stratified approach. The rationale for this decision was the expected software development and maintenance efforts. The second fundamental design decision was to take a *unified algorithmic approach*, with a similar rationale. This means that all problem instances are basically resolved by the same algorithmic machinery. As we shall see below, this does not mean that SPIDER refrains from taking advantage of characteristics of the instance at hand. Third, we selected a *metaheuristic approach* for VRP resolution. A heuristic version of the exact method of column generation would arguably be a realistic alternative, but the choice was made mainly due to the high flexibility and robustness towards model extensions offered by metaheuristics.

Given our choice of algorithmic approach, a mathematical formulation of the rich VRP model has not been central in the development of SPIDER[15]. Local Search and metaheuristics operate on the combinatorial object in question. In the development of SPIDER, conceptual modeling of the VRP "world" is crucial. A conceptual model, covering both the problem and the solution part, is needed. We have used the Unified Modeling Language (UML) as our vehicle for conceptual modeling. The model and the optimization algorithms are implemented in the object-oriented language C++. Below, we give an

[14] In a few small-sized or simple cases, the best known results are optimal.

[15] A mathematical formulation is not uninteresting, as it will add to precision and provide opportunities for obtaining lower bounds.

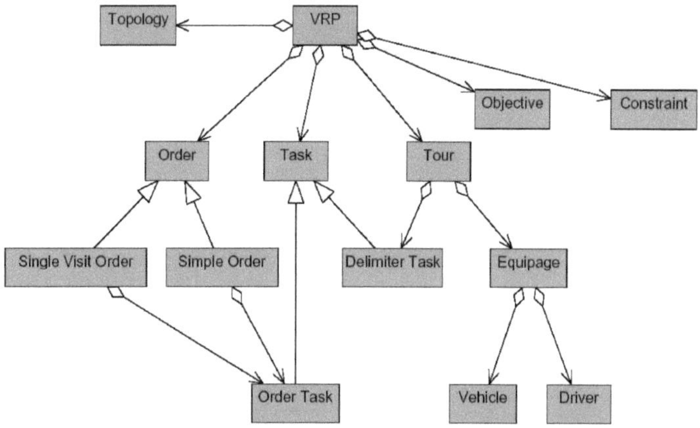

Fig. 3. The problem encoding part of the SPIDER conceptual model.

overall description of the SPIDER Conceptual Model, and the extensions of the classical CVRP that it accommodates.

4.3 The SPIDER Model

Figures 3 and 4 show SPIDER's conceptual model. The optimization problem is represented as a VRP, which consists of tasks, orders, tours, constraints, an objective, and a topology (e.g., Euclidean or road network). There are two types of orders: Simple and Single Visit. A simple order has two tasks; one for loading and one for unloading. A single visit order contains only a single service task. Each task represents the requirement to visit a certain location and possibly load/unload something. A location is either a node location, corresponding to a point (node) in the road topology, or an edge location that corresponds to an edge or arc with two nodes. Edge locations are used to model arc routing problems and for modeling aggregate orders.

The transportation resources are represented as a set of tours. In each tour, a given vehicle and driver—collectively known as an equipage—are available for serving orders. The tour contains two delimiter tasks that represent the start and end of the tour.

A solution consists of a set of tour instances and task instances, one for each tour and task in the VRP. A tour instance specifies the ordered sequence of tasks that are performed during the tour. A task instance contains more specific information about one task, such as the arrival and departure times and vehicle load at those times.

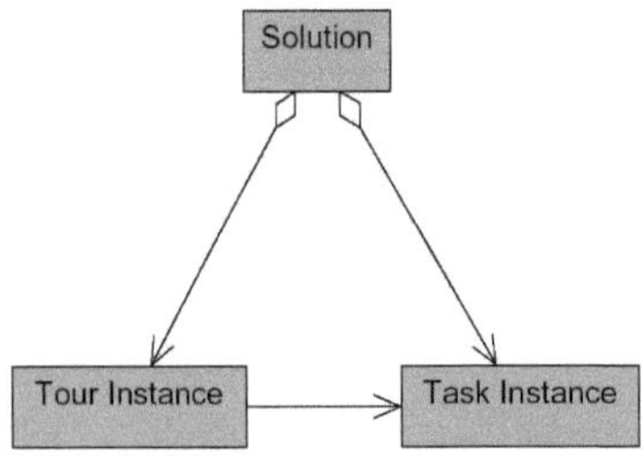

Fig. 4. The solution part of the SPIDER conceptual model.

Extended Modeling Capabilities

The SPIDER model was designed from the outset to be rich, flexible and general. Requirements from novel applications have motivated several extensions. Below we briefly describe some major SPIDER model extensions relative to the CVRP. We refer to Section 3 for a motivation of these extensions.

In addition to the ability to model a large variety of real-world routing problems, the generic VRP model generalizes many VRP variants described in the literature, e.g., VRPTW, PDPTW, and Multiple Depot VRP with Time Windows (MDVRPTW). However, it is easy to find interesting VRP variants that are not covered by our model, for instance split delivery VRPs.

Types of Order

There are four types of order: *Delivery*, *Pickup*, *Direct*, and *Single Visit*. A Single Visit order does not have a size. It is used to model service (repairman) orders. Pickup, Delivery, and Direct orders each have two tasks: one for loading and one for unloading. Single visit orders have a single (service) task each. For delivery type orders, we take advantage of the fact that the loading task is fixed at the depot. As the SPIDER model currently does not model problems with loading constraints, the sequence of loading tasks at the depot is irrelevant. Hence, computational effort is saved in insertion. Similarly, we take advantage of the fact that pickup orders all have their unloading task at the depot and their sequence is immaterial.

Direct orders are used to model pickup and delivery problems, where the vehicle will not visit the depot between pickup and delivery of the same goods. There is a causal precedence constraint between the pickup an delivery task of a given direct order. The pickup and delivery task must be executed by the same vehicle (pairing constraint).

Time Windows and Alternative Periods

Orders (through their tasks) and tours have single or multiple hard time windows. Customer service must commence within (one of) the time window(s) for the order. Tours must start and finish within its time window.

The model allows for specification of periodic, e.g., weekly orders. A periodic order has alternative periods (e.g., Tuesday, or Thursday). Couplings between alternative periods for a set of orders may be expressed. Within a period, (multiple) time windows may be defined.

Service Times

In many routing applications, total service time takes a substantial share of the total time, even more than 50 %. An adequate service time model is critical to overall representational adequacy. Each order has a service time that can depend on the location of the preceding order and the size of the order. This is useful, for instance to model that the delivery of the first order to a given customer takes a certain time, but any additional orders to the same customer will take less time.

Capacity Dimensions

A virtually unlimited number of capacity dimensions can be defined, e.g., volume, weight, and pallet places.

Vehicles

An arbitrary but fixed number of vehicles can be defined. The fleet can be homogeneous or heterogeneous. Heterogeneity will result from differences in cost structure, capacity, or equipment. If there is a substantial initial cost on vehicles, the optimization procedures described below will reduce fleet size by optimizing the sum of initial costs and driving costs. For a heterogeneous fleet, there will be a tendency to allocate empty tours to the least useful vehicles.

Tours

Vehicles are allocated to Tours. Tours have arbitrary start and end locations. In many cases, all tour start and end locations are identical, i.e., there is a single depot. In a solution, a subset of the predefined tours will be populated with tasks (belonging to orders) in a certain sequence. Some predefined tours will typically be empty. In addition to single tours, a chain of multiple tours may be predefined.

Alternative locations

Orders (through their tasks) can have alternative locations. Tours may have alternative start and end locations, for instance alternative depots. As explained above, a location is either a node or an edge.

Topologies

Travel times, distances and costs are defined through alternative types of topology: Euclidean topology, Tabular topology (distance, time, and cost tables), or an electronic road network with various restrictions and speed models generated from Geographical Information Systems. The road network topology handles travel times that vary with time, also on an individual link basis.

Cost Model

A variety of cost elements exist, including travel costs on distance and time, waiting time cost, unserviced costs on orders, initialization costs on tours, cost per order in a tour, and costs for breaking working time regulations.

Additional Constraints

In addition to the standard capacity constraints in CVRP, constraints on total capacity over a set of tours are supported. There are compatibility constraints between tour / order and tour / location.

Locks

Parts of a solution can be locked, for instance to freeze history in dynamic routing. As mentioned in Section 4.1, the VRP Solver is currently being extended with facilities for solving dynamic VRPs using knowledge on uncertain information.

4.4 The SPIDER Unified Algorithmic Approach

The CVRP is an **NP**-hard problem. For the VRPTW, finding a feasible solution is **NP**-complete. As our goal is to be able to solve a wide variety of more general VRP variants, and also large size instances, we have selected a heuristic method. Moreover, we have chosen to follow a *uniform algorithmic approach*, i.e., we basically use the same algorithm for all types of instances. This approach has obvious pros regarding software engineering issues. However, there are obvious cons related with the asymptotic behavior of algorithms, as one generally cannot fully exploit the structure of "simple" instances. In a competition with academic VRP research, SPIDER has a handicap because demanding computations are performed also for instances where they are not necessary[16].

The SPIDER heuristic approach is based on Local Search. There are three phases: *Construction* of Initial Solutions, *Tour Depletion*, and *Iterative Improvement.*

[16] The propagation of time window constraints is a good example. Special code to recognize special cases could easily be included, but code volume and maintainability would suffer.

Construction

The goal of the Construction phase is to create a starting point for iterative improvement. The SPIDER Iterative Improvement mechanisms are robust in the sense that the end result is not highly dependent on the initial solution. However, convergence to a high quality solution will generally be faster with a better starting point. In this phase, one attempts to generate an initial solution with as high quality as possible in a short time. For this, fairly greedy heuristics are used, at least for other than small sized problems. Alternative solution constructors with different strengths and quality vs. response time characteristics are available in SPIDER. The construction phase is optional, as the iterative improvement mechanism contains an INSERT neighborhood that will attempt to insert unserviced orders into a partial or even empty solution.

Tour Depletion

If the objective has a vehicle acquisition / depreciation cost component, the reduction of tours is important as these costs will generally constitute a major part of the total objective. An initial solution from the Construction phase will have a large potential for reduction of tours. Tour Depletion by use of relatively simple heuristics will be harder after a period of iterative improvement, because the solutions will be "tighter". There will be less room for adding orders in tours that have already been optimized. Often it pays off to invoke a focused Tour Depletion phase immediately after the Construction phase. SPIDER has the option to invoke a Tour Depletion phase. Tour Depletion is also an operator for the Iterative Improvement phase.

Iterative Improvement

The goal of the Iterative Improvement phase is to focus search efforts in an optimal way to achieve higher quality solutions. In SPIDER, search varies between sub-phases of intensification and diversification. Intensification of search towards a deep local optimum is achieved through Variable Neighborhood Descent [31]. When a local optimum is found, alternative diversification mechanisms exist to "jump" to an unexplored, promising part of the search space. The overall procedure is an example of Iterated Local Search [36].

Some more detail on the three phases follow below, but space constraints and confidentiality concerns forbid us to go into full detail.

Construction of Initial Solutions

SPIDER offers extensions of the classical construction heuristic of Clarke and Wright ("Savings") [15] and Solomon's construction heuristic I1 [43]. These heuristics were developed for the CVRP and VRPTW, respectively. A considerable effort has been made in order to generalize these heuristics so they will accommodate the rich VRP model of SPIDER. In addition, we have developed a *regret-based* cheapest insertion heuristic (see also Section 2.4). The regret value for an unserviced order is the difference between the cost of inserting it in its second best position and the cost of inserting it in its best position. The cheapest insertion heuristic will insert each unserviced order at its best position in the routing plan in a certain sequence[17], namely the sequence of decreasing regret values. For more on regret-based insertion heuristics we refer to [39].

Despite that generalizations were made to the classical construction heuristics to accommodate model extensions, we observed serious shortcomings, mainly related with heterogeneous fleet problem instances. Hence we developed a novel construction heuristic called *SPIDER Constructor* for generating one or more initial solutions in the Construction Phase. It builds upon ideas from classical construction heuristics (see for instance [7]) but includes non-trivial extensions to address heterogeneous fleet, multiple depots, and tours with different start and end locations. The construction heuristic has a structure similar to the I1 heuristic [43], namely sequential best insertion of the most critical unserviced order, but the heuristics for determining best insertion and most critical order are different. An instance analysis is used to determine whether the problem is heterogeneous, and, in case, to determine a good sequence in which additional tours will be opened. Tour preference values, based on certain metrics, are used to heuristically reduce the number of orders to be considered for insertion.

Tour Depletion

In the Tour Depletion Phase, a greedy tour removal heuristic is invoked. A single tour is depleted, and insertion of the unassigned orders in the remaining tours is attempted. The new solution is accepted if all unassigned orders are successfully inserted in the remaining tours. Optionally, the tours that have been changed will be optimized by the intra-tour improvement operators 2-OPT or 3-OPT. Such depletion of a single tour is attempted on all tours in sequence, and the process is repeated until quiescence.

The depletion procedure is also implemented as a local search operator to be used in the Iterative Improvement phase.

Iterative Improvement

The Iterative Improvement phase is based on Variable Neighborhood Descent (VND) [31], using a selection of several intra-tour, inter-tour, and special operators. The operators are mostly extensions of well-known operators for the VRPTW. The intra-tour operators are: 2-OPT, OR-OPT, 3-OPT, and

[17] The insertion sequence is critical to the quality of a cheapest insertion heuristic.

RELOCATE. INSERT, RELOCATE, CROSS (a.k.a. 2-OPT*), EXCHANGE, SEQUENCE EXCHANGE, and TOUR DEPLETION constitute the inter-tour operators. CHANGE ALTERNATIVE PERIOD, and CHANGE ALTERNATIVE LOCATION are special operators that search over alternative time periods and locations.

Operators have been extended to accommodate the SPIDER model. Some of these extensions are non-trivial and require extra search effort, e.g., the extension to accommodate direct orders. These orders have two tasks that are positioned independently, except that the selected positioning must adhere to causal precedence and pairing (same tour) constraints.

Exact and heuristic filters on neighborhoods are applied to increase speed. For some operators, additional opportunistic search focus is added by analyzing the current solution and exploring promising moves only. For the EXCHANGE operator, promising segment end points are identified by analysis of arc length, and only segments with end points corresponding to long arcs are included in the exchange neighborhood. The above mechanisms are candidate list strategies, as suggested by Glover in the context of Tabu Search. Similar work that draws upon arc lengths to reduce neighborhood size has been suggested by Toth and Vigo [48]. Irnich et al. [34] discuss important local search speedup techniques for the VRP.

When VND reaches a local optimum, a diversification mechanism is employed. *Very Large Neighborhood Search* [2] is used to generate alternative local optima using the incumbent solution. In more detail, a number of orders are retracted from the incumbent solution and re-inserted using a regret-based insertion procedure, see [39] and [42]. The overall strategy is a hybrid of VND and Iterated Local Search [36].

Empirical Investigation

In the world of Discrete Optimization Problems, and for VRP in particular, exact methods have a limited scope in terms of instance size. Often, no computationally effective approximation schemes exist, and one must resort to heuristics that do not come with a performance guarantee. To assess the practical performance of heuristic optimization algorithms, empirical investigation is needed. Scientists and applied researchers in the OR community develop test instances for the problems they study. Often, such instances are made available to other researchers, e.g., through web pages, and become standard test cases. Published results, either in journal articles, reports, or on web pages, constitute benchmarks for experimental investigation. Some instances have been solved to optimality. Most instances have not. As we have briefly discussed in Section 2.1 above, exact solutions to relaxations will provide bounds, although they might be weak. Together, a heuristic feasible solution and the exact solution to a relaxation provide an upper bound on the *error*, i.e., the relative difference between the value of a candidate solution and the optimal value.

Empirical investigation is an important part of VRP research. There are a number of pitfalls. A set of standard test instances does not necessarily span the variety of instances for a given problem variant. There is a danger of over-fitting in the sense that a proposed algorithm may be tuned to perform optimally on a small and possibly non-representative set of test instances, and perform relatively worse on other instances. Also, performance must be qualified with response time, and response times will vary between computers. For a stochastic optimization method, a sufficient number of independent runs should be performed to ensure statistical significance, and variance should be reported. Despite these dangers, empirical investigation is an indispensable part of research methodology in discrete optimization.

For the VRP, there are test sets for many variants. The CVRP and DVRP have been the most studied problems, and there are benchmarks by Christofides et al. [13], [14], Augerat et al. [3], and others. For the recently much studied VRPTW, there is a classical benchmark by Solomon 56 instances with 100 customers [43]. The Solomon benchmark is divided into 6 subsets: C1, C2, R1, R2, RC1, RC2. C denotes instances where customer locations are clustered. In the R subsets locations are randomly distributed, and in RC there is a combination of clusters and randomly distributed customers. In Type 1 instances the time horizon is short and tours have few customers, whereas the time horizon is longer with a resulting larger number of customers per tour for Type 2 instances.

The Solomon set has been extended with test sets of 60 instances for 200, 400, 600, 800, and 1000 customers by Homberger and Gehring [32]. Li and Lim [35] have utilized the Solomon and Homberger and Gehring instances to generate a benchmark for the Pickup and Delivery Problem with Time Windows (PDPTW). The Li and Lim benchmark has the same structure of 6 subsets as the Solomon benchmark. The subsets are denoted LC1, LC2, LR1, LR2, LRC1, LRC2.

SPIDER has undergone comprehensive empirical investigations based on test instances for many VRP variants taken from the literature, as well as real-life instances from industry. The investigation shows that the uniform algorithmic approach produces solutions of high quality in reasonable time for the CVRP, DVRP, VRPTW, PDPTW, and FSMVRPTW. At the time of writing (August 2006, see [46] for updates), SPIDER has produced the best known solutions until now for 46 out of the 354 PDPTW test cases provided by Li and Lim, and many ties for previous best known results.

Tables 1–6 in Appendix A compare the best SPIDER results for a single run with a single parameter setting, with a collection of the best known results reported (as of August 2006) for each instance of the Li and Lim PTPTW benchmark with 100 orders. The objective is hierarchical: first minimize the number of vehicles used, then minimize total distance. The Authors column refers to the authors that first reported the best known result, where Li&Lim refers to [35], BVH refers to [6], and SAM refers to unpublished work based on SPIDER by the Group of Optimization at SINTEF Applied Mathematics.

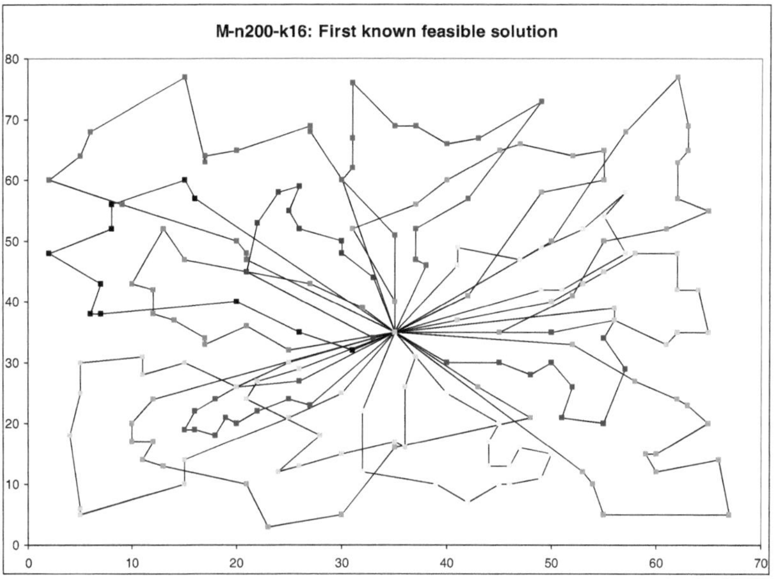

Fig. 5. The first known feasible solution to the CVRP instance M-N200-K16.

SPIDER results in bold are best known results and ties. The tables show that in only four of the 56 instances, SPIDER does not yield the best known solution.

SPIDER is written in C++ and the experiments were run on a Pentium IV 1 GHz PC with 512 Mb of memory. CPU time is only shown for SPIDER, as some authors have not reported computing time. Typically, SPIDER will be less computationally efficient than academic solvers due to its rich model and uniform algorithmic approach. Research solvers focused on an particular idealized VRP model can take advantage of problem structure to reduce computational complexity. In principle, a solver for rich VRPs can recognize special types of instance and take advantage of them, but for an industrial solver one has to worry about the size and complexity of code as well.

Interestingly enough, SPIDER also performs quite well on some test instances for the most basic and extensively studied CVRP and DVRP variants. Figure 1 in Section 2.1 shows a new best solution for the instance G-n262-k25 with 261 customers and a limit of 25 vehicles, more than 7 % better than the previous best known solution. For the M-n200-k16 instance with 200 customers and a limit of 16 vehicles of Christofides et al. [14], no feasible solution was previously known until SPIDER found one in 2003 (see Figure 5). Table 7 in Appendix A compares the best known results for 14 classical CVRP / DVRP instances defined by Christofides et al. [14]. For this benchmark, a maximum number of vehicles is given for each instance, and the objective is total distance. The best out of four SPIDER runs with different parameter settings is compared with the best known results as of 2004. We refer to [17] for details on the best known results. The SPIDER experiment was run on a 1.7 GHz Pentium IV with 512 Mb memory. SPIDER results in bold are ties with the best known result. The average distance error is 0.7 %.

There are many unexplored ideas for improving the SPIDER algorithmic approach. This suggests that there is still a potential for improvement of state-of-the-art in resolution of classical VRPs.

For a web resource on VRP benchmarks, and a record of "best until now" results for VRPTW and PDPTW test instances, we refer to the TOP web pages [46].

5 Summary and Conclusions

In this chapter, we have described the industrially highly relevant and scientifically very challenging Vehicle Routing Problem. The development of methods for the VRP is considered one of the great successes of Operations Research. A sizable routing tool industry has been built upon scientific innovations that have followed extensive research efforts since the introduction of the VRP by the seminal paper of Dantzig and Ramser in 1959 [18]. Today, routing technology based on VRP models and optimization algorithms is implemented at an increasing rate across many industries. Tool vendors claim typical cost savings of between 5 % and 30 %. Still, current technologies face serious challenges regarding modeling flexibility and optimization performance for many application types. In particular, optimization performance seems to degrade ungracefully for current routing technology when the number of orders increases above a few hundred. An improvement of scalability of VRP optimization algorithms is needed. The introduction of Enterprise Resource Planning systems, mobile communication technologies, geographical information systems, and positioning technology has enabled a new level of dynamic control in transportation management. This trend motivates an increased study of stochastic and dynamic VRPs, and commercial exploitation of results.

Although the "simpler" VRP variants are regarded as being solved from a pragmatic stance, there are clear indications that there is still a fair potential for improvement of optimization performance, even for the CVRP and DVRP. As the general power of computers has increased, and novel VRP optimization methods have improved our solution capabilities considerably, the scientific VRP community now takes on new challenges in the form of richer VRP variants and larger size problems. It is our belief that novel hybrid methods that combine exact, mathematical approaches, metaheuristics, and methods from constraint satisfaction is a promising way to go. In metaheuristics for the VRP, the predominant approach is based on exact exploration of neighborhoods. More work in the direction of focused, opportunistic investigation of neighborhoods that grow faster than linearly with problem size is needed. Also, parallel computing methods and collaborating searches at different levels of abstraction are interesting to increase performance of heuristic VRP methods.

At SINTEF Applied Mathematics, we have studied rich variants of the VRP for more than a decade. We have developed a generic, rich VRP model that accommodates many of the idiosyncrasies found in real-world routing problems. The model generalizes many of the extended VRP variants found in the literature. A uniform, heuristic approach consisting of Construction, Tour Depletion, and Iterative Improvement phases is used for resolution. The Iterative Improvement phase utilizes a hybrid meta-heuristic search strategy that combines Variable Neighborhood Descent [31] and Iterated Local Search [36]. The rationale for selecting Variable Neighborhood Descent is the goal of reaching a high quality local optimum fast. By employing a rich set of neighborhoods in an intensification phase one generally finds a high quality solution that is a local optimum with respect to all neighborhoods employed relatively quickly. Diversification with fast improvement is achieved through a Very Large Neighborhood Search approach, using a Ruin and Recreate [42] strategy. Empirical investigation on test problems from the literature and the industry has shown that the overall approach is robust and efficient over a large variety of VRPs.

There is still room for improvement. In particular, future work will include development of candidate list strategies for non-exact exploration of large neighborhoods, more informed selection of promising moves, novel diversification mechanisms, compound neighborhood operators, opportunistic selection of search operators, as well as aggregation and decomposition heuristics to achieve high optimization performance even for huge-scale, rich VRP instances.

There is no reason to believe that VRP research will be less scientifically interesting or industrially irrelevant in the future. The road from methodological improvements in discrete optimization and VRP resolution methods to economical and environmental savings in transportation logistics will be short.

Acknowledgement. The development of SPIDER has very much been a team effort. Over the last decade, some 20 researchers at SINTEF Applied Mathematics, students, and visiting researchers have been active in VRP research and SPIDER software development. We thank everyone, without naming them. Many thanks also to the people involved from numerous end user companies and SPIDER Solutions AS for providing requirements, test data, and feedback.

The writing of this chapter, and substantial parts of SINTEF's VRP research efforts, were funded by the TOP strategic institute program financed by the Research Council of Norway. Earlier and present research and technology development activities have been partly funded by Telenor Mobil AS, the European Commission, and the Research Council of Norway. All support is gratefully acknowledged.

Also, we thank two anonymous referees for their thorough work and constructive criticism.

References

1. E. H. L. Aarts and J. K. Lenstra (eds). *Local Search in Combinatorial Optimization.* Wiley-Interscience Series in Discrete Mathematics and Optimization, ISBN 0-471-94822-5, 1997.
2. R. Ahuja, O. Ergun, J. Orlin, and A. Punnen. A Survey of very large-scale neighborhood search techniques. *Discrete Appl. Math.*, 1(23):75–102, 2002.
3. P. Augerat, J. M. Belenguer, E. Benavent, A. Corbéran, D. Naddef. Separating capacity constraints in the CVRP using tabu search. *European J. Oper. Res.*, 106:546–557, 1998.
4. C. Barnhart, E. Johnson, G. Nemhauser, M. Savelsbergh, and P. Vance. Branch-and-price: Column generation for solving huge integer programs. *Oper. Res.*, 46(3):316–329, 1998.
5. M. Bellmore and S. Hong. Transformation of the multisalesmen problem to the standard traveling salesman problem. *J. ACM*, 21(3), 500–504, 1974.
6. R. Bent and P. Van Hentenryck. A two-stage hybrid algorithm for pickup and delivery vehicle routing problems with time windows. In *Proceedings of the International Conference on Constraint Programming (CP-2003) Kinsale, Ireland, September 2003.* Springer Verlag, 2003.
7. O. Bräysy and M. Gendreau. Vehicle routing problem with time windows, part I: Route construction and local search algorithms. *Transport. Sci.*, 39(1):104–118, 2005.
8. O. Bräysy and M. Gendreau. Vehicle routing problem with time windows, part II: Metaheuristics. *Transport. Sci.*, 39(1):119–139, 2005.
9. O. Bräysy, M. Gendreau, G. Hasle, and A. Løkketangen. A Survey of Heuristics for the Vehicle Routing Problem, Part I: Basic Problems and Supply Side Extensions. *SINTEF Report*, Oslo, Norway, 2005.
10. O. Bräysy, M. Gendreau, G. Hasle, and A. Løkketangen. A Survey of Heuristics for the Vehicle Routing Problem, Part II: Demand Side Extensions. *SINTEF Report*, Oslo, Norway, 2005.
11. A. M. Campbell and M. Savelsbergh. Efficient insertion heuristics for vehicle routing and scheduling problems. *Transport. Sci.*, 38(3):369–378, 2004.
12. M. Christiansen, K. Fagerholt, and D. Ronen. Ship routing and scheduling: Status and perspectives. *Transport. Sci.*, 38(1):1–18, 2004.
13. N. Christofides and S. Eilon. An algorithm for the vehicle dispatching problem. *Oper. Res. Q.*, 20(3):309–318, 1969.
14. N. Christofides, A. Mingozzi, P. Toth, and C. Sandi (eds). Chapter 11 in *Combinatorial optimization.* John Wiley, Chichester, 1979.
15. G. Clarke and J. Wright. Scheduling of vehicles from a central depot to a number of delivery points. *Oper. Res.*, 12(4):568–581, 1964.
16. J. F. Cordeau, G. Desaulniers, J. Desrosiers, M. M. Solomon, and F. Soumis. *VRP with Time Windows.* Chapter 7 in [47].
17. J.-F. Cordeau, M. Gendreau, A. Hertz, G. Laporte, and J.-S. Sormany. New heuristics for the vehicle routing problem. *Technical Report G-2004-33*, GERAD, Montreal, Canada 2004.
18. G. B. Dantzig and J. H. Ramser. The truck dispatching problem. *Manag. Sci.*, 6:80, 1959.
19. G. Desaulniers, J. Desrosiers, I. Ioachim, M. Solomon, F. Soumis, and D. Villeneuve. A unified framework for deterministic time constrained vehicle routing

and crew scheduling problems. In T. Crainic and G. Laporte (editors): *Fleet Management and Logistics*, Chapter 3, 57–93, Kluwer 1998.
20. G. Desaulniers, J. Desrosiers, and M. Solomon (editors). *Column Generation.* Number 5 in GERAD 25th Anniversary Series. Springer 2005.
21. T. Flatberg, G. Hasle, O. Kloster, E. J. Nilssen, and A. Riise. *Dynamic and Stochastic Aspects in Vehicle Routing – A Literature Survey.* SINTEF Report STF90A05413, ISBN 82-14-02843-4, Oslo, Norway, 2005.
22. T. Flatberg, G. Hasle, O. Kloster, E. J. Nilssen, and A. Riise. Dynamic and Stochastic Vehicle Routing in Practice. Accepted as Chapter 3 in V. S. Zeimpekis, G. M. Giaglis, and C. D. Tarantillis (editors). *Dynamic Fleet Management: Concepts, Systems, Algorithms and Case Studies.* To be published by Springer Verlag.
23. M. Dror (editor). *Arc Routing: Theory, Solutions and Applications.* Kluwer, ISBN-0792378989, 2000.
24. W. Dullaert, G. K. Janssens, K. Sörensen, and B. Vernimmen. New heuristics for the Fleet Size and Mix Vehicle Routing Problem with Time Windows. *J. Oper. Res. Soc. 1*, 53(11):1232–1238, 2002.
25. Y. Dumas, F. Soumis and J. Desrosiers. Optimizing the schedule for a fixed vehicle path with convex inconvenience costs. *Transport. Sci.* 24(2): 145–152, 1990.
26. European Commission. *European Transport Policy for 2010: time to decide.* White Paper. Office for official publications of the European Communities, Luxemburg, ISBN 92-894-0341-1, 2001.
27. M. R. Garey and D. S. Johnson. *Computers and Intractability: A Guide to the Theory of NP-Completeness.* Freeman, San Francisco, CA, USA, 1979.
28. M. Gendreau, M. Iori, G. Laporte, and S. Martello. A tabu search algorithm for a routing and container loading problem. *Transport. Sci.*, 40(3):342–350, August 2006.
29. B. E. Gillet and J. G. Johnson. Multi-terminal vehicle-dispatch algorithm. *Omega 4*, 711–718, 1976.
30. F. Glover and G. Kochenberger (eds). *Handbook of Metaheuristics.* International Series in Operations Research & Management Science, vol. 57. Kluwer Academic Publishers, Norwell, MA, USA, ISBN 1-4020-7263-5, 2003.
31. P. Hansen and N. Mladenovic. An Introduction to Variable Neighborhood Search. In S. Voss et al. (eds): *Metaheuristics, Advances and Trends in Local Search Paradigms for Optimization.* 433–458, Kluwer, Dordrecht, 1999.
32. J. Homberger and H. Gehring. Two evolutionary meta-heuristics for the vehicle routing problem with time windows. *INFOR*, 37:297–318, 1999.
33. H. H. Hoos and T. Stützle. *Stochastic Local Search – Foundations and Applications.* Elsevier, 2005.
34. S. Irnich, B. Funke, and T. Grünert. Sequential search and its application to vehicle-routing problems. *Comput. Oper. Res.*, 33(8):2405–2429, 2006.
35. H. Li and A. Lim. A Metaheuristic for the Pickup and Delivery Problem with Time Windows. *Proceedings of the 13th IEEE International Conference on Tools with Artificial Intelligence (ICTAI'01)*, ISBN 0-7695-1417-0, 2001.
36. H. R. Lourenço, O. C. Martin, and T. Stützle. *Iterated Local Search.* Chapter 11 in [30].
37. G. Nemhauser and L. A. Wolsey. *Integer and Combinatorial Optimization.* Wiley, 1988.

38. J. Oppen and A. Løkketangen. The Livestock Collection Problem. *Working paper 2006:3*. Molde University College.
39. D. Pisinger and S. Ropke. A general heuristic for vehicle routing problems. *Comput. Oper. Res.*, 2005. doi:10.1016/j.cor.2005.09.012.
40. J.Y. Potvin and J. M. Rousseau. A parallel route building algorithm for the vehicle routing and scheduling problem with time windows. *European J. Oper. Res.*, 66, 331-340, 1993.
41. M. Rao. A note on multiple travelling salesman problem. *Oper. Res.*, 28:628–632, 1980.
42. G. Schrimpf, J. Schneider, H. Stamm-Wilbrandt, and G. Dueck. Record breaking optimization results using the ruin and recreate principle. *J. Comput. Phys.*, 159:254–265, 2000.
43. M. M. Solomon. Algorithms for the vehicle routing and scheduling problems with time window constraints. *Oper. Res.*, 35:254–265, 1987.
44. M. Stølevik, G. Hasle, and O. Kloster. Solving the Long-Term Forest Treatment Scheduling Problem. In *this book*.
45. Statistisk Sentralbyrå. Lastebilundersøkelsen 2002 (in Norwegian).
46. TOP web pages. `http://www.top.sintef.no`
47. P. Toth and D. Vigo (eds). *The Vehicle Routing Problem*. SIAM Monographs on Discrete Mathematics and Applications, Philadelphia, 2002.
48. P. Toth and D. Vigo. The granular tabu search and its application to the vehicle-routing problem. *INFORMS Journal on Computing*, 15(4):333–346, 2003.
49. L. A. Wolsey. *Integer Programming*. Wiley, 1998.

A Experimental Results

Table 1. SPIDER Performance on the PDPTW LC1 instances.

Instance	Best Known			SPIDER		
	Vehicles	Distance	Authors	Vehicles	Distance	CPU (s)
lc101	10	828.94	Li&Lim	**10**	**828.94**	0
lc102	10	828.94	Li&Lim	**10**	**828.94**	5
lc103	9	1035.35	BVH	9	1052.33	8452
lc104	9	860.01	SAM	**9**	**860.01**	3336
lc105	10	828.94	Li&Lim	**10**	**828.94**	0
lc106	10	828.94	Li&Lim	**10**	**828.94**	0
lc107	10	828.94	Li&Lim	**10**	**828.94**	0
lc108	10	826.44	Li&Lim	**10**	**826.44**	0
lc109	9	1000.60	BVH	10	827.82	0

Table 2. SPIDER Performance on the PDPTW LC2 instances.

Instance	Best Known			SPIDER		
	Vehicles	Distance	Authors	Vehicles	Distance	CPU (s)
lc201	3	591.56	Li&Lim	**3**	**591.56**	0
lc202	3	591.56	Li&Lim	**3**	**591.56**	0
lc203	3	585.56	Li&Lim	**3**	591.17	147
lc204	3	590.60	SAM	**3**	**590.60**	1223
lc205	3	588.88	Li&Lim	**3**	**588.88**	0
lc206	3	588.49	Li&Lim	**3**	**588.49**	7
lc207	3	588.29	Li&Lim	**3**	**588.29**	0
lc208	3	588.32	Li&Lim	**3**	**588.32**	8

Table 3. SPIDER Performance on the PDPTW LR1 instances.

Instance	**Best Known**			**SPIDER**		
	Vehicles	Distance	Authors	Vehicles	Distance	CPU (s)
lr101	19	1650.80	Li&Lim	**19**	**1650.80**	19
lr102	17	1487.57	Li&Lim	**17**	**1487.57**	17
lr103	13	1292.68	Li&Lim	**13**	**1292.68**	13
lr104	9	1013.39	Li&Lim	**9**	**1013.39**	9
lr105	14	1377.11	Li&Lim	**14**	**1377.11**	14
lr106	12	1252.62	Li&Lim	**12**	**1252.62**	12
lr107	10	1111.31	Li&Lim	**10**	**1111.31**	10
lr108	9	968.97	Li&Lim	**9**	**968.97**	9
lr109	11	1208.96	SAM	**11**	**1208.96**	11
lr110	10	1159.35	Li&Lim	**10**	**1159.35**	10
lr111	10	1108.90	Li&Lim	**10**	**1108.90**	10
lr112	9	1003.77	Li&Lim	**9**	**1003.77**	9

Table 4. SPIDER Performance on the PDPTW LR2 instances.

Instance	**Best Known**			**SPIDER**		
	Vehicles	Distance	Authors	Vehicles	Distance	CPU (s)
lr201	4	1253.23	SAM	**4**	**1253.23**	77
lr202	3	1197.67	Li&Lim	**3**	**1197.67**	723
lr203	3	949.40	Li&Lim	**3**	**949.40**	314
lr204	2	849.05	Li&Lim	**2**	**849.05**	237
lr205	3	1054.02	Li&Lim	**3**	**1054.02**	20
lr206	3	931.63	Li&Lim	**3**	**931.63**	30
lr207	2	903.06	Li&Lim	**2**	**903.06**	153
lr208	2	734.85	Li&Lim	**2**	**734.85**	593
lr209	3	930.59	SAM	**3**	**930.59**	374
lr210	3	964.22	Li&Lim	**3**	**964.22**	240
lr211	2	911.52	SAM	3	884.294	258

Table 5. SPIDER Performance on the PDPTW LRC1 instances.

Instance	**Best Known**			**SPIDER**		
	Vehicles	Distance	Authors	Vehicles	Distance	CPU (s)
lrc101	14	1708.80	Li&Lim	**14**	**1708.80**	83
lrc102	12	1558.07	SAM	**12**	**1558.07**	16
lrc103	11	1258.74	Li&Lim	**11**	**1258.74**	120
lrc104	10	1128.40	Li&Lim	**10**	**1128.40**	74
lrc105	13	1637.62	Li&Lim	**13**	**1637.62**	559
lrc106	11	1424.73	SAM	**11**	**1424.73**	587
lrc107	11	1230.15	Li&Lim	**11**	**1230.15**	486
lrc108	10	1147.43	SAM	**10**	**1147.43**	88

Table 6. SPIDER Performance on the PDPTW LRC2 instances.

Instance	Best Known			SPIDER		
	Vehicles	Distance	Authors	Vehicles	Distance	CPU (s)
lrc201	4	1406.94	SAM	**4**	**1406.94**	3004
lrc202	3	1374.27	Li&Lim	**3**	**1374.27**	3137
lrc203	3	1089.07	Li&Lim	**3**	**1089.07**	548
lrc204	3	818.66	SAM	**3**	**818.66**	756
lrc205	4	1302.20	Li&Lim	**4**	**1302.20**	113
lrc206	3	1159.03	SAM	**3**	**1159.03**	796
lrc207	3	1062.05	SAM	**3**	**1062.05**	5
lrc208	3	852.76	Li&Lim	**3**	**852.76**	356

Table 7. SPIDER Performance on a CVRP / DVRP benchmark.

Instance			Best Known	SPIDER	
Name	Vehicles	Customers	Distance	Distance	CPU (s)
P01 (E051-05e)	5	50	524.61	**524.61**	170
P02 (E076-10e)	10	75	835.26	844.80	436
P03 (E101-08e)	8	101	826.14	829.63	578
P04 (E151-12c)	12	150	1028.42	1040.65	1784
P05 (E200-17c)	17	199	1291.45	1320.57	899
P06 (D051-06c)	6	50	555.43	**555.43**	49
P07 (D076-11c)	11	75	909.68	**909.68**	344
P08 (D101-09c)	9	100	865.94	**865.94**	187
P09 (D151-14c)	14	150	1162.55	1175.33	1441
P10 (D200-18c)	18	199	1395.85	1421.97	1757
P11 (E121-07c)	7	120	1042.11	1042.12	1602
P12 (E101-10c)	10	100	819.56	**819.56**	48
P13 (D121-11c)	11	120	1541.14	1544.51	687
P14 (D101-11c)	11	100	866.37	**866.37**	123
TOTAL	149	1588	13761.17	13796.91	10105

Solving the Long-Term Forest Treatment Scheduling Problem

Martin Stølevik, Geir Hasle, and Oddvar Kloster

Summary. The Long-Term Forest Treatment Scheduling Problem (LTFTSP) is the task of allocating treatments in a forest such that both sustainability and economic outcome is maximized. Solving such problems is demanded in more and more countries and the task is increasingly more complex because one must adhere to local legislation, environmental issues, and public interests. To be able to handle most aspects of the LTFTSP with adjacency constraints (which is the problem we solve), a rich, spatial model which is parameterized, is required. We present a model defined on discrete land units and time points, where the treatments to perform are parameterized. Many of the most commonly used criteria in the form of constraints and objective components in long-term forestry scheduling are included. Such criteria may be defined for the complete forest region in question, or for specific sub-regions.

The complexity of the model requires a robust solution method. We have selected a heuristic approach based on Tabu Search. An initial solution is constructed by composition of economically optimal schedules for each land unit. This solution is made feasible by a greedy heuristic. The initial solution is iteratively improved by Tabu Search. Two different types of move are used in the Tabu Search procedure: Shifting a treatment to another time point, and exchanging one treatment program for another treatment program.

The solution method is implemented in the software tool Ecoplan. Empirical results have been produced for a 1,541 stand case from Norway. The results show that when more than one objective is included in the objective function, the quality of the solution with respect to the individual objectives may be considerably reduced. Some of the quality loss, especially with regards to the "old forest" objective component may be explained by the initial state of the forest.

Key words: forest harvest scheduling, rich model, spatial constraints.

1 Introduction

The long-term planning of sustainable forest treatment at the landscape level is an increasingly complex task. Local treatment schedules, pertaining to homogeneous sub-areas called stands, must be developed over a time horizon of 50-150 years. Thousands of local schedules must be coordinated to satisfy hard constraints, and balance soft constraints and optimization criteria.

Forestry provide services that effect human welfare, the global public, the local public, forest workers, forest owners, and the forest industry. Sustainable forestry is about trying to take all those different kinds of services from the forest and forestry into account when planning how to manage the forests. The increasing demands from today's society regarding environment and nature issues have made forest management planning a much more challenging and demanding task. Examples are: New legislation that constrains the maximum size of clear-cut areas, different elements of nature that need special care, and an increasing focus on recreational values.

The aim of a long-term forest treatment schedule is twofold. Over the near time horizon, it must provide clear instructions for forest treatment. In addition, sustainability over the full horizon must be demonstrated. Many Operations Research (OR) based approaches have been suggested for different types of problems within long-term forest planning when considering the spatial characteristics of the problem. Simulated Annealing has been used to model harvest scheduling problem [17, 8]. Tabu Search, another well known and popular metaheuristic, has been used for instance in [22] (which also used Simulated Annealing), [5] and [3]. In [6], an indirect search algorithm is used to solve an LTFTSP. There has also been some use of genetic algorithms for solving this problem [21]. Most articles in this field describe research on only one or a few of the features in a real-world LTFTSP. We have not been able to find articles that describe a rich model of the LTFTSP.

We describe a model for long-term forest treatment scheduling built on explicit constraint descriptions and a flexible memory-based Local Search procedure. Based on the described model we have implemented the software Ecoplan. Ecoplan relies on close interaction with a stand simulator, which provides the forestry knowledge necessary to guide the scheduling process. The software also facilitates user interaction in the planning process, the ability to lock specific parts of the plan, or to alter key factors in the plan in order to set up "what-if" scenarios.

This chapter is organized as follows: In Section 2 we describe our model of the long-term forest treatment scheduling problem. Section 3 describes the solution procedure and in Section 4 we describe results from a computational study based on a case in Norway. In Section 5 we list possible algorithmic improvements, while Section 6 is the conclusion.

2 Modeling

The Long-Term Forest Treatment Scheduling Problem (LTFTSP) deals with the allocation of elementary forest treatment *actions* over time to different regions of the forest. The forest is usually divided into *stands*; local, homogeneous, spatial units. Actions may include clear-cutting, thinning, fertilizing and reforestation. The special action "do nothing" is also available. The set of actions allocated over time to a specific forest stand is called a *Local Forest Treatment Schedule, LFTS*. A plan, or a solution to the LTFTSP, is the sum of all the LFTS-es, one for each stand in the problem region. The optimal treatment plan for a single forest stand (when regarded in isolation) depends on the current state of the forest, including the condition of the soil, as well as economical factors. Locally optimal plans need to be coordinated due to local (geographic neighbor) and global constraints. A feasible forest treatment schedule must adhere to several types of constraints, originating from forestry knowledge, concerns on biodiversity, and human recreational (social) values. Moreover, the schedule must ensure sustainable forest harvesting and provide a sound and fair economy for the forest owner(s) involved. The resulting set of possible combinations of forest treatment actions can be prohibitively large.

2.1 A Simple Model

A simple forest treatment scheduling problem may be described by: "How does one harvest timber at different times so as to maximize the harvested volume?". Formulated like this, the problem is easy to solve and can be decomposed into N sub-problems if there are N timber units. An LTFTSP is interesting and challenging, and not to forget, realistic, when the harvest of each unit is dependent on the harvest in other units. Consider the problem: "How does one harvest timber at different times so as to maximize the harvested volume such that adjacent units are not harvested in the same time period?". This problem (which is a spatial problem) is much harder, since harvests will influence each other. The simple problem above may be described mathematically by:

$$\begin{aligned} &\text{Maximize} \\ &\qquad V = \sum_{t=1}^{T} v_t = \sum_{t=1}^{T}\sum_{i=1}^{N} X_{it} v_{it} \\ &\text{Subject to} \\ &\text{(i)} \qquad X_{it} + X_{jt} \le 1 \quad \forall t, i, j \in U(i) \\ &\text{(ii)} \qquad \sum_{t=1}^{T} X_{it} \le 1 \quad \forall i \end{aligned} \tag{1}$$

Here:

t: is the index of a time period.
T: is the total number of time periods.
v_t: is the volume harvested in time period t.
i: is the index of a stand.
N: is the total number of stands.
X_{it}: is a binary decision variable that is 1 if unit i is harvested in time period t and 0 otherwise.

v_{it}: is the volume harvested from stand i in time period t.
$U(i)$: is the set of neighboring stands for stand i.

In words: "Maximize the harvested volume V from the stands over all time periods t such that: (i) no neighboring stands are harvested in the same time period; and (ii) each stand is harvested at most once". The harvested volume v_t in each time period t may be computed by summing the volume of all stands that are harvested that period.

Another way of coupling the harvests could be to formulate the problem like this: "How does one harvest such that the variation in harvested volume between time periods is minimized?". Again, harvesting in one stand will influence harvests in other stands. All stands may not be harvested in the optimum time period, since even harvesting between time periods is required.

2.2 A Rich Model

Simple models like the one described in the previous section are used to illustrate a point, but are not really useful in the real world. Forest owners are generally not interested in solving the simple problem modeled above. Instead, they want to solve the problems at hand. These are complex problems that require a rich model.

The model must be able to represent real-world problems in an adequate way. We cannot simplify too much. It would be convenient to restrict the number of action types to two; clear-cut, and do nothing. It would also be convenient to set the planning horizon to less than one rotation period[1], as in the simple model in Section 2.1, so that each stand is treated at most once. Choosing these two simplifications would render the model much easier to solve. However, we can not afford this luxury. Our mission is to describe a model that solves real-world problems. To accommodate this, we must develop a rich model that describes many important aspects of the problem.

The modeling must allow for different objective functions and constraints to be used. This is an important aspect since different users have different needs. The modeling must support special regional constraints like when a wood grouse mating game area is to be protected, or when an area being particularly exposed visually from some viewpoint in the landscape is to be saved from the most glaring clear-cutting.

The model must also provide site-specific constraints on a stand or at regional level. The approach to the planning process can also be a stepwise procedure. One must be able to define critical areas, for which a planning process with a specified planning horizon is to be performed. Such a critical area may correspond to the before-mentioned wood grouse mating area. The resulting plan can be locked, and an optimized plan for the surrounding area can be generated.

[1]The "rotation period" is the approximate time between clear-cuts.

The task when solving the long-term forest treatment scheduling problem is to find the best, or at least a good enough, *plan* for the planning area over the planning horizon. The "best" in this context means the plan that gives the best (minimum or maximum) value when evaluated by some *objective function*. A plan consists of a collection of one LFTS for every stand in the planning area. The number of possible plans is huge. We illustrate this by a small example. Use the simple model of Section 2.1 with $T = 5$ and $N = 25$. Assume that there are two basic treatments ("do nothing" and "clear-cut"). Furthermore note that an LFTS may only contain one (or no) clear-cut actions (constraint 1-ii). Then there are six different possible LFTS-es for each stand; the clear-cut may be assigned one of the five time periods or all time periods may be set to "do nothing". The number of possible plans is thus all combination of such LFTS-es for the 25 stands: $6^{25} = 2.84 \times 10^{19}$. In theory, all these plans must be evaluated, checking constraint 1-i and computing the objective function V, to find the optimal, legal plan. The simplest conceivable optimization approach is *explicit enumeration*: generate all possible solutions, evaluate each of them, and keep the best, feasible one[2]. Even if a computer could evaluate a billion plans per second, it would still take more than 900 years to evaluate all plans using explicit enumeration.

Real-world problems are much larger than this. A typical real-world problem may contain 5,000 – 10,000 stands and 10 – 20 treatments, rendering the number of theoretical possible plans enormous. One can easily see that in order to find the best (or even a good) solution amongst all these plans, one has to use better algorithms than explicit enumeration. Typically, problem knowledge may help in reducing the number of plans to evaluate. Let us revisit the simple model (1) but assume no adjacency constraints (i.e., "maximize V subject to the constraints 1-ii only"). In general, stands may need to be left uncut due to constraints and time period length, but in model 1 with only the constraints 1-ii, it is always better to harvest in some period than never to harvest at all. Consequently one does not need to check plans containing stands that are never harvested. The number of plans to check is reduced to $5^{25} = 2.98 \times 10^{17}$, which could be checked in a little more than nine years (on a billion plans per second-computer). Using simple problem knowledge, the problem size is reduced by a factor 100, but still, explicit enumeration is hardly a practical approach.

As we have seen, compared to explicit enumeration, a clever solution method may reduce the number of plans one needs to evaluate. The LTFTSP is an example of a discrete optimization problem (DOP). Several solution methods may be used to solve DOPs. Which algorithm to utilize, depends partly on the structure of the problem, as defined by its constraints and objectives, the size of instances to solve, and the available response time. In general, exact methods (mathematical programming, or explicit enumeration) can be used

[2]Explicit enumeration is also called *Generate and Test* or *the British Museum Procedure*.

if the instance size is relatively small, or the problem has a nice structure that renders it solvable in polynomial time as a function of instance size. For the LTFTSP, if the instance size is realistic, and real-world constraints are used, exact methods are not applicable due to the computing time required. Solution methods are further discussed in Section 3.

2.3 Stands

To be able to assign different treatments to different parts of the forest, the forest must be divided into smaller parts – treatment units. Traditionally, dividing the forest into homogeneous polygons called *stands* has been the way of doing this. The stands are assumed to be static (i.e., have permanent borders) over the planning horizon. A forest stand is (or may be considered) a homogeneous area of forest, regarding growth conditions, the age of the forest, the distribution of tree species and the conditions for forestry operations. The idea is that when a treatment is performed, the treatment is performed in the entire stand. The normal size of forest stands in Scandinavia is between one and twenty hectare (ha). For each stand, one usually has information about the mean age and height of trees, tree species, density and volume of trees, site index, etc. In our model we will use the index i when indexing stands. Each stand has a number of other stands neighboring it. The set of neighboring stands for stand i is denoted $U(i)$.

The above described model for stands is referred to as the Unit Restriction Model (URM) [23]. In recent years, researchers have developed planning models where stands can be grouped together for harvest, in what is termed harvest blocks. Instead of using predefined blocks, one defines the stands that can be grouped together, either a priori or dynamically during optimization. This is called the Area Restriction Model (ARM) [23]. ARM models can be used when the size of stands are significantly below the maximum area that can be clear-cut in the same time period: The smaller units are grouped together to form contiguous areas that may be clear-cut in the same time period since the joint area is less than the maximum area restriction. The ARM-model increases problem size, because one must also decide which small units should be grouped together. On the other hand, it allows more freedom in solving the problem. For problems that are difficult to solve because adjacency constraints restrict the degree of freedom, this approach may be successful. ARM-models are usually solved by heuristics, but there has also been some research on exact methods for ARM [12].

2.4 Time

We model time using a discretized time line made up of *time points* $\{t_j\}$, $j \in [0, \ldots, T]$. The plans are to be set up within the planning horizon $\{t_j\}$, $j \in [0, \ldots, T_H]$, $T_H \ll T$. The time points are uniformly spaced along the time line. The time period length (the time between two consecutive time points) is typically 5 or 10 years, but this is not fixed in the model. The time periods are indexed using t such that time period $t = 1$ is the period between t_0 and t_1 and time period $t = T$ is the period between t_{T-1} and t_T. The shorter the time period, the more difficult the model will be to solve. To put it another way: The shorter the time period, the more accurate results are obtained, at the cost of more CPU power.

2.5 State

A *state* s is an n-tuple of state attributes and is a description of the state of a stand at some time point t_j:

$$s(t_j) = (s_1(t_j), \ldots, s_n(t_j)) \text{ where } s_m(t_j) \in D_m \text{ for } m \in [1, \ldots, n].$$

D_m is the set of all possible values for the attribute s_m. Examples of attributes are age, height, average tree volume, etc. In general, the state at any time τ (not necessarily a time point) may be found by a forest simulator (see Section 2.7).

2.6 Actions and Stand Treatment Programs

An *action* is a human intervention in a stand. An action changes the state and is modeled to happen instantaneously. A *Stand Treatment Program (STP)* is a description of the actions to perform in a given stand between two time points; the start time point t_s and end time point t_e of the STP. The collection of all STPs for one stand, which is also an STP, is a Local Forest Treatment Schedule (LFTS). STPs are built in a hierarchical manner so that two or more STPs can be combined to give another STP. A *basic STP* is an STP between two time points with one or no actions. The number of basic STPs in an STP is referred to as the complexity of the STP. Fig. 1 shows an LFTS of complexity seven, in three different levels. The complexity of STP1 is two, and the complexity of STP2 is six.

Actions may in general occur in between time points. At what point between two consecutive time points (t_j and t_{j+1}) of an STP an action occurs, is defined by the *action fraction*, $f_a \in [\,0, 1\rangle$, which is a global parameter in our model. The time at which the action occurs is $\tau_a = t_j + (t_{j+1} - t_j)f_a$.

Actions are assumed to have zero time duration. Let $s^-(\tau_a)$ and $s^+(\tau_a)$ denote the states immediately before and immediately after time point τ_a. Let $A = \{a_i\}$, $i \in [0, \ldots, N_A]$ be the discrete set of actions. There exists an action state transformation function $g : D_1 \times \ldots \times D_N \times A \rightarrow D_1 \times \ldots \times D_N$ such that $s^+(\tau_a) = g(s^-(\tau_a), a)$, $a \in A$. The action "do nothing", denoted a_0, has the following property: $s^+(\tau_{a_0}) = g(s^-(\tau_{a_0}), a_0) = s^-(\tau_{a_0})$, i.e., the states immediately before and after the "do nothing" action has been performed, are equal. This action is typically used in let-grow type STPs.

An LFTS has information about the start and end points of the STPs. For instance, say that the time period length is five years and there are 21

- LFTS
  - STP1
    - Let-grow (Basic STP)
    - Clear-cut (Basic STP)
  - STP2
    - 30% thinning (Basic STP)
    - Let-grow(Basic STP)
    - 60% thinning (Basic STP)
    - Let-grow(Basic STP)
    - Clear-cut(Basic STP)

Fig. 1. Hierarchical configuration of an LFTS.

time points in the planning horizon (i.e., a time span of 100 years). Then the above LFTS might be illustrated as in Fig. 2 (where t_0 is time 0 and t_{T_H} is 100 years).

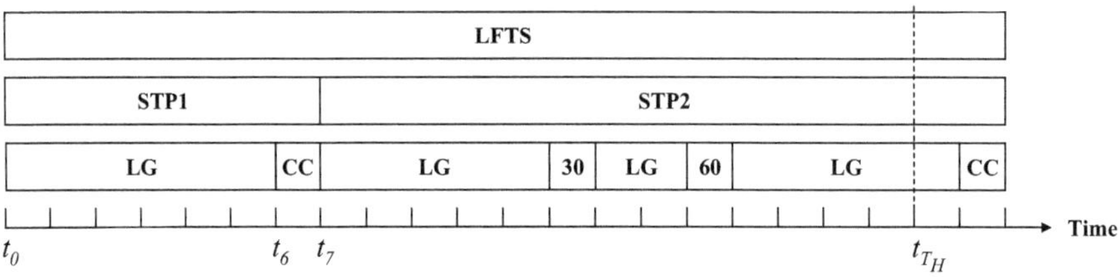

Fig. 2. Visualization of the hierarchical configuration of an LFTS.

In Fig. 2 'LG' means let-grow, 'CC' means clear-cut, '30' means 30% thinning and '60' means 60% thinning. As can be seen from the figure, the end time of an LFTS is usually outside the end of the planning horizon, t_{T_H}. This is to ensure that the treatments performed in the forest are sustainable also for the periods immediately after the time horizon.

2.7 Forest Simulator

State changes due to natural processes happen continuously "in between" the time points. These state changes are computed by a simulator. We model the simulator as a prediction function e. The simulator can compute the natural evolution of the forest from a given initial state $s(t_j)$ over a time span $t_2 - t_1$ between two time points ($t_2 > t_1$):

$$e : D_1 \times \ldots \times D_n \times \mathbf{R}^+ \rightarrow D_1 \times \ldots \times D_n \text{ such that } s(t_2) = e(s(t_1), t_2 - t_1).$$

The forest simulator is used extensively throughout the computations. Its task is to compute what effect the different actions will cause in the forest over time. Since the simulator is used so much, it is important that the software implementation of the simulator do the computations quickly.

In most articles describing the LTFTSP, the forest simulator is replaced by either:

1) Simple (piecewise) linear growth functions that, given the time since the last clear-cut and some parameter describing the soil (like site quality), compute relevant state attributes like standing volume (per area) and height.
2) Growth tables that facilitate look-up of the value of important state attributes.

The advantage of replacing the simulator with such simplifications is that during optimization one spends considerably less time on determining the state of the forest. The time spent can instead be used on the search for better solutions. Replacing the simulator is typically done for the purpose of demonstrating and comparing different solution methods. The disadvantage of replacing the simulator by simplifications is that the computations of forest states are less correct. This may lead to decisions that can be sub-optimal. However, many real-world applications of harvest scheduling models use some form of pre-defined yield tables.

The forest simulator does numerous computations. Some of the computations are identical, but performed at different times. To avoid unnecessary computations, all simulations performed by the simulator may be cached. Before a computation is performed, the simulator checks if an identical computation has already been performed. If so, the cached value is returned. In large cases, the number of computations avoided by such caching may be extensive.

2.8 Regions

When planning long-term treatments in a forest, the forest area under consideration may be large, both in terms of number of stands and area. In some parts of the forest, special considerations must be taken. To model this we introduce *regions*. A region may be as small as one stand, and as large as all stands in the planning area. In general, a region is a sub-part of the planning area. Regions may overlap. Constraints and objectives (Sections 2.9 and 2.10) are linked to a region – *the applicable region*. This allows the overall optimization process to take special considerations in some regions.

2.9 Constraints

In the following paragraphs we list some of the most common constraints used when solving the LTFTSP. A constraint may be considered a *hard constraint*, i.e., it may not be violated in a solution under any circumstance. A constraint may also be used as a *soft constraint*, in which case the constraint may be violated, but at a cost. The motivation for using both hard and soft constraints is that some constraints may not be satisfied in any solution at all or it may be very difficult to obtain even one feasible solution because of some hard constraint(s). To be able to find feasible solutions, such constraints are made soft. The cost, or *penalty*, is added to the objective function. In a minimization problem, the more a soft constraint is violated, the more penalty is incurred. To be able to use a constraint as a soft constraint one must define a function F_c that evaluates the penalty of the constraint $c \in C_S$ (where C_S is the set of soft constraints). The penalty must increase as the violation to the constraint increases (i.e., the penalty function F_c is monotonically non-decreasing). The value of F_c is zero if the constraint is not violated. If the objective function is called F, the original minimization problem can be stated:

$$\begin{aligned} &\text{Minimize} \\ &\qquad F \\ &\text{Subject to} \\ &\qquad c_i \in C \quad (i = 1, \ldots, n) \end{aligned} \tag{2}$$

Now assume that the constraints C are partitioned into m hard constraints (C_H) and $n - m$ soft constraints (C_S) such that $C = C_S \cap C_H$. Then the minimization problem is transformed to:

$$\begin{aligned} &\text{Minimize} \\ &\qquad \tilde{F} = F + \sum_{c=1}^{n-m} \alpha_c F_c \\ &\text{subject to} \\ &\qquad c_j \in C_H \quad (j = 1, \ldots, m) \end{aligned} \tag{3}$$

In the new objective function ($\tilde{F}$) the number α_c is a scaling factor that assure that the term $\sum \alpha_c F_c$ is so large that the solution method will focus on removing violations if any of the F_c components become too large. The optimal solution to problem (3) will be a lower bound on the optimal solution to the original problem. In addition, this bound will be tighter than optimizing F while discarding the soft constraints [2]:

$$\begin{aligned} &\text{Minimize} \\ &\qquad F \\ &\text{Subject to} \\ &\qquad c_j \in C_H \quad (j = 1, \ldots, m) \end{aligned} \tag{4}$$

In our model, the constraints may generally be defined between two time points t_s and t_e, in which case all actions that take place in the half open interval $[t_s, t_e\rangle$ are considered.

All constraints are posted in an applicable region (see Section 2.8). If the constraint is posted on all stands, the special region containing all stands (the global region) is set as the constraint's applicable region. Applying constraints in regions allows for instance modeling of preservation of old forest in only specific parts of the forest. Several constraints may be defined in the same region and different constraints may be defined in regions that are not equal, but overlap. In the description of constraints below, it is assumed that only stands in the constraint's region are considered. Summing over the constraint's region is indicated by a sum over i. Likewise summing over t for a constraint means summing over the applicable time interval. In the following sub-sections we go through the different constraints and describe their penalty functions.

The x-Meter Constraint

This constraint (or similar constraints) is also known as the "green-up" constraint, the "neighborhood" constraint or the "adjacency" constraint. The constraint is applied to avoid large continuous clear-cut areas and may be formulated as "a stand is not allowed to be clear-cut in the same time period as a neighboring stand". We use the formulation "a stand may not be clear-cut before the average tree height of all neighboring stands is at least x meters". If the trees grow fast or the time periods are long, this may lead to the same effect as the first formulation. However, in real-world problems this is usually not the case, thus the second formulation is richer and more flexible and as a consequence it demands more computing power. Our formulation is also more demanding because, in general, a clear-cut in one stand prevents clear-cutting of neighboring stands for more than one time period. The number of periods one has to wait before neighboring stands can be cut is dependent on the speed at which the trees grow, which is simulated by the forest simulator. The formulation used here is actually too restrictive. The usual restriction for clear-cutting is that neighboring stands may be clear-cut as long as the contiguous area of clear-cut forest is less than some area limit [23].

Penalty

Let the height of stand i at time point t be h_{it}(meters). Let the critical height in meters (the "x") of stand i be H_i^{min} and let

$$x_{it}^h = \begin{cases} 1 & \text{if } h_{it} < H_i^{min}, \\ 0 & \text{otherwise.} \end{cases}$$

Then the penalty incurred by the x-Meter constraint if it is used as a soft constraint is

$$F_{x\text{-}Meter} = \sum_i \sum_t x_{it}^h \left(\sum_{j \in U(i)} x_{jt}^h \left(H_j^{min} - h_{jt}\right)^2 \right)$$

In some areas around Oslo (Norway), in Nordmarka, $H_i^{min} = 2 \, \forall \, i$.

Maximum Total Area of Clear-Cut Forest

This is a constraint for limiting the area of clear-cut (or newly clear-cut) forest. The clear-cut area that is limited is not contiguous. The definition of clear-cut is the same as for the x-Meter constraint; if the height of the stand is less than H_i^{min}, then the stand is considered clear-cut.

Penalty

Let the area of stand i (in m^2) be A_i. Let the maximum area (in m^2) of forest that may be clear-cut in a given region at time point t be A_t^{max}. The total area of clear-cut forest (i.e., forest below the critical height) in time point t is $a_t = \sum_i x_{it}^h A_i$. Let

$$x_t^a = \begin{cases} 1 & \text{if } a_t > A_t^{max}, \\ 0 & \text{otherwise.} \end{cases}$$

The penalty incurred by this constraint if it is used as a soft constraint is

$$F_{MaxCC} = \sum_t x_t^a \left(a_t - A_t^{max}\right)^2 = \sum_t x_t^a \left(\left(\sum_i x_{it}^h A_i\right) - A_t^{max}\right)^2$$

Minimum Amount of Old Forest

The Old Forest constraint ensures that the area of old forest at all time points is at a minimum level. The (average) age of a stand is given initially and updated by the growth simulator. A stand is said to be old if the age is above a certain age. The "old age" may vary depending on the site quality of the stand. This constraint ensures that the area of old forest should be at least some given percent of the total area in the applicable region at all time points.

The Old Forest constraint is also used as a constraint on the maximum amount of young forest in a region since "maximum x% young forest" is the same as "minimum $(100 - x)$% old forest", where "old" forest in the latter formulation is the same as forest that is "not young" in the first formulation.

Penalty

Let the age of stand i at time point t be y_{it}. The age at which the forest in stand i is considered old is Y_i^{old}. This age depends on different parameters for the stand, including the site quality. The age of a stand after a clear-cut is zero. Let

$$x_{it}^y = \begin{cases} 1 & \text{if } y_{it} > Y_i^{old}, \\ 0 & \text{otherwise.} \end{cases}$$

Then the area of old forest in time period t is $a_t^y = \sum_i x_{it}^y A_i$. The value A_t^{min} is the minimum area of old forest required at time point t. Now, let

$$x_t^y = \begin{cases} 1 & \text{if } a_t^y < A_t^{min}, \\ 0 & \text{otherwise.} \end{cases}$$

The penalty is

$$F_{OF} = \sum_t x_t^y \left(A_t^{min} - a_t^y\right)^2 = \sum_t x_t^y \left(A_t^{min} - \sum_i x_{it}^y A_i\right)^2$$

Minimum Produced Volume

This constraint puts limits on minimum harvested volume of timber.

Penalty

Let the total volume harvested (from thinning, clear-cuts and any other actions that produce timber) in stand i at time point t be v_{it}. Let the minimum harvested volume required at time point t be V_t^{min}. The total volume harvested in time point t is $v_t = \sum_i v_{it}$. Let

$$x_t^v = \begin{cases} 1 & \text{if } v_t < V_t^{min}, \\ 0 & \text{otherwise.} \end{cases}$$

The penalty is computed by

$$F_{PV} = \sum_t x_t^v \left(V_t^{min} - v_t\right)^2 = \sum_t x_t^v \left(V_t^{min} - \sum_i v_{it}\right)^2$$

Minimum Income

This constraint puts limits on the minimum income. The minimum level may be different for different periods. Note that since the timber has different value depending on quality, age etc. this constraint can not be modeled using the Minimum Produced Volume constraint. E.g., consider two scenarios; 1) You harvest a lot of cheap timber (low quality) and 2) you harvest the same amount of expensive timber. Using the Minimum Produced Volume constraint above, both scenarios would either be feasible or infeasible (since the same amount is harvested). Using the Minimum Income constraint, the two situations may be treated differently, allowing the latter, but disallowing the first.

Penalty

The penalty is the same as the penalty for the Minimum Produced Volume constraint if one replaces v_{it} by w_{it}, the income from all harvested timer in stand i at time point t, and replaces V_t^{min} by W_t^{min}, the minimum required income in time period t. Then the penalty is:

$$F_I = \sum_t x_t^w \left(W_t^{min} - w_t\right)^2 = \sum_t x_t^w \left(W_t^{min} - \sum_i w_{it}\right)^2$$

where x_t^w is defined in the same manner as x_t^v.

2.10 Objective Components

Objective components (also called "objectives") evaluate a solution and return a real number. As mentioned above, a constraint may be treated as a soft constraint, in which case it also works as an objective component.

The penalty incurred by the soft constraints has the property that the better the solution is, the less penalty is incurred. The objective components must have the same property. The goal is to minimize the total value (i.e., the value from objective components + the penalty from soft constraints).

In the following paragraphs we present the different objective components. Typically, the objective value is summed over some stands, indexed with i, and time periods, indexed with t, depending on which applicable region and time span the objective has been defined for.

Net Present Value (NPV)

This objective component evaluates the economic value. The value of the objective is the sum of net present value for all harvested timber in the stands and time periods multiplied by minus one.

Value

The net present value of all harvested timber in a solution is:

$$F_{NPV} = -\sum_t \left(\frac{\sum_i w_{it}}{(1 + \frac{R}{100})^{y(t)}} \right)$$

Where R is the yearly interest rate and $y(t)$ is the number of years between now and the time point t. The formula for net present value increases as the solution gets better – more harvest yields higher value on w_{it}, which is the income from stand i in time period t (see 2.9). To be able to use the net present value together with other objective components and soft constraints in a minimization problem, the formula for net present value is multiplied by minus one. Thus the NPV objective component always takes on negative objective values.

Match Profile (Income and Production Volume)

The Match Profile objective is a generic component that is used to evaluate to what degree a (calculated) quantity matches a target profile over time. The objective is a set of target levels for some or all time points. It then measures the discrepancy between the target levels and the calculated quantity in a plan at the given time points. This component is used in two objectives; the Match Income Profile objective and Match Produced Volume Profile objective. The first evaluates how much the income during the periods resembles income from a predefined (required) income profile, while the latter evaluates how much the harvested volume during the periods resembles a predefined (required) production volume profile.

Value

The required profile is set by giving the numbers L_t^v for production profile and/or L_t^w for the income profile, for time points t. One wishes to be as close to the required profile as possible. The value of the objective is the squared deviation from the required profile:

$$F_{MP_v} = \sum_t \left(L_t^v - \sum_i v_{it} \right)^2$$

$$F_{MP_w} = \sum_t \left(L_t^w - \sum_i w_{it} \right)^2$$

Age Distribution

This is an objective component similar to the penalty from the Old Forest soft constraint. It specifies a desired percentage of area that should be in a given age interval. E.g., "20% of the forest should have an age in the interval $[40, 60\rangle$ years". An age interval is a half open interval $[y_1, y_2\rangle$, where y_1 and y_2 are ages. The objective is to minimize the squared difference between the actual area percentage and the desired percentage in each age interval. When evaluating this objective component for a given age interval and time point, the total forest area of the age interval is computed at the given time. The quotient of this value and the total area of the region gives the percentage which is compared with the desired percentage.

Value

Assume that the age intervals, indexed $k = 1, .., K$, do not overlap. Let a_t^k be the area of all stands at time point t that are in age class k and A_k be the area required for this age class. Then, the value of this component is

$$F_{AD} = \sum_t \left(\sum_{k=1}^{K} \left(a_t^k - A^k\right)^2 \right)$$

Note that $\sum_k a_t^k \leq A^{tot}\ \forall t$, where A^{tot} is the area of all stands. If the K age intervals cover all possible ages, then $\sum_k a_t^k = A^{tot}\ \forall t$.

2.11 Summery of the Model

The model consists of minimizing the sum of objective components

$$\begin{aligned} F &= \alpha_{NPV} F_{NPV} + \alpha_{MP_v} F_{MP_v} + \alpha_{MP_w} F_{MP_w} + \alpha_{AD} F_{AD} \\ &= \sum_t \Bigg(-\alpha_{NPV} \frac{\sum_i w_{it}}{(1+\frac{R}{100})^{y(t)}} + \alpha_{MP_v} \Big(L_t^v - \sum_i v_{it} \Big)^2 \\ &\qquad + \alpha_{MP_w} \Big(L_t^w - \sum_i w_{it} \Big)^2 + \alpha_{AD} \sum_{k=1}^{K} \Big(a_t^k - A^k \Big)^2 \Bigg) \end{aligned} \tag{5}$$

subject to the (hard) constraints:

$x_{it}^h \sum_{j \in U(i)} x_{jt}^h = 0\ \forall t, i$	As long as the height of stand i is less than H_i^{min} ($x_{it}^h = 1$), then the height of all neighbors, j, must be more than H_j^{min} ($x_{jt}^h = 0\ \forall j \in U(i)$).
$x_t^a = 0\ \forall t$	Can not harvest an area larger than A_t^{max} at any time point t.
$x_t^y = 0\ \forall t$	The area of old forest at time point t must be at least A_t^{min}.
$x_t^v = 0\ \forall t$	The harvested volume at time point t must be at least V_t^{min}.
$x_t^w = 0\ \forall t$	The income at time point t must be at least W_t^{min}.

Where we have the following constants:

α_i:	is a positive constant, $i \in \{NPV, MP_v, MP_w, AD\}$. The value must be set so that the influence of the objectives is approximately at the same level.
A_i:	is the area of stand i.
A_k:	is the desired area required for age class k.
A_t^{max}:	is the maximum area of forest that may be clear-cut at time point t.
A_t^{min}:	is the minimum area of old forest desired at time point t.
H_i^{min}:	is the critical height of stand i.
L_t^v:	is the required level for harvested volume at time point t.
L_t^w:	is the required level for income in stand at time point t.
R:	is the interest rate (in %).

$U(i)$: is the set of neighboring stands of stand i.
V_t^{min}: is the minimum harvested volume desired at time point t.
W_t^{min}: is the minimum desired income in time period t.
Y_i^{old}: is age at which the forest in stand i is considered old.
$y(t)$: is the number of years until time point t.

And the following variables:

a_t^k: is the area of all stands at time point t that are in age class k.
h_{it}: is the height of stand i at time point t.
v_{it}: is harvested volume in stand i at time point t.
w_{it}: is the income from harvested timber in stand i at time point t.
y_{it}: is the age of stand i at time point t.
x_{it}^h: is binary and 1 if $h_{it} < H_i^{min}$ at time point t, 0 otherwise.
x_{it}^y: is binary and 1 if $y_{it} > Y_i^{old}$ at time point t, 0 otherwise.
x_t^a: is binary and 1 if $A_t^{max} < \sum_i x_{it}^h A_i$ at time point t, 0 otherwise.
x_t^v: is binary and 1 if $\sum_i v_{it} < V_t^{min}$ at time point t, 0 otherwise.
x_t^w: is binary and 1 if $\sum_i w_{it} < W_t^{min}$ at time point t, 0 otherwise.
x_t^y: is binary and 1 if $\sum_i x_{it}^y A_i < A_t^{min}$ at time point t, 0 otherwise.

The model does not contain explicit decision variables. This is not necessary because we solve the problem by applying a heuristic procedure. There is no explicit linkage between the x-variables used in the constraints (e.g. x_{it}^h, x_{it}^y, ...) and the objective function. The x-variables are not decision variables. The linkage between the constraints and the objective function is via the other variables. E.g.: The volume produced in stand i at time t, v_{it}, is used in the objective function. The constraint $x_t^v = 0$ ensure a minimum amount of produced volume at time t: $v_t = \sum_i v_{it} \geq V_t^{min}$ which is a limit on the sum of v_{it} for all i. The moves (see Section 3.2) of the local search procedure will alter the value of some of the variables.

Weighting is a difficult task when optimising an objective function with several objective components. How does one choose a value for the constants α_i? The simplest way around this problem is to let the user set the weights. This may be a good solution if the user is experienced. Another way around the problem is to employ multi-objective optimisation. If one uses one objective function for each objective component, there is no need to weigh. On the other hand, muli-objective optimisation consumes a lot of CPU power and does not find one "best" solution but rather a set of solutions (the pareto set) in which no solution can be said to be better than any other.

If a constraint is too hard to satisfy, or the problem solver deems it right, the constraint is relaxed by penalizing it in the objective function (see equation 3 in Section 2.9). The penalty terms for the (hard) constraints when used as soft constraints are listed in Section 2.9.

The problem description given here does not cover all possible aspects of an LTFTSP, but it does cover a wide range of problems encountered in Norwegian and international forestry.

3 The Solution Procedure

Over the years there have been many suggestions as to what kind of solution procedure one should choose for solving the LTFTSP. Mathematical programming [27, 28] has been widely used for solving different types of planning problems in forestry, including the LTFTSP. The advantage of mathematical programming is that the solution found is the optimal solution. The disadvantage is that as the problem size increases the solution time increases drastically. This is mainly due to the discrete nature of the problem. If response time is limited, as it often is, one may utilize mathematical programming as an approximation method and return the best feasible solution found within the time limit. In case, an upper bound on the error is provided. However, the time to find the first feasible solution may be very high and also hard to predict. Moreover, non-linearities in the model will cause serious challenges to a mathematical programming approach. In an industrial setting, it is important that the choice of optimization method is robust towards model changes.

In contrast with *exact optimization methods* that guarantee to find the optimal solution, *heuristic optimization methods* focus on finding good solutions in reasonable time. In general, they are search techniques that do not come with a strong performance guarantee. However, a heuristic method is a robust choice for a rich model of real-world LTFTSPs with many variables, adjacency constraints and other non-linearities, as we shall discuss below.

Many heuristics are built on the basic method of *Local Search*, a simple and efficient way of finding reasonably good solutions to discrete optimization problems. A problem with Local Search is that it may get stuck in a local optimum with an objective value that is substantially worse than a global optimum. *Metaheuristics* are search strategies that avoid getting trapped in local optima and explore the search space more thoroughly than Local Search.

We have developed an optimization algorithm for the LTFTSP based on the *Tabu Search* metaheuristic. Before we describe our algorithmic approach in detail, we discuss alternative methods, motivate our methodological choice, and give a short introduction to Local Search and Tabu Search.

3.1 Alternative methods for solving the LTFTSP

Whether to choose an exact or a heuristic method for the LTFTSP highly depends on the overall goal, how rich the model is, and the problem size. The literature shows that researchers have used exact methods such as dynamic programming [16], Mixed Integer Linear Programming (MILP) [19], [7], and column generation [25] to study LTFTSPs. Models of limited richness and instances of limited size have been solved to optimality.

Since our overall goal is to develop an industrial decision-support system, we have developed a rich, spatial model that takes into account most of the aspects that forest planners meet in practice. An important usability issue for decision-support systems is predictability of response time. In decision support for planning, users want the optimal solution instantaneously. If they cannot have that, it is important to provide them with a good solution within an acceptable response time. Unpredictability in response time is very frustrating to users. An equally important issue is extendibility: The algorithmic approach must not be based on assumptions that may be invalidated by future model changes. Both issues are central to our choice of optimization method.

A rich LTFTSP model will become very large if modeled using linear programming [17, 26]. Some of the most central constraints, such as the x-Meter constraint, lead to binary variables, X_{it}, which is one if stand i is clear-cut in time period t and zero otherwise. A simplified version of the constraint may specify that two neighboring stands may not be clear-cut in the same time period. This leads to the constraints: $X_{it} + X_{jt} \leq 1 \; \forall \, t, i, j \in U(i)$, where $U(i)$ is the neighboring stands of stand i. This makes the problem a MILP. It is well known that solving large MILPs may take too much time, unless the formulation has a special structure. Typically, exact methods can only solve instances with a limited number of integer variables in reasonable time.

Linear and mixed integer programming are useful in studying forest scheduling problems, for two main reasons. First, one can study the properties of the problem and solve instances of limited size to optimality. Second, in combination with heuristic methods, mathematical programming is useful for acquiring upper bounds on the optimality gap. When applying a heuristic, one does not know whether the optimal solution is found, or how large the gap is. Any feasible solution will give an upper bound (when minimizing) on the optimal value. To find a lower bound, and hence an upper bound on the optimality gap, an exact method for a relaxed or reduced problem may be used. In this way, performance guarantees may be arrived at through a combination of exact and heuristic methods.

As mentioned above, exact methods from mathematical programming can be used to find approximate solutions [18]. MILP-solvers can be set to solve a problem within a given solution tolerance gap. For many MILP instances it is too time consuming to find an optimal solution. Allowing the solution to differ from the optimal (unknown) solution within a given tolerance will usually reduce the solution time dramatically. It has been shown that solution times may be reduced by as much as 90% for approximately 50% of over-mature and old-growth forest cases [18] by increasing the tolerance gap from 0.01% to 1%[3]. Still, reducing the solution time by 90% may not be enough, since as the problem size increases, the number of variables and/or constraints increases disproportionately.

Assume we have an LTFTSP instance of 1,000 stands and five time periods. This instance takes, say, two hours to solve using mixed integer linear programming. It will be solved in twelve minutes if the solution time is reduced by 90% (by increasing the tolerance gap from 0.1% to 1%). This may be acceptable in a given situation for a given user. But, due to the combinatorial

[3]The results were obtained on a LTFTS problem of three time periods.

nature of LTFTSPs, a slightly larger instance, say 1,200 stands, may take forty hours. The instance might be solved in four hours by increasing the tolerance gap to 1%, but there is no guarantee that the speed improvement is linear. Moreover, the time for finding the first feasible solution is unpredictable. This is a serious issue for the design of an industrial decision support system, even for strategic, long-term planning.

Unless a heuristic method is accompanied by an efficient method for determining high quality lower bounds, it gives no non-trivial bound on the optimality gap. An advantage of Local Search based heuristics is that the method for constructing the initial solution often has low computational complexity. A feasible solution will be constructed in limited time[4]. After the initial solution has been constructed, normally in a short time, the user may at any time stop the optimization process and get the best solution found until now. This *anytime property* has considerable value in practical problem solving. The property is shared by the approximation methods described above, but the unpredictable response time for the first feasible solution is often a serious issue.

Exact methods has been combined with heuristics to solve the LTFTSP. Good results are reported for a hybrid method that utilizes a Simulated Annealing metaheuristic for solving spatial considerations and linear programming for non-spatial aspects [24].

We have selected a heuristic algorithmic approach for providing approximate, practical solutions to LTFTSP instances in a decision-support setting. The better anytime property, the added flexibility for handling of non-linearities in a rich model, and the robustness towards model extensions are our main reasons. Our selection of a single solution type metaheuristic based on Tabu Search was motivated by the fact that Tabu Search has shown high performance on a number of hard, discrete optimization problems. Also, a basic Tabu Search variant is simple to implement. A population based metaheuristic would probably face problems with memory consumption for real-life LTFTSPs.

3.2 Local Search and Metaheuristics

We now turn to a short, general introduction to Local Search and metaheuristics. For a thorough treatment of Local Search, we refer to the book edited by Aarts and Lenstra [1]. For a survey of metaheuristics, we refer to the book edited by Glover and Kochenberger [10].

---

[4]In highly constrained problem instances, a feasible solution may be hard to find using simple, greedy construction heuristics.

Local Search for Discrete Optimization Problems

We define a *Discrete Optimization Problem* (DOP) in the following way. A (usually finite) set of *solutions* $\mathcal{S}'$ is given. A function $f : \mathcal{S}' \to \mathbf{R}$, called *the objective*, is defined. A constraint, $c : \mathcal{S}' \to \mathbf{Z_2}$ has the property:

$$c(\mathbf{s}') = \begin{cases} 0 & \text{if the solution } \mathbf{s}' \text{ is infeasible with respect to } c, \\ 1 & \text{if the solution } \mathbf{s}' \text{ is feasible with respect to } c. \end{cases}$$

Let $\mathcal{C}$ be the set of constraints and the set $\mathcal{S} \subseteq \mathcal{S}'$ be the set of feasible (legal) solutions:

$$\mathcal{S} = \{s' \in \mathcal{S}' \mid c(s') = 1 \ \forall c \in \mathcal{C}\}$$

The goal is to find a *global optimum*, i.e., a *feasible* solution such that the objective f is minimized[5].

Hence, a DOP may be defined by a pair $(\mathcal{S}, f)$. Normally, $\mathcal{S}$ is not given explicitly but defined implicitly through *variables* and their *domains*, together with *relations* (constraints) between these variables. $\mathcal{S}$ is called the *search space*. We note that the definition of a DOP covers (pure) integer programs. With discretization of time, our LTFTSP model is a DOP.

A solution $\mathbf{s}^* \in \mathcal{S}$ is a *global optimum* of the DOP $(\mathcal{S}, f)$ if the following holds:

$$f(\mathbf{s}^*) \leq f(\mathbf{s}), \forall \mathbf{s} \in \mathcal{S}$$

Local Search (LS), also called *Neighborhood Search*, is based on the idea of improvement through small modifications of a current solution. More formally, a neighborhood $\mathcal{N}$ is defined as a function $\mathcal{N} : \mathcal{S} \to 2^{\mathcal{S}}$, where $2^{\mathcal{S}}$ denotes the set of all subsets of $\mathcal{S}$. Given a solution $\mathbf{s}$, $\mathcal{N}(\mathbf{s}) \subseteq \mathcal{S}$ is called the *Neighborhood* of $\mathbf{s}$. Normally, but not necessarily, a neighborhood is a small subset of the set of all possible solutions.

A neighborhood is typically defined on the basis of a certain type of operation, defined by an *operator* on the solution. An example from the LTFTSP is a 1-OPT operator on STPs. Given a current solution, it will create a set of solutions that differ from the original in that exactly one STP in the solution has been exchanged for another. The 1-OPT neighborhood then consists of all feasible solutions that are generated by this operator. The size of this neighborhood is dependent on the total number of STPs in the current plan, and the number of alternative STPs for each of them. Usually, several operators are defined for a given DOP. For the LTFTSP we may for instance define a k-OPT operator that selects k STPs in a solution and replaces them with k others. The exchange of local treatment schedules between two stands is another example.

We define a neighborhood *move* as the acceptance of a neighboring solution. The set of solutions generated by an operator represents potential moves.

[5] A minimization problem may easily be transformed to a maximization problem.

It is a superset of the corresponding neighborhood because it may contain infeasible solutions. Members of this set must be checked for feasibility and improvement over the best solution so far, *the incumbent*, before a move is selected. Checks for feasibility and improvement is often a computational bottleneck in Local Search.

Let $\mathcal{N}$ be a Neighborhood function for the DOP $(\mathcal{S}, f)$. We define a solution $\hat{\mathbf{s}}$ to be a *local optimum* if the following holds:

$$f(\hat{\mathbf{s}}) \leq f(\mathbf{s}), \forall \mathbf{s} \in \mathcal{N}(\hat{\mathbf{s}})$$

Observe that a local optimum is defined relative to a specific neighborhood. Different neighborhoods will typically give rise to different sets of local optima.

Local Search is an iterative improvement procedure. It needs an *initial solution* $\mathbf{s}_0 \in \mathcal{S}'$ to start. Normally the initial solution must be feasible ($\mathbf{s}_0 \in \mathcal{S}$). There are alternative general methods for producing an initial solution. These include random generation, greedy construction, exact resolution of a reduced or relaxed problem, and perturbation of a previously found solution (e.g., a local optimum). A common approach is greedy construction. Here, one starts with an empty solution and uses a more or less myopic and computationally cheap heuristic to incrementally build a solution.

Starting with the initial solution $\mathbf{s}_0$, Local Search moves iteratively from one solution to a better solution. In the kth iteration, it searches the neighborhood of the current solution $\mathbf{s}_{k-1}$ for an improving solution. The iterative search process stops when there are no improving solutions in the neighborhood of the current solution. It follows from the definition above that the final solution is a local optimum relative to the neighborhood used. If the search space $\mathcal{S}$ is finite, Local Search stops in some iteration K, where $K < |\mathcal{S}|$.

Note that the above does not precisely define Local Search. There may be many improving solutions in the neighborhood of the current solution. A *strategy* for the acceptance of improving neighbors is needed. The most common acceptance strategy is *Best Improving*: the best (or one of the best) improving neighbor is accepted. Alternative strategies may be employed. For instance, one may accept the *First Improving* neighbor. Given an initial solution, Local Search has the anytime property. It may be interrupted at any time and still provide a useful solution.

Pseudo-code for Local Search with the Best Improving strategy is shown in Listing 1.

Metaheuristics

As we have mentioned, the main problem with Local Search is that it gets stuck in a local optimum. The topology of the search space defined by the neighborhood relations, the neighborhood employed, the topography defined by the objective function, the initial solution, and the selected acceptance strategy defines which local optimum a Local Search process will end up in. The process will typically be confined to a small part of the search space.

Listing 1. Local Search with Best Improving strategy

```
Procedure LocalSearch(s0,f,N, C)
begin
  current:=s0; // This is the initial solution
  localopt:=false;
  while not localopt do
  begin
    (current,localopt):=
        SearchNeighborhood(current,N(current),f, C);
    if localopt then return current;
  end
end

Procedure SearchNeighborhood(current,Neighbors,f,Constraints)
begin
  bestneighbor:=current;
  for n in Neighbors do
  begin
    feasible:=CheckFeasibility(n,Constraints);
    if feasible and f(n) < f(bestneighbor) then
    begin
      bestneighbor:=n
    end
  end
  return (bestneighbor,bestneighbor=current)
end

Procedure CheckFeasibility(solution,Constraints)
begin
  for c in C do
  begin
    if c(solution)=0 then return false;
  end
  return true;
end
```

Metaheuristics are search strategies that are able to escape from local optima and spread the search effort to different, promising areas of the search space. Glover and Kochenberger [10] give the following informal definition:

> *Metaheuristics, in their original definition, are solution methods that orchestrate an interaction between local improvement procedures and higher level strategies to create a process capable of escaping from local optima and performing a robust search of a solution space.*

We may distinguish between two main classes of metaheuristics: *single solution based metaheuristics*, and *population based metaheuristics*. Single solution based metaheuristics maintain a single current solution and utilize some variant of Local Search for local improvement. There are two fundamental mechanisms in single solution based metaheuristics: acceptance of non-improving moves, and restart.

Arguably, the simplest metaheuristic is to restart Local Search from a different initial solution. Other general techniques in metaheuristics are use of randomness, and short term memory to avoid looping and revisiting similar solutions. Long-term memory structures may be used both to diversify search to visit new regions of the search space, and to intensify search around solutions with properties that seem good. Some metaheuristics change the search space topography by augmenting the objective function, thus escaping from local optima. Others draw upon the fact that local optima will differ between alternative neighborhoods. They systematically change neighborhood when a local optimum has been found.

In contrast, population based metaheuristics keep a set of solutions during search. In general, their iterative procedures utilize information from two or more solutions to generate new solutions. Modern, population based metaheuristics typically also use Local Search for improving individuals in the population.

To conclude, the general introduction to Local Search and metaheuristics, we give a short description of the single solution based metaheuristic called *Tabu Search* that is the basis of our algorithmic approach for the LTFTSP.

Tabu Search

Tabu Search (TS) is a single solution metaheuristic that is based more on use of memory structures than use of randomness. Its fundament is Local Search as defined above with a small strategy modification that makes it more "aggressive": the best move in the neighborhood is always accepted even if it is not improving. This is enough to enable Tabu Search to escape from local optima, but at the expense of a high probability of looping back to a previously visited solution. To avoid looping, *short term memory* is used to maintain certain *attributes* of the solutions visited or moves made over the past iterations.

In principle, the short term memory may contain full solutions, but memory and time requirements will typically render this strategy impractical. Instead, essential attributes of solutions, or the moves that lead to them, are stored in short term memory. Attributes are typically defined as basic structures in the solution representation. An LTFTSP example of an attribute is the STP structure of the current plan. Another example is the LFTSes that were changed by a move. A *tabu criterion* is defined on the chosen attribute. Again, an LTFTSP example would be that, when a given STP in the solution is changed by an accepted move, it is forbidden (tabu) to make a move that changes it again for a certain number of Local Search iterations.

Aspiration criteria are used in Tabu Search to override the tabu status of certain moves. The obvious and most commonly used aspiration criterion is improvement over the incumbent, i.e., the move will lead to the best solution found so far. The duration of tabu status is called the *tabu tenure*, which is a critical parameter to the Tabu Search method. Generally, one may enhance performance of a Tabu Search procedure by varying tabu tenure during search.

A basic Tabu Search procedure with short term memory and a simple aspiration criterion is shown in Listing 2.

Tabu Search is arguably a framework for memory-based metaheuristics rather than a concrete search technique. Tabu Search based methods has shown good performance over a large variety of DOPs. There is a huge literature on Tabu Search. The interested reader is referred to [9], [11], [15], and the chapter "Introduction to Tabu Search" in [10].

3.3 Initial Solution for the LTFTSP

Iterative improvement techniques like Tabu Search are dependent on an initial solution. The better the initial solution is, the better solutions will in general be found. Or put another way: the same solution as with a worse initial solution will be found in less time.

The strategy we have chosen for creating a good, feasible initial solution is this: Create a first solution $\mathbf{s}_f$ by ignoring adjacency constraints. Focus on each stand, one by one, and create a harvest schedule with as many clear-cuts and as many thinnings as possible, given the stands' site quality and age. The thinnings and clear-cuts will be scheduled at ages where it is possible to move the treatments both forward and backward in time (see Section 3.4 about neighborhoods). The first solution will have treatment schedules that are locally optimal for each stand. Depending on which (hard) constraints the user has chosen, the first solution $\mathbf{s}_f$ will typically be infeasible ($\mathbf{s}_f \in \mathcal{S}'$). It must be made feasible to serve as an initial solution. This is fixed in our approach by performing pre-optimization: For every constraint c in the set of hard constraints $\mathcal{C}$ define a penalty function F_c as described in Section 2.9. Then we make $\mathbf{s}_f$ feasible by applying the algorithm shown in Listing 3.

After this procedure, the solution returned from the procedure "Create-FeasibleInitialSolution", $\mathbf{s}_0$, will satisfy all constraints in $\mathcal{C}$. Thus, $\mathbf{s}_0$ is feasible and a possible starting point for the tabu search procedure. The proposed procedure has one weakness; The call to the procedure "TabuSearch" may never return solution for which $F_c = 0$. There might be two reasons for this:

a) There *does not exist* a solution for which $F_c = 0$.
b) The chosen solution technique (Tabu Search) is not able to find a solution for which $F_c = 0$.

Listing 2. Basic Tabu Search

```
Procedure TabuSearch(Init,f,N,Stop(),C)
begin
  incumbent:=current:=Init; // incumbent holds best solution
  InitializeTabuList (current);
  while not Stop() do
  begin
    current:=SearchNeighborhood(current,N(current),
                                f,incumbent,C);
    if f(current) < f(incumbent) then
    begin
      incumbent:=current;
    end
  end
  return incumbent;
end


Procedure SearchNeighborhood(current,Neighbors,f,incumbent,
                             Constraints)
begin
  bestneighbor:=ReallyBadSolution();
  Neighbors=N(current);
  for n in Neighbors do
  begin
    feasible :=CheckFeasibility(n,Constraints);
    if feasible and f(n) < f(bestneighbor) and
      (not Tabu(n,TabuList) or f(n) < f(incumbent)) then
    begin
      bestneighbor:=n;
    end
  end
  UpdateTabuList(bestneighbor);
  return bestneighbor;
end
```

If a) is the case, then the user should be informed about this fact. If b) is the case, then another solution technique could be chosen and the above procedure repeated. A compromise in case of b) is to inform about which constraint that could not be satisfied, and ask (the user) if it can be relaxed. "Relaxed" means to make the constraint easier to satisfy (typically by adjusting some of the parameters in the constraint).

An alternative to the above procedure might be to handle all (hard) constraints together as in Listing 4.

Listing 3. Create a feasible initial solution

```
Procedure CreateFeasibleInitialSolution(sf,N,C)
begin
  Feasible:=sf;
  Constraints:=Empty();
  for c in C do
  begin
    Constraints:=Constraints+c;
    Feasible:=TabuSearch(Feasible,Fc,N,StopWhenFcIsZero(),
                         Constraints);
  end
  return Feasible;
end
```

Listing 4. Create a feasible initial solution (2)

```
Procedure CreateFeasibleInitialSolution(sf,N,C)
begin
  Feasible:=sf;
  Constraints:=Empty();
  F=0;
  for c in C do
  begin
    F:=F+Fc;
  end
  return TabuSearch(Feasible,F,N,StopWhenFIsZero(),C);
end
```

Again, the same problem may occur when calling "TabuSearch": this procedure may never find a solution for which $F = \sum_c F_c = 0$.

3.4 Local Search Neighborhoods for the LTFTSP

We have defined two Local Search neighborhood operators (see Section 3.2 above) for our model of the LTFTSP: A "time adjustment operator" and an "exchange STP operator". In the following, we shall denote them TIME and 1-OPT, respectively. Both operators, their corresponding neighborhoods and moves are described in the following paragraphs.

A move in the TIME neighborhood corresponds to an adjustment of the length of a single, adjustable STP. The move is carried out by changing the end time of an STP. A TIME move can be *local* or *global* depending on how the change in end time affects the later STPs. In a local TIME move the change in the STP that is adjusted will be absorbed by the next adjustable STP. In

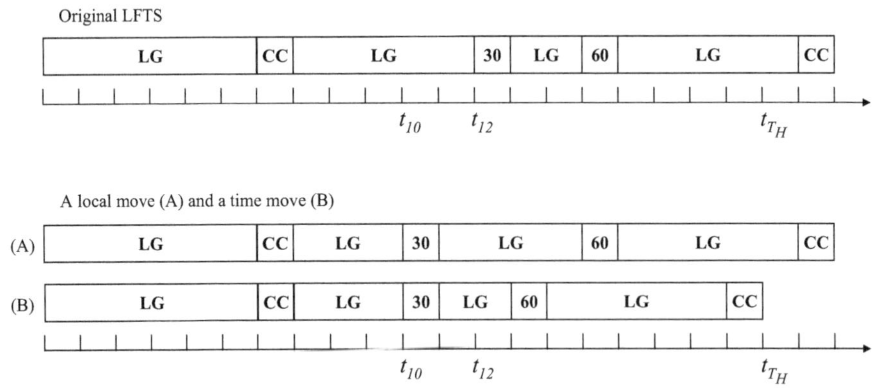

Fig. 3. Two types of TIME moves, "local" and "global", in an LFTS.

a global TIME move all consecutive STPs are shifted by the same amount of time. The two types of TIME moves are illustrated in Fig. 3. In the figure the local move (A) shortens the length of the second let-grow period. The move is absorbed by the next let-grow period, which increases its length. The program is unchanged from the 60% thinning and later. For the global move, the start and end times for all STPs in the program after the second let-grow period is shifted by the same amount of time $t_{12} - t_{10}$.

The 1-OPT neighborhood corresponds to moves that exchange a single STP with another pre-defined STP. Basic STPs are defined together with a set of non-basic STPs. The non-basic STPs are collections of basic STPs (see Section 2.6). A 1-OPT move will exchange exactly one STP (basic or non-basic) with another STP (basic or non-basic). A 1-OPT move is illustrated in Fig. 4 where the STP between t_a and t_b – "STP2" – is to be exchanged.

The 1-OPT move is more complicated than the TIME move and requires substantially more computation (mainly by the simulator). The strategy used in our solution procedure implies a more frequently use of the TIME operator than the 1-OPT operator.

The size of a local TIME neighborhood is equal to the number of new time points which all end time point can be set to. Let us look at one specific end time point; Consider the original LFTS in Fig. 3. If the end time of the second let-grow period (t_{12}) is chosen as the end time to adjust, then there are at most four time points to the left t_8 through t_{11} to consider. To the right of t_{12} there is only one possible time point since the third let-grow period can not have a length equal to zero. Thus the total number of neighbors for a local TIME move that will change the second let grow period is five.

The size of a global TIME neighborhood in the above example has the same limit if the adjustment is to decrease the length of the let-grow period (i.e., size equal to four). If the second let-grow period is to be enlarged, the number of possible, new end points is restricted by limits on when the following 60%-

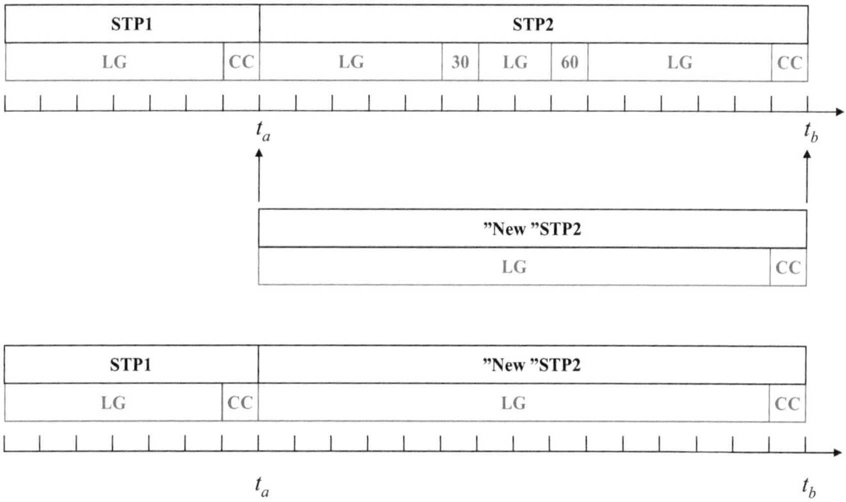

Fig. 4. Example of a 1-OPT move.

thinning and clear-cut may occur. The size of the global TIME neighborhood is generally larger than the size of the local TIME neighborhood.

A 1-OPT move exchanges a subpart of an LFTS. The size of the 1-OPT neighborhood is thus dependent on the number of available STPs that has been defined and can legally be inserted in the same time interval as the removed STP.

The TIME and 1-OPT neighborhoods may both be characterized as *intra-LFTS* neighborhoods, in the sense that the local treatment schedule of *one stand only* is changed by each move. Some articles [4, 14] describe an *inter-LFTS* neighborhood, where a move will make changes in two stands. One advantage of the inter-LFTS neighborhood is that the objective value of many objective components may be conserved by such moves.

3.5 Focal Points for the LTFTSP

The general idea behind *focal points* is to reduce the search effort. Instead of searching full neighborhoods, the search focuses on critical areas of the current solution that seem the most promising for improvement. In the LTFTSP context, we define focal points as descriptions of where (which stand) and when (time point) in the current forest treatment plan there seems to be a high potential for improvement.

We return again to the simple model in Section 2.1. Assume there are 25 stands, five time periods, and two treatments – clear-cut and do nothing. Each LFTS contains one clear-cut. The number of neighbors for a simple time move will be[6] $25 \times 4 = 100$. If we were using focal points, the harvest objective might be able to generate a few focal points indicating in which stands the harvested volume is small and would increase much if the harvest was moved to a later time (so that the forest will grow). For example, if the number of focal points were three, the number of neighbors would be twelve. The number of neighbors to evaluate will be reduced by 88% at the cost of letting the objective component do the computations to generate the focal points. In general, we save a lot of computing effort by utilizing focal points.

Focal points are similar to the *Candidate list* concept in Tabu Search [11] in that one avoids generating and assessing neighbors that are not promising. Hence, one reduces the computational expense of evaluating all moves in the neighborhood. Of course one thus loses the exactness of neighborhood evaluation. If the focal point identification heuristics are good, this reduction of neighborhoods will lead to an overall speedup without missing high quality solutions.

3.6 Tabu Search for the LTFTSP

Our implementation of Tabu Search for the LTFTSP is quite basic, except for the use of focal points. It utilizes the basic LFTS structure defined for each stand as attribute. The tabu criterion is defined as follows: if a given LFTS has been changed by an accepted move, it is tabu to accept a move that changes it again, for the next n iterations. The tabu tenure n is a parameter to the algorithm.

The tabu criterion is implemented by a short term memory that identifies the last n stands that have had their LFTS changed. We use the standard aspiration criterion that prescribes acceptance of a tabu move if it gives an improvement over the incumbent.

3.7 Software Implementation

Ecoplan [20, 13] is a software implementation of the model described in the previous sections. SINTEF started developing this software in 1995.

In Ecoplan the user may exploit flexibility in variables and STPs. Any number of variables may be specified. Variables are specified in a text file that is read during start-up. STPs may be defined by specifying criteria for the STP and the result the STP has. Both criteria and result are specified through the use of variables; criteria by giving one or more variables and setting legal values or intervals for them, result by specifying the change in variables as a result of the STP. STPs are specified in a text file that is read during start-up.

[6] The clear-cut is performed in one of the five periods and can thus (in every stand) be moved to one of the other four periods.

The above features may be easily changed without changing the software. But it is also possible to extend Ecoplan with e.g., a new simulator or new constraints and objectives. The software is object oriented, so in order to implement e.g., a new simulator, a defined interface must be implemented by the simulator. Once this is done, variables and STPs may take advantage of the new features of the added simulator.

Ecoplan is the software that has been used to produce the results presented in Section 4.

4 Illustrative Example and Experimental Results

The following section contains experimental results for a real 1,541 stand case from Norway.

4.1 Description of the case

The case contains 1,541 stands with an average area of 2.23 ha. An important number for the x-Meter constraint is the average number of neighbors for each stand, which is 5.2. The initial state of the forest is as follows: The age of the stands are in the interval $[0, 148]$ years, with an average of 43.4 years. The area contains mostly spruce (49.4%), but also birch (27.5%) and pine (23.11%). The average height of conifers (spruce and pine) is 8.45 m, the average height of broadleaves (birch) is 7.84 m. The time horizon of the problem is set to be 100 years, divided into 20 five-year periods.

Four objective components are used in this case. The applicable region of all components is the global region of all stands. All time periods are considered. The two first are pure objective components (F_{NPV} and F_{MP_v}), while the two other (F_{OF} and $F_{x\text{-}Meter}$) are soft constraints. The last constraint, the x-Meter constraint, is also used as hard constraints in some runs.

F_{NPV}: Net present value (objective component).

F_{MP_v}: Match profile of production volume (objective component). The periodical production volume is set to 1800 m3/ha per period in order to achieve an even production volume.

F_{OF}: Minimum amount of old forest (constraint/objective component). The minimum amount of area that should be old forest is set to 30% in each time period. Initially, the amount of old forest is 20.6%. A stand is considered old depending on site index and dominating tree specie, according to the following table:

Site index	"Old" age
0 – 9	46
10 – 12	56
13 – 15	61
16 – 18	71
> 19	76

Since the old area initially is 20.6%, the penalty will be rather large until more forest grow old. The initial old area is the reason why this component is a soft constraint and not a hard constraint.

$F_{x\text{-}Meter}$: The x-Meter constraint with the critical height set to two meters for all stands.

Eighteen different basic STPs are defined. The large number is necessary because there are different basic clear-cut STPs for each combination of tree specie and site index groups (i.e., there is one clear-cut STP defined for spruce stands where the site index is in the interval $[1, 10\rangle$, one for spruce stands where the site index is in the interval $[10, 13\rangle$ and so on. In addition, two thinning STPs are defined; 15% and 20% thinning.

4.2 The Impact of the Different Objectives

We compare eight different runs on the case. In each run we start from scratch (i.e., no initial solution is provided) and run for 100,000 iterations. The runs are named "Npv", "Match", "Oldforest" and "Xmeter". When the name of a run is followed by "C_xmeter" it means that the x-Meter component is used as a (hard) constraint. So for instance; "Npv_C_xmeter" means that the objective function is "maximize net present value[7]", while obeying the x-Meter constraint. The objective function consist of only one component in each run except for the last run (minimize the deviation from the required production volume, maximize net present value, minimize the missing amount of old forest and minimize the violations of the x-Meter constraint, respectively). In the last line of the table an objective function with all the four objective components has been run. The different objective functions evaluate the solutions at very different values. To be able to add together several objectives one must choose the α_c-s defined in Section 2.11. This is done by experience and by trial and error. For the last run in the table, "Npv_Match_Oldforest_xmeter", the objective function was set to:

$$F = 1.5 \times 10^{-5} F_{NPV} + 3.0 \times 10^{-5} F_{MP_v} + 2.0 \times 10^{-4} F_{OF} + 0.75 F_{x\text{-}Meter}$$

The results from all runs are compared in Table 1. The table contains four columns measuring the effect of the four different objectives:

NPV: The Net Present Value with 5% interest rate measured in 10^6NOK.
Dev. Prod.: Deviation from the production profile measured in $10^5(\mathrm{m}^3/\mathrm{ha})^2$.
Dev. Old forest: Deviation from 30% old forest measured in 10^3ha.
x-Meter viol.: Violation of the x-Meter constraint in meters (m).

Note that only the two first columns contain objective values (F_{NPV} and F_{MP_v}, respectively), the two last columns contain values that are closely related to the corresponding objective component (F_{OF} and $F_{x\text{-}Meter}$).

[7] This actually mean that we minimize F_{NPV}

Table 1. The effect of the four different objectives.

Name of run	NPV	Dev. Prod.	Dev. Old forest	x-Meter viol.
Npv	38.0	454	4.99	210
Npv_C_xmeter	35.7	259	1.06	—
Match	23.8	0.186	1.25	76.9
Match_C_ xmeter	24.1	4.33	1.32	—
Oldforest	23.7	210	0.922	251
Oldforest_C_xmeter	23.6	151	0.967	—
Xmeter	27.5	92.9	1.66	0.00
Npv_Match_Oldforest_xmeter	26.6	10.7	1.29	0.15

4.3 Observations

As expected, net present value is highest in the run where the objective function was "maximize net present value" ("Npv"). The deviation from the 1,800 m^3/ha-profile was lowest in the "minimize the deviation from the desired production volume"-run ("Match"). The deviation from 30% old forest in each period is smallest when "minimize the missing amount of old forest" ("Oldforest") is used as objective function. And when "minimize the violations of the x-Meter constraint" ("Xmeter") is used as objective function, the violations are smallest (no violations at all).

Among the three runs with other objective functions than the x-Meter, and where the x-Meter is not a hard constraint ("Npv", "Match" and "Oldforest"), the violation to the x-Meter constraint is smallest for the "Match" run. This seems logical because in this run harvesting will be evenly divided amongst the periods to minimize the objective, thus also (on average) minimizing violations to the x-Meter constraint. The objective functions F_{MP_v} and $F_{x\text{-}Meter}$ are less conflicting than for instance F_{MP_v} and F_{NPV}. The latter will typically try to place as much harvesting as possible in the early periods because this gives higher net present value (see Section 2.10). That is in direct conflict with the F_{MP_v} objective function, which evens out the harvested volume.

The missing old forest area in all runs is missing area in the early periods. The initial old forest area is 709 ha, which is 20.6% of the total area. This means that no matter how long one optimizes, there will be a deviation from 30% old forest in the early periods. In the "Oldforest"-run, the old area in the first four time points is 20.6%, 22.1%, 23.0%, 27.4%, respectively. After that, there is enough old forest (more than 30%).

The objective value displays a typical pattern. At first it improves much per time unit, but after some time the improvements are fewer. This is illustrated in Fig. 5 for the "Npv_Match_Oldforest_xmeter" run. The objective value can

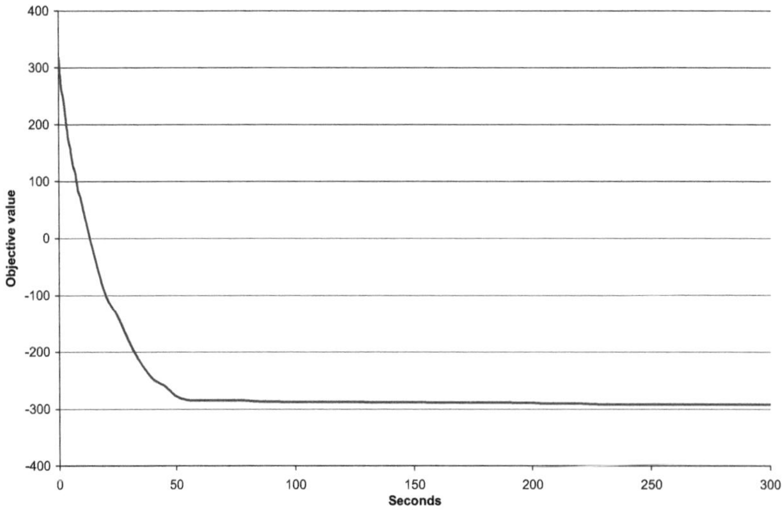

Fig. 5. Improvement in objective value as a function of time for the run "Npv_Match_Oldforest_xmeter".

become negative because the objective component net present value F_{NPV} takes negative values (see Section 2.10).

Development of harvested volume for different number of iterations for the match production profile ("Match") is shown in Fig. 6. As one can see, the adjustment of the profile to the level of 1,800 m^3/ha becomes better and better as the number of iterations increases. The results for this plot are from the run "Match". The relatively bad match to 1800 m^3/ha in the first periods is due to the fact that there is a lot of old forest that must be harvested in the first periods (because of limits on how old forest can be when clear-cut is performed).

5 Improvements

5.1 Two-opt Moves

Today the software described contains two moves; a time move (TIME) which changes the duration of an adjustable STP and an exchange move (1-OPT) which exchanges part of the program of a stand with another part. Used together, these moves can improve and solve many forest scheduling problems. Still, one would benefit from a wider range of moves. For certain objective functions, a two-opt move would be useful. Two-opt in the setting of the described model would mean to swap part of a program in one stand with part of a program of another stand [5]. In [4] it is showed that using two-opt in conjunction with one-opt instead of one-opt alone gives better results. The

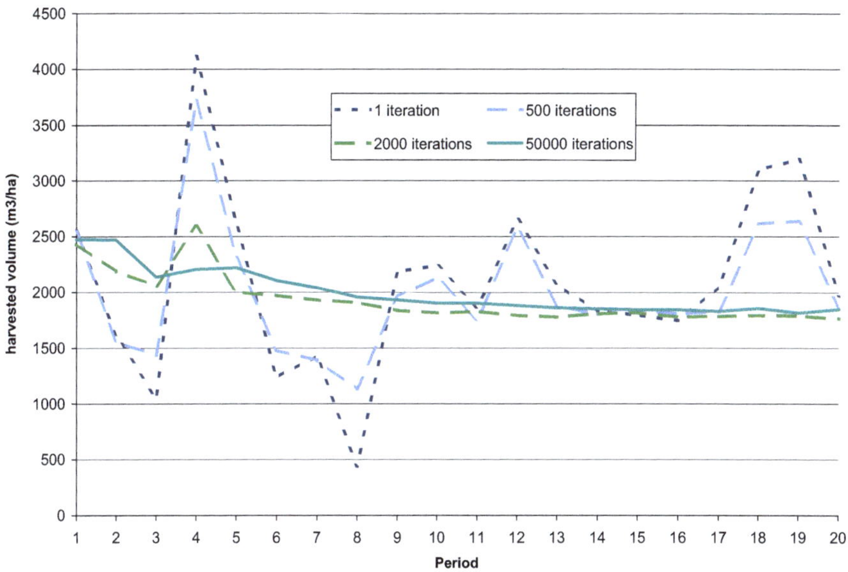

Fig. 6. Harvested volume for different number of iterations for the run "Match".

two-opt moves described in [4] is simpler than a two-opt move would be in our framework. Still we believe that such a move would improve the quality of the search.

5.2 Focal Points

A focal point is a descriptions of where (stand) and when (time period) the current plan may be improved. The idea behind focal points is that instead of searching all possible combinations of where and when to improve, the search focuses on the most promising focal points. For some of the objective components it is difficult to describe focal points, while for other objective components it is easy. We illustrate this by two examples.

The x-Meter objective component. Focal points can be found by checking all neighboring strands of a stand that is clear-cut. If the height of neighboring stands at the time of clear-cut is less than the critical height, then this can be identified as a focal point (i.e., a change at this time in either the stand or the neighbor, could improve the objective value of this component).

Net present value. To identify a focal point one would have to identify where a change in the plan could increase the net present value. There is no obvious way to do this without actually performing a move and evaluating the effect. Perhaps it would be possible to find stands where a clear-cut is not in the "optimal age interval". But for this to work one needs to find the "optimal age interval" which is dependent on tree growth functions and how prices on timber are calculated. One will soon end up spending so much time finding the focal points that the time one was suppose to save, is lost.

6 Conclusions

We have described a rich model for solving long-term forest treatment scheduling problems. The model is flexible and capable of describing many forest treatment scheduling problems.

The heuristic optimization technique based on Tabu Search that we have proposed allows "any" kind of constraint/objective to be implemented. The size of the problems that can be handled in reasonable time is large. No exact solutions can be guaranteed, but solutions may be produced in reasonable time for moderate problem sizes. The quality of solutions will decrease as the size of the problem instance increases.

The model and our suggested optimization procedure are implemented in a decision support system called Ecoplan, to be used in real-life management of forests. In our opinion, Ecoplan strikes the right balance between the types of problems it can solve, the quality of the solutions produced, and the time it takes to reach an acceptable solution.

References

1. E. Aarts and J. K. Lenstra, editors. *Local Search in Combinatorial Optimization.* John Wiley & Sons Ltd, Chichester, England, 1997.
2. T. Back, D. B. Fogel, and Z. Michalewicz, editors. *Handbook of Evolutionary Computation*, chapter Penalty Functions - Section C 5.2. Oxford Univ. Press, 1995.
3. P. Bettinger, K. Boston, Y.-H. Kim, and J. Zhu. Landscape-level optimization using tabu search and stand density-related forest management prescriptions. *European J. Oper. Res.*, 2006, doi:10.1016/j.ejor.2005.09.025.
4. P. Bettinger, K. Boston, and J. Sessions. Intensifying a heuristic forest harvest scheduling search procedure with 2-opt decision choices. *Can. J. For. Res.*, 29(11):1784–1792, 1999.
5. F. Caro, M. Constantino, I. Martins, and A. Weintraub. A 2-opt tabu search procedure for the multiperiod forest harvesting problem with adjacency, greenup, old growth and even flow constraints. *For. Sci.*, 49(5):738–751, 2003.
6. K. Crowe and J. Nelson. An indirect search algorithm for harvest-scheduling under adjacency constraints. *For. Sci.*, 49(1):1–11, 2003.
7. K. Crowe, J. Nelson, and M. Boyland. Solving the area-restricted harvest-scheduling model using the branch and bound algorithm. *Can. J. For. Res.*, 33(9):1804–1814, 2003.
8. K. A. Crowe and J. D. Nelson. An evaluation of the simulated annealing algorithm for solving the area-restricted harvest-scheduling model against optimal benchmarks. *Can. J. For. Res.*, 35(10):2500–2509, 2005.
9. F. Glover. Tabu search - a tutorial. *Interfaces*, 20(4):74–94, 1990.

10. F. Glover and G. Kochenberger, editors. *Handbook of Metaheuristics*. International Series in Operations Research & Management Science, vol. 57. Kluwer Academic Publishers, Norwell, MA, USA, 2003.
11. F. Glover and M. Laguna. *Tabu search*. Kluwer Academic Publishers, 1997.
12. M. Goycoolea, A. T. Murray, F. Barahona, R. Epstein, and A. Weintraub. Harvest scheduling subject to maximum area restrictions: Exploring exact approaches. *Oper. Res.*, 53(3):490–500, 2005.
13. G. Hasle, J. Haavardtun, O. Kloster, and A. Løkketangen. Interactive planning for sustainable forest management. *Ann. Oper. Res.*, 95(1-4):19–40, 2000.
14. T. Heinonen and T. Pukkala. A comparison of one- and two-compartment neighbourhoods in heuristic search with spatial forest management goals. *Silva Fennica*, 38(3):319–332, 2004.
15. M. Hindsberger and R. V. V. Vidal. Tabu search - a guided tour. *Control Cybernet.*, 29(3):631–651, 2000.
16. M. Hoganson and G Borges. Using dynamic programming and overlapping subproblems to address adjacency in large harvest scheduling problems. *For. Sci.*, 44(4):526–538, 1998.
17. C. Lockwood and T. Moore. Harvest scheduling with spatial constraints - a simulated annealing approach. *Can. J. For. Res.*, 23(3):468–478, 1993.
18. M. E. McDill and J. Braze. Using the branch and bound algorithm to solve forest planning problems with adjacency constraints. *For. Sci.*, 47(3):403–418, 2001.
19. M. E. McDill, S. A. Rebain, and J. Braze. Harvest scheduling with area-based adjacency constraints. *For. Sci.*, 48(4):631–642, 2002.
20. G. Misund, G. Hasle, and B. S. Johansen. Solving the clear-cut scheduling problem with geographical information technology and constraint reasoning. In *Proceedings of the Third International Conference/Workshop on Integrating GIS and Environmental Modelling*, 1996.
21. D. S. Mullen and R. M. Butler. The design of a genetic algorithm based spatially constrained timber harvest scheduling model. In *Proceedings of the Seventh Symposium on Systems Analysis in Forest Resources*, pages 57–65, 1997.
22. A. T. Murray and R. L. Church. Heuristic solution approaches to operational forest planning problems. *OR Spectrum*, 17(2-3):193–203, 1995.
23. A.T. Murray. Spatial restrictions in harvest scheduling. *For. Sci.*, 45(1):45–52, 1999.
24. K. Öhman and L. O. Eriksson. Allowing for spatial consideration in long-term forest planning by linking linear programming with simulated annealing. *For. Ecol. Manage.*, 161(1-3):221–230, 2002.
25. A. Weintraub, F. Baharona, and R. Epstein. A column generation algorithm for solving general forest planning problems with adjacency constraints. *For. Sci.*, 40(1):142–161, 1994.
26. A. Weintraub, G. Saez, and M. Yadlin. Aggregation procedures in forest management planning using cluster analysis. *For. Sci.*, 43(2):274–285, 1997.
27. H. P. Williams, editor. *Model building in Mathematical Programming*. John Wiley & Sons Ltd, Chichester, England, 1978.
28. H. P. Williams, editor. *Model Solving in Mathematical Programming*. John Wiley & Sons Ltd, Chichester, England, 1983.

An Integer Programming Approach to Image Segmentation and Reconstruction Problems

Geir Dahl and Truls Flatberg

Summary. This paper discusses segmentation and reconstruction problems using a integer linear programming approach. These problems have important applications in remote sensing, medical image analysis and industrial inspection. We focus on methods that produce optimal or near-optimal solutions for the corresponding optimization problems. We show that for the two problems one may use similar ideas in both modeling and solution methods. These methods are based on Lagrangian decomposition and dynamic programming for certain subproblems (associated with lines in the image). Some computational experiences are also reported.

Key words: Image analysis, segmentation, reconstruction, integer programming.

1 Introduction

Nowadays digitized images are routinely recorded in a huge number of applications, e.g., in remote sensing, medical image analysis and industrial inspection. In several of these applications the observed image may be viewed as a matrix of pixels and in each pixel one has an observed value (or vector). There are several important problems that arise in the analysis of such images. One such problem is *image segmentation* which is a method for labeling each pixel in order to describe the content of the image. These labels are typically chosen from a small discrete set; in medical images the labels correspond to tissue types. In other applications the actual image of some object may be unknown and it is observed indirectly by measuring projections using e.g., x-rays in different directions. This is, for instance, the case in some medical and industrial applications where computerized tomography (CT) is used for scanning objects. In these applications one is confronted with *reconstructions problems* where the goal is to reconstruct an image (or object) based on density projections along parallel lines in a given set of directions.

Both image segmentation and reconstruction problems may be considered as mathematical optimization problems, actually as discrete optimization problems. The main goal of this paper is to present this approach and discuss modelling issues and corresponding optimization methods. Moreover some related theoretical results are considered. For image segmentation the basis is a statistical model, a Markov Random Field model, and the problem is rewritten into an integer linear programming problem with special structure which is investigated and exploited in solution methods. The reconstruction problems are formulated in terms of binary matrices and, again, integer programming.

Image analysis, segmentation and reconstruction represent very active research areas where many different mathematical and engineering approaches are used. For instance, the recent research area of *discrete tomography* is largely motivated by image reconstruction problems, see [17] for a survey of this area. Still, the use of integer programming in these areas is very recent and this chapter presents main ideas and results here. For some references on this approach, see [8], [9], [13], [22], [16], [11] and [24].

2 A General Framework

In this section we present some problems in image analysis that we focus on in this paper. We also introduce some mathematical terminology and a rather general framework for discussion of these problems.

We restrict the attention to two-dimensional discrete images. However, both ideas and methods presented here may be adopted to multi-dimensional images, although the corresponding optimization problems then are more difficult to solve. For our purposes an *image* is a function

$$f : \mathcal{D} \rightarrow \mathcal{K}$$

where the domain is $\mathcal{D} = \{1, 2, \ldots, M\} \times \{1, 2, \ldots, N\} = \{(i, j) \in \mathbb{Z}^2 : 1 \leq i \leq M,\ 1 \leq j \leq N\}$ and $\mathcal{K} = \{0, 1, \ldots, K\}$. We consider $\mathcal{D}$ and $\mathcal{K}$ as fixed. Thus, our notion of image corresponds to two-dimensional discrete images with M (horizontal) rows and N (vertical) columns containing a total of $M \cdot N$ pixels (picture elements). Moreover, in each pixel (i, j), there is a value $f(i, j)$ which is an integer between 0 and K.

Remark. Often in image analysis and tomography one meets problems involving images viewed as real-valued functions defined on some (connected) domain in $\mathbb{Z}^d$. This is different from our setting: we have both a discrete domain and a discrete-valued function.

It is natural to view, or represent, an image f as a matrix $X = [x_{ij}]$ of size $M \times N$ where the entries are all integers and satisfy $0 \leq x_{ij} \leq K$. If $K = 1$, i.e., $\mathcal{K} = \{0, 1\}$, then we have a *binary image*. In connection with binary images one also uses the term *lattice set* which is simply a subset S of $\mathbb{Z}^2$ (we may here assume that $S \subseteq \mathcal{D}$). Thus, there is a correspondence between such lattice sets and binary images, and this correspondence is given by

$$S = \text{supp}(f) := \{(i,j) \in \mathcal{D} : f(i,j) = 1\}.$$

Thus, supp(f) is the *support* of f; the set of elements in the domain of f with nonzero function value. In applications, the set S corresponds to some kind of object of interest.

A broad class of problems concerning images have the following form which we shall refer to as the *generic image construction problem*:

For fixed sets $\mathcal{D}$ and $\mathcal{K}$ construct, if possible, an image f having "specified properties".

These "specified properties" represent partial knowledge or information about the image to be found. For instance, this may be a degraded version of the image, some kind of "cross-section" or geometrical knowledge of the underlying image (or object). Let us give some important examples of problems that fall into this generic framework.

1. **Reconstruction problems for binary images based on projections.** A major class of problems in discrete tomography consist in finding a binary image f with given *projections*. The *horizontal projection* of an image f is the M-dimensional vector $\mathcal{P}_{e_1}(f)$ whose ith component is given by
$$(\mathcal{P}_{e_1}(f))_i = \sum_{j=1}^{N} f(i,j) \qquad (i \leq M).$$
This corresponds to finding the cardinality of the intersection between supp(f) and each of the lattice lines parallel to the coordinate vector e_1 (the "line" $\{(i,j) : j \in \mathbb{Z}\}$). The *vertical projection* of f is defined similarly, and actually, one may consider projections in more general directions. Thus, the problem is to reconstruct an unknown binary image from its projections. Here we assume that the projections are known without any error.
Two basic problems in this area are
   - *Consistency:* determine if there exists an image f with the given projections.
   - *Uniqueness:* if such an image exists, is it unique?

   A lot of work has been done on these problems, see for instance [17]. In all these problems, whenever the solution is nonunique, it may be of interest to find a solution with some additional property. For instance, one might want the object to be found to have some sort of geometrical property. We return to this later. Other variants are obtained if the projections are not known exactly, but with some (small) errors.
2. **Image segmentation problems.** In many applications, as in medical imaging, the underlying "true" image f is not directly observed, but is observed through a degraded image g. These images are related via a statistical model of the blurring process which specifies the conditional

probability distribution of the observed image g given that the underlying image is f. In the Bayesian approach one combines this conditional distribution with an a priori distribution of f which reflects partial knowledge about the true image. One then wants to find a maximum a posteriori (MAP) solution f which maximizes the conditional distribution of f given the observed image g. Thus, one looks for an image f with the additional property of maximizing a certain likelihood function.

The generic image construction problem therefore contains some challenging problems in image analysis. Several computational methods have been suggested for these problems. We are here concerned with the *integer programming approach* to the generic image construction problem. The starting point is to introduce decision variables that describe the image f. As above f may be represented by an $M \times N$ matrix X with entries $x_{ij} \in \mathcal{K}$. This leads to an optimization problem of the form

$$\begin{array}{ll} \text{maximize} & F(X) \\ \text{subject to} & \\ & X \in \mathcal{P} \\ & 0 \leq x_{ij} \leq K \; (i \leq M, j \leq N) \\ & x_{ij} \text{ integer} \;\; (i \leq M, j \leq N). \end{array} \tag{1}$$

Here $F(X)$ is an objective function to be maximized and together with the abstract constraint $X \in \mathcal{P}$ they represent the specific properties that we require the image X to have. In the concrete examples we discuss in the remaining part of the paper, this function F is linear and the constraint $X \in \mathcal{P}$ may be represented by linear inequalities. We then have a *integer linear programming problem.* For instance, linear inequalities along with some extra variables may be used to model that neighboring positions are assigned the same value. This technique will be illustrated later. Although such problems are theoretically hard to solve, the specific structure of the problem (encoded in F and $\mathcal{P}$) may open up for efficient algorithms in practice. For instance, one has the possibility to solve the corresponding linear programming (LP) relaxation (the problem obtained by dropping the integrality constraints) and develop some kind of approximation method from this. This can be a a very useful approach since large-scale linear programming problems may be solved fast using available LP software (as e.g., CPLEX). But several other general techniques may also come into play such as a decomposition technique called *Lagrangian relaxation* which we shall be using in the next few sections.

Finally, we should say that the integer programming approach does not "solve" such a broad class of problems as discussed here. However, it provides a starting point where the idea is to exploit a vast range of modelling techniques and computational methods that are available in the area of integer programming in order to solve the image analysis problem of interest.

3 The Integer Linear Programming Models

Now, having introduced our general framework, we proceed to present two concrete examples of the integer programming approach. The following two problems will be treated:

1. **(CON)**: Binary *hv*-convex images with *hv*-projections.
2. **(SEG)**: Image segmentation (where $K \geq 1$).

Both problems (CON) and (SEG) belong to the two broad classes of problems discussed in the previous section. The first problem may be seen as a combinatorial optimization problem whereas the second is a statistical estimation problem that may be transformed into a integer linear programming problem. In Section 4 we discuss some solution methods for these problems and focus on some common techniques that have turned out to be useful.

In these models it is convenient to introduce a certain graph G, called the *image graph*. The vertex set of G is $\mathcal{D}$ (the set of all positions in the matrix), and the edge set consists of an edge e for each pair of neighbor positions, i.e., consecutive positions in the same row or column. Thus, E consists of the edges $\{(1,1),(1,2)\}$, $\{(1,2),(2,3)\}$ etc. (In some application one may also want to consider "diagonal neighbors", e.g., so that $(1,1)$ and $(2,2)$ are neighbors, but this is not done in this paper.)

3.1 (CON) and Integer Linear Programming

We are now concerned with the reconstruction of binary images based on given horizontal and vertical projections. Consider a binary image X, here viewed as a $(0,1)$-matrix of size $M \times N$. Define its *row-sum vector* $R(X) = (r_1, \ldots, r_M)$ where $r_i = \sum_{j=1}^{N} x_{ij}$ $(i \leq M)$ and its *column-sum vector* $S(X) = (s_1, s_2, \ldots, s_N)$ where $s_j = \sum_{i=1}^{M} x_{ij}$ $(i \leq M)$. We now assume that the row-sum and column-sum vectors are known, say equal to $R = (r_1, r_2, \ldots, r_M)$ and $S = (s_1, s_2, \ldots, s_N)$, respectively. These two vectors have nonnegative integral components and $\tau := \sum_i r_i = \sum_j s_j$. The last condition, as well as $r_i \leq N$ and $s_j \leq M$ for all i, j, are clearly necessary for an image with these projections to exist. We therefore assume that these conditions hold.

Consider now the set

$$\mathcal{A}(R,S) = \{X \in \{0,1\}^{M \times N} : R(X) = R, S(X) = S\}$$

which consists of all binary images having row-sum vector and column-sum vector equal to R and S, respectively.

A basic question is to characterize whenever $\mathcal{A}(R,S)$ is nonempty. It was shown independently by Gale [14] and Ryser [21] that $\mathcal{A}(R,S)$ is nonempty if and only for each $k \leq n-1$ the sum of the k largest components in S does not exceed $\sum_{j=1}^{k} R_j^*$. Here R_j^* denotes the sum of column j in the matrix whose ith row consists of r_i ones followed by $N - r_i$ zeros. This condition may be expressed nicely in terms of the notions of majorization and conjugate vectors, see e.g., Marshall and Olkin [19].

We shall consider the more difficult problem of finding a matrix in $\mathcal{A}(R,S)$ satisfying an additional constraint of a geometrical nature. A $(0,1)$-matrix X is h-convex (v-convex) if the ones in each row (column) form a contiguous interval, and X is called hv-convex if it is both h- and v-convex. It is known that it is computationally difficult (in precise terms, NP-complete) to decide if there exists a hv-convex binary matrix with given row and column sums, see [23]. Similar results hold for h-convexity and v-convexity, see [2]. However, [9] suggested an optimization approach to these problems which we now explain briefly.

The idea is to try to find a matrix in $\mathcal{A}(R,S)$ with a maximum number of neighboring ones. For this purpose we consider two positions (in the matrix) as neighbors if they occur consecutively in either a row or a column, i.e., the positions are adjacent in the image graph. This problem may be expressed as a integer linear programming problem and the structure of this problem leads to an algorithm which decomposes the problem into much simpler subproblems.

If $X \in \mathcal{A}(R,S)$, then the number of neighboring ones in X is given by the function

$$\eta(X) = \sum_{i=1}^{M-1}\sum_{j=1}^{N} \min\{x_{ij}, x_{i+1,j}\} + \sum_{j=1}^{N-1}\sum_{i=1}^{M} \min\{x_{ij}, x_{i,j+1}\}.$$

The mentioned problem therefore leads to the following optimization problem (CON):

$$\max\{\eta(X) : X \in \mathcal{A}(R,S)\}. \tag{2}$$

Recall that $G = (\mathcal{D}, E)$ is the image graph. Our problem may be formulated as the following integer linear programming problem

$$\begin{array}{lll} \max & \sum_{e\in E} y_e & \\ \text{subject to} & & \\ \text{(i)} & \sum_{j=1}^{N} x_{ij} = r_i & (i \le M) \\ \text{(ii)} & \sum_{i=1}^{M} x_{ij} = s_j & (j \le N) \\ \text{(iii)} & y_e \le x_{ij} & (e \in E, (i,j) \in e) \\ \text{(iv)} & x_{ij}, y_e \in \{0,1\} & ((i,j) \in \mathcal{D}, e \in E). \end{array} \tag{3}$$

The variables here consist of the *image variables* x_{ij} and the *neighbor variables* y_e ($e \in E$) that correspond to each pair e of neighbors. Constraints (iv) say that all variables are binary while constraints (i) and (ii) are the *projection constraints*. They ensure that the matrix X has the correct row and column sums so it lies in $\mathcal{A}(R,S)$. Constraint (iii) relates the two sets of variables and ensures that the variable y_e can be set to one only if both neighboring entries in e are set to one, too. Note that constraint (iii) represents two constraints for each edge.

We have now formulated our problem in the way we want, and the next step is to come up with a solution method for this problem. Unfortunately, the problem (3) is *NP*-hard since a special case is the mentioned problem of checking if $\mathcal{A}(R,S)$ contains a *hv*-convex matrix. Thus, we must expect that any algorithm for solving (3) exactly will have computational time growing exponentially as a function of the problem input size. We propose an algorithm in Section 4 and present some computational results that indicate the size of problem instances that currently can be solved using this method.

Finally we briefly mention some results concerning bounds related to model (3), for further details see [9]. It is easy to see that an upper bound for $\eta(X)$ is given by

$$\eta(X) \leq \hat{v} = \sum_{i=1}^{M-1} \min\{r_i, r_{i+1}\} + \sum_{j=1}^{N-1} \min\{s_j, s_{j+1}\}.$$

Thus, this bound is easy to calculate. A stronger bound may be obtained by solving the linear programming relaxation of (3); let $v(LP)$ denote the optimal value in this problem. Then it may be shown that these bounds are related as follows

$$\max_X \eta(X) \leq v(LP) \leq \hat{v}.$$

Some empirical testing indicates that the bound $v(LP)$ may be quite good, although a general theoretical justification for this is missing.

3.2 (SEG) and Integer Linear Programming

We now consider an image segmentation problem which arises in applications in remote sensing, medical image analysis and industrial inspection. Usually the observed image consists of a (possible vector-valued) observation in every pixel (picture element, i.e., the position in the matrix). A part of the analysis of such images is *segmentation* where one determines a labeling of each pixel by a class-label describing the content in the image. For instance, in medical images, the labels represent tissue types. Usually, the segmentation process is complicated by noise in the observed image. One therefore tries to incorporate other information in the segmentation and a popular stochastic model is called *Markov Random Field (MRF)*. This model is flexible (as prior knowledge may be exploited) and some kind of smoothness in the image is introduced. Combining the MRF model with a Bayesian approach leads to the (SEG) problem which is to find a maximum a posteriori solution. Further details on the stochastic model are given below. Our presentation is mainly based on [13] and [22]. A closely related model is investigated in [24] and some efficient network flow based algorithms are found.

The main idea in the model is to find the most probable configuration of labels based on the observed image. Note that the number of pixels, i.e., $M \cdot N$, is usually quite large, typically $M = N = 256$. The number K of classes may vary from 2 up to say 40-50, but in some medical images there may be 5-6 tissue types ($= K$) under consideration. Thus, one is confronted with a large-scale computational problem. Some popular methods for solving (SEG) are stochastic simulation and simulated annealing [15] or by using some sort of heuristic like the ICM algorithm [3]. Efficient approximation algorithms based on minimum cuts in graphs also exist, see [5]. We will discuss an alternative approach based on integer programming.

We first present the model in an intuitive manner since it can be explained without going into details of the statistical model. However, the model involves some parameters that relate to the underlying statistical model so we present the statistical framework after a discussion of the integer programming model.

Instead of representing the image with integer variables $x_{ij} \in \mathcal{K}$ we prefer to use binary variables. This means that the integral variable x_{ij} is replaced by $K+1$ binary variables x_{ij}^k, called *class variables*, with the interpretation that $x_{ij}^k = 1$ means that $x_{ij} = k$. In the model we then need the equation $\sum_k x_{ij}^k = 1$ in order to ensure that exactly one integer is assigned to position (i,j) in the image. The model contains some parameters:

- d_{ij}^k: measures how likely it is that the true image contains class k in position (i,j) for each $(i,j) \in \mathcal{D}, k \in \mathcal{K}$.
- β: a single weight parameter reflecting smoothness.

The parameters d_{ij}^k are determined by the observed image and the statistical model. The details are provided at the end of the section. The goal of the model is to find a best possible balance between a measure of the distance to the observed image and a term reflecting the smoothness in the image. This objective is reflected in an objective function (with two terms) that we wish to maximize. In the following model we let $e = \{(i,j),(i',j')\}$ in constraints (4)(i), (i').

$$\begin{array}{lll}
\max \sum_{(i,j)\in\mathcal{D}} \sum_{k\in\mathcal{K}} d_{ij}^k x_{ij}^k + \beta \sum_{e\in E} y_e & & \\
\text{s.t.} & & \\
\text{(i)} \quad x_{ij}^k - x_{i'j'}^k + y_e \le 1 & (k \in \mathcal{K}, e \in E) & \\
\text{(i')} \quad -x_{ij}^k + x_{i'j'}^k + y_e \le 1 & (k \in \mathcal{K}, e \in E) & \\
\text{(ii)} \quad \sum_{k\in\mathcal{K}} x_{ij}^k = 1 & ((i,j) \in \mathcal{D}) & \\
\text{(iii)} \quad x_{ij}^k, y_e \in \{0,1\} & ((i,j) \in \mathcal{D}, e \in E, k \in \mathcal{K}). &
\end{array} \tag{4}$$

In addition to the class variables x_{ij}^k we have the *neighbor variables* y_e. They have a similar function as the variables y_e have in (3), they model when neighbors are assigned the same class. Thus, if we assign some class k to two neighbor positions (i,j) and (i',j'), then the corresponding neighbor variable y_e (where $e = \{(i,j),(i',j')\}$ may be set to 1 and this gives a contribution of β in the objective function. Thus, the second term of the objective function is increased by letting many neighbor pixels be assigned to the same class. The first term in the objective function measure, in a probabilistic sense, the distance between the selected image and the observed image. If $\beta = 0$, an optimal solution is obtained by assigning a class k to pixel (i,j) for which d_{ij}^k is maximal. For a $\beta > 0$, however, we are confronted with finding a solution which balances the two terms in an optimal manner. We observe that in an optimal solution of (4) each neighbor variable y_e (where $e = \{(i,j),(i',j')\}$) may be expressed in terms of x as follows

$$y_e = 1 - \max_{k \in \mathcal{K}} |x_{ij}^k - x_{i'j'}^k|. \tag{5}$$

There is another natural way of modeling that neighbors are assigned the same class. Instead of using a single variable y_e for each edge e one may use variables y_e^k that indicate whether the corresponding neighbors are both assigned class k. This leads to the following alternative linear integer programming model of (SEG)

$$\begin{array}{lll} \max & \sum_{(i,j)\in\mathcal{D}} \sum_{k\in\mathcal{K}} d_{ij}^k x_{ij}^k + \beta \sum_{e\in E} \sum_{k\in\mathcal{K}} y_e^k & \\ \text{subject to} & & \\ \text{(i)} & y_e^k \le x_{ij}^k \quad (k \in K, e \in E, (i,j) \in e) & \\ \text{(ii)} & \sum_{k\in\mathcal{K}} x_{ij}^k = 1 \;\; ((i,j) \in \mathcal{D}) & \\ \text{(iii)} & x_{ij}^k, y_e^k \in \{0,1\} \;\; ((i,j) \in \mathcal{D}, e \in E, k \in \mathcal{K}). & \end{array} \tag{6}$$

This model contains more variables than model (4), and again one may express y_e^k in terms of x in an optimal solution. There are different algorithms that may be developed for either of these models. It turns out that using suitable relaxation techniques one gets nice combinatorial subproblems for both models. However, these subproblems are quite different. This is discussed in Section 4.2.

The remaining part of this subsection discusses some further issues related to (4). A more detailed discussion and further references are found in [22].

- **A relation to the multi-terminal cut problem.** Consider a graph $G' = (V', E')$ and let $t_1, \ldots, t_r$ be specified (terminal) vertices in this graph. Let nonnegative weights w_e for $e \in E'$ be associated with the edges in G'. The *multi-terminal cut problem* asks for a partition $V_1', \ldots, V_r'$ of the vertex set V' satisfying $t_j \in V_j'$ for $j = 1, \ldots, r$, and such that one maximizes the total sum of the weights of all edges with both end-vertices in the same set of the partition. The problem arises in many applications, see e.g., [7]. The multi-terminal cut problem is *NP*-hard for every fixed $r \geq 3$ (even when all weights are equal), see [10]. For $r = 2$ the multi-terminal cut problem specializes into the min-cut problem which is polynomial (if weights are arbitrary we get the *NP*-hard max-cut problem). Consider again (4) and construct a weighted graph G from its image graph G in the following way. Add $K + 1$ new vertices $t_0, \ldots, t_K$ to G and for each vertex t_k and each vertex $(i,j) \in \mathcal{D}$ we add an edge $[t_k, (i,j)]$ and define its weight to be d^k_{ij}. Each original edge $[u, v]$ in G has weight β. Let G' denote the resulting graph. It is easy to see that the multi-terminal cut problem in G' (with terminal vertices $t_0, \ldots, t_K$) is equivalent to (4). It follows that if one allows an arbitrary image graph (where any pairs of pixels is possibly adjacent), then (4) is *NP*-hard. Another consequence is that when $K = 1$, i.e, for binary images, (4) reduces to the minimum cut problem so it may be solved efficiently. Unfortunately, we do not know the complexity of (4) when the image graph is defined as above with only horizontal/vertical neighbors.

- **More about the statistical background.** We assume that the image $X = [x_{ij}]$ is not directly observed, but is observed through a degraded image $Z = [z_{ij}]$. The connection between X and Z is given through a statistical model which contains the conditional (probability) distribution $f(Z|X)$ of Z given X. We here assume conditional independence between pixels in the sense $f(Z|X) = \prod_{(i,j)} f(z_{ij}|x_{ij})$. Usually, some prior information about the true image Z is available and a simple and popular class of models for representing this information is the Markov Random Field (MRF). It builds in smoothness properties of the image. One such model (the Potts model) is given by

$$\pi(Z) = \frac{1}{s} \exp\Big\{\sum_{(i,j)} \alpha_{z_{ij}} + \beta \sum_{[(i,j),(i',j')]\in E} I(z_{ij} = z_{i'j'})\Big\} \tag{7}$$

where α_k defines the prior probabilities for the different classes k, $\beta > 0$ is a parameter giving the degree of smoothness, s is a normalization constant making $\pi(\cdot)$ a proper distribution (i.e., total probability is 1) and the indicator function $I(z_{ij} = z_{i'j'})$ equals 1 if $z_{ij} = z_{i'j'}$ and 0 otherwise. More generally, one may let β be dependent on e.g., the classes z_{ij} and $z_{i'j'}$. The *posterior* distribution for X given Z may be found through Bayes' formula to be $\pi(X|Z) = \phi(Z)\pi(X)f(Z|X)$ where ϕ is a suitable normalization function. Consider now Z as fixed; it is the observed image. Thus $\phi(Z)$ is a constant. The Bayesian paradigm [4] is to base all inference on this posterior distribution. In particular, one estimate of X is the maximum a posteriori (MAP) solution $\hat{X}$ which maximizes $\pi(X|Z)$, or equivalently, is an optimal solution of

$$\max_X \pi(X) f(Z|X) \tag{8}$$

where X runs through the set of $M \times N$ matrices with entries in $\mathcal{K}$. It is more convenient to maximize the logarithm of the posterior distribution, or equivalently, after removing a constant (depending only on Z), the following function

$$U(X) = \sum_{(i,j)} \sum_{k} (\log f(z_{ij}|k) + \alpha_k I(x_{ij} = k) + \beta \sum_{E} I(z_{ij} = z_{i'j'})$$

where the last summation is over all $[(i,j),(i',j')] \in E$. In the image analysis literature (the negative of) U is usually referred to as the *energy function*. By defining $d_{v,k} = \log f(z_v|k) + \alpha_k$ for all $(i,j) \in \mathcal{D}$ and $k \in \mathcal{K}$ we obtain the integer linear programming problem (4).

4 Solution Methods

In this section we discuss the main ideas in solution methods for the two integer linear programming problems in the previous section. The two models have some similarity and this is reflected in the methods we now present.

Briefly the main ideas are the following

1. Both (3) and (4) can be solved efficiently if the image has a single row or column.
2. Using Lagrangian relaxation, a general decomposition technique, each of the problems (3) and (4) is decomposed into subproblems, one for each row and column in the image. These subproblems are solved fast using the algorithm for the special case in 1.
3. An outer optimization routine modifies some Lagrangian parameters (multipliers) that reflect the interaction between neighboring rows, and one returns to step 2.

In the following subsections we explain how to treat the special cases with a single row or column. As expected the problems (3) and (4) are easier to solve if the image has a single row, i.e., $M = 1$. (The case where $N = 1$ is treated similarly.) The main reason for this is that we then obtain one-dimensional problems that can be solved using dynamic programming.

In Subsection 4.3 we discuss the Lagrangian relaxation approach in more detail.

4.1 (CON) in Case of a Single Row

Consider (CON) with $M = 1$. Of course, a similar approach applies to the case $N = 1$. For reasons that will be clear in the next subsection, we allow a more general objective function for the x-variables. The problem has the form

$$\begin{aligned} \max \quad & \textstyle\sum_{j=1}^{N-1} y_j + \sum_{j=1}^{N} c_j x_j \\ \text{subject to} \quad & \\ & \textstyle\sum_j x_j = r \\ & y_j \le x_j, \quad y_j \le x_{j+1} \text{ (for all } j) \\ & x_j, y_j \in \{0,1\} \qquad \text{(for all } j). \end{aligned} \tag{9}$$

where x_i and y_i correspond to the variables x_{1i} and y_e where $e = \{(1,i),(1,i+1)\}$ in the model (3), and $r = r_1$. This problem may viewed as a longest (simple) path problem with exactly $k+1$ edges in a certain directed graph D defined next. Let D have vertices $x_1, x_2, \ldots, x_n, s, t$, i.e., the vertices are the

x-variables plus two extra vertices s (source vertex) and t (terminal vertex). The arcs in D and their associated weights are

- (s, x_j) with weight c_j
- (x_{j-1}, x_j) with weight $c_j + 1$
- (x_i, x_j) for $i < j - 1$ with weight c_j
- (x_j, t) with weight 0.

The construction is illustrated in Figure 1. A maximum weight path from s to t in D with exactly $r+1$ edges corresponds to an optimal solution of problem (9) by setting x_j equal to one if vertex x_j is used in the longest path and zero otherwise. Similarly, we set y_j to one if the longest path visits both x_j and x_{j+1} and zero otherwise.

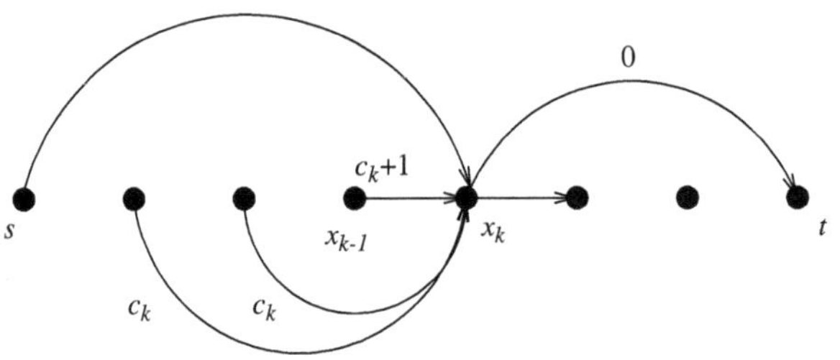

Fig. 1. Arcs going to and from a vertex x_j.

The maximum weight path problem with a prescribed number of arcs in an acyclic graph can be solved efficiently by dynamic programming. Let $z(x_i, p)$ denote the value of the maximum weight path from vertex s to vertex x_i using exactly p arcs. The values of $z(x_i, p)$ can be calculated recursively from

$$z(x_i, 1) = c_i$$
$$z(x_i, p) = \max\{z(x_{i-1}, p-1) + c_i + 1, \max_{j<i-1}[z(x_j, p-1) + c_i]\}.$$

The optimal value is then $z(t, r+1) = \max_i z(x_i, r)$ and the corresponding solution can be found by retracing the path in the graph. The algorithm has running time $O(rn^2)$.

4.2 (SEG) in Case of a Single Row

We now consider (SEG) in the case with $M = 1$. This problem may be treated rather similar to our approach for (CON) in this special case. (And, again, the case with $N = 1$ follow the same pattern.) The difference, however, is that we now have more image variables x, there are K variables in each position in the image.

Consider again (4). Recall that the y_e-variables may be eliminated using (5); this fact is exploited in the construction below. We denote the x-variables in (4) by x_j^k (corresponding to x_{1j}^k) for $j \leq N$.

Construct a directed graph D with vertices x_j^k ($j \leq N$, $k \leq K$) along with two extra vertices s (source vertex) and t (terminal vertex). D contains the following arcs: (i) an arc from x_j^k to $x_{j+1}^{k'}$ for $j \leq N-1$ and $k, k' \leq K$, and (ii) an arc from s to x_1^k and an arc from x_N^k to t for $k \leq K$. The digraph D therefore has $N+2$ layers $\{s\}$, $\{x_1^k : k \leq K\}, \ldots, \{x_N^k : k \leq K\}$ and $\{t\}$, and arcs go from each vertex in a layer to each vertex in the next layer. We define the weight of the arcs in the following way. The weight of the arc $(x_i^k, x_{i+1}^{k'})$ is $d_{i+1}^{k'} + \beta I(k = k')$ ($i \leq N-1$) where $I(k = k') = 1$ if $k = k'$ and $I(k = k') = 0$ otherwise. For $k \leq K$ we define the weight of (s, x_1^k) to be d_1^k and the weight of each edge incident to t is zero. We let the weight of an arc (u, v) be denoted by $w(u, v)$.

There is a one-to-one correspondence between the feasible solutions of (4) that satisfy (5) and the directed st-paths in D. Moreover, the objective function value of the solution x and the weight of the corresponding path coincide. Thus, one may solve (4) by finding a maximum weight st-path in D. Since D is acyclic, this may be viewed as a shortest path problem. A maximum weight path from s to a vertex $x_{i+1}^{k'}$ consists of a maximum weight path from s to a some vertex x_i^k plus the arc from x_i^k to $x_{i+1}^{k'}$. Let $z(v)$ denote the maximum weight of a path from s to vertex v, and from the Bellmann–Ford equations we obtain

$$\begin{aligned} z(x_{i+1}^{k'}) &= \max\nolimits_{k \leq K}[z(x_i^k) + w(x_i^k, x_{i+1}^{k'})] \\ &= \max\nolimits_{k \leq K}[z(x_i^k) + d_{i+1}^{k'} + \beta I(k = k')] \\ &= d_{i+1}^{k'} + \max\{z(x_i^{k'}) + \beta, \max\nolimits_{k \neq k'} z(x_i^k)\}. \end{aligned} \tag{10}$$

The algorithm may calculate the maximum weight path lengths layer by layer as follows:

Step 1: Initialize z by $z(x_1^k) = d_1^k$ for $k \leq K$ and let $i = 1$

Step 2: Determine the largest number s_1 and the second largest number s_2 in the set $\{z(x_i^k) : k \leq K\}$, and also determine the "maximizers" $M := \{k \leq K : z(x_i^k) = s_1\}$

Step 3: For each $k' \leq K$ do the following : if $M = \{k'\}$ let $z(x_{i+1}^{k'}) = d_{i+1}^{k'} + \max\{z(x_i^{k'}) + \beta, s_2\}$, otherwise let $z(x_{i+1}^{k'}) = d_{i+1}^{k'} + \max\{z(x_i^{k'}) + \beta, s_1\}$

Step 4: Let $i := i + 1$ and return to Step 2.

The algorithm terminates, of course, when the vertex t is reached (in iteration $i = n+1$, in which just finding the smallest of the numbers $z(x_n^k)$ is required). The complexity of this specialized Bellman-Ford algorithm is $O(nK)$.

The so-called *Viterbi algorithm* [12] is known in the image analysis literature for solving the (IMS) problem with a single row. It is easy to see that the Viterbi algorithm is equivalent to the algorithm above.

4.3 Lagrangian Decomposition

Both (CON) and (SEG) problems arising in practice are large-scale so one needs special-purpose algorithms for solving the problems exploiting their structural properties. This may be done in many ways, but a common theme would be to decompose the problem somehow. As shown in the previous section, both problems can be solved efficiently in case of a single row (or column). In the case of (CON) and (SEG) a decomposition into single row (and column) subproblems can be achieved using a technique called Lagrangian decomposition.

The edge set E of the image graph G may be partitioned by $E = E^r \cup E^c$ where E^r are all the "horizontal edges" joining two adjacent pixels in the same row and E^c are all the "vertical edges" joining two adjacent pixels in the same column. We introduce two set of x variable vectors x^r and x^c to represent the "row solution" and "column solution" respectively. The x^r variables will be used in constraints involving edges in E^r while the x^c variables will be used in constraints involving edges in E^c. By introducing a new constraint that imposes equality of the two sets we will obtain new formulations of (3) and (4). We will show the construction for (CON). The model is now

$$\begin{array}{lll}
\max & \sum_{e\in E} y_e & \\
\text{subject to} & & \\
\text{(i)} & \sum_{j=1}^{N} x^r_{ij} = r_i & (i \le M) \\
\text{(ii)} & \sum_{i=1}^{M} x^c_{ij} = s_j & (j \le N) \\
\text{(iiir)} & y_e \le x^r_{ij} & (e \in E^r, (i,j) \in e) \\
\text{(iiic)} & y_e \le x^c_{ij} & (e \in E^c, (i,j) \in e) \\
\text{(iv)} & x^r_{ij} = x^c_{ij} & (i,j) \in \mathcal{D} \\
\text{(v)} & x_{ij}, y_e \in \{0,1\} & ((i,j) \in \mathcal{D}, e \in E).
\end{array} \tag{11}$$

Note that constraints (iv) ensure that this formulation is equivalent with (3). If we now relax constraints (iv) using Lagrangian multipliers λ_{ij} for each $(i,j) \in \mathcal{D}$, we obtain the following Lagrangian subproblem:

$$\begin{array}{lll}
\max & \sum_{e\in E} y_e + \sum_{(i,j)\in\mathcal{D}} \lambda_{ij}(x^r_{ij} - x^c_{ij}) & \\
\text{subject to} & & \\
\text{(i)} & \sum_{j=1}^{N} x^r_{ij} = r_i & (i \le M) \\
\text{(ii)} & \sum_{i=1}^{M} x^c_{ij} = s_j & (j \le N) \\
\text{(iiir)} & y_e \le x^r_{ij} & (e \in E^r, (i,j) \in e) \\
\text{(iiic)} & y_e \le x^c_{ij} & (e \in E^c, (i,j) \in e) \\
\text{(v)} & x_{ij}, y_e \in \{0,1\} & ((i,j) \in \mathcal{D}, e \in E).
\end{array} \tag{12}$$

The resulting Lagrangian subproblem decomposes into a "row problem" involving only the x^r-variables and a "column problem" involving only the x^c-

variables. Furthermore, in the row problem there are no constraints between variables from different rows, making it possible to solve for each row separately using the shortest path procedure. The column problem may be treated similarly.

A similar approach may be used for the (SEG) problem based on the integer linear programming model (4). Thus, by introducing x^r-variables and x^c-variables and relaxing the corresponding equations between these, the overall problem decomposes into nice row problems and column problems. These subproblems are solved efficiently as described in Subsection 4.2. For further details we refer to [22].

4.4 The Subgradient Procedure

In the previous section we presented a Lagrangian decomposition technique in connection with (CON) and (SEG). The resulting subproblems give, for each Lagrangian multiplier vector λ, an upper bound $z(\lambda)$ on the optimal value of the corresponding integer programs. The Lagrangian dual problem of finding the best (i.e., smallest) upper bound by varying λ may be solved using the subgradient algorithm, see [20] for a general description. We briefly describe the principle in this setting. Consider the Lagrangian subproblem of (12) when $x^r = x^c$ is relaxed using the multiplier vector λ^s; this is in iteration s. Let $\bar{x}^r_{v,k}$ and $\bar{x}^c_{v,k}$ denote the x^r and x^c variables in an optimal solution of the subproblem corresponding to λ^s. Then $g := \bar{x}^r - \bar{x}^c$ is a subgradient of the (piecewise linear and) convex function $\lambda \to z(\lambda)$ at the point λ^s. The new multiplier λ^{s+1} is obtained by taking a step from λ^s in the direction of the subgradient, that is

$$\lambda^{s+1}_{v,k} = \lambda^s_{v,k} - \theta_s(\bar{x}^r_{v,k} - \bar{x}^c_{v,k})$$

where $\theta_s > 0$ is the step-length. We decrease the step-lengths according to $\theta_s = \mu^s(z(\lambda_s) - z_L)/\|g|^2$ where z_L is a lower bound on the optimal value $\min_\lambda z(\lambda)$ and μ^s is a scaling factor. A problem in connection with relaxing equalities, is that the solutions "flip around" too much and then the subgradients oscillates. A good idea is then to stabilize the process by moving λ in a direction given by some convex combination of the new and the previous subgradients (with the larger weight on the new subgradient). This was done in the algorithms reported in the next section. A detailed discussion of these aspects are found in [6].

The final point we mention is that we used simple heuristics for turning the nonfeasible solutions in the subproblems into feasible ones. In the case of (SEG) this may be done by letting $x^c := x^r$, and defining the variable y accordingly (see (5)). The same heuristics will fail for (CON) as using only the horizontal variables may lead to incorrect projection values vertically. Instead, we solve the following network flow problem based on the solution vectors x^r and x^s; we let $\alpha_{ij} := x^r_{ij} + x^c_{ij}$:

$$\begin{array}{ll} \max & \sum_{(i,j)\in\mathcal{D}} \alpha_{ij}x_{ij} \\ \text{subject to} & \\ & \sum_j x_{ij} = r_i \quad (i \le M) \\ & \sum_i x_{ij} = s_j \quad (j \le N) \\ & 0 \le x_{ij} \le 1. \end{array}$$

This may be solved efficiently using e.g., the network simplex method, see [1]. For integer data, the vertices of the corresponding polytope are integral. Any method that finds an optimal vertex solution, will give an integer solution and the matrix $X = [x_{ij}]$ will be a member of $\mathcal{A}(R, S)$.

Our lower bound z_L may be updated when a better integer solution has been found and the new bound is used in the step-length calculation.

5 Computational Results

In this section we report computational results and experiences with the two algorithms discussed in Section 4.

5.1 (CON) and Image Reconstruction

The main motivation for formulating problem (CON) was to reconstruct *hv*-convex matrices, but the method can also work for matrices that are approximately *hv*-convex. The following example illustrates the method on a test-case where the original image was not *hv*-convex.

We used the image on the left in Figure 2 to generate row and column sum vectors as input to our algorithm. The image on the right shows the result of solving the reconstruction problem using a subgradient procedure to solve the Lagrangian dual problem. Feasible solutions were obtained with the network flow based heuristics. The graph shows the upper bound from the subgradient procedure and the lower bound from the heuristic as a function of the number of subgradient iterations.

Some further test results are given in Table 1. There are three classes of test data with varying degree of *hv*-convexity. All data were generated by producing (0,1)-matrices and then calculating row and column-sum vectors. For each case the table shows the dimension (M and N), upper and lower bound (v^{UB} and v^{LB}) with the corresponding gap in percent (g) and the time used for the calculations using the Lagrangian heuristic. The two last columns show the gap and time for solving the problem (3) using the commercial integer linear programming solver ILOG CPLEX 9.0 [18]. Each case was run with default settings for a maximum of 3600 seconds. A $*$ for the gap is used for those cases where CPLEX where unable to find a feasible integer solution.

Even though the number of test cases are limited, the results seem to indicate that the algorithm is more successful with the cases having a large

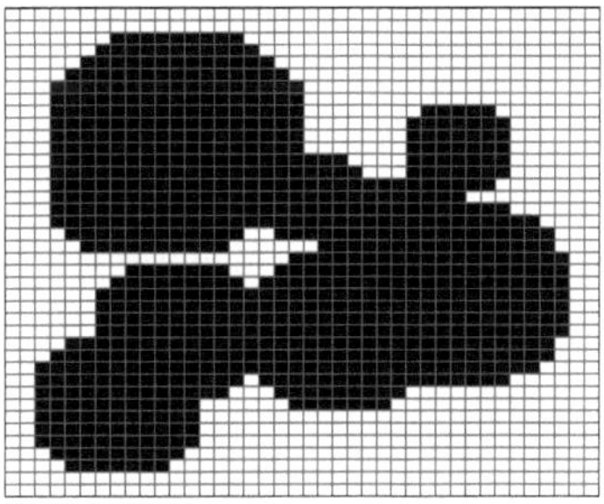

(a) Original figure

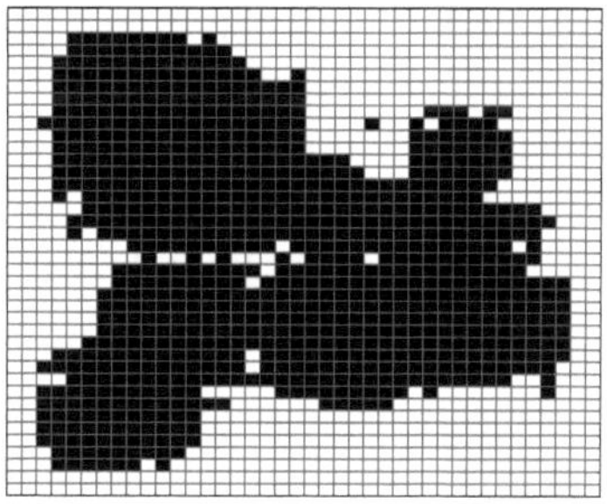

(b) Best feasible solution

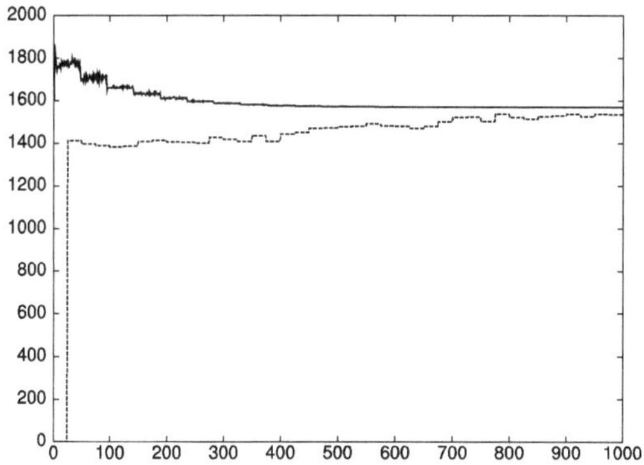

(c) Upper and lower bounds

Fig. 2. Reconstruction example, $M = N = 40$

Table 1. Computational results.

Type	M	N	v^{UB}	v^{LB}	g (%)	T (sec.)	g_{cplex}	T_{cplex}
hv-convex	14	16	98	98	0	0.77	0	8.55
	34	36	432	432	0	3.67	6.17	3600
	56	54	1140	1140	0	15.1	20.0	3600
	73	77	1748	1748	0	38.2	*	3600
	100	90	3094	3094	0	103	*	3600
Two circles	20	20	426	424	0.47	2.59	0	2700
	40	40	2144	2144	0	11.6	0	1055
	60	60	4176	4158	0.43	72.7	1.57	3600
	80	80	8180	8180	0	190	*	3600
	100	100	11781	11657	1.05	500	*	3600
Random circles	20	20	265	265	0	1.08	0	10.3
	40	40	1674	1674	0	7.84	0	629
	60	60	3072	2914	5.14	59.7	3.46	3600
	80	80	8483	8408	0.884	253	*	3600
	100	100	9423	8347	11.4	456	*	3600

degree of hv-convexity in the original image. For the hv-convex data, optimal solutions were produced for all cases. There is a considerable increase in running time as the size of the problems increases. Thus, the algorithm in its current form may not be able to handle large problems. Still, the results show the advantage of using a specialized solver compared to a general solver for integer linear programs.

The algorithm can be generalized to handle problems with more than two projection directions and general neighborhood relations. A forthcoming paper [8] treats these extensions in detail.

5.2 (SEG) and Image Segmentation

In Section 4 we described a Lagrangian relaxation approach for the (SEG) problem. This method was based on model (4) and the idea of exploiting that the (SEG) problem for a single row or column may be solved efficiently using dynamic programming. We denote this algorithm by *Algorithm SEG-line*.

It is also possible to develop algorithms based on the other model (6) presented in Subsection 3.2. Again it is natural to relax some of the complication constraints in order to obtain easier subproblems. It turns out that the best results are obtained by relaxing the equations $\sum_{k \in \mathcal{K}} x_{ij}^k = 1$ $((i,j) \in \mathcal{D})$. In this case one can show that the resulting subproblems can be solved as minimum st-cut problems in a suitable graph. This graph is an extension of the image graph (two extra nodes s and t) and the edge weights are derived form the model parameters and the Lagrangian multipliers. The overall algorithm uses a subgradient procedure to update the multipliers and in each iteration one solves $K + 1$ minimum st-cut problems, one for each $k \in \mathcal{K}$. We denote this by *Algorithm SEG-cut*. Further details are found in [13] (see also [22]).

We now briefly discuss some experiences with these two algorithms. In Figure 3 (left panel) there is a real Magnetic Resonance (MR) image of the brain (the image is a so-called T2-weighted band). The set $\mathcal{K}$ correspond to $K = 5$ classes, namely *fat, air/bone, connective tissue, cerebrospinal fluid* and *brain parenchyma*. The size of the image was $M \times N$ where $M = N = 256$, so model (4) then contained nearly half a million variables while model (6) had almost one million variables. Both algorithms found an optimal solution of the (SEG) problem. Algorithm SEG-cut needed 25 subgradient iterations and the CPU time was 1267.5 seconds, while Algorithm SEG-line used 63 subgradient iterations and the CPU-time was 121.5 seconds. Thus, Algorithm SEG-cut converged much faster in terms of number of subgradient iterations, but Algorithm SEG-line was superior in actual CPU time (see [13] for information about the computer hardware, today the computation would be significantly faster). Figure 4 displays the upper and lower bounds obtained from Algorithm SEG-line and Algorithm SEG-cut as functions of CPU time. The figure also displays results from a more commonly used algorithm based on simulated annealing [15] (we show the best out of many runs using different initial temperature and temperature schedules). The speed of Algorithm

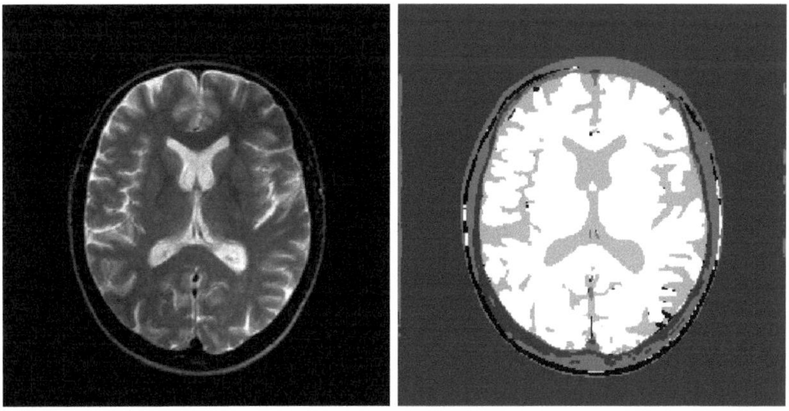

Fig. 3. To the left, $T2$-weighted Magnetic Resonance image of the brain. The gray values correspond to values in the range from 0 (black) to 255 (white). To the right, the MAP solution obtained by Algorithm SEG-line. The classes are *fat, air/bone, connective tissue, cerebrospinal fluid* and *brain parenchyma*, displayed as grey-levels with the first class being white and the last one being black.

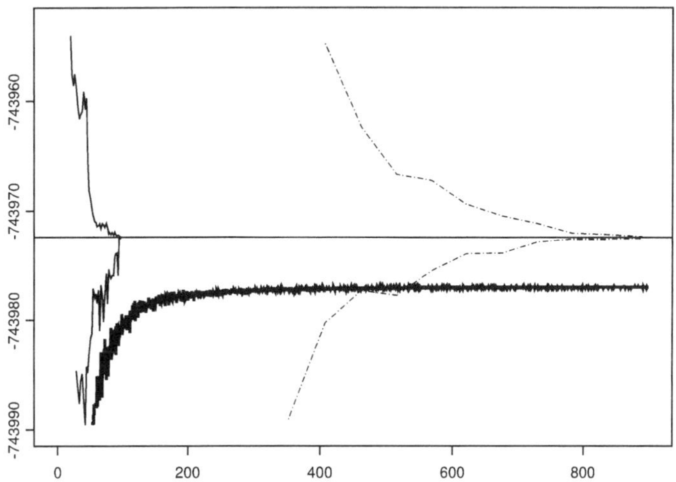

Fig. 4. Upper and lower bounds obtained by using Algorithm SEG-line (solid lines) and Algorithm SEG-cut (dashed lines) for the brain image in Fig. 3. The bounds are plotted as functions of CPU time. The thick solid line show the values obtained by a simulated annealing algorithm. The horizontal line displays the optimal value obtained.

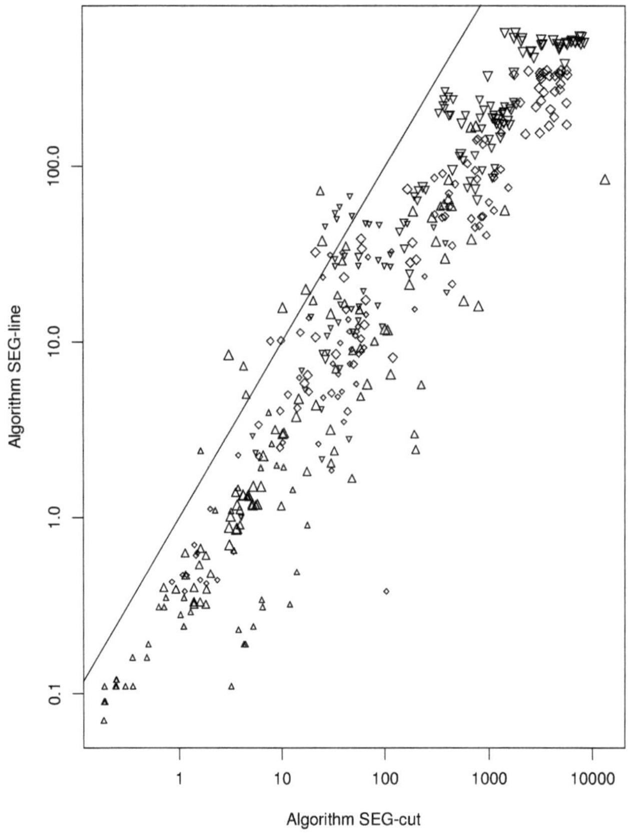

Fig. 5. Plot of time in seconds with Algorithm SEG-cut on the x-axis and Algorithm SEG-line on the y-axis. Symbols $\triangle$, $\diamond$ and $\bigtriangledown$ are used for $K = 1, 3$ and 5, respectively. The sizes of the symbols are proportional to the sizes of the images. Both the x- and y-axes are on logarithmic scales. The line represents $y = x$.

SEG-line is clearly seen. The optimal image obtained by the SEG-algorithm is shown in the right panel of Figure 3, giving a very reasonable estimate of the true distribution of classes.

Some further computational tests were conducted in [13] for images of size $M \times N$ where M, N varied in the range 20-60. Moreover, the number of classes varied between 2 and 6. Random problem instances were generated and Algorithm SEG-line and Algorithm SEG-cut were run for each of these instances. The results are displayed in Figure 5, the CPU time (in seconds) for Algorithm SEG-line and Algorithm SEG-cut are plotted against each other. (For points on the linear curve $y = x$, the times are equal). The computations were done on a SPARC 20 computer. As expected the CPU time increases with K. Moreover, Algorithm SEG-line is consistently faster than Algorithm

SEG-cut. However, it turned out that the distinction between the algorithms in terms of the number of subgradient iterations is much less clear. Thus, the superiority of Algorithm SEG-line is mainly due to faster subproblems (the shortest path problems compared to the minimum st-cut problems in Algorithm SEG-cut). It could be that there is potential for improvement on Algorithm SEG-cut by specializing the minimum st-cut algorithm somehow.

A further comparison of different algorithms, including the SEG-algorithm and simulated annealing may be found in [13] and [22] where solution quality and convergence speed is analysed. A main advantage of the SEG-algorithm is that it produces both lower and upper bounds on the optimal value, even if it is terminated before convergence.

6 Concluding Remarks

In this paper we have discussed segmentation and reconstruction problems for digitized images using a linear integer programming approach. Although these problems are rather different we have explained a common approach based on efficiently solvable one-dimensional subproblems and Lagrangian relaxation. The resulting algorithms are promising and have the attractive feature of producing both lower and upper bounds on the optimal value. Thus, the quality of a proposed solution may be reported. We believe that the idea of applying an integer programming approach to problems in image analysis is interesting and promising.

References

1. R.K. Ahuja, T.L. Magnati, and J.B. Orlin. *Network flows : theory, algorithms, and applications.* Prentice-Hall, 1993.
2. E. Barcucci, A. Del Lungo, M. Nivat, and R. Pinzani. Reconstructing convex polyominoes from horizontal and vertical projections. *Theoret. Comput. Sci.*, 155:321–347, 1996.
3. J. Besag. On the statistical analysis of dirty pictures. *J. R. Stat. Soc. Ser. B Stat. Methodol.*, 48(3):259–302, 1986.
4. J. Besag. Towards Bayesian image analysis. *J. Appl. Stat.*, 16(3):395–407, 1989.
5. Y. Boykov, O. Veksler, and R. Zabih. Fast approximate energy minimization via graph cuts. *IEEE Trans. Pattern Anal. Mach. Intell.*, 23(11):1222–1239, 2001.
6. T.G. Crainic, A. Frangioni, and B. Gendron. Bundle-base relaxation methods for multicommodity capacitated fixed charge network design. *Discrete Appl. Math.*, 112:73–99, 2001.
7. W. Cunningham. The optimal multiterminal cut problem. *DIMACS Ser. Discrete Math. Theor. Comput. Sci.*, 5:105–120, 1991.
8. G. Dahl and T. Flatberg. Reconstructing (0,1)-matrices from projections using integer programming. Working paper.
9. G. Dahl and T. Flatberg. Optimization and reconstruction of hv-convex $(0,1)$-matrices. *Discrete Appl. Math.*, 151(1-3):93–105, 2005.

10. E. Dahlhaus, D.S. Johnson, C.H. Papadimitriou, P.D. Seymour, and M. Yannakakis. The complexity of multiterminal cuts. *SIAM J. Comput.*, 23:864–894, 1994.
11. P. Fishburn, P. Schwander, L. Shepp, and R.J. Vanderbei. The discrete Radon transform and its approximate inversion via linear programming. *Discrete Appl. Math.*, 75:39–61, 1997.
12. G.D. Forney. The Viterbi algorithm. *Proc. of the IEEE*, 3(61):268–278, 1973.
13. G. Storvik G. Dahl and A. Fadnes. Large-scale integer programs in image analysis. *Operations Research*, 50 (3):490–500, 2002.
14. D. Gale. A theorem on flows in networks. *Pacific J. Math.*, 7:1073–1082, 1957.
15. S. Geman and D. Geman. Stochastic relaxation, Gibbs distribution, and Bayesian restoration of images. *IEEE Trans. Pattern Anal. Mach. Intell.*, 6(6):721–741, 1984.
16. P. Gritzmann, S. de Vries, and M. Wiegelmann. Approximating binary images from discrete x-rays. *SIAM J. Comput.*, 11(2):522–546, 2000.
17. G.T. Herman and A. Kuba, editors. *Discrete Tomography: Foundations, Algorithms and Applications*. Birkhäuser, 1999.
18. ILOG CPLEX. http://www.ilog.com/products/cplex.
19. A.W. Marshall and I. Olkin. *Inequalities : Theory of Majorization and Its Applications*. Academic Press, 1979.
20. G.L. Nemhauser and L.A. Wolsey. *Integer and Combinatorial Optimization*. Wiley, 1988.
21. H.J. Ryser. Combinatorial properties of matrices of zeros and ones. *Canad. J. Math.*, 9:371–377, 1957.
22. G. Storvik and G. Dahl. Lagrangian-based methods for finding MAP solutions for MRF models. *IEEE Trans. Image Process.*, 9(3):469–479, 2000.
23. G.J. Woeginger. The reconstruction of polyominoes from their orthogonal projections. *Inform. Process. Lett.*, 77:225–229, 2001.
24. B.A. Zalesky. Network flow optimization for restoration of images. *J. Appl. Math.*, 2 (4):199–218, 2002.

The Impacts of By-products on Optimal Supply Chain Design

Marielle Christiansen and Roar Grønhaug

Summary. During the last half decade, the metal industry has been in a harsh situation seeing their profit margins squeezed to an extreme. Many companies have been forced to close down furnaces and plants. To help a major metal producing company manage this process, we developed a strategic mixed integer programming model. The main decisions addressed by the model involve the future plant structure and production capacities, the production portfolio at each plant and the by-product production. Here, we present the underlying MIP-model and give computational results. In addition, we show how the valuable by-product production can have impact on the optimal supply chain design.

Key words: Facilities/Equipment Planning: Capacity Expansion, Design, Location; Industries: Mining/Metals; Manufacturing: Strategy

1 Introduction

The last decade has been characterized by an increasing globalization resulting in international groups of companies collaborating and merging. This has also been the case in the metal industry. Together with the global economy slowdown that started some years ago, this industry has needed to emphasize supply chain redesign. In this paper we focus on how to improve the efficiency of the supply chain network, and show how these decisions depend on the production of finished products and important by-product production.

There are very few contributions in the literature focusing on by-product production, but in general more and more corporations are realizing the importance of using mathematical programming models to support their supply chain decisions. Already in 1993, Ballou and Masters [2] survey the use of commercial optimization software for location analysis in the USA. They find that about 51 % of the companies responding use commercial software to support their network design. Furthermore, Geoffrion and Powers [11], in discussing the possible future development of OR tools for the design of production and distribution networks, have proved right in believing that supply chain will continue "to gain in scope and influence as a corporate function".

The metal company Elkem ASA is the largest supplier of silicon metal in the world. The difficult situation in this industry has resulted in a clear corporate objective to improve the efficiency of the supply chain and evaluate the product portfolio. To help Elkem in this process, a mathematical programming model was developed [20]. The decisions addressed by the model pertain to future plant structure including possible closures, new plant acquisitions and investments in production equipment. The silicon division has made extensive use of the model and its scenario analysis capabilities, resulting in important organizational benefits and an expected significant relative improvement in yearly revenue.

Silicon and ferrosilicon are produced by mixing different raw materials in large furnaces at high temperatures. From the furnaces, the molten metal goes through production to become different qualities of silicon and ferrosilicon metals. In addition, some smoke comes out of the furnaces during heating. This smoke used to be disposed as waste until it was discovered that the smoke could be collected by filtering. The resulting product is called microsilica, and today, microsilica is an important by-product from the production of silicon and ferrosilicon metals. The production process, furnace and raw materials have great impact on the quality of microsilica. Some plants produce high-quality microsilica, and the revenue from the microsilica production can be important for the further existence of the plant. In the existing strategic optimization model [20] the impact of by-product on the supply chain is not considered. Elkem has expressed a need for an extension of that model to address this issue.

This paper presents a supply chain design in the metal working industry where by-product production is focused and gives a mathematical model of the problem. In addition, the aim is to show the effects on the plant structure and production mix by different conditions of by-product production.

The rest of this paper is organized as follows: In Section 2, we describe the supply chain design with emphasis on the by-product production. Comparisons to related studies in the literature are presented in Section 3. Both facility location and supply chain studies are addressed. Section 4 is dedicated to the description of the underlying mixed-integer programming model. We will also in this section point out some modeling challenges. Computational results are given and discussed in Section 5. Finally, some concluding remarks follow in Section 6.

2 Problem Description

Elkem is the largest producer of silicon metal in the world, and the western world's largest producer of ferrosilicon products. The silicon division has nine plants located in Canada, Iceland, Norway and USA. The division's main products are silicon and ferrosilicon metals in a wide range of qualities. Several by-products result from the smelting process when producing silicon and ferrosilicon metals. The most valuable by-product is silica fume, called microsilica. Elkem is also the market leader of microsilica with a market share of 50%.

The slowdown of the global economy that started in 2000 had a great impact on the metal production industry. Indeed, Elkem's silicon division experienced a serious weakening in the end market in 2001 and 2002, compared with 2000. For example, by the end of 2001, spot prices for silicon and ferrosilicon metals in the European market were reduced to half those seen in the mid-90s. This resulted in widespread restructuring in the entire industry as a number of companies were forced to close down furnaces and plants. Nevertheless, the global production capacity was still greater than market demand.

To ensure that Elkem could keep the position as the world's leading silicon manufacturer, its management evaluated the supply chain of the silicon division. By the end of 2001, Elkem had closed down five furnaces at four different plants. It became necessary to improve the supply chain and evaluate the product portfolio. Closing a production plant might be the best alternative if the plant is not competitive enough, although this decision can be expensive. Elkem must demolish the plant and rehabilitate the environment to the same condition it had before the plant was built. Moreover, the company has to compensate the redundant employees. Another alternative is to temporarily close down the plant. This is less expensive than a permanent closure, but it is not allowed for a long period of time.

Elkem has studied its supply chain for the silicon division since 2001. [20] describes the development of an optimization package to support restructuring the silicon division. Although this optimization model has been of great value for Elkem in restructuring, further analyses have brought up the question of how the production and sale of by-products can affect the of supply chain. The model neither treats the production and the demand for the by-products explicitly, nor satisfactorily treats conversions between different furnace technologies. Because the impacts of plant location decisions are drastic, not only for stockholders' value, but also for the thousands of affected employees and their families, the management wants to ensure that they are making the right decisions. The management expresses a need to analyze if and how the decisions of by-product allocation between plants affect the supply chain design of the silicon division.

The rest of the section is organized as follows: In Section 2.1 we will discuss Elkem's supply chain, while the production process is described in Section 2.2.

2.1 Elkem's Supply Chain

These products have a complex production process, and reach the customers at the end of a global supply chain. The supply chain is illustrated in Figure 1.

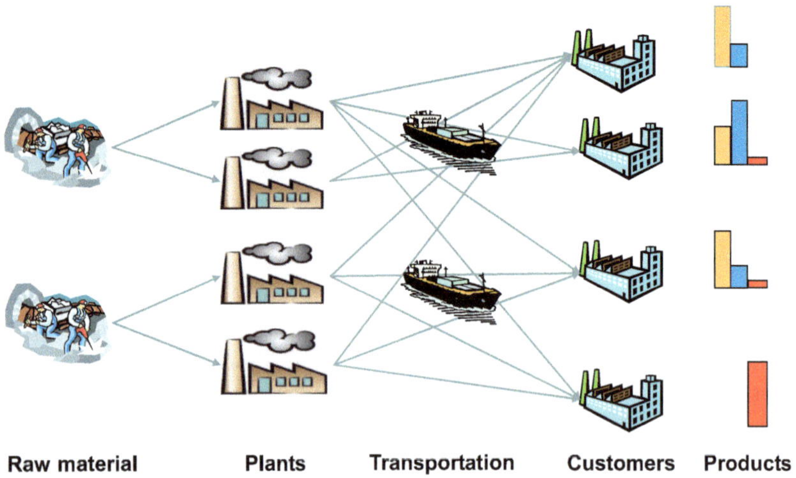

Fig. 1. Elkem's supply chain for silicon, ferrosilicon and microsilica. Raw materials are transported from coal and coke mines to the production plants where the silicon and ferrosilicon metals are produced. The finished products are transported mainly at sea to the end-customers. The bars at right illustrate different product mixes of silicon and ferrosilicon metals sent to end-customers.

Raw materials are sourced from mines in Europe, Asia and South America. Because Elkem has either a long-term relationship with or ownership of its suppliers, smelting plants have few raw materials suppliers to choose among.

Smelting plants have different costs due to economies of scale. Elkem plants differ substantial in size with production output ranging from 25 to 115 thousand metric tons per year. Production capacity is mostly limited by the number and capacity of furnaces at the plants. Each furnace operates at a relatively fixed production level close to its' maximum capacity. The crew size to operate a furnace is similar for small and large furnaces. Thus, small furnaces are more expensive to operate per ton of product output. Nevertheless, they are still necessary as some products can only be made on small furnaces. Maintenance costs and reinvestments depend on the state of the equipment in terms of age and previous restoration work.

Elkem's main customers are located in Europe, the United States and Japan. For ferrosilicon products there are a large number of medium-sized customers, while for both silicon metal and microsilica products there are a few large- and many small-sized customers. Elkem is particularly capable of producing high quality silicon metal. For example, the company supplies 50% of the total demand for silicon to the highly exacting Japanese electronics industry. The finished products are usually transported by ship to the customers. Microsilica and special grades of silicon metal are shipped in containers while standard products are transported in bulk. Because the production facilities are far away from most of the customers, transportation costs are considerable: for standard ferrosilicon products transportation costs can be up to 15% of the sales prices.

2.2 Elkem's Production Process

To identify components assembled into a product assembly industries use a bill-of-materials which is often referred to as a recipe in the process industries. A recipe states the amount of each raw material required to produce a finished product. Raw materials purchase costs, inbound transportation costs, and handling costs are often denoted recipe cost. The production costs depend on the products; in general it is twice as expensive to produce silicon metal as ferrosilicon metal. The costs related to recipe and energy are the major cost components. Recipe costs make up 35-50% of the sales prices depending on plant and product, while energy costs can be up to 25% of the sales prices. At each plant the raw material is fed into the furnace, and heated by electric power until it smelts at 1800° C before it is tapped into big ladles. To increase quality, the metal can be refined in the ladles by adding slag-forming elements and blowing gas through the metal. After this, the refined metal is cast into ingots. Alternately, when it is not necessary to refine the metal, the molten metal is cast directly from the furnace into ingots. Furthermore, the ingots are crushed into pebbles in grinding machines. Finally, the finished products are transported to the customers by either bulk- or container freight in bags. The production process is illustrated in Figure 2.

Microsilica was earlier considered as waste. As a consequence of strict Norwegian environmental laws, Elkem was forced to clean the smoke from production. This led to research on how to use microsilica as an additive in other products. Elkem has developed several different application areas for microsilica, and offers a range of different qualities for sale. For example, microsilica is used as an additive in concrete, fiber cement and refractory materials. Microsilica is generated from intermediate products in the smelting process. The purity, or quality, of microsilica is dependent on the main product the furnace is producing. Microsilica is collected by filtering smoke from the furnace. Production of 1 ton of main product generates 0.15-0.5 ton of microsilica, depending on the furnace characteristics and the purity of the main product. Because microsilica from this filtering has low density, it is often densified before it is packed into bags. Some plants offer microsilica mixed with water to slurry, which can be transported by tankers to the customers. Because the plants are obligated to clean the smoke from the production of ferrosilicon and silicon metals, the marginal cost of producing microsilica is very low. The price obtained for microsilica depends heavily on its quality. There is a high demand for low-quality microsilica, but the demand is limited for the best quality. Since there are several suppliers of low-quality microsilica, the prices are not attractive. On the other hand it is hard to produce high-quality microsilica, so there are few suppliers of high-quality microsilica. Elkem, one of

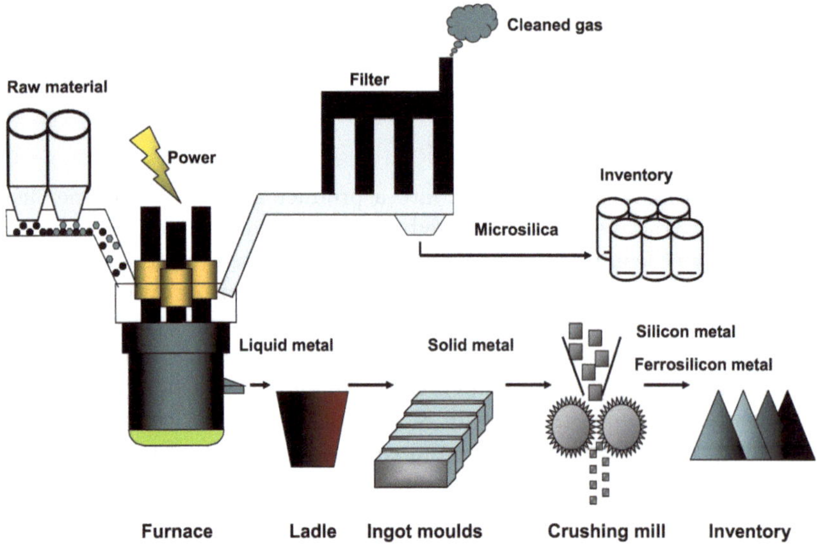

Fig. 2. Silicon, ferrosilicon and microsilica production. Raw materials are fed into a furnace where they melt and transform to liquid metal. The liquid metal is poured into big ladles before it goes through several post-smelting activities. Finally, the finished product is stored. Microsilica is collected by filtering the smoke from the furnace and packed in bags before it is stored.

the few suppliers capable of producing this quality, can obtain relatively high prices for it.

The throughput of silicon and ferrosilicon from a plant is limited by furnaces' capacity, in addition to refining and casting capacities. Because building new furnaces is very expensive, this is seldom an alternative, but a closed furnace might be reopened at a reasonable cost. Expanding the refining or grinding capacities is also possible to increase the plant's total throughput.

There are several different furnace technologies which put limitations on the product range. For instance, a furnace producing high quality silicon metal needs electrodes which do not liberate iron to the liquid metal. Iron from the electrodes will contaminate the silicon metal, resulting in lower purity. Converting a furnace to another technology costs 10 – 20 million USD. Furthermore, conversion of a furnace to better technology will also lead to higher operating costs, although it will also lead to more flexibility by increasing the furnace's product range. The choice of furnace technology will also affect the production of by-products.

Each Elkem plant consumes enormous quantities of energy: the furnaces at Elkem's plants consume between 10 and 45 MW. Earlier, the Norwegian Government supported industrial development by awarding good electricity contracts to plants. Many of those contracts are due to expire in the near future. However Elkem can often negotiate new first-rate contracts with local providers, and there is a range of contract agreements. Some contracts state that Elkem can only buy energy for use in metal production, while others allow Elkem to buy specified amounts and to sell any surplus electricity in the spot market. A third type of contract is called take-or-pay; meaning that Elkem must pay a given amount for the energy even if they do not use it. In a market with high energy prices and low metal prices, it is sometimes profitable to close down production temporarily and sell the energy instead.

3 Related Literature

In this section we will discuss the theory, optimization models and real cases presented in the literature that are related to the supply chain modeling at Elkem's silicon division. The rest of the section is organized as follows: First, in Section 3.1 we address facility location models, while Section 3.2 is devoted to supply chain models. Finally, in Section 3.3 we will discuss supply chain planning in the process industry.

3.1 Facility Location Models

The operations research literature on facility location and supply chain design is extensive. Problem formulations and solution methods for multi-commodity facility location problems have been published since the early seventies. One of the earliest implementation of strategic facility design is given by Geoffrion and Graves [10]. They present a single-period mixed integer linear program for multi-commodity distribution design. Warszawski [23] presents model formulations for both the single-period multi-commodity location problem and the multistage location problem with several time periods. A more recent study of the multi-period location problem is presented by Canel and Das [5]. They present a multi-period location problem with profit maximization as the objective. The demand for different products can be served by different plants, although the model can choose to not fill all demand. The decisions to be made are where to locate facilities and when to invest in them; furthermore, the problem formulation also allows plant closing at no cost. Although most of the facility location literature assumes linear costs and revenues, this assumption might not always be realistic due to economies of scale. Harkness and ReVelle [13] discuss several model formulations for the facility location problem with convex production costs.

Choice of capacity and technology can often influence the decision of where to locate manufacturing facilities; for example the decision between the numbers of production plants versus their sizes. Verter [21] presents an integrated model for simultaneously solving the facility location, capacity acquisition and technology selection problems. Although the author considers technology selection he does not include conversion of existing equipment. Verter and Dasci [22] have expanded this work to include both product-dedicated and flexible technology alternatives. Because flexible technology alternative can manufacture the entire product range, it offers economics of scale. Both models are single-period multi-commodity optimization models. When the plant locations are predetermined, the problems from [21] and [22] reduce to the plant loading problem. This problem is concerned with the issue of plant product mix, sourcing and market allocation in the context of a firm's supply chain for production and distribution. A model formulation of this problem with fixed facility costs and concave production costs is given by Cohen and Moon [7]. Corporations that want to evaluate their locations usually already have an existing network of facilities. The reverse location problem handles such problems where the main objective is to improve this network efficiently within a limited budget [26].

3.2 Supply Chain Models

The increased emphasis on improving and optimizing the supply chain, from both the business strategists and logistics practitioners, has given an opportunity and motivation for the operations research community to develop mathematical programming models for supply chain optimization [17]. Such modeling techniques are well suited to support decision making for integrated planning in the corporations' supply chain. Shapiro, Singhal and Wagner [17] describe a mixed integer program for strategic redesign of the supply chain for a large corporation which had just recently acquired another company. The main issue in their model is to decide which plants to operate and which plants to close. The model also considers plant expansion, and deployment and utilization of equipment among the operative plants. An implementation of a decision support system for supply chain management at tactical level is given by Brown, Graves and Honczarenko [4]. They have developed a mixed integer program designed for facility selection, equipment location and utilization, manufacturing and distribution of products for the biscuit producer Nabisco.

Global supply chains have a number of issues that make them different from single-country supply chains [8]. Some of these differences are taxes and duties, currency exchange rate fluctuation, market penetration strategies, product design differences between countries, and handling of multinational supply chain as global system to achieve economics of scale. Cohen, Fisher and Jaikumar [8] present an optimization model which has incorporated many of these international issues. The model is a mixed integer program where the objective is to maximize the total after-tax profits for the corporation. One of the main contributions of that paper is the inclusion of vendor supply contracts. A well documented application of global strategic supply chain optimization is the redesign of Digital Equipment Corporation in the 1990's [1]. The rapid changes in the computer industry in the beginning of the nineties lead to a shift in demand from large central mainframes to networks of smaller, less expensive computers. To face this challenge Digital developed a model called Global Supply Chain Model. The goal was to evaluate and redesign their worldwide manufacturing and distribution strategy. The decision support system was designed to handle questions like how many plants and distribution centers are needed, which capacity and technology should be acquired at each plant, how many plants should be involved in producing a product, and how do tax and transportation affect the optimal supply chain design.

Yan, Yu and Cheng [25] present a multi-commodity and multi-echelon single period strategic supply chain model. The model's objective is to minimize total cost and they include bill-of-material considerations in the form of logical constraints. The bill-of-materials constraints are used to express the relationship between product, suppliers and producers.

Activity-based costing (ABC) is a valuable tool in determining product and customer costs for managerial decision support [16]. ABC is a descriptive tool for allocation of costs to different cost drivers. Shapiro [16] discusses synergies between mathematical programming models for strategic supply chain design and ABC. The author argues that ABC analysis can be a valuable tool for developing much of the necessary cost data needed for a strategic optimization model, moreover accurate product and consumer costs can only be extracted from an optimal solution of an optimization model.

3.3 Supply Chain Planning in the Process Industry

Recently, there have been many contributions on optimization in forestry in the literature, which from a modeling point of view have some similarities with modeling the silicon metal industry. Carlsson and Rönnqvist [6] give an overview of supply chain optimization projects in the Swedish forest industry, while Bredström et al. [3] present a tactical decision support tool for short-term planning of daily supply chain decisions, developed for a major manufacturer of market pulp. Another model for strategic and tactical supply chain planning in the paper industry is given by Philpott and Everett [14]. For a general overview of supply chain models for strategic, tactical and operational levels in forestry, see Rönnqvist [15]. By-products from forestry can be used as bio-energy fuel by conversion of by-products from sawmills and forest residue to forest fuel. Gunnarsson, Rönnqvist and Lundgren [12] consider strategic analysis and tactical planning for a forest fuel supply chain.

There is some supply chain optimization literature in the smelting industry. Sinha et al. [18] describe the development of a mixed integer program for strategic decision support and planning at Tata Steel. The mathematical program was developed to give decision support for optimal allocation of steel-making capacity in order to maximize profit. The model considers the use of scarce power, operating hours, capacity of processes and upper and lower bounds on demand. Timpe and Kallrath [19] discuss production planning on a tactical level in process industry. They apply a rolling horizon principle and use time periods with different lengths. Much of the focus is on turnover between different mode of operation, e.g. between production of different products. Furthermore, Wolsey [24] presents a more general framework for modeling production changeover in planning models.

Ulstein et al. [20] discuss supply chain redesign in the silicon metal industry. They have developed a strategic optimization model which determinates optimal plant structure for Elkem. The main features include plant acquisitions, closings and expansions. The model also takes into consideration the product mix at each plant and decisions regarding the total product portfolio at Elkem's silicon division.

4 The Strategic Planning Model

The strategic planning problem described in Section 2 will be formulated as a deterministic problem where the objective function is to maximize the net present value of the future incomes and costs. All incomes and cost data given are discounted and an appropriate discount rate is used. Here we will present one model, but to facilitate the presentation we have split the model into three parts. First, we present the supply chain network in Section 4.1. Then in Section 4.2, the conditions related to the production of silicon and ferrosilicon products are described. Furthermore, the by-product constraints are presented in Section 4.3, and in Section 4.4 the overall objective function is given. Finally, in Section 4.5 we discuss some of the challenges we faced during the modeling process.

We present here a simplified version of the strategic planning model for Elkem. Some of the model features are left out to enhance the readability and understanding of the model. The features mentioned in the text that are not included in this formulation are: different types of electricity contracts and the separate investments in refining and casting equipment; instead we just consider a general type of equipment. Trade barriers and duties are omitted as well.

The notation is based on the use of lower-case letters to represent subscripts and decision variables, and capital letters to represent constants. Normally, the variables and constraints are only defined for some combination of subscripts. To limit the level of details in the mathematical formulation, we introduce only the necessary sets in the model.

4.1 The Supply Chain

The main decisions for the company regarding the supply chain are the operating status of the plants during the planning horizon. An existing plant can either be operating during the entire planning horizon or be closed down in any of the time periods. Due to the seriousness of a closure decision, the company cannot re-open a plant which has been closed down. However, the company can invest in new plants during the planning horizon.

Another important decision for the supply chain is the allocation to customers of products produced at particular plants. The supply chain from the raw material vendors to the plants is not included in the model, because for each plant the raw material vendors are given and there is no reason to include this issue into the model.

First, we describe the indices, sets, parameters and variables for the supply chain part of the problem. Figure 3 gives an overview of the variables in the supply chain.

Indices

c customer
i production plant
p product
q by-product quality
t time period

Sets

I^N the set of candidate plants

Parameters

CCL_{it} cost of closing plant i in period t
CIP_{it} cost of investing at plant i in period t
COP_{it} cost of operating plant i in period t
CT_{ipct} unit transportation cost for product p from plant i to customer c in period t
CMT_{iqct} unit transportation cost for by-product quality q from plant i to customer c in period t
F_{pct} fixed contract volume of product p for customer c in period t
FN_{ipct} fixed contract volume for product p at potential plant i for customer c in period t
MD_{qct} demand for by-product quality q for customer c in period t
$P0_i$ equals 1 if plant i is operating at the beginning of the planning period, and 0 otherwise
PM_{qct} unit sale price for by-product quality q to customer c in period t
PS_{pct} unit sale price for product p to customer c in period t
S_{pct} spot volume of product p for customer c in period t
SN_{ipct} spot volume of product p at new plant i for customer c in period t

Variables

pin_{it}	equals 1 if there is an investment at plant i in period t, and 0 otherwise
po_{it}	equals 1 if plant i is operating in time period t, and 0 otherwise
xs_{ipct}	quantity of product p produced at plant i and sold to customer c in period t
ys_{iqct}	quantity of by-product quality q produced at plant i and sold to customer c in period t
zsc	the objective function related to the supply chain network

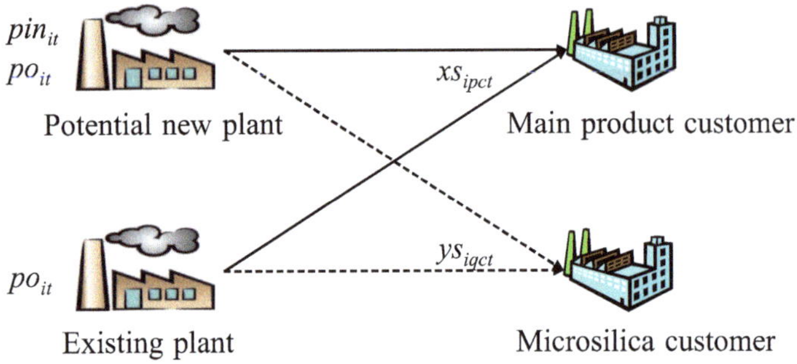

Fig. 3. Variables in the supply chain. Solid-drawn lines indicate transportation of silicon and ferrosilicon between plants and customers while the dashed lines indicate transportation of microsilica between plants and customers.

Model Formulation

The part of the model regarding the supply chain can then be written as:

$$\begin{aligned} \max zsc = & \sum_i \sum_p \sum_c \sum_t (PS_{pct} - CT_{ipct})\, xs_{ipct} \\ & + \sum_i \sum_q \sum_c \sum_t (PM_{qct} - CMT_{iqct})\, ys_{iqct} \\ & - \sum_i \sum_t COP_{it}\, po_{it} - \sum_i \sum_t CIP_{it}\, pin_{it} \\ & - \sum_i \sum_t CCL_{it}\, (po_{i(t-1)} - po_{it} + pin_{it}) \end{aligned} \quad (1)$$

subject to:

$$po_{it} = P0_i, \quad \forall i, t = 0, \quad (2)$$

$$po_{it} - po_{i(t-1)} - pin_{it} \leq 0, \quad \forall i, t, \tag{3}$$

$$\sum_i xs_{ipct} - \sum_{i \in I^N} \sum_{\tau \leq t} FN_{ipct}\, pin_{i\tau} \geq F_{pct}, \quad \forall p, c, t, \tag{4}$$

$$\sum_i xs_{ipct} - \sum_{i \in I^N} \sum_{\tau \leq t} (SN_{ipct} + FN_{ipct})\, pin_{i\tau} \leq S_{pct} + F_{pct}, \quad \forall p, c, t, \tag{5}$$

$$\sum_i ys_{iqct} \leq MD_{qct}, \quad \forall q, c, t, \tag{6}$$

$$xs_{ipct}, ys_{iqct} \geq 0, \quad \forall i, p, q, c, t, \tag{7}$$

$$pin_{it}, po_{it} \in \{0, 1\}, \quad \forall i, t. \tag{8}$$

The objective function (1) expresses the profit from selling products less the associated costs. The variable costs include variable operating costs at the plants and transportation costs from the plants to customers. The fixed costs include the costs to operate plants and potential investment and closure costs at plants. Constraints (2) represent the initial operating status of existing and potential plants. Constraints (3) impose that a plant can only operate if the plant was operating in the previous time period or if the company has invested in the plant at the beginning of the time period. The company cannot re-open a plant which has been closed down. Constraints (4) make sure that the sales to a customer in a given time period have to be at least equal to the amount required by the fixed order contract. Constraints (5) require that sales to a customer in a given time period cannot exceed the sum of fixed and spot orders. In constraints (6), we ensure that the amount of by-products sold to a customer do not exceed the demand. Finally, constraints (7) and (8) handle the non-negativity and binary requirements.

4.2 The Production

Production at each plant starts with a given recipe for each product type and electricity as input to the furnaces. The smelt from the furnaces is constrained by furnace and equipment capacities. Furnaces operating with one furnace technology can be converted to another technology producing different products. Capacity can be increased by equipment investment. Figure 4 gives an overview of the variables in the production process.

The indices, sets, parameters and variables not defined in Section 4.1, will be defined in the following.

Indices

d furnace technology
e equipment
f furnace

Sets

P^e the set of products that require production equipment e
P^d the set of products that can be produced with furnace technology d

Parameters

$CCO_{ifdd\prime t}$	cost of converting furnace f at plant i from technology d to technology $d\prime$ in period t
CE_{iet}	cost of using equipment e at plant i in period t
CEL_{it}	unit cost of buying spot electricity for plant i in period t
CIE_{iet}	cost of investing in equipment e at plant i in period t
COF_{ift}	cost of operating furnace f at plant i in period t
CR_{ifpt}	recipe costs for producing product p in furnace f at plant i in period t
EC_{iet}	capacity of equipment e at plant i in period t
ECI_{iet}	capacity increase from investing in equipment e at plant i in period t
ELC_{ifpt}	amount of electricity consumed to produce one unit of product p in furnace f at plant i in period t
ELF_{it}	amount of electricity stipulated in the fixed contract for plant i in period t
$FCAP_{ifdpt}$	capacity of furnace f with furnace technology d at plant i for the producing product p in period t
$F0_{ifd}$	equals 1 if plant i has a furnace f with furnace technology d at the beginning of the planning period, and 0 otherwise
PEL_{it}	unit sale price for electricity from plant i in period t

Variables

ein_{iet}	equals 1 if an investment in equipment e at plant i is made at the beginning of time period t, and 0 otherwise
elb_{it}	amount of electricity bought in the spot market for plant i in period t
els_{it}	amount of electricity sold in the spot market from plant i in period t
$fclos_{ifdt}$	equals 1 if furnace f with furnace technology d at plant i is closed down in period t, and 0 otherwise
$fconv_{ifdd\prime t}$	equals 1 if furnace f with furnace technology d is converting to furnace technology $d\prime$ at plant i in period t, and 0 otherwise
fo_{ifdt}	equals 1 if furnace f with furnace technology d at plant i is operating in period t, and 0 otherwise
xm_{ifdpt}	quantity of smelted product p tapped from furnace f with furnace technology d at plant i in period t
zpp	the objective function related to the production process

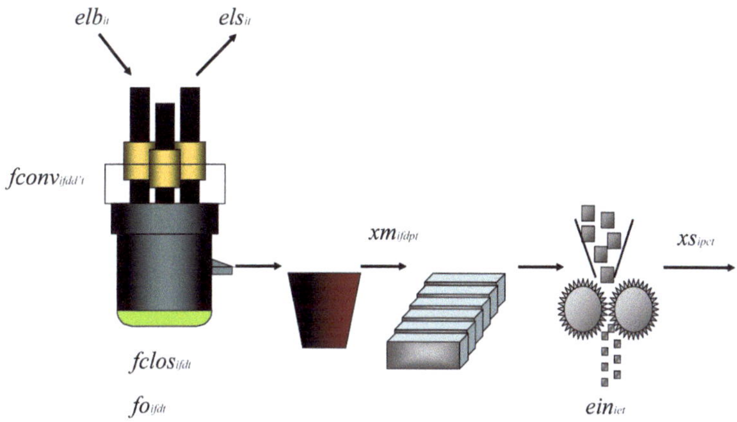

Fig. 4. Variables in the production process.

Model Formulation

The part of the model regarding the production can then be written as:

$$
\begin{aligned}
\max\ zpp = & \sum_i \sum_t PEL_{it}\, els_{it} \\
& - \sum_i \sum_f \sum_d \sum_p \sum_t CR_{ifpt}\, xm_{ifdpt} - \sum_i \sum_t CEL_{it}\, elb_{it} \\
& - \sum_i \sum_f \sum_d \sum_t COF_{ift}\, fo_{ifdt} \\
& - \sum_i \sum_f \sum_d \sum_{d'} \sum_t CCO_{ifdd't}\, fconv_{ifdd't} \\
& - \sum_i \sum_f \sum_d \sum_e \sum_{p \in P^e} \sum_t CE_{iet}\, xm_{ifdpt} \\
& - \sum_i \sum_e \sum_t CIE_{iet}\, ein_{iet}
\end{aligned} \tag{9}
$$

subject to:

$$po_{it} - \sum_d fo_{ifdt} \geq 0, \quad \forall i, f, t, \tag{10}$$

$$\sum_d fo_{ifdt} + \sum_d fclos_{ifdt} = 1, \quad \forall i, f, t, \tag{11}$$

$$
\begin{aligned}
& fo_{ifdt} + fclos_{ifdt} \\
& \quad + \sum_{d'} fconv_{ifdd't} - \sum_{d'} fconv_{ifd'dt} \\
& = F0_{ifd}|_{t=1} + (fo_{ifd(t-1)} + fclos_{ifd(t-1)})|_{t>1}, \quad \forall i, f, d, t,
\end{aligned} \tag{12}
$$

$$fo_{ifdt} - \sum_{p} \frac{1}{FCAP_{ifdpt}} xm_{ifdpt} \geq 0, \quad \forall i, f, d, t, \tag{13}$$

$$\sum_{f} \sum_{d} \sum_{p \in P^e} xm_{ifdpt} - \sum_{\tau \leq t} ECI_{iet}\, ein_{ie\tau} \leq EC_{iet}, \quad \forall i, e, t, \tag{14}$$

$$\sum_{f} \sum_{d} \sum_{p} ELC_{ifpt} xm_{ifdpt} - elb_{it} + els_{it} \leq ELF_{it}, \quad \forall i, t, \tag{15}$$

$$\sum_{f} \sum_{d} xm_{ifdpt} - \sum_{c} xs_{ipct} = 0, \quad \forall i, p, t, \tag{16}$$

$$xm_{ifdpt} \geq 0, \quad \forall i, f, d, p \in P^d, t, \tag{17}$$

$$elb_{it}, els_{it} \geq 0, \quad \forall i, t, \tag{18}$$

$$ein_{iet}, fclos_{ifdt}, fconv_{ifdd't}, fo_{ifdt} \in \{0, 1\}, \quad \forall i, e, f, d, d', t. \tag{19}$$

The objective function (9) expresses the profit from selling electricity less the associated costs. Variable costs include recipe costs, electricity costs and costs for using the equipment. The fixed costs include the costs of operating the furnaces, converting a furnace from one technology to another and investment costs in refining and casting equipment. Constraints (10) ensure that furnaces at a plant can only operate if the plant is in operation. Constraints (11) impose that each furnace is either operating or closed down in a time period. The status of each furnace in each time period is calculated in constraints (12), and they concern both the furnace technology and operation status. In each time period the furnace technology should be known, even if the furnace is closed down. From one period to the next, a furnace can operate or be closed with the same furnace technology. Alternately, the furnace can convert to another furnace technology. These furnace technology conservation constraints are given in (12). Constraints (13) state that the amount of all products produced in a furnace cannot exceed its capacity. Constraints (14) force total output not to exceed the total capacity of any equipment used. The capacity of a type of equipment in a given time period is equal to its initial capacity and the capacity added by investments in all earlier time periods. Constraints (15) restrict total electricity consumed or sold to that obtained from fixed contracts or purchased on the spot market. The mass balance constraints (16) describe the relationship between the smelted products and the sales products transported to the customers. Finally, constraints (17), (18) and (19) handle the non-negativity and binary requirements not declared in Section 4.1.

4.3 By-product Production

The smelt from the furnaces is split in a given ratio between main products and by-products. The marginal costs of producing the by-product are very low and are therefore ignored in the model. Figure 5 gives an overview of the variables in the production of the by-product.

The indices, sets, parameters and variables not defined in Sections 4.1 and 4.2, will be defined in the following.

Sets

W the set containing the lowest quality of the by-product

Parameters

Y_{ifpq} the portion of by-product with quality q generated due to production of product p at plant i and furnace f

$YT_{qq'}$ equals 1 if quality q' can be transferred and sold as quality q,and 0 otherwise. For each quality q' just one transfer is possible

Variables

ym_{iqt} quantity of by-product of quality q produced at plant i in time period t

y^{+}_{ipt} quantity of by-product of quality q that is in surplus and can be sold as a lower quality at plant i in time period t

yw_{iqt} quantity of by-product of the lowest quality q that is in surplus and must be considered as waste at plant i in time period t

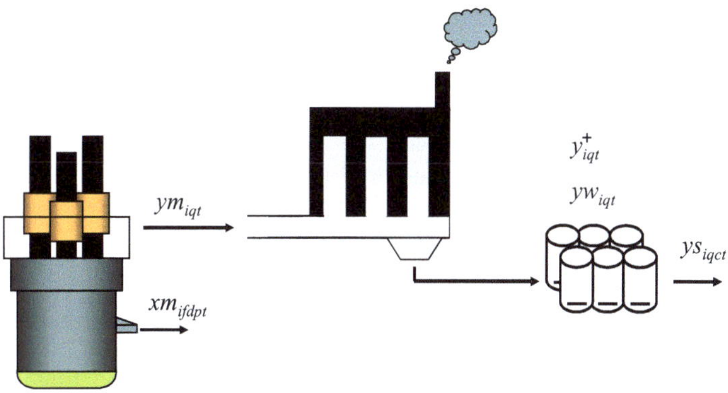

Fig. 5. Variables in the production of the by-product.

Model Formulation

The profit from the by-product production is included in the objective function term for the supply chain network. Because the marginal costs of producing the by-product are ignored in the model, we have no contribution to the objective function from the by-product production. The constraints regarding the by-product production can then be written as:

$$ym_{iqt} - \sum_{f}\sum_{d}\sum_{p} Y_{ifpq}\, xm_{ifdpt} = 0, \quad \forall i, q, t, \tag{20}$$

$$\begin{aligned} ym_{iqt} + \sum_{q'} YT_{qq'}\, y^{+}_{iq't} - \sum_{q'} YT_{q'q}\, y^{+}_{iqt} \\ -yw_{iqt} - \sum_{c} ys_{iqct} = 0, \quad \forall i, q, t, \end{aligned} \tag{21}$$

$$ym_{iqt}, y^{+}_{iqt} \geq 0, \quad \forall i, q, t, \tag{22}$$

$$yw_{iqt} \geq 0, \quad \forall i, q \in W, t. \tag{23}$$

The amount of by-products generated in the furnaces depends on the products produced and the particular furnace, and this dependency is described in constraints (20). The mass balance of the by-product produced and sold is given by constraints (21). The amount by-product sold of a particular quality is equal to the amount produced adjusted for by-products sold as a lower quality than produced. With a surplus of the lowest quality, this amount of by-product is wasted. Finally, constraints (22) and (23) handle the non-negativity requirements not declared in Sections 4.1 and 4.2.

4.4 The Objective Function

The resulting objective function consists of the terms from the supply chain network and the production process and becomes:

$$\max\, z = zsc + zpp \tag{24}$$

4.5 Some Modeling Challenges

The first and perhaps the most important challenge in the modeling is to find a level of detail which corresponds to the decision problem at hand. For instance, it was always clear that furnace operations had to be modeled, but it soon became evident that other major equipment types also had to be included. Furthermore, given the size of the problem, we had to aggregate products of similar type into groups and customers into clusters. The aggregation of time into time periods was implemented so the user can easily change the aggregation. This enables the user to trade-off run-time and detail output of data in each run.

Another challenge was extracting and breaking down economic information about the operations of each plant. Prior to the development of the optimization tool, top managers were accustomed to comparing more aggregated economic information. The mathematical model made it necessary to derive the costs with the financial directors at the plants in a way that disaggregated the costs of operating plants into product specific activity costs per equipment and fixed and variable maintenance and administration costs.

5 Computational Results

An advanced version of the optimization model has been implemented and solved using Xpress-MP version 14 [9] and run on a computer with a 3GHz processor and 1GB RAM. Furthermore, the model has been tested with real data from Elkem's silicon division. Xpress-MP takes less then 30 seconds to find optimal solutions for the optimization model with 5 time periods.

Some of the plants in this division have fixed contracts and/or are producing very specialized products that other plants cannot produce. Because the future operating status of these plants does not affect the rest of the plants, they are not considered in this optimization model. As a result, the model has been tested for six plants using a rolling horizon of 11 years. Furthermore, the time periods are not equal in length. We have split the time horizon into 5 distinct time periods where the first three periods are 1 year long and the last time period is 6 years long. The choice of splitting up the planning horizon in different length of each time period is mainly due to the fact that decisions early in the planning horizon are more critical for the company.

The development and the sales of silicon and ferrosilicon metals are carried out by the silicon division itself. On the other hand, a separate business unit in Elkem has been responsible for the development, marketing and sales of the various microsilica products. This business unit is called Elkem Materials. Because the production and sales of microsilica are handled by two separate units in Elkem, there have been some discussions regarding the level of the transfer prices for microsilica. These transfer prices are currently 2/3 of the market prices. Elkem's corporate management has expressed a need for a model that can assure that the plant structure is optimal and is not affected by the chosen level of transfer prices for microsilica.

The rest of the section is organized as follows: In Section 5.1 we will discuss the optimal operational status for the plants. Furthermore, in Section 5.2 we will examine furnace conversions and optimal furnace technology for the plants, and finally, in Section 5.3 product allocations between plants will be discussed.

5.1 Operating Status of the Plants

Table 1 shows the transfer prices between Elkem's silicon division and Elkem Materials. A price level of 100% equals the current level of the transfer prices for microsilica, while a price level of 150% is the current market price. A price level of 0% means that the silicon division does not get any revenues from the microsilica sales. The expected net present value of the cash flow for the different scenario is standardized to be 100 for scenario F, when the transfer prices are held at the current level. The net present value increases considerably when the microsilica prices increase. The reason for this is that there are no marginal costs for the microsilica production and therefore the sale of microsilica will only give positive contribution to the profit. As we can see from Table 1, the operating status of plant 1 is dependent on the price obtained for microsilica. If the microsilica prices are cut to 60% of the current transfer prices or lower, it is optimal to close plant 1.

Table 1. Optimal plant operating status for different microsilica price scenarios. A price level of 100% reflect the current level of transfer prices for microsilica and a price level of 150% is the current market prices for microsilica. MS price = microsilica price, NPV = net present value.

Scenario	MS Price	NPV	Plant 1	Plant 2	Plant 3	Plant 4	Plant 5	Plant 6
A	0 %	59	Closed	Open	Open	Closed	Open	Closed
B	20 %	65	Closed	Open	Open	Closed	Open	Closed
C	40 %	73	Closed	Open	Open	Closed	Open	Closed
D	60 %	80	Closed	Open	Open	Closed	Open	Closed
E	80 %	90	Open	Open	Open	Closed	Open	Closed
F	100 %	100	Open	Open	Open	Closed	Open	Closed
G	150 %	126	Open	Open	Open	Closed	Open	Closed
H	500 %	305	Open	Open	Open	Closed	Open	Closed
I	800 %	463	Open	Open	Open	1 period	Open	Closed

The plant operating status is much more robust for increases in the microsilica prices than for price reductions if the demand is held at a constant level. Indeed, the microsilica prices must increase up to a level of 800% before it is optimal to let plant 4 stay open in one period. The analyses show that plant 6 and plant 4 should close down in all scenarios. In fact, if management decides to keep plant 6 open in the entire planning horizon, the expected total profit for the silicon division will cut by 10 %.

The sales department at Elkem Materials is constantly developing the end-market for microsilica. As a consequence, the market for high quality microsilica is likely to increase during the next decade. A linear annual growth of 20% in demand is a well-supported assumption. Table 2 shows the optimal operating status for the plants under these conditions. The plant operating status is very robust for changes in microsilica demand, and a higher demand for microsilica does not affect the future plant status.

Table 2. Optimal plant operating status with 20% linear annual growth in microsilica demand. MS = microsilica.

Scenario	MS Price	MS Demand	NPV	Plant 1	Plant 2	Plant 3	Plant 4	Plant 5	Plant 6
J	60 %	120 %	83	Closed	Open	Open	Closed	Open	Closed
K	80 %	120 %	93	Open	Open	Open	Closed	Open	Closed
L	100 %	120 %	105	Open	Open	Open	Closed	Open	Closed
M	150 %	120 %	133	Open	Open	Open	Closed	Open	Closed
N	200 %	120 %	161	Open	Open	Open	Closed	Open	Closed

5.2 Furnace Technology Conversions

To produce high-quality microsilica, a special furnace technology has to be installed at the smelting plants. Such furnace conversion is expensive, and will only be considered if Elkem can sell this high-quality microsilica with a positive profit. Moreover, plants that are only capable of producing ferrosilicon products have to convert their furnaces in order to produce silicon products if profitable. These conversions are even more expensive than converting furnaces to produce high-quality microsilica.

Table 3 lists the optimal furnace technology at each open plant in scenario A-I. The table shows that the price level of microsilica affects the product range at Elkem's smelting plants. Today, only plant 2 is capable of producing the finest microsilica quality. On the other hand, plant 3 has a lower cost than plant 2, and Elkem can increase its profitability by reallocating the high quality microsilica products to plant 3 in all scenarios except in scenario A. Although it is profitable to reallocate this microsilica quality from plant 2, it is still optimal to leave plant 2 open producing silicon metal. Both plants 1 and 5 are today producing ferrosilicon products, but by converting the furnaces, the plants can also produce silicon products. A conversion of furnaces at plant 1 is profitable as long as the microsilica prices do not reduce to below 30% of the current transfer price. Furthermore, if the microsilica price is 30% of the current transfer price or lower, it is more profitable to close plant 1 and instead convert plant 5 to handle the demand for silicon products. Note that plant 5 is open in all of these scenarios, but the product mix at the plant is dependent on the price obtained for the by-products.

5.3 Product Allocation between Plants

As a consequence of furnace technology conversions and plant closings, the portfolio of the main products will also change. The optimal overall product mix when the silicon division obtains prices at the current transfer rate for the microsilica production (scenario F) is 65% ferrosilicon products and 35% silicon products. On the other hand, if there are no revenues from the microsilica production (scenario A), some of the ferrosilicon contracts will no

Table 3. Optimal furnace technology for the plants. FeSi = Ferrosilicon products, Si = Silicon products, hqMS = high quality microsilica, * = plant is open first period and closed in the other periods.

Scenario	Plant 1	Plant 2	Plant 3	Plant 4	Plant 5	Plant 6
A	Closed	Si, hqMS	Si	Closed	Si, FeSi	Closed
B	Closed	Si, hqMS	Si, FeSi, hqMS	Closed	Si, FeSi	Closed
C	Closed	Si, hqMS	Si, FeSi, hqMS	Closed	Si, FeSi	Closed
D	Closed	Si, hqMS	Si, FeSi, hqMS	Closed	Si, FeSi	Closed
E	FeSi, Si	Si, hqMS	Si, FeSi, hqMS	Closed	FeSi	Closed
F	FeSi, Si	Si, hqMS	Si, FeSi, hqMS	Closed	FeSi	Closed
G	FeSi, Si	Si, hqMS	Si, FeSi, hqMS	Closed	FeSi	Closed
H	FeSi, Si	Si, hqMS	Si, FeSi, hqMS	Closed	FeSi	Closed
I	FeSi, Si	Si, hqMS	Si, FeSi, hqMS	Si*	FeSi	Closed

longer be attractive and Elkem should reduce this production to a level of 47% of the main products from the silicon division. This means that even if Elkem chooses to close down a smelting plant producing silicon products when the microsilica market collapses, Elkem should also convert furnaces at other plants to secure the total production of silicon metal. The reason for this is that the silicon products have higher margins than the ferrosilicon products.

Another issue is how the product mix at the plants depend on the microsilica prices. Plant 5 is today a ferrosilicon plant, and as long as plant 1 is open, plant 5 will continue to produce ferrosilicon products. On the other hand, when it is optimal to close plant 1, Elkem should convert one furnace to produce silicon metal at plant 5. When there are no revenues from microsilica production (scenario A), the optimal product mix for plant 5 is 25% silicon metal and 75% ferrosilicon metal.

The optimal production mix changes as a consequence of reduced production volume for ferrosilicon metal when the microsilica transfer prices are cut down. The reason for this is that the sales revenues from selling microsilica with no marginal production costs contribute to a positive margin for the low-margin products. If the revenues from microsilica sales are removed, these low-margin products will not be profitable anymore, and the plants producing these low margin products will have to close down. Note that neither the optimal product mix nor optimal production volume is affected by the current microsilica transfer price rate.

6 Concluding Remarks

In this paper we have presented a mixed integer programming model for strategic supply chain optimization. The model is tailor-made for actors in the metal industry, and handles the production and sales of by-products in addition to the production and sales of the main products. Moreover, the model optimizes the supply chain design and allocates the production of both main products and by-products at the different plants. It also considers plant and furnace closings, plant acquisitions, capacity extensions, and technology conversions.

The model has been tested with real data from Elkem, the largest producer of silicon metal in the world and the western world's largest producer of ferrosilicon products. The testing of the model for different scenarios shows that the production and sales of by-products influence the optimal supply chain configuration. In fact, for a smelting plant the sales of by-products can make the difference between negative and positive profit. As a consequence, the sales of by-products may help the plants to survive in a harsh situation. Furthermore, the sales of by-products influence the product mix at the smelting plants and the total corporate product portfolio.

The optimization model described in this paper is an extension of the model described in [20], and the optimization tool from [20] is used in Elkem today. Due to the importance of the strategic restructuring process, Elkem employed optimization experts from our research modeling team to support further model developments and use. These optimization experts have triggered the extension of the model presented in this paper. The new version of the model is about to be implemented in the company.

References

1. B.C. Arntzen, G.G. Brown, T.P. Harrison, and L.L. Trafton. Global supply chain management at Digital Equipment Corporation. *Interfaces*, 25(1):69–93, 1995.
2. R.H. Ballou and J.M. Masters. Commercial software for locating warehouses and other facilities. *J. Bus. Logist.*, 14(2):71–107, 1993.
3. D. Bredstrom, J.T. Lundgren, M. Ronnqvist, D. Carlsson, and A. Mason. Supply chain optimization in the pulp mill industry – IP models, column generation and novel constraint branches. *European J. Oper. Res.*, 156(1):2–22, 2004.
4. G.G. Brown, G.W. Graves, and M.D. Honczarenko. Design and operation of a multicommodity production/distribution system using primal goal decomposition. *Manag. Sci.*, 33(11):1469–1480, 1987.
5. C. Canel and S.R. Das. The uncapacitated multi-period facilities location problem with profit maximization. *Int. J. Phys. Distrib. Logist. Manag.*, 29(6):409–433, 1999.
6. D. Carlsson and M. Ronnqvist. Supply chain management in forestry–case studies at Sodra Cell ab. *European J. Oper. Res.*, 163(3):589–616, 2005.
7. M.A. Cohen and S. Moon. An integrated plant loading model with economies of scale and scope. *European J. of Oper. Res.*, 50(3):266–279, 1991.
8. M.M. Cohen, M. Fisher, and R. Jaikumar. International manufacturing and distribution design networks: A normative model framework. In K. Ferdows, editor, *Managing International Manufacturing*, pages 67–93. Elsevier Science Publishers B.V (North-Holland), 1989.
9. Dash Optimization. *Xpress-Optimizer reference manual, release 14*, 2002.

10. A.M. Geoffrion and G.W. Graves. Multicommodity distribution system design by Benders Decomposition. *Manag. Sci.*, 5(5):822–844, 1974.
11. A.M. Geoffrion and R.F. Powers. Twenty years of strategic distribution system design: An evolutionary perspective. *Interfaces*, 25(5):105–128, 1995.
12. H. Gunnarsson, M. Ronnqvist, and J.T. Lundgren. Supply chain modelling of forest fuel. *European J. Oper. Res.*, 158(1):103–123, 2004.
13. J. Harkness and C. ReVelle. Facility location with increasing production costs. *European J. Oper. Res.*, 145(1):1–13, 2003.
14. A. Philpott and G. Everett. Supply chain optimisation in the paper industry. *Ann. Oper. Res.*, 108(1 - 4):225–237, 2001.
15. M. Ronnqvist. Optimization in forestry. *Math. Program.*, 97(1 - 2):267–284, 2003.
16. J.F. Shapiro. On the connections among activity-based costing, mathematical programming models for analyzing strategic decisions, and the resource-based view of the firm. *European J. Oper. Res.*, 118(2):295–314, 1999.
17. J.F. Shapiro, Vijay M. Singhal, and S.N. Wagner. Optimizing the value chain. *Interfaces*, 23(2):102–117, 1993.
18. G.L. Sinha, B.S. Chandrasekaran, N. Mitter, G. Dutta, S.B. Singh, A.R. Choudhury, and P.N. Roy. Strategic and operational management with optimization at Tata Steel. *Interfaces*, 25(1):6–19, 1995.
19. C.H. Timpe and J. Kallrath. Optimal planning in large multi-site production networks. *European J. Oper. Res.*, 126(2):422–435, 2000.
20. N.L. Ulstein, M. Christiansen, R. Gronhaug, N. Magnussen, and M.M. Solomon. Elkem uses optimization in redesigning its supply chain. *Interfaces* 34(4):314–325, 2006.
21. V. Verter. An integrated model for facility location and technology acquisition. *Comput. Oper. Res.*, 29(6):583–592, 2002.
22. V. Verter and A. Dasci. The plant location and flexible technology acquisition problem. *European J. Oper. Res.*, 136(2):366–382, 2002.
23. A. Warszawski. Multi-dimensional location problems. *Oper. Res. Q.*, 24(2):165–179, 1973.
24. L.A. Wolsey. MIP modelling of changeovers in production planning and scheduling problems. *European J. Oper. Res.*, 99(1):154–165, 1997.
25. H. Yan, Z. Yu, and T.C. Edwin Cheng. A strategic model for supply chain design with logical constraints: formulation and solution. *Comput. Oper. Res.*, 30(14):2135–2155, 2003.
26. J. Zhang, Z. Liu, and Z. Ma. Some reverse location problems. *European J. Oper. Res.*, 124(1):77–88, 2000.

Optimization Models for the Natural Gas Value Chain

Asgeir Tomasgard, Frode Rømo, Marte Fodstad, and Kjetil Midthun

Summary. In this paper we give an introduction to modelling the natural gas value chain including production, transportation, processing, contracts, and markets. The presentation gives insight in the complexity of planning in the natural gas supply chain and how optimization can help decision makers in a natural gas company coordinate the different activities. We present an integrated view from the perspective of an upstream company. The paper starts with decribing how to model natural gas transportation and storage, and at the end we present a stochastic portfolio optimization model for the natural gas value chain in a liberalized market.

Key words: Natural gas markets, Gas value chain, Stochastic programming, Portfolio optimization, Gas transport.

1 Introduction

The models in this paper are based on the authors experience from making decision support tools for the Norwegian gas industry. We present an integrated view on how to model the natural gas value chain in optimization models. The paper gives insight in the complexity of planning in the natural gas supply chain and how optimization can help decision makers in a natural gas company coordinate the different activities. Our focus is on describing modeling techniques and important technological issues, rather than a very detailed representation needed for commercial models.

We study the natural gas value chain seen from the point of view of an upstream company with a portfolio of production fields. Such a company should plan its operations considering long term contract obligations, the short term markets and transportation capacity booking. In particular we describe how the operations and planning are influenced by the existence of spot markets and forward markets. For the models to make sense it is also critical to include the technological characteristics of natural gas transportation and processing.

We therefore give a set of models where the interplay between the technological characteristics of natural gas and the markets are highlighted. In these models the economical content and the understanding of gas markets is essential.

We structure the paper by gradually introducing the different levels of the supply chain. We start by describing the most important components of the natural gas value chain in Section 2. Then in Section 3 we focus on how to model natural gas transportation in a steady-state situation. This is the type of transportation models suitable for planning problems with time resolution weeks, months or years. In Section 4 we introduce gas storages, and in Section 5 we describe a portfolio perspective and start investigating the integrated supply chain view. Here we introduce short term markets. In Section 6 we see how the spot-markets can be used to price natural gas storage capacity and indicate how to estimate the terminal value of natural gas still in storages or in reservoirs at the end of the planning horizon using the concept of an Expected Gas Value Function. An appendix describing all notation used in the paper is included at the end. Together, these sections will give a supply chain optimization model with an integrated view of the value chain, from production, via transportation and processing to contract management and gas sales.

2 The Natural Gas Value Chain

Here we give a brief description of the different elements of the natural gas value chain on the Norwegian continental shelf: production, transportation, processing, contract management and sales. The first action is to transport the natural gas from production fields to processing plants or transportation hubs where gas from different fields is mixed. Rich gas components are extracted and sold in separate markets. The remaining dry gas is transported to the import terminals in UK or on the European continent. In these hubs bilateral contracts and spot-trades are settled. Also upstream markets exist, where the gas is sold before it is transported to the import terminals. We focus on the value chain of a producing company, hence the issues of transmission and distribution to end customers are not considered.

In Figure 1 we show the main components of the natural gas export value chain. Before we go in detail on these we give a short summary of the main effects of liberalization and regulation in the European gas market.

2.1 Production

Production of natural gas takes place in production fields. Often these fields have several owners, and each owner has production rights that are regulated by lifting agreements. Typically a producer's rights allow him to produce between a minimum level of production and a maximum level of production

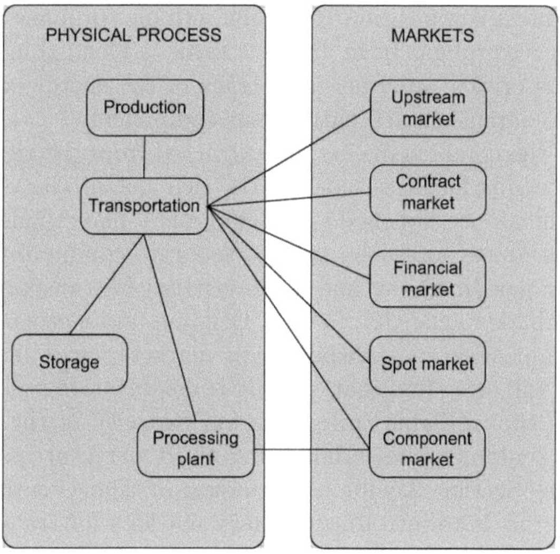

Fig. 1. Important objects in the natural gas value chain

within a set of time periods of different length. This production band may be flexible so that gas can be transferred between periods within predefined limits. Normally such production intervals are defined for single days, for years, and for intermediate periods in between like weeks and months.

Much of the natural gas produced is traditionally committed to take-or pay contracts where the buyer has agreed to take a volume in a given import terminal for a sequence of years. Again there is flexibility on when to take the gas within a year (or other time periods) and typically the daily offtake is within a minimum and maximum level. The customer nominates volumes within the take-or-pay agreements, and the producer has to deliver. These nominations are often done weekly, with final nomination the day before production. In take-or-pay contracts the price is usually indexed to other commodities like oil, to temperature and several other parameters.

2.2 Transportation and Processing

Natural gas is transported in pipelines by using compressors to create a higher pressure in the originating end of a pipeline, so that molecules will flow towards the end. Several pipelines may meet in a node in the transportation network. They may have different pressure at the end of the pipeline, but the input pressure of all pipelines going out of a transportation node must be smaller than the smallest end pressure of pipelines coming into the node, unless there is a compressor in the node.

An example of an export network for natural gas is the one you find at the Norwegian continental shelf which consist of 6600 km of pipelines. Here natural gas from different fields have different quality, in terms of energy content and its chemical composition (methane, ethane, propane and several more). Hence when natural gas from different fields is blended in the transportation network, it is critical to either keep track of the energy content of the blend or the total content of each natural gas component.

Some of the components can be extracted from the rich gas in processing plants. Processing facilities separate the rich gas into its various components. The components are liquefied petroleum gases like ethane, propane and butanes, which are exported by ship to separate commodity markets. The remaining dry gas (methane and some ethane) is transported in pipelines to import terminals in the UK, France, Belgium and Germany.

The organization of transportation markets varies a lot from region to region. One will often find that existing transportation rights already accounts for much of the available transportation capacity in the network. The Gas directive [6] enforces undiscriminating third party access to the remaining capacity (see Section 2.5 for a discussion of The Gas directive). One way of resolving this is to introduce primary markets for transportation capacity where capacity can be booked. In some cases a fixed tariff is used for zones or for pipelines, in other cases bids are given for capacity and the market settled by some auction mechanism. In all cases the market is cleared and capacity is allocated by given transparent rules. In a secondary market with shorter time horizons transportation capacity is balanced with transportation needs for the different shippers.

In this paper we will only focus on the utilization of transportation capacity, while the capacity allocation regime and tariff regime is not discussed. For a further discussion on these topics see [3].

2.3 Storage

There exist several types of natural gas storages. Abandoned oil and gas fields have high capacity and thereby a cost advantage. They also have low risk as geological data are known. In aquifers water is replaced with gas. They have higher risk as seismic investigation is necessary. Salt caverns are underground storage tanks washed out from salt layers. They typically have high costs. Injection rates, capacities, withdrawal rates, and characteristics depending on filling rate vary between the types. Storages are important in planning models because they allow us to store natural gas close to the market and thereby use them to exploit spot-market variations. They also allow producers to produce in time periods where demand is low and thereby utilize the available transportation capacity. Also they can be used as seasonal storages to smooth out seasonal effects. Whether storage is used as to avoid bottlenecks in the system in high demand periods or to utilize market possibilities, today's storage capacity is very limited when compared to the total production volumes.

2.4 Import Terminals and Markets

The import terminals are landing facilities for natural gas where the export pipelines end. Natural gas is delivered here according to specification on minimum and maximum pressure and energy content. These characteristics are often specified by the contracts as terms of delivery. Further transportation from the import terminals are taken on by the buyer using a transmission network to distribute the gas to the end customers.

Originally these terminals were the points of deliverance for the many take-or pay contracts. Recently these terminals have also been the location of the growing spot markets for natural gas and for financial derivatives on the spot market. The leading European hubs in terms of liquidity are the National Balancing Point in Great Britain, TTF in the Netherlands, and Zeebrugge in Belgium.

A different type of markets is also emerging upstream in the pipeline network. The main idea here is to have standardized trading mechanisms for natural gas at some important locations in the network to be able to provide an additional flexibility for the producers. Upstream markets are used to perform trades of natural gas before transportation takes place. They are useful because there is a need for having standardized mechanisms to exchange gas between producers. They include additional flexibility for producers in terms of being able to stay within the limits of their own lifting agreements, transportation capacities, and contract commitments in case of unexpected events or in case overselling or underselling of natural gas has occurred. The buyer of gas upstream also has the responsibility to transport the gas to downstream markets. Upstream markets are not as well developed as the other markets. Still the idea is old and the former variant was the less standardized bilateral long term swing agreements between different producers, allowing fields with little flexibility an option to draw gas from fields with more flexibility in volumes.

2.5 Liberalization and Regulation

The European natural gas industry has developed rapidly over the past thirty years. The European Commission has worked toward strengthening the competition within the complete gas- and energy markets value chain. A breakthrough in this process came on the 22nd of June 1998 when the gas directive was passed in the European Commission [5]. In the directive a stepwise liberalization of the European gas market is described. The key components of the gas directive are third party access to all transportation installations, division of activities within the firms in the value chain (physically or by accounting) and the possibility for certain consumers to obtain their gas from the supplier of their choice. The directive was followed by a second gas directive in 2003 [7] which moved another step towards liberalization.

Another implication of the gas directive and of EU competition laws was the 2002 closing down of the Gas Negotiation Committee (GFU), the forum for coordinated gas sales from the Norwegian continental shelf. The GFU formerly coordinated the supply of Norwegian natural gas producers Statoil and Hydro. Now the sales are company based and rarely linked to a specific production field.

An expected result from these changes is that short-term markets will evolve for natural gas. Though liquidity is still low, there are already clear signs indicating that short-term contracts and spot trades will play an important role in the future. The prior market structure is dominated by long-term agreements and thus minimizes the uncertainty for the participants. In practice the producers take the price risk, as prices are fixed towards various indexes, while the buyers take the volume risks by going into long-term agreements. The new markets will include short-term bilateral contracts, spot markets and financial markets. The introduction of short-term markets will most likely also lead to higher volatility and thus higher uncertainty.

Abolishment of the GFU-system and the introduction of a system where the individual companies are responsible for disposal of their own gas reserves called for a new access and tariff regime in the transportation network. The first step in the Norwegian transportation system was taken with the creation of Gassco AS in May 2001 under the provisions of a Norwegian White Paper. Gassco is assigned all the operator's responsibilities warranted in the Norwegian Petroleum Law and related Regulations. As a State owned company, Gassco AS should operate independently and impartially and offer equal services to all shippers. Systems operated by Gassco are the rich and dry gas systems previously operated by Statoil, Norsk Hydro, and TotalFinaElf.

The models presented in this paper are simplified variants of models developed in co-operation with Gassco and Statoil to deal with the changes mentioned above.

3 A Natural Gas Transportation Model

When modeling natural gas pipeline flow it is important to have a conscious view on how time and the dynamics of gas flow should be handled. The modeling of natural gas flow in continuous time has clear links to the process control paradigm [8]. Within this paradigm one normally uses active control, to operate the system according to a predetermined load and supply, finding a sequence of control actions which leads the system to a target state. The control regime often focuses on single processes or single components in the network. For our purpose we need to model a system of pipelines with a set of production fields, processing plants and markets. The natural choice is to look at mixed integer programming models from the modeling paradigm of mathematical programming. Here time is discretized. If the resolution of time periods is minutes or hours there is a need to model the transient behavior of natural gas. Some attempts on optimizing the transient behavior of a system of natural gas pipelines are [23] and [13], but only systems of limited size and complexity can be handled. To be able to handle the complexity needed for our models, we leave the concept of modeling the transient behavior of natural gas and approximate the time dimension by discrete time periods of such length that steady-state descriptions of the flow will be adequate. When the time resolution of the model are months, weeks, and maybe days, rather than minutes and hours, we can assume that the system is in a steady-state in each time period. The mathematical optimization models used to describe natural gas flow in the case of steady-state models are typical non-linear and non-convex. An approach using non-linear formulation of the mathematical models is illustrated in [4]. We present here a linearized model based on mixed integer programming to optimize routing of natural gas in pipeline networks. We base our presentation on work done on linearization from [18]. Several examples on linearization of natural gas flow exist in the literature. For a recent PhD thesis on linearization of natural gas flow see [22].

In this paper we describe the essential constraints needed to model the technological characteristics of natural gas flow in a steady-state setting. Issues like pressure, gas quality and gas components are dealt with from a pipeline transportation perspective. More detailed models very similar to the one we present here are today in use by Statoil and Gassco in the software package GassOpt developed by SINTEF. GassOpt is mainly used by the operator of the gas transportation system in the Norwegian sector of the North Sea. They are obliged to verify the delivery capabilities and robustness of the pipeline system transporting natural gas to European markets.

In the model presented in this section, we will focus on the transportation alone with the main purpose to meet demand for transportation generated by planned production profiles for the different fields. This typically represents the situation facing the neutral operator. The pipeline system is a natural monopoly, and is controlled by the authorities. This verification is also of strategic importance for the independent producers and customers in Germany, Belgium and France. The security of supply will influence the price possible to achieve for long term contracts, and contribute to infrastructure investment decisions, and GassOpt is one of the tools used to ensure maximum utilization of the infrastructure.

GassOpt itself focuses on analyzes of transportation possibilities. It can be used for optimal routing decisions from a flow maximization perspective. Also it is used to reroute natural gas when unexpected incidents lead to reduced capacity (in production units or pipeline). Thirdly, it can be applied at more tactical/operational level by a commercial player in capacity planning and capacity booking.

In this section we present a static model of one period. Demand for natural gas in the import terminal is assumed to be aggregated over the contracts in the terminals and planned production volumes given as constants to represent the license holders' production plans. So the main task of this model is to operate the transportation network to make sure demand is met by the planned production. In Section 4 we extend the model with several time periods and storage capabilities. In Section 5 we include contracts, markets and a portfolio perspective on managing the natural gas supply chain with stochastic prices and demand.

3.1 The GassOpt Modeling Interface

In GassOpt, the underlying physical network is represented in a graphical modeling environment with nodes and arcs. The modeling tool is hierarchical and applies to general network-configurations. Figure 2 indicates the network complexity for the North Sea network. The squared nodes contain subsystems with further nodes and pipelines. When modeling the North Sea system we need approximately 75 nodes and 100 arcs to represent the network.

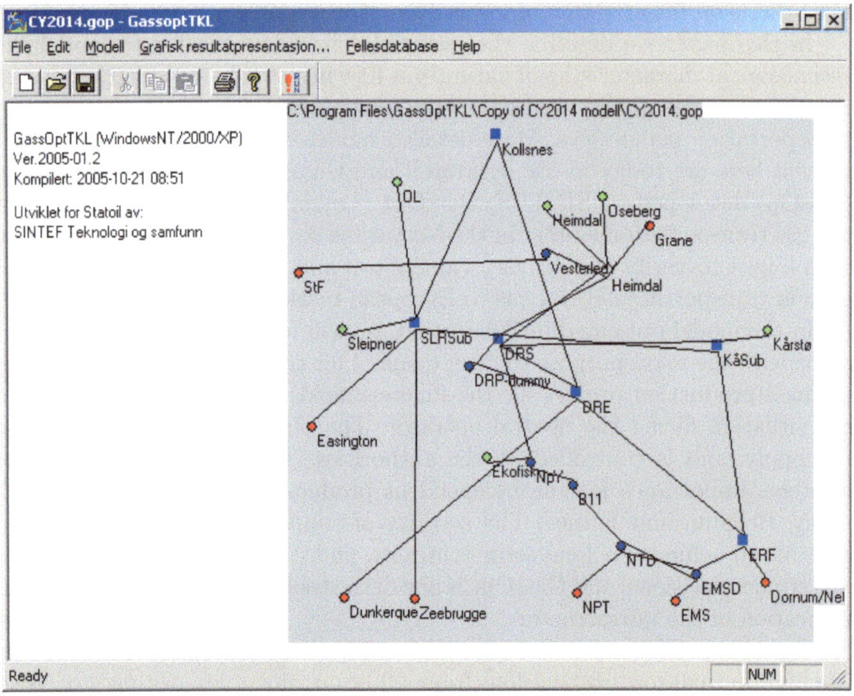

Fig. 2. Network presentation in GassOpt

3.2 The GassOpt Mathematical Model

This network model includes flow balances, blending of different gas qualities from different fields, processing nodes for extracting components of the natural gas, compressor nodes, node pressures and the nonlinear nature of pressure drop in pipelines. The model describes a steady-state situation where the network is in equilibrium in terms of pressures and natural gas mix. It is typically the kind of model used to model situations where flows are aggregated over a given time period. When the time period gets short enough, for example hours or minutes, this steady-state description will not be good enough because of the need to describe the transient behavior of natural gas flow. The objective for the optimization model is to ensure optimal routing and mixing of natural gas.

The model should make sure the nominated volumes are delivered to the import terminals within a time period. This objective can be achieved in several ways. Penalties are introduced in the objective function to influence the impact of the following goals:

1. Maintain planned production from the producers, where this is physically possible.
2. Deliver natural gas which meets quality requirements in terms of energy content.
3. Deliver within the pressure requirements in the contracts.
4. Minimize the use of energy needed in order to deliver the natural gas to the customers by minimizing the pressure variables.

A typical optimization case describes a specified state of the network, including expected production and demand (characterized by volume and quality), shutdown situations and turn-up capacity (additional available but unplanned production capacity) from production fields. In a normal situation, there will be several possible strategies to deliver the maximum amount of gas to the customers. To make the model generate and report these realistic flows, we have introduced penalty costs in the objective function on deviation from planned production, quality requirements, pressure agreements and the energy use. These penalty costs can of course theoretically interfere with and prevent us to achieve the main goal, to deliver in accordance with the demand of the customers. The tests we have performed on the full North Sea network, show that this 'multi-criteria' aspect does not sacrifice much of the maximal flow potential, but is rather used to choose between alternative solutions with about the same flow. In a fault situation, for example if a field or pipeline is down, the model will prioritize to deliver the nominated volumes in the import terminals. For more information about multi-criteria decision making, see for instance [16].

Seen from an operator's point of view the model tries to meet the customer's requirements for a given state of the network: either by optimal routing of gas or by turning up production in fields with flexibility on the production side. In the last case we say that we use turn-up capacity, which is available in some fields with flexible production characteristics.

3.3 Sets

Below the sets used in the mathematical description of the model is presented.

$\mathcal{N}$ The set of all nodes in the network.
$\mathcal{B}$ The set of nodes where gas flows are splitted into two or more pipelines.
$\mathcal{M}$ Nodes with buyers of natural gas: typically import terminals.
$\mathcal{I}(n)$ The set of nodes with pipelines going into node n.
$\mathcal{O}(n)$ The set of nodes with pipelines going out of node n.
$\mathcal{R}$ The set of nodes with processing capabilities.
$\mathcal{S}$ The set of nodes with storage facilities.
$\mathcal{K}(b)$ The set of contracts in node $b \in \mathcal{B}$.
$\mathcal{C}$ The set of components defining the chemical content of the natural gas.
$\mathcal{T}$ The set of time periods included in the model.
$\mathcal{L}$ The set of breakpoints used to linearize the Weymouth equation.
$\mathcal{Z}$ The set of split percentages used to discretize possible split fractions in split-nodes of the network.
$\mathcal{Y}$ The number of discretized storage and injection rate levels used to linearize storage characteristics.

3.4 Objective Function

Our goal is to route the gas flow through the network, in order to meet demand in accordance with contractual obligations (volume, quality and pressure). In the formulation given below, variable f_{im} is the flow of gas from node i into market node m, p_{ij}^{in} is the pressure into the pipeline going from node i to j, ϵ_m^+ and ϵ_m^- is the positive and negative deviation from the contracted pressure level respectively, Δ_g^+ and Δ_g^- represents underproduction and the use of turn-up in relation to the planned production in field g, δ_m^{l-} is the negative deviation from the lower quality level limit, and δ_m^{u+} is the positive deviation from the upper quality level limit in market node. The value of the flow to the customer nodes is given by the constant ω. Furthermore, κ is the penalty cost for pressure level, ϖ is the penalty cost for deviation from contracted pressure level, χ is the penalty cost for deviation from contracted quality to customers and ι for use of turn-up.

$$\begin{aligned} \max Z = & \sum_{i\in\mathcal{I}(m)} \sum_{m\in\mathcal{M}} \omega_m f_{im} - \sum_{i\in\mathcal{N}} \sum_{j\in\mathcal{N}} \kappa p_{ij}^{in} - \sum_{m\in\mathcal{M}} \varpi \left(\epsilon_m^+ + \epsilon_m^-\right) \\ & - \sum_{g\in\mathcal{G}} \iota \left(\Delta_g^+ + \Delta_g^-\right) - \sum_{m\in\mathcal{M}} \chi \left(\delta_m^{l-} + \delta_m^{u+}\right) \end{aligned} \quad (1)$$

Energy consumption for transporting the natural gas is minimized through making the penalty cost (κ) insignificant in size as compared to the value of the natural gas transported. This contributes to reduce the necessary build up of pressure to a minimum, without interfering with the correct volume, quality and pressure to the customer terminals. The penalty on using turn-up capacity will make sure that planned production in the fields is prioritized first, as long as it does not influence the throughput of the pipeline system. For most practical cases the contracted pressure level is not a soft constraint, and will then rather be put into a hard constraint instead of being penalized in the objective function.

3.5 Constraints

Production capacity

The following constraint says that the total flow out of a production node g cannot exceed the planned production of the field in that node. Here f_{gj} is the flow from production field g to node j:

$$\sum_{j \in \mathcal{O}(g)} f_{gj} \leq G_g, \; g \in \mathcal{G}. \tag{2}$$

Demand

This constraint says that the total flow into a node with customers for natural gas must not exceed the demand of that node:

$$\sum_{j \in \mathcal{I}(m)} f_{jm} \leq D_m, \quad m \in \mathcal{M}. \tag{3}$$

Mass balance for node j

The following constraint ensures the mass balance in the transportation network. What flows into node j must also flow out of node j:

$$\sum_{i \in \mathcal{I}(j)} f_{ij} = \sum_{n \in \mathcal{O}(j)} f_{jn}, \; j \in \mathcal{N}. \tag{4}$$

Pressure constraints for pipelines

Offshore transportation networks often consist of very long pipelines without compression, where it is crucial to describe the pressure drops in the pipeline system. We use the Weymouth equation to describe the flow in a pipeline as a function of input and output pressure. The Weymouth equation is described in e.g., [2]. In the Weymouth equation $W_{ij}(p_{ij}^{in}, p_{ij}^{out})$ is the flow through a pipeline going from node i to node j as a consequence of the pressure difference between p_{ij}^{in} and p_{ij}^{out}:

$$W_{ij}(p_{ij}^{in}, p_{ij}^{out}) = K_{ij}^{W} \sqrt{{p_{ij}^{in}}^2 - {p_{ij}^{out}}^2}, \qquad j \in \mathcal{N}, i \in \mathcal{I}(j). \tag{5}$$

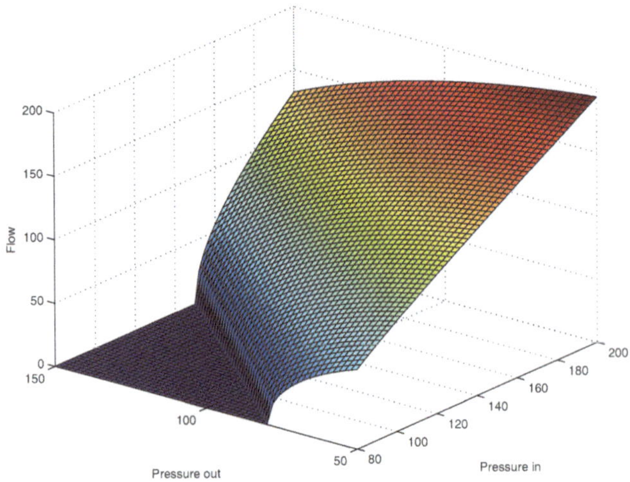

Fig. 3. A three-dimensional illustration of how the Weymouth relates pressure at the inlet and outlet points to the capacity in the pipeline.

Here K_{ij}^W is the Weymouth constant for the pipeline going from i to j. This constant depends among others on the pipelines length and its diameter and is used to relate the correct theoretical flow to the characteristics of the specific pipeline. Figure 3 illustrates the Weymouth equation. The figure shows that the function in the interesting area (positive pressure levels) is one fourth of a cone. The cone starts in origo, and is limited by the inlet pressure axis, and the 45° line between the inlet pressure and outlet pressure axes.

Through Taylor series expansion it is possible to linearize Equation (5) around a point (PI, PO) representing fixed pressure into the pipeline and fixed pressure out of the pipeline respectively:

$$\begin{aligned} W_{ij}(p_{ij}^{in}, p_{ij}^{out}) \leq & W_{ij}(PI, PO) + \frac{\partial W_{ij}}{\partial p_{ij}^{in}}(p_{ij}^{in} - PI) \\ & + \frac{\partial W_{ij}}{\partial p_{ij}^{out}}(p_{ij}^{out} - PO), \qquad j \in \mathcal{N}, i \in \mathcal{I}(j). \end{aligned} \tag{6}$$

We introduce a set of points to linearize this expression, (PI_l, PO_l), where $l = 1, \ldots, L$. Then we replace for each pipeline the nonlinear function (5) with L linear constraints of the type:

$$\begin{aligned} f_{ij} \leq & K_{ij}^W \frac{PI_l}{\sqrt{PI_l^2 - PO_l^2}} p_{ij}^{in} \\ & - K_{ij}^W \frac{PO_l}{\sqrt{PI_l^2 - PO_l^2}} p_{ij}^{out}, \qquad j \in \mathcal{N}, i \in \mathcal{I}(j), l = 1, \ldots, L. \end{aligned} \tag{7}$$

For any given pipeline flow, only one of these L constraints will be binding, namely the one that approximates the flow best. The planes described in (7) will be tangent to the cone at the line where the ratio between pressure in and out of the pipeline is equal to the ratio between PI_l and PO_l. Together the planes give an outer approximation of the cone. This approximation will consist of triangular shapes defined by these planes.

Pipelines without pressure drop

For physical pipelines between nodes where the distances are very limited it is not necessary to model pressure drops by the Weymouth equation. In this case a simple maxflow restriction is:

$$f_{ij} \leq F_{ij}, \quad j \in \mathcal{N}, i \in \mathcal{I}(j), \tag{8}$$

where F_{ij} is the capacity. In this case there is no pressure drop, so:

$$p_{ij}^{out} = p_{ij}^{in}, \quad j \in \mathcal{N}, i \in \mathcal{I}(j). \tag{9}$$

Relationship between pressures into a node and out of a node

To achieve a relevant flow pattern, it is sometimes preferable to model the pressure out of all the pipelines going into the same node homogenously:

$$p_{in}^{out} = p_{jn}^{out}, \quad n \in \mathcal{N}, i \in \mathcal{I}(n), j \in \mathcal{I}(n). \tag{10}$$

Another important issue is the relationship between pressure in ingoing pipelines and the outgoing. In general for a node n the input pressure of all pipelines going out of n must be lower than the lowest pressure out of any pipeline going into node n, see Figure 4. There is one exception, and that is the case where a pipeline into the node has 0 flow. The end pressure of this arc is neglected. In the Equation (11) the variable ρ_{ij} is 0 for pipelines without flow and 1 for the others. M is a number which is large enough to not restrict the pressures when the flows are 0. Then the following constraints make sure that the input pressure of a pipeline leaving n is less than the output pressure of a pipeline ending in n as long as both pipelines have a flow larger than 0.

$$p_{nj}^{in} - p_{in}^{out} + M(\rho_{nj} + \rho_{in} - 1) \leq M, \qquad n \in \mathcal{N}, i \in \mathcal{I}(n), j \in \mathcal{O}(n) \tag{11}$$

$$f_{nj} \leq M\rho_{nj}, \quad n \in \mathcal{N}, j \in \mathcal{O}(n) \tag{12}$$

$$\rho_{nj} = \begin{cases} i & \text{if flow from node } n \text{ to node } j \\ 0 & \text{otherwise.} \end{cases} \tag{13}$$

The Weymouth equation used gives an upper bound on the flow in a pipeline. This means that even if there is a pressure difference in a pipeline the flow can be zero. Because of this property it is not necessary to explicitly model the possibility of shutting down a pipeline. The model can simply put the flow to zero, and still keep the desired pressure. If omitting the constraints presented above one has to be aware of this when interpreting the results from the model.

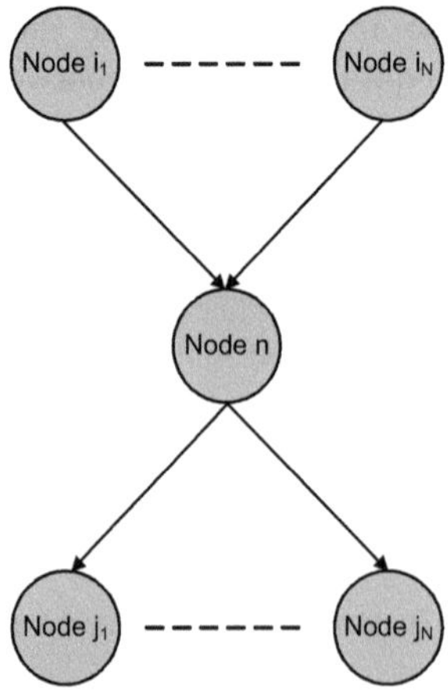

Fig. 4. Example of a split node with the possibility to shut down operation of one of the pipelines. The upper nodes, i, have pipelines going to node n. The lower nodes, j, have pipelines coming from node n. The index on i and j goes from 1 to N, where N is the total amount of nodes.

Modeling bidirectional pipelines

For pipelines designed to handle flows in both directions, the ρ_{ij} variable defined in the previous paragraph is used to determine the direction of flow. Equations (14) and (15) make sure that there only flows gas in one direction in the pipeline.

$$f_{ij} \le M\rho_{ij}, \qquad i \in \mathcal{N}, j \in \mathcal{O}(i), \tag{14}$$

$$\rho_{jn} = 1 - \rho_{nj}, \quad n \in \mathcal{I}(j), j \in \mathcal{I}(n). \tag{15}$$

Nodes with compression or pressure drop

In some cases we allow the pressure to increase in a node by using a compressor, or we force a pressure drop in the node. We here present a simplified formulation for modeling compression nodes where pressure can be build up or forced down. The compressor characteristics includes a compressor factor Γ used to limit how much the gas can be compressed in a node. If there is no compressor, this factor is 1. If there is a compressor, this Γ is a function of the flow $f_n = \sum_{j \in \mathcal{I}(n)} f_{jn}$ into the node:

$$\Gamma_n(f_n) = \left(\frac{W^{max}\eta(K_a - 1)}{100 K_a f_n} + 1 \right)^{\frac{K_a}{K_a - 1}}, \quad n \in \mathcal{N} \tag{16}$$

In this expression, the parameter K_a is the adiabatic constant for a certain gas type, W^{max} is the power output capacity of the compressor, and η is the compressor efficiency, [2]. Here we simplify this by using a constant compression factor independent of the flow. Then the pressure out of the compressor node n is limited by the compressor factor times the pressure into the node n:

$$\Gamma_n p_{jn}^{out} \geq p_{ni}^{in}, \qquad n \in \mathcal{N}, j \in \mathcal{I}(n), i \in \mathcal{O}(n). \tag{17}$$

Pressure drop is modeled in the same way, but with a reduction factor Θ_n instead of a compressor factor:

$$\Theta_n p_{jn}^{out} \geq p_{ni}^{in}, \qquad n \in \mathcal{N}, j \in \mathcal{I}(n), i \in \mathcal{O}(n). \tag{18}$$

Here Θ_n and Γ_n are constants, where $0 < \Theta_n \leq 1$ and $1 \leq \Gamma_n$. The formulation is only meaningful if at most one of the factors is different from 1 in a node.

Contracted pressure

It may be necessary to model the contracted pressure in nodes with customers. Most import terminals have a limited range around a target pressure P_m which they accept for incoming gas:

$$p_{im}^{out} + \epsilon_m^- - \epsilon_m^+ = P_m, \qquad m \in \mathcal{M}, i \in \mathcal{I}(m). \tag{19}$$

Here ϵ_m^- and ϵ_m^+ are negative and positive deviations from the target pressure. These deviations are penalized in the objective at a level reflecting how hard the pressure constraint is in practice.

It is also possible to specify restrictions for each pipeline for example for the pressure into and out of a given pipeline. Pressure restrictions often apply to nodes with compression or nodes where processing of the gas is being performed. These constraints are called technical pressure constraints. Examples are minimum and maximum pressure out of pipeline (represented by (20) and (21) respectively).

$$p_{ij}^{out} \geq P_{ij}^{min}, \qquad j \in \mathcal{N}, i \in \mathcal{I}(j). \tag{20}$$

$$p_{ij}^{in} \leq P_{ij}^{max}, \qquad j \in \mathcal{N}, i \in \mathcal{I}(j). \tag{21}$$

Gas quality and energy content

In this model, gas quality can be specified in two different ways, focusing on combustion value (GCV) of the natural gas, or the content of CO_2. These properties are both technically and economically important for the customer. When dealing with CO_2, the customer accept a maximum content in terms of [mol %]. This is typically due to environmental taxes or to requirements related to avoiding corrosion in pipelines. If we focus on GCV, the customer accepts deliveries between a minimum and maximum combustion value. High GCV is in itself tractable as the energy content is higher, but in practice the plants using the natural gas are technically calibrated for a certain GCV-range. The quality is then measured in $[MJ/Sm^3]$. Here we only give the formulation for GCV:

$$Q_m^{min} \leq q_{im} \leq Q_m^{max}, \quad m \in \mathcal{M}, i \in \mathcal{I}(m), \tag{22}$$

where q_{im} is gas quality (GCV) in a pipeline going from node i to market node m. In practice we need more flexibility in the model by allowing reduced quality in order to increase the flow. Modeling this as hard constraints could lead to situations where unexpected shutdowns of production fields or pipelines may lead to a complete stop in deliveries to a customer due to the contractual quality. If it is an alternative to get some deliverances, outside the contracted limits, but within what is technically acceptable the latter will be chosen. This tradeoff will be valued in economical terms as reduction in the customer price. We need the variables δ_m^{l+} and δ_m^{l-} to indicate the positive and negative deviation from the lower quality limit Q_m^{min} of customer node m. Likewise we need δ_m^{u+} and δ_m^{u-} to indicate the positive and negative deviation from the upper quality limit Q_m^{max}:

$$q_{im} + \delta_m^{l-} - \delta_m^{l+} = Q_m^{min}, \quad m \in \mathcal{M}, \ i \in \mathcal{I}(m), \tag{23}$$

$$q_{im} + \delta_m^{u-} - \delta_m^{u+} = Q_m^{max}, \quad m \in \mathcal{M}, \ i \in \mathcal{I}(m). \tag{24}$$

Gas quality and blending

Gas quality is a complicating element because we have to keep track of the quality in every node and pipeline, and this depends on the flow. Where two flows meet, the gas quality out of the node to the downstream pipelines depends on flow and quality from all the pipelines going into the node. The flow in each pipeline is a decision variable in the model, and so is the quality out of each node. We assume that the resulting blending quality is common for all the downstream pipelines being connected to a node, and that it is decided by the convex combination of inflow qualities to the node:

$$q_{ij} = \frac{\sum_{n \in \mathcal{N}} q_{ni} f_{ni}}{\sum_{n \in \mathcal{N}} f_{ni}}, \qquad i \in \mathcal{N}, j \in \mathcal{O}(i), \tag{25}$$

or:

$$q_{ij} \sum_{n \in \mathcal{N}} f_{ni} - \sum_{n \in \mathcal{N}} q_{ni} f_{ni} = 0, \quad i \in \mathcal{N}, j \in \mathcal{O}(i). \tag{26}$$

This equation has two quadratic terms on the form $q_{ni} f_{ni}$. These terms can easily be reformulated in the following way: Define $\alpha = q_{ni} - f_{ni}$ and $\beta = q_{ni} + f_{ni}$. Then $q_{ni} f_{ni} = 1/4(\alpha^2 - \beta^2)$. Linearizing α^2 and β^2 is straightforward using Special Ordered Sets of type 2 (SOS2) [24]. In the SOS2 set at most two variables can be non-zero, and the two variables must be adjacent. Still this means that we need to move into solution techniques from integer programming, in particular branch and bound, so solution time will increase exponentially with the numbers of SOS2 sets needed.

Modeling multi component flows

If we model the flow of C components of the natural gas we require that the split fractions of the components going into the different pipelines out of the node n is equal for all components. For simplicity let us assume we always have only two pipelines out of a split node $n \in \mathcal{N}$ going to node j_1 and j_2 (see Figure 5). Let us also denote the first component in the set $\mathcal{C}$ of components for c_1. All components are indexed from $c_1, \ldots, c_C$. Then the relation of the volume split between j_1 and j_2 is equal for all components:

$$\frac{f_{nj_1}^{c_1}}{f_{nj_2}^{c_1}} = \frac{f_{nj_1}^{c}}{f_{nj_2}^{c}}, \qquad n \in \mathcal{N}, c \in \mathcal{C}. \tag{27}$$

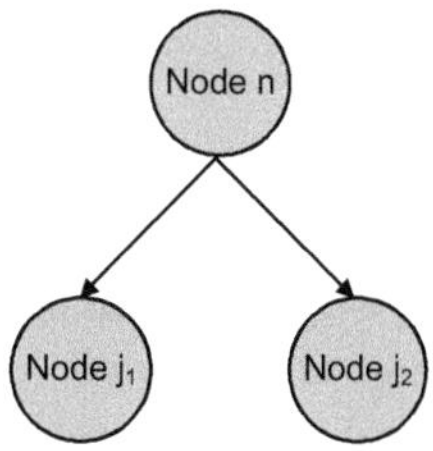

Fig. 5. The flow is split in node n to node j_1 and j_2.

This is a quadratic expression, and we reformulate it using the equations (28) to (32). We need a set of binary variables ϑ_{nz} where $z = 1, \ldots, Z$, each representing the choice of a split percentage for the share of natural gas going to node j_1. The ϑ_{nz} variable is modeled as a special ordered set of type 1 (SOS1), where only one variable can be non-zero [24]. For each ϑ_{nz} we define a constant E_z giving the percentage related to the z. We also define a new variable e_{nz} representing the flow through node n of component c if $\vartheta_{nz} = 1$.

The first constraint says that the flow from n to j_1 of component c equals the percentage E_z multiplied with the total flow through node n of the component c.

$$f_{nj_1}^{c} = \sum_{z=1}^{Z} E_z e_{nz}^{c}, \qquad n \in \mathcal{B}. \tag{28}$$

The set $\mathcal{B}$ consists of all split nodes in the network. Then we need to restrict the formulation so that only one ϑ_{nz} is positive for each node:

$$\sum_{z=1}^{Z} \vartheta_{nz} = 1, \qquad z \in \{1, \ldots, Z\}, n \in \mathcal{B}. \tag{29}$$

The e_{nz}^{c} variables giving the flow through the node of each component is constrained by the capacity of the node, corresponding to the active ϑ_{nz}.

$$\sum_{c \in \mathcal{C}} e^c_{nz} \leq F_n \vartheta_{nz}, \quad z \in \{1, \ldots, Z\}, n \in \mathcal{B}. \tag{30}$$

We also require that what flows through the node of each component either goes to node j_1 or to node j_2:

$$\sum_{z=1}^{Z} e^c_{nz} = f^c_{nj_1} + f^c_{nj_2}, \qquad n \in \mathcal{B}, c \in \mathcal{C}. \tag{31}$$

And to make sure that there does not flow more out of the node of each component than what comes in:

$$f^c_{nj_1} + f^c_{nj_2} = \sum_{i \in \mathcal{N}} f^c_{in}, \quad c \in \mathcal{C}, n \in \mathcal{B}. \tag{32}$$

Processing plants

Some of the gas components are extracted and sold in separate component markets. The extraction is handled in processing plants in the network. In the modeling of this process it is assumed that the volume of each component extracted is a constant fraction of the total volume of that component in a processing plant (A^c_r). Hence, no decision on the configuration of the processing plant is made, but pressures and gas flows through a processing plant can be modeled by several processing nodes in sequence or parallel. This is expressed in equation (33). The mass balance for the processing plant nodes can then be formulated as in equation (34). The variable a^c_r is used to keep track of how much of component c is extracted from the flow in processing plant r.

$$a^c_r = A^c_r \sum_{i \in \mathcal{N}} f^c_{ir}, \quad c \in \mathcal{C}, r \in \mathcal{R} \tag{33}$$

$$\sum_{i \in \mathcal{N}} f^c_{ir} = \sum_{j \in \mathcal{N}} f^c_{rj} + a^c_r \tag{34}$$

Modeling turn-up: flexibility in the production fields

Turn-up is an expression used for the flexibility present in some production fields. For example reduced transport capacity due to a shutdown in one part of the network may be compensated by turning up the planned production from other gas fields not directly affected by the reduced capacity. When modeling this turn-up capacity it is important to keep in mind that even if one are free to utilize this flexibility, it is not acceptable from a practical point of view that the model presents a flow allocation where fields with significant turn-up capacity will take over production from minor fields, which basically is not affected by the shutdown. The turn-up is only used to take over production from fields that for some reason are prevented to deliver. Hence, our first priority is to meet demand in the network and our second priority is to produce in accordance with the planned production at the fields.

We model this by adding a penalty cost for using turn-up in the objective to avoid turn-up to be used at the expense of normal production capacity in other fields. This works because not delivering gas to customers would generate a loss which is considerably higher than the small penalty put on using turn-up capacity.

The variables Δ_g^- and Δ_g^+ represent underproduction and the use of turn-up in relation to the planned production of G_g for field g. As before f_{gj} is the flow from g to j:

$$\sum_{j \in \mathcal{O}(g)} f_{gj} + \Delta_g^- - \Delta_g^+ = G_g, \qquad g \in \mathcal{G} \tag{35}$$

4 Management of Natural Gas Storages

As a consequence of the liberalization process in the natural gas industry, the natural gas markets have become more dynamic. The spot markets and the possibility to trade gas in forward markets have increased the importance of gas storages. In this section we discuss models for gas storage operations in a market with uncertain demand.

In order to discuss the management of natural gas storages, a couple of terms need to be established (see Figure 6 for an illustration of the terms):

Storage capacity gives the maximal volume of natural gas in the storages facility. The storages capacity is limited by the physical properties of the storage.

Volume of natural gas in the storage is the total volume of natural gas in a given storage at a given time.

Cushion gas is the amount of gas needed to create necessary pressure in order to lift gas from the storage. The amount of cushion gas needed varies with the type of storage and the geological conditions at the storage location. For some types of storages the cushion gas requirement is as high as 80% of the total gas volume in the storage.

Working gas is the gas volume available during normal operation of the storage. This corresponds to the total amount of gas in the storage subtracted the cushion gas.

4.1 Storage Facilities

The most common storage facilities are abandoned oil- and gas reservoirs, aquifers, salt caverns and LNG-storages. In the following, a short overview of advantages and disadvantages of these possibilities will be given. For further discussion of storage facilities, see [1].

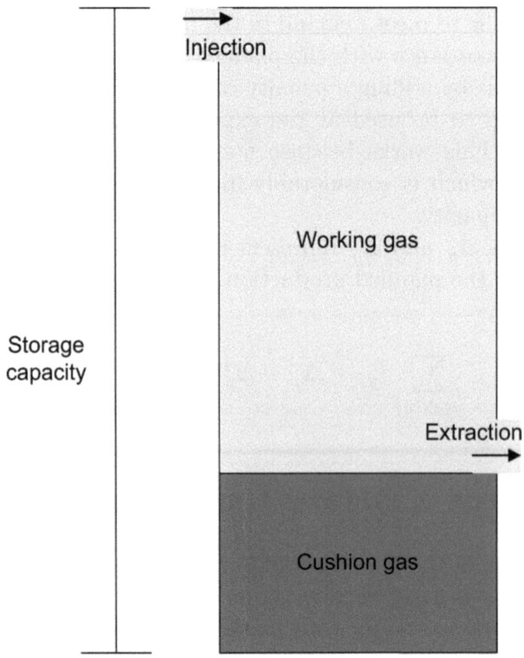

Fig. 6. The complete square is the total storage capacity. The lower part of the figure is the cushion gas needed for operation of the storage, and the upper part of the figure is the gas currently available for extraction from the storage.

Abandoned oil- and gas reservoirs are the most common storage facility. One reason for this is the relatively low startup costs. The storage facility is already in place, and so is most of the surface installations needed. Another advantage of this type of storage is the fact that infrastructure is normally already in place. One major drawback is the amount of cushion gas needed for operation.

Aquifer is a porous, underground water-bearing layer which can be transformed into a storage facility by replacing the water with natural gas. When using abandoned oil- and gas reservoirs the geological properties are known, this is not the case when using aquifers. This adds risk to the development of this type of storages. Cushion gas in the amount of 80 to 90 % is needed for operation, and the development takes time and is costly. These storages are normally only used in locations where no oil- and gas reservoirs are available. One advantage of this type of storage is the relatively high delivery rate.

Caverns are created from underground salt or rock formations. In the salt caverns, water is used to dissolve halite and to shape cavities in natural salt formations. These cavities have the properties of a high-pressure gas container, with impenetrable walls. The storages have a high delivery capacity, and a cushion gas requirement of only approximately 25 %. The process of dissolving halite and shaping the cavities makes this alternative more expensive than the previous two alternatives.

LNG-storages are, in contrast to the previously presented alternatives, above-ground facilities. These storages consist of tanks containing liquefied natural gas (LNG) or liquefied petroleum gas (LPG). The capacity of these tanks is normally very limited compared to the other alternatives presented.

4.2 Motivation for Utilization of Storages

The possibility of storing natural gas gives the participants increased flexibility with regards to production and transportation decisions. One important use of natural gas storages is to take advantage of the strong seasonal pattern in prices. Since the primary use of natural gas is for heating and production of electricity, the fundamental price determinant in the markets is the weather conditions. The demand is normally higher in winter than in summer, and the production capacity is also lower than the peak demand. This means that the monthly demand for natural gas may be much higher than the possible changes in production level can satisfy. The difference between production capacity and peak demand can to a certain degree be satisfied through utilization of storages. The use of storages can substitute for investments in new production fields and transportation capacity. Traditionally the storages have been used in order to ensure a high security of supply. When problems occurred either in the production or transportation facilities, storages could be used to supply the downstream participants. The storages operate as a security buffer in this case. With the development of short-term markets and volatile spot prices, the storages will be important for participants wanting to utilize the price fluctuations. Especially for gas producers not having a reservoir close to the market this will be important. It can take several days before a decision to change production level at the field will result in increased delivery in the market.

4.3 Modeling Storages

The maximum in- and outflow rates of the storage varies with the current storage level. The maximal injection rate is a strictly decreasing convex function of the storage level. Likewise the outflow rate can be given as a strictly increasing convex function of the storage level. To be able to realistically represent the in- and outflow rates, the use of special ordered sets of type 2 is chosen [24]. An illustration of the implementation of the SOS2 is shown in Figure 7 for the injection rate. The storage levels are discretized by a set of constants $X_1, \ldots, X_Y$, the corresponding injection rates are $H_1, \ldots, H_Y$ and the variables $\nu_1, \ldots, \nu_Y$ are used to give a convex combination of two of the

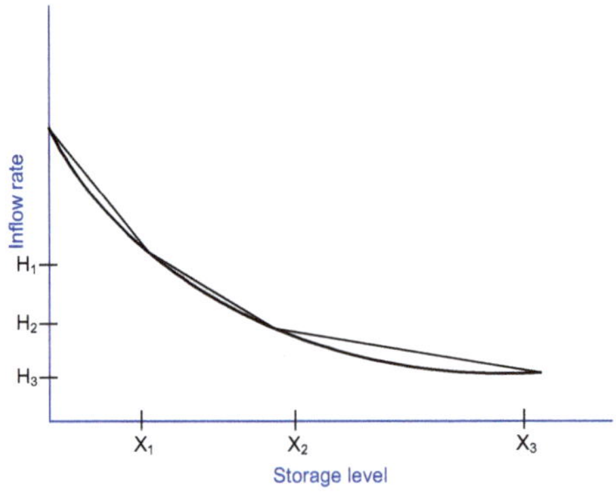

Fig. 7. Illustration of the linearization of the injection rate of a storage.

points. This means that if ν_y has a value different from 0, then only one additional variable can be non-zero. The only two candidates in this case are ν_{y-1} or ν_{y+1}. The storage level at a given time t is represented by x_s^t.

$$\sum_i f_{is}^t \le \sum_{y=1}^{Y} \nu_{ys}^t H_{ys}, \qquad s \in \mathcal{S}, \tag{36}$$

$$\sum_{y=1}^{Y} \nu_{ys}^t = 1, \; SOS2, \qquad s \in \mathcal{S}, \tag{37}$$

$$x_s^t = \sum_{y=1}^{Y} \nu_{ys}^t X_y, \qquad s \in \mathcal{S}, \tag{38}$$

$$x_s^t = x_s^{t-1} + \sum_{i \in \mathcal{I}(s)} f_{is}^t - \sum_{i \in \mathcal{O}(s)} f_{si}^t, \qquad s \in \mathcal{S}. \tag{39}$$

The maximum and minimum levels of storage are modeled implicitly with the representation given. The maximal level (equal to the total capacity of the storage) is restricted by the inflow function. When the storage reaches the upper capacity level, the associated inflow rate is equal to zero. The minimum level (coming from the requirement of a certain level of cushion gas in the storage) is handled in a similar way: when the minimum storage level is reached, the associated outflow rate will be equal to zero.

5 Value Chain Optimization and Portfolio Management

We will here give a short description on how to include markets and portfolio optimization in the natural gas value chain. For more advanced models on portfolio optimization in the natural gas value chain see [17] from which most of the ideas presented here originate. Other relevant references are [12] which considers portfolio optimization for oil and gas fields in a strategic horizon and [21] which considers tactical value chain coordination, but without stochasticity and without multi commodity flows.

5.1 Different Levels of Portfolio and Value Chain Integration

The models presented here have both a portfolio and a value chain perspective. These are important properties of a natural gas optimization model. The importance of these perspectives can be realized when considering the complexity of the transportation system. Due to the technical nature of the gas network, several physical and technical threshold-values exist. If such values are trespassed, only minor incremental deliveries in one part can cause significant unintended reductions elsewhere. The bottlenecks in the transportation system make the flexibility incorporated in a system perspective valuable. We will not give all the previous formulations of the transportation network again, but each time period t in a value chain model will include transportation network constraints and variables like the ones from Section 3 with an additional index t on all variables.

The motivation behind the portfolio and value chain perspectives can be summarized by considering four levels of planning:

1. *Traditional production planning*: In this first level the model ensures balancing of the production portfolio with the contract portfolio. Stochastic demands and prices that are not perfectly correlated motivate a portfolio perspective on the planning, as the portfolio variation will be lower than the variation of the stochastic parameters of separate fields or contracts.
2. *Production and market optimization*: At this level markets are used to supplement the physical production in order to gain more from the physical production capabilities. The market can be used to resolve bottlenecks in the transportation network or on the production side. The purpose is to maximize the profit from production and contract obligations using also spot markets. At this level geographical swaps and time swaps of gas can be performed using the market, and they are used to fully utilize the flexibility in the system.
3. *Trading:* At this level contracts and financial instruments are traded independently of the physical production and contract obligations based on market opportunities. The trading is similar to the previous level in terms of using the spot market and financial instruments like futures and options, but the motivation is now speculation, not solving bottleneck problems. These trades are in no way connected to the physical production and contract obligations, unless the producer has market power.
4. *Risk management:* So far we have assumed the producer is risk neutral and tries to maximize expected profit. In that case it is enough to supplement physical production with trades in the spot market at level 2. If the producer is risk averse hedging the portfolio outcome using futures, forwards or options may be optimal.

The distinction between level 2 and 3 is clear in theory, but in practice the transition will be gradual.

5.2 Utilization of Short-Term Markets in Value Chain Optimization

The use of short-term markets allows for considerable flexibility in the system. Consider the network in Figure 8. In a situation where field B needs to produce and the company has an obligation to deliver in a bilateral contract in Emden several possibilities exist:

- Field A supplies Emden, while field B sells spot in Zeebrugge
- The company may buy spot in Emden and the production from field B can be sold in the spot market in Zeebrugge.
- The company buys spot in Emden, while it sells the production from B spot in the upstream market.
- Storage might be used to supply Emden, while the production from field B is sold elsewhere.

These simple geographical swaps makes the system more flexible and gives the company the possibility to maximize the flow of natural gas (and the value of their production) beyond what traditional transportation planning would have done. For example bottlenecks in the production or in the transportation may be resolved or moved using the markets actively.

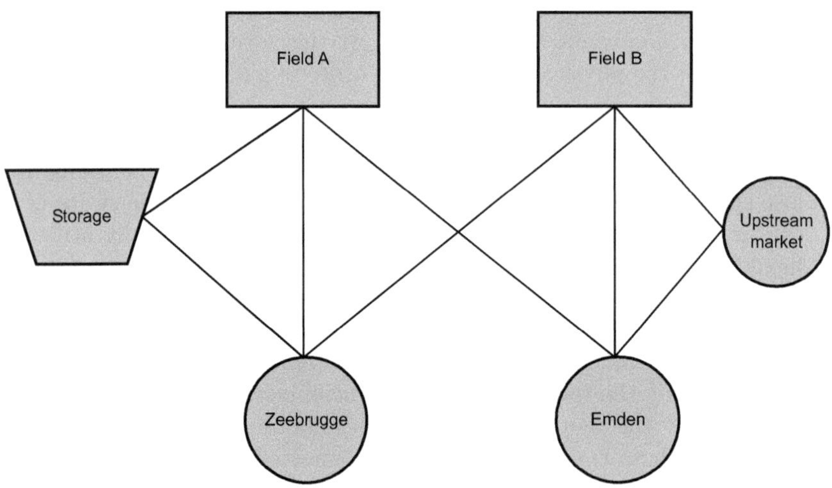

Fig. 8. Example of a natural gas network

A different reason to use the markets is time swaps. Consider Figure 8 again. This time field B needs to produce in time 1, and the company has an obligation to deliver in time 2. Several options are then available to the company:

- In period 1 field B may supply storage, and in period 2 the storage supplies Emden.
- In period 1 field B can sell spot in Zeebrugge, and in period 2 either use a forward contract or buy spot in Emden.
- In period 1 field B can sell spot upstream, and then use either a forward contract or the spot market to supply Emden

This is just some of many possibilities that exist for geographical swaps and time swaps. The network considered is also very small. When expanding the network to, for instance, 20 fields, 80 pipelines and 10 markets, the number of possible routing decisions gets very large and the flexibility increases. It is this flexibility we try to capture when modeling the portfolios of production fields and contracts. The flexibility further increases when perfect spot markets are added. The need for flexibility comes from the fact that demands and prices are stochastic. The gain from portfolio thinking increases because they are not perfectly correlated. We assume the company is a price taker. For simplicity of notation, we assume there is only one company in the markets. If not, we would also need to model the other companies' transportation needs.

5.3 Including Markets and Contracts

In Section 3 only aggregated deliveries to take-or-pay contracts in the different customer nodes $m \in \mathcal{M}$ were considered. When including market transactions in the model a representation of the uncertainty in the price process is important. Based on this representation scenarios describing the uncertainty can be generated and optimal decisions in the interaction between the physical system and the market can be made. In this description some simplifications have been made. Only one company is considered, so no upstream market exists, the possibility of delaying production through lifting agreements will be disregarded, and only trades in the spot market will be considered. The possibility of trading forward contracts is only interesting for a risk adverse company. This will be discussed shortly at the end of this section.

Figure 9 illustrates how the market nodes are included in the model. The arrows show that gas might flow from the transportation network to the market. There is no flow from the market to the network (as would be the case for an upstream market). In addition, transactions within the market node can be performed. In the spot market the company can purchase or sell volumes of natural gas. Obligations in the take-or-pay contracts can be fulfilled either by flow from the network to the market node, or by transactions within the market node.

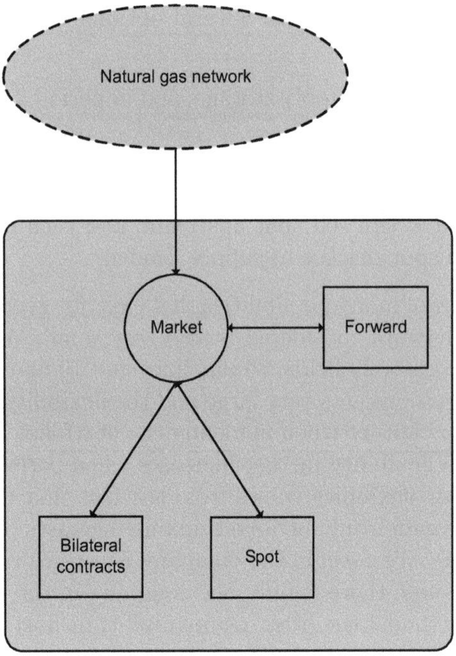

Fig. 9. The market node

5.4 Modeling Stochasticity

We use the modeling paradigm of stochastic programming to represent uncertainty in the models, see for example [10]. Uncertainty is then represented in a scenario tree, see Figure 10. The nodes in the scenario tree represent decision points, and uncertainty is resolved along the arcs going out of a node with several branches. In practice decisions are only made when new information becomes known. A stage is the set of time periods elapsing between each time information is learned by the decision maker. Each stage in the tree typically consists of several time periods, but only nodes after a branching are decision points, as they are the only time periods when new information about the future is resolved. Still, decision variables are present in time periods where information is not resolved, hence the time periodization using time periods t reflect in which time period the decision has effect. In Figure 10 there are 3 time periods. Time periods 1 and 2 are in stage 1, starting with the decision in node 0 and ending just before the new decisions at stage 2 (in nodes 2, 6 and 10).

In a two-stage stochastic programming model we define a set of time periods $t \in \mathcal{T}_1 = \{t_1, \dots, T_1\}$ belonging to the first stage where information is deterministic, and a set of time periods $t \in \mathcal{T}_2 = \{T_1 + 1, \dots, T_2\}$ where some parameters are stochastic (as seen from $t \in \mathcal{T}_1$). When time passes on and one enters the first $t \in \mathcal{T}_2$, uncertainty is resolved and also the remaining time periods can be considered deterministic.

In the model presented here we use a two-stage formulation for ease of notation. Several parameters are stochastic in reality. We will consider stochasticity in: contractual demands, contract prices and spot prices. We denote the stochastic contract price for contract k in customer node m at time period $t \in \mathcal{T}_2$ as $\tilde{\phi}_{mk}^t$. Stochastic demand for contract k in customer node m at time period t is $\tilde{\mu}_{mk}^t$. The stochastic spot price is represented with $\tilde{\psi}_m^t$. The vector of all stochastic variables in time period t is $\tilde{\xi} = (\tilde{\psi}^t, \tilde{\phi}^t, \tilde{\mu}^t)$.

We use a tilde over the variable to reflect that it is stochastic (as seen from $t \in \mathcal{T}_1$) and remove the tilde when the variable is deterministic. We then get the following:

$\tilde{\xi}_m^t$ Stochastic variables for customer node m in time period $t \in \mathcal{T}_2$ seen from a point in time $t' \in \mathcal{T}_1$.

ξ_m^t Deterministic parameters for customer node m in $t \in \mathcal{T}_1$, or $t \in \mathcal{T}_2$ after uncertainty is resolved (Seen from a point in time t' where $t \in \mathcal{T}_2$).

A scenario tree can be constructed for example using price processes for natural gas or descriptions of the dynamic aspects of stochastic demand. We will not go in detail on how to do this here, but assume the scenario tree exists in the remaining part of this paper.

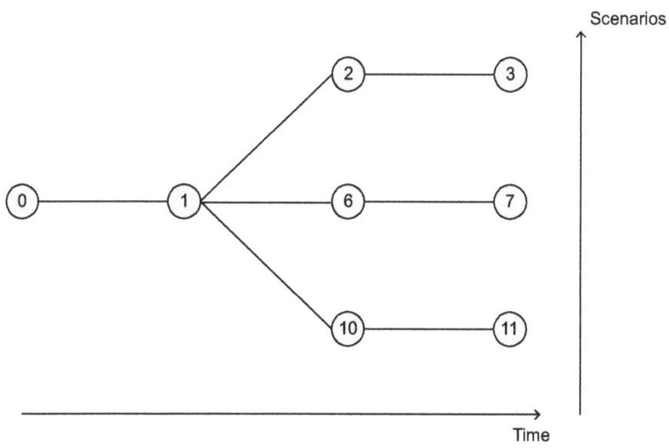

Fig. 10. An example of a scenario tree

5.5 The Objective

We describe the supply chain portfolio optimization model as a two-stage stochastic program with relatively complete recourse [10]. The only stochasticity that is present is in the right hand side and in the objective. The typical length of a time period for a tactical planning model is one month, and the planning horizon would typically be 12 months, where for example the first 6 months would belong to $\mathcal{T}_1$ and the last 6 months to $\mathcal{T}_2$ in a two-stage formulation. The objective is to maximize expected profit taken into consideration cash flows and shortfall costs. Hence the objective can be described by summarizing the expected cash flow of the time periods. The cash flow of each time period t can be described as a function $\Pi^t(x^{t-1};\xi^t)$ (or $\tilde{\xi}^t$ if stochastic) where x^{t-1} is the storage level in the start of the time period. The decision variables and constraints are equal in all time periods, except for initialization in time period 0 where only initial storage levels are defined x^0_{ns} and for the terminal conditions at the end of the model horizon. We use the vector x^0 to denote the initial level of all storages and x^t to denote the level of all storages in time period t. The profit function for time period $t \in \mathcal{T}_1 \cup \mathcal{T}_2$ can be formulated as:

$$\Pi^t(x^{t-1};\xi^t) = \sum_{m\in\mathcal{M}}\sum_{k\in\mathcal{K}} \phi^t_{mk}\mu^t_{mk} + \sum_{m\in\mathcal{M}} \psi^t_m(\zeta^{t-}_m - \zeta^{t+}_m), \tag{40}$$

where the first term is the income from delivery in contract k in market m at time t and the second term gives the profit from trading in the spot market in node m in time period t.

The two-stage stochastic program with fixed relatively complete recourse is:

$$\max \sum_{t\in\mathcal{T}_1} \Pi^t(x^{t-1}) + \mathcal{Q}(x^{\mathcal{T}_1}), \tag{41}$$

where

$$\mathcal{Q}(x^{\mathcal{T}_1}) = \max E_{\tilde{\xi}}[\sum_{t\in\mathcal{T}_2} \Pi^t(x^{t-1};\tilde{\xi}^t) + EGV\big(x^{\mathcal{T}_2}\big)], \tag{42}$$

subject to a set of constrains representing transportation, production and markets. These constraints are mainly the constraints descried earlier in this paper, but we will look closer at the ones changing because of the introduction of markets and contracts. The constraint sets are identical for all time periods $t \in \mathcal{T}_1 \cup \mathcal{T}_2$. For the last time period the objective includes the terminal value of the natural gas in storages expressed by the Expected Gas Value function, $EGV(x^{\mathcal{T}_2})$. This function is described in more detail in Section 6.

The solution resulting from maximizing expected profits will normally be different from the solution reached with the objective function presented in Section 3.4. This means that the solution does not necessarily maximize the throughput in the network, or minimize the cost of achieving a given throughput. The solution will however show how the network should be managed in order to achieve the highest possible expected profit.

5.6 Constraints Including Markets and Contracts

The mass balance in the market node for each time period and each scenario is expressed as:

$$\sum_{i \in \mathcal{I}(m)} f_{im}^{t} + \zeta_{m}^{t+} = \zeta_{m}^{t-} + \sum_{k \in \mathcal{K}(m)} \mu_{mk}^{t}, \quad \forall m \in \mathcal{M}, \ \forall t \in \mathcal{T}. \tag{43}$$

In (43), ζ_m^t represent transactions in the spot market in node m in time period t. The $+$ sign indicates purchases of natural gas whilst the $-$ sign indicates sales. Delivery in contract type k in the node m in time period t are included in μ_{mk}^t. The mass balance equation illustrates the flexibility gained by including markets in the model. It is no longer necessary to ship the gas to the market node in order to fulfill the contractual agreements, since the spot market can be utilized for this. This means that geographical and time swaps are now available to the company.

Risk Aversion

In the discussion so far only the possibility for trading natural gas through the spot market has been discussed. For a risk neutral company that is maximizing expected profits this is an adequate approach. Since the forward price is a good predictor of the expected future spot price, trading in the forward market would on average be approximately equal to trading on the spot market (this is based on a simple arbitrage argument, see for instance [9]. The fact that natural gas is a commodity makes the argument less obvious, but under some assumptions still valid. In the case where the company is risk averse however the situation changes and some tools to handle risk management are needed. The inclusion of a forward market then gives the company the possibility to hedge, that is: to reduce the risk of their position. By trading forward contracts a given price can be locked in on advance.

In this case the company will no longer maximize expected profits from their operations, but rather maximize a utility function that incorporates the risk aversion of the company. Another way of doing this is to introduce a penalty function that will put extra cost in the objective function on deviations from some target profit value. In addition to the change in the objective function, the mass balance in the market node (see (43)) will be changed to incorporate the possibility to trade in the forward market.

Solution Times

The complexity of the models introduced in this paper to a large extent depends on the modeling of the gas components. The inclusion of gas components adds a large number of integer variables to the problem. When excluding the gas components, a stochastic model with a network consisting of approximately 80 nodes and 1000 scenarios, can be solved within an hour. This problem will have approximately one million rows, one and a half million columns, four million non-zero elements and fourteen thousand binary variables. When including gas components the solution time increase significantly, and it is difficult to find an optimal solution. For a physical system similar to the one above, with 100 scenarios and 10 breakpoints (see Section 3.5), a solution with an integrality gap of 4% to 5 % typically can be reached within 12 hours. If the objective is only to maximize flow in a static model, solution times are within minutes when components are omitted and increases correspondingly when components are added.

6 The Expected Gas Value Function (*EGV*)

So far no considerations have been made with respect to how the final period in the planning horizon will be handled. The model presented so far will most likely end up with a very low storage level, and the production might also be higher than optimal when considering a longer horizon (since the value of the gas still contained in the reservoirs is neglected).

In order to handle the end-of-horizon problem, several possibilities exist. One way of handling the storage problem is to set a target value for the storage level at the end-of-horizon, for instance the starting level.

$$x_s^T \geq x_s^0 \tag{44}$$

This level might also be made dependent on various factors, such as the season in which the end-of-period belongs. This way of modeling the end-of-period however allows for limited flexibility and also neglects the true value of the gas contained in the storage. A way of determining the optimal level for the storages in the last period is by using the expected gas value function.

The Expected Gas Value function (EGV) gives an estimate of the value of a unit of gas in storage at some point in time t, based on expectations for the future development of the spot price of natural gas. When the EGV is used as a boundary value, the alternative value of the natural gas in storage is thereby included. This alternative value comes from the opportunities present after the end of the model horizon. Hence for each end-of-horizon storage level, the EGV must reflect the value of an optimal out-of-horizon strategy for injecting gas in the storage and selling gas from the storage.

If high prices are expected in the future, the EGV will encourage a high storage level in final time period T_2, whilst if lower prices are expected the optimal level in period T_2 may be lower. Figure 11 illustrates how the EGV is included in the existing optimization model. As the figure shows, the estimation of EGV is performed independently from the value chain model and the purpose is to give a boundary condition for the value of gas.

An important element in the model used to estimate EGV is the development of the natural gas spot price represented through spot price curves. These can be modeled using stochastic processes. Several versions of such models exist, for an overview of some of them, see [20]. Based on the chosen price model, scenarios describing possible future outcomes can be constructed (see Figure 12). Hence, for any given initial storage level a strategy is found

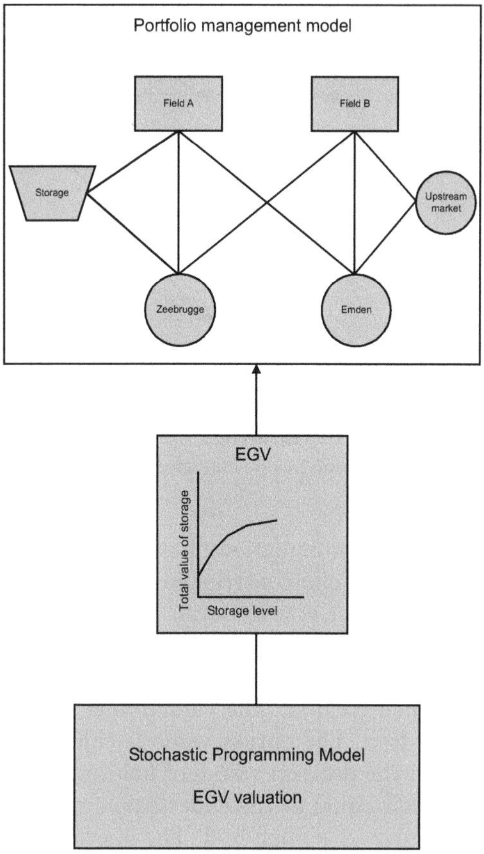

Fig. 11. The estimation of the EGV is performed in a stochastic optimization model that is independent of the existing optimization model. The EGV is then used in the value-chain model as a boundary value on the gas in storage and reservoirs.

for injection and withdrawal of natural gas based on a stochastic process for the gas price. In practice this is a real-options approach used to value the value of gas in the storage. The option value in gas storages comes from the operational flexibility. The company can switch between injection, withdrawal or idle modes, depending on the price development. For references on real-options, see for instance [9]. It is possible to do this estimation both for individual storages, and also for the combination of all or some of the storages in the network. In the latter case a more complicated model is needed for estimation of the EGV.

In the following, an example of how the EGV can be calculated is given. The procedure is based on [19] and [11], and use a stochastic dynamic programming framework. For references to similar work in hydro power, see for instance [14, 15]. After choosing a stochastic model to represent the price

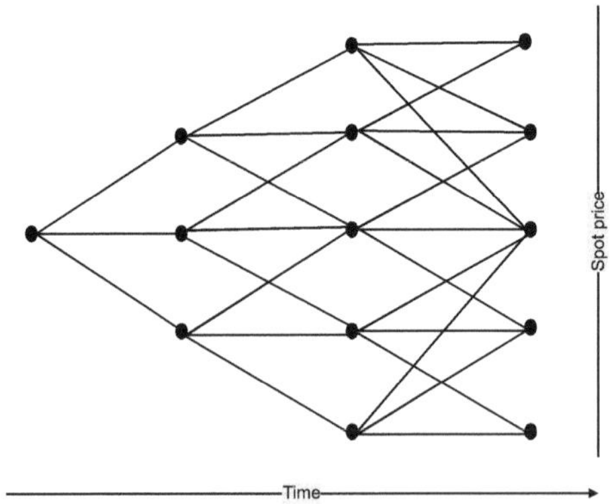

Fig. 12. Representation of the development of the spot price of natural gas. In this case a recombining trinomial tree. The arcs in the figure represent price movements, while the nodes represent different price scenarios.

of natural gas, a discrete approximation of the storage facility state space is made. A tree similar to the one constructed for the spot price (Figure 12) can be constructed also for the storage level. In this case the nodes represent different storage levels, while the arcs represent injection and withdrawal of natural gas in the storage. A multilevel tree representing the spot price and the amount in storage is then created. The valuation is performed by backward induction through the tree. The option value is calculated in each node by taking the maximum of the decision values of hold, injection and withdrawal. The hold decision value is equal to the expectation of the option value of the next steps, when storage level is unaltered. The injection value is the negative value of gas injected in this period, plus the expected value of increased storage level in future nodes. The withdrawal value is then the value of releasing gas in this period, plus the expectation of option values of decreased storage levels in coming nodes. This can be illustrated by (45), which shows the value in a given node in the tree:

$$I^t(\tau^t) = \pi^t\left(\varphi^t\right) + I^{t+1}\left(\tau^{t+1}\right). \tag{45}$$

$I^t(\tau^t)$ is the value of storage level τ in time period t in the node considered. This is determined by the value of flow $\pi^t\left(\varphi^t\right)$, (where φ^t is the volume injected or withdrawn in period t) in the node in period t plus value of the storage level τ^{t+1} in the next time period (in nodes following the considered one). The storage level is updated according to (46):

$$\tau^{t+1} = \tau^t + \varphi^t. \tag{46}$$

An illustration of a gas value function is given in Figure 13. The challenge is in finding the appropriate total value for each level of storage, as well as finding the breakpoints.

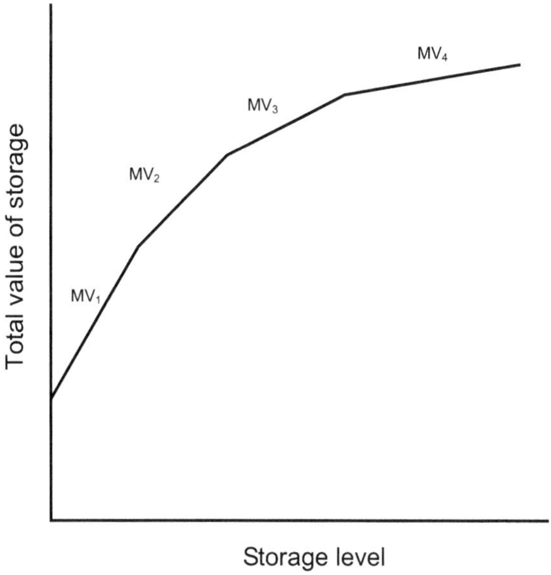

Fig. 13. An example of an expected gas value function. The MVs shows the expected marginal value of gas for various levels of storage. This is the additional value of one more unit of natural gas in the storage.

Even though short-term markets for natural gas are developing in Europe, the liquidity in these markets is still very limited. This lack of liquidity makes estimation of the spot-price process difficult, and therefore also estimation of the EGV difficult. Given that a spot-price model can be modeled for any given time horizon, a time-horizon of a couple of years may be appropriate for estimating the EGV. As the time-horizon for estimation is increased, the discount rate will make the gas value in the last periods decrease strongly.

7 Conclusions

In this paper we have gradually introduced the complexity of a stochastic optimization model for the natural gas value chain. We focus on coordination of the different levels of the chain and on a portfolio perspective. We started out by defining necessary constraints for a steady-state formulation of the underlying transportation network, supporting multi commodity flows and pressures. Next we introduced the time aspect and the use of storages. Thereafter we introduced stochasticity in demands and prices and gave a stochastic programming formulation for a portfolio optimization model. Natural extensions of this model would be contract selection and more advanced modeling of the production flexibility reflected by lifting agreements. Finally we defined the Expected Gas Value function and explained its use for giving the terminal value of stored natural gas and indicated how to calculate it.

Most of the model formulations presented here are simplified variants of models that are implemented for commercial use on the Norwegian continental shelf. In this paper we have taken the position of a large producer, but many of the formulations would be relevant for more general models focusing on other parts of the natural gas value chain.

References

1. Energy Information Administration. The basics of underground natural gas storage, 2002. http://www.eia.doe.gov/pub/oil_gas/natural_gas/analysis_publications/storagebasics/storagebasics.pdf.
2. J.M. Campbell. *The Equipment Modules.* Gas Conditioning and Processing. Norman, Okla., 7 edition, 1992.
3. H.J. Dahl, F. Rømo, and A. Tomasgard. An optimisation model for rationing-efficient allocation of capacity in a natural gas transportation network. In *Conference Proceedings: IAEE Praha*, 2003.
4. D. De Wolf and Y. Smeers. The gas transmission problem solved by an extension of the simplex algorithm. *Manag. Sci.*, 46(11):1454–1465, 2000.
5. Directorate-General for Energy European Commission and Transport. Opening up to choice: Launching the single european gas market. Office for Official Publications of the European Communities, July 2002.
6. Directive 98/30/EC of the european parliament and of the council, 22 June 1998.
7. Directive 2003/55/EC of the european parliament and of the council, 26 June 2003.
8. K. Hofsten. *Model-Based Dynamic Control and Optimization of Gas Networks.* PhD thesis, NTNU, Trondheim, 2000.
9. J. Hull. *Options, Futures and Other Derivatives.* Prentice Hall International, Upper Saddle River, N.J, fifth edition, 2003.
10. P. Kall and S. W. Wallace. *Stochastic Programming.* John Wiley & Sons, Chichester, 1994.
11. M. Manoliu. Storage options valuation using multilevel trees and calendar spreads. *Int. J. Theor. Appl. Finance*, 7(4):425–464, 2004.
12. B. Nygreen, M. Christiansen, T. Bjørkvoll, K. Haugen, and Ø. Kristiansen. Modelling Norwegian petroleum production and transportation. *Ann. Oper. Res.*, 82:251–267, 1998.
13. M. Nowak and M. Westphalen. A linear model for transient gas flow. In *Proceedings of ECOS*, pages 1751–1758, 2003. available as SINTEF-report STF 38S03601.
14. M. Pereira, N. Campodónico, and R. Kelman. Application of stochastic dual DP and extensions to hydrothermal scheduling. Technical report, PSRI, 012/99.
15. M.V.F Pereira and L.M.V.G. Pinto. Multi-stage stochastic optimization applied to energy planning. *Math. Program.*, 52:359–375, 1991.
16. R. L. Rardin. *Optimization in Operations Research.* Prentice Hall, Upper Saddle River, New Jersey, 1998.
17. F. Rømo, A. Tomasgard, M. Fodstad, and K. Midthun. Stochastic optimization of the natural gas value chain in a liberalized market. Technical report, SINTEF, Trondheim, Norway, 2004.

18. F. Rømo, A. Tomasgard, M.P. Nowak, M. Fodstad and L. Hellemo. An optimization system for gas transportation on the Norwegian continental shelf. Technical report, SINTEF, Trondheim, Norway, 2006.
19. T. Scott, H. Brown, and N. Perry. Storing true value? *Energy and Power Risk Management*, March 2000.
20. E.S. Schwarz. The stochastic behavior of commodity prices: Implications for valuation and hedging. *J. Finance*, 52(3):923 – 973, 1997.
21. N.L. Ulstein, B. Nygreen, and J.R. Sagli. Tactical planning of offshore petroleum production. Working paper, NTNU, Trondheim, 2004.
22. T. Van der Hoeven. *Math in Gas and the Art of Linearization.* PhD thesis, International business school and research center for natural gas, Groningen, The Netherlands, 2004.
23. M. Westphalen. *Anwendungen der Stochastischen Optimierung im Stromhandel und Gastransport.* PhD thesis, University Duisburg-Essen (Germany), 2004.
24. H. P. Williams. *Model Building in Mathematical Programming.* John Wiley & Sons, Chichester, 4th edition, 1999.

A Notation and Definitions

A.1 Sets

$\mathcal{N}$ The set of all nodes in the network.
$\mathcal{G}$ The set of nodes in the network with production fields .
$\mathcal{B}$ The set of nodes where gas flows are splitted into two or more pieplines.
$\mathcal{M}$ Nodes with buyers of natural gas: typically import terminals.
$\mathcal{I}(n)$ The set of nodes with pipelines going into node n.
$\mathcal{O}(n)$ The set of nodes with pipelines going out of node n.
$\mathcal{R}$ The set of nodes with processing capabilities.
$\mathcal{S}$ The set of nodes with storage facilities.
$\mathcal{K}(b)$ The set of contracts in node $b \in \mathcal{B}$.
$\mathcal{C}$ The set of components defining the chemical content of the natural gas.
$\mathcal{T}$ The set of time periods included in the model.
$\mathcal{L}$ The set of breakpoints used to linearize the Weymouth equation.
$\mathcal{Z}$ The set of split percentages used to discretize possible split fraction in split-node of the network.
$\mathcal{Y}$ The number of discretized storage and injection rate levels used to linearize storage characteristics.

A.2 Indexes

n Used for nodes in general. $n \in \mathcal{N}$. When more indexes are needed, i and j will be used.
g Used for nodes with production fields, $g \in \mathcal{G}$.

b Split nodes, $m \in \mathcal{M}$.
m Customer nodes $b \in \mathcal{B}$.
r Used for nodes with processing plants.
s Storage facility $s \in \mathcal{S}$.
k Contract $k \in \mathcal{K}$.
c Component $c \in \mathcal{C}$.
t Time period $t \in \mathcal{T}$.
l Breakpoits in linearized Weymouth restrictions.
z Breakpoints in linearization of split percentages in split nodes.
y Breakpoints for linearization of injection rate levels in storages.

A.3 Constants

G_g Planned production $[Sm^3/|t|]$ in field $g \in \mathcal{G}$.
F_{ij} Upper limit for flow through the pipeline from node $i \in \mathcal{N}$ to node $j \in \mathcal{N}$.
F_n Upper limit for flow through node $n \in \mathcal{N}$.
P_n^{max} Max pressure [bar]into a node $n \in \mathcal{N}$.
P_{ij}^{max} Max pressure [bar] into the pipeline from node $i \in \mathcal{N}$ to node $j \in \mathcal{N}$.
P_{ij}^{min} Min pressure [bar] out of the pipeline from node $i \in \mathcal{N}$ to node $j \in \mathcal{N}$.
P_b Target pressure [bar] for deliverances to a customer node $b \in \mathcal{B}$.
Q_n^{max} Max energy content requirement for gas deliverances to node $n \in \mathcal{N}$. Energy quality is given by (GCV or CO_2) or $\frac{[MJ/Sm^3]}{[mol\%]}$, where GCV is the (Gross Caloric Value).
Q_n^{min} Min energy content requirement for gas deliverances to node $n \in \mathcal{N}$.
D_b Demand in standard cubic meter pr time unit $[Sm^3/|t|]$ for natural gas in node $b \in \mathcal{N}$.
S_l Storage capacity $[Sm^3/|t|]$in node $s \in \mathcal{S}$.
K_{ij}^W The Weymouth constant is used as a constant in an empirical expression for linking flow and pressure in pipelines.
A_r^c Fraction of component c in processing plant r that is extracted from the flow.
PI Fixed point for pressure into a pipeline.
PO Fixed point for pressure out of a pipeline.
Γ_n Compressor factor in node $n \in \mathcal{N}$.
Θ_n Pressure reduction factor in node $n \in \mathcal{N}$.
η Compressor efficiency.
K_a Adiabatic constant for a certain gas type.
W^{max} Power output capacity of the compressor.
ω_b Value of gas to customer b.
κ Penalty cost for pressure level.
ϖ Penalty cost for deviation from contracted pressure level.

ι Penalty cost for use of turn-up.
χ Penalty cost for deviation from contracted quality to customers.
E_z Gives the split percentage related to a given z in linearization of split nodes.
X_y Discrete representations of storage level in linearization of storages.
H_y Discrete representations of injection rates in storages.

A.4 Decision Variables

f_{ij}^c Flow from node $i \in \mathcal{N}$ to node $j \in \mathcal{N}$ of component c. In some cases index c is omitted when we do not consider multi commodity flow.
f_n Total flow into node n.
e_{nz} Flow through node n of component c used for linearization in splitting nodes $m \in \mathcal{M}$.
p_{ij}^{in} Pressure [bar] into the pipeline going from node i to node j.
p_{ij}^{out} Pressure [bar] out of the pipeline going from node i to node j.
q_{ij} Gas quality (GCV or CO_2) in pipeline going from node i to node j.
ν_y Give convex combinations of X_y and H_y.
σ_{in} Equal to 1 if flow from i to n, otherwise 0.
ϑ_{nz} Binary variable representing split percentage in node n.
a_r^c Amount extracted of component c in plant r.
ρ_{ij} Equal to 1 if flow goes from i to j, otherwise 0.
ζ_m^{t-} Volume sold in spot market m in time period t.
ζ_m^{t+} Volume bought in spot market m in time period t.
δ_b^{l+} Positive deviation from the lower quality limit Q_b^{min} of customer node b.
δ_b^{l-} Negative deviation from the lower quality limit Q_b^{min} of customer node b.
δ_b^{u+} Positive deviation from the upper quality limit Q_b^{max} of customer node b.
δ_b^{u-} Negative deviation from the upper quality limit Q_b^{max} of customer node b.
x_s^t The storage level at a given time t in a storage $s \in \mathcal{S}$.
ϵ_b^+ Positive deviation from the contracted pressure to customer b.
ϵ_b^- Negative deviation from the contracted pressure to customer b.
Δ_g^+ Positive deviation from the planned production in field g.
Δ_g^- Negative deviation from the planned production in field g.

A.5 Functions

$EGV^t(x_s)$ Expected gas value in time period t as a function of the storage level in storage s.
$W_{ij}(PI, PO)$ Flow resulting from pressure difference between Pressure in,

	PI and pressure out, PO, of a pipeline according to the Weymouth equation.
$\Gamma(f_n)$	Compressor factor as a function of flow f_n into the node $n \in \mathcal{N}$.

A.6 Stochastic Variables

$\tilde{\phi}^t_{bk}$ Contract price for contract k in customer node b in time period t.
$\tilde{\mu}^t_{bk}$ Demand for contract k in customer node b in time period t.
$\tilde{\psi}^t_m$ The spot price in market m in time period t.
$\tilde{\xi}^t$ The vector of all stochastic variables $\tilde{\phi}^t$, $\tilde{\mu}^t$ and $\tilde{\psi}^t$.

In time periods where these parameters are not stochastic or where uncertainty is resolved, the tilde is dropped in the variable name.

MIX
Papier aus verantwortungsvollen Quellen
Paper from responsible sources
FSC® C105338
FSC
www.fsc.org

If you have any concerns about our products,
you can contact us on
ProductSafety@springernature.com

In case Publisher is established outside the EU,
the EU authorized representative is:
Springer Nature Customer Service Center GmbH
Europaplatz 3, 69115 Heidelberg, Germany

Printed by Libri Plureos GmbH
in Hamburg, Germany